R00219 67770

R0021967770

REF
QD
945
.B57
Cop.1

Blundell, T. L.

Protein crystal-
lography

DATE DUE

REF
QD
945
.B57
Cop.1

FORM 125M

Business/Science/Technology
Division

The Chicago Public Library

Received _____ NOV 1 9 1977

PROTEIN
CRYSTALLOGRAPHY

MOLECULAR BIOLOGY

An International Series of Monographs and Textbooks

Editors: BERNARD HORECKER, NATHAN O. KAPLAN, JULIUS MARMUR, AND HAROLD A. SCHERAGA

A complete list of titles in this series appears at the end of this volume.

Protein Crystals. a) monoclinic aldolase. b) orthorhombic triosephosphate isomerase. c) trigonal insulin. d) tetragonal lysozyme. e) hexagonal aldolase. f) cubic glucagon.

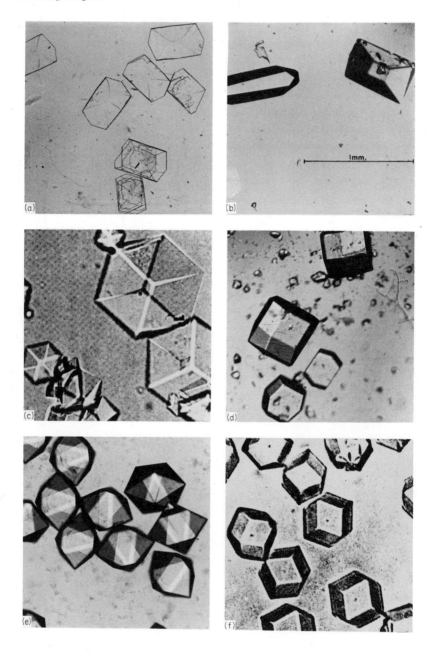

PROTEIN CRYSTALLOGRAPHY

T.L. BLUNDELL
School of Biological Sciences, University of Sussex
Falmer, Brighton, Sussex, England

and

L.N. JOHNSON
Laboratory of Molecular Biophysics, University of Oxford
South Parks Road, Oxford, England

 1976

ACADEMIC PRESS
NEW YORK · LONDON · SAN FRANCISCO
A Subsidiary of Harcourt Brace Jovanovich, Publishers

ACADEMIC PRESS INC. (LONDON) LTD.
24/28 Oval Road,
London NW1

United States Edition published by
ACADEMIC PRESS INC.
111 Fifth Avenue
New York, New York 10003

Copyright © 1976 by
ACADEMIC PRESS INC. (LONDON) LTD.

All Rights Reserved
No part of this book may be reproduced in any form by photostat, microfilm, or any other means, without written permission from the publishers

Library of Congress Catalog Card Number: 76–1068
ISBN: 0 12 108350–0

Typeset in IBM Press Roman by PREFACE LTD., SALISBURY
and printed in Great Britain by
WHITSTABLE LITHO LTD.,
WHITSTABLE, KENT

PREFACE

The scope of this book is simply defined: we have tried to cover all aspects of the X-ray diffraction analysis of protein crystals, with special emphasis on the differences between the analysis of protein crystals and the analysis of crystals of small molecules. In several cases we have had a difficult task in selection of material. In general we have limited our detailed descriptions to those procedures which have contributed to successful structure determinations. We also include chapters on neutron diffraction, γ-ray resonance and electron microscopy. The information obtained from the application of these techniques to protein crystals is complementary to that obtained by X-ray analysis and in recent years there have been considerable advances in these fields.

T. L. BLUNDELL

April, 1976   L. N. JOHNSON

CONTENTS

Preface vii
Acknowledgements xv

1. **Introduction** 1

    1.1. Introduction 1
    1.2. The Nature and Function of Globular Proteins . . 2
    1.3. The Development of Protein Crystallography and its Limitations 10
    1.4. Steps in the X-ray Analysis of Crystalline Proteins . 13

2. **The Principles of Protein Structure** 18

    2.1. Introduction 18
    2.2. Amino Acids – the Building Blocks 18
    2.3. The Primary Structure 25
    2.4. The Nature of the Peptide Link 25
    2.5. Peptide Structure 28
    2.6. Secondary Structure of Polypeptides 31
    2.7. Protein Tertiary Structure 45
    2.8. Quaternary Structure 51
    2.9. Protein Interactions with Small Molecules and Ions . 57

3. **Crystallization of Proteins** 59

    3.1. Introduction 59
    3.2. Factors affecting the Solubility of Proteins 60
    3.3. Factors affecting Crystal Nucleation and Growth . 66
    3.4. Techniques for Crystallization 68
    3.5. Protein Crystals: the Solvent of Crystallization . . 79
    3.6. Crystal Mounting 81

4. **Symmetry, Space Groups and Optical Properties of Crystals** 83

    4.1. Introduction 83
    4.2. The Thirty-two Classes of Symmetry 86
    4.3. The Seven Crystal Systems 90
    4.4. The Fourteen Bravais Lattices 90

	4.5.	Internal Symmetry Elements 93
	4.6.	The two-hundred-and-thirty Space Groups 94
	4.7.	Optical Properties of Crystals 97
	4.8.	The Use of the Polarizing Microscope by the Protein Crystallographer 103
	4.9.	The Determination of Molecular Weight from Protein Crystals 104

5. The Principles of X-ray Diffraction 107

	5.1.	Introduction – the Wave Nature of X-rays 107
	5.2.	Bragg's Law 109
	5.3.	The Analogy with Light Scattering 110
	5.4.	Resolution 117
	5.5.	The Scattering of X-rays by an Atom 119
	5.6.	The Scattering of X-rays by a Molecule 123
	5.7.	Scattering of X-rays by a Crystal 124
	5.8.	The Relationship between the Laue Conditions for Diffraction and Bragg's Law. The Reciprocal Lattice 130
	5.9.	The Ewald Construction 132
	5.10.	Friedel's Law 133
	5.11.	The Electron Density Equation 134
	5.12.	The Phase Problem 136
	5.13.	The Patterson Function 137
	5.14.	Direct Methods 140
	5.15.	Refinement 144
	5.16.	Summary 149

6. Isomorphous Replacement 151

	6.1.	Introduction 151
	6.2.	Single Isomorphous Replacement 153
	6.3.	Centric Zones 159
	6.4.	Multiple Isomorphous Replacement 160
	6.5.	The Size of the Heavy Atom Contribution 161
	6.6.	Tests for Isomorphism 162

7. Anomalous Scattering 165

	7.1.	Introduction 165
	7.2.	The Dispersion and Absorption Effects 165
	7.3.	The Use of Anomalous Dispersion to Simulate Isomorphism 168

CONTENTS xi

	7.4.	The Breakdown of Friedel's Law 170
	7.5.	Fourier Syntheses for Structures with an Anomalous Scatterer 175
	7.6.	The Use of Anomalous Scattering Differences in the Determination of Phases 177
	7.7.	Choosing the Absolute Configuration using Anomalous Scattering 180

8. **Preparation of Heavy Atom Derivatives** 183

    8.1. Introduction 183
    8.2. Metal Ion Replacement in Metalloproteins 184
    8.3. Heavy Atom Labelled Specific Inhibitors of Enzymes and Carrier Proteins 188
    8.4. Replacement of an Amino Acid with a Heavy Atom Labelled Analogue 191
    8.5. Direct Binding of the Heavy Atom Salts 192
    8.6. Result of Direct Binding of Heavy Atom Reagents . 199
    8.7. The Introduction of New Functional Groups as Heavy Atom Binding Sites 235
    8.8. Techniques for Carrying out Modification of Protein Crystals 236
    8.9. Warning 238

9. **Data Collection** 240

    9.1. Introduction 240
    9.2. X-Rays 242
    9.3. Factors which affect the Measurement of Integrated Intensities 246
    9.4. Cylindrical Polar Co-ordinates 257
    9.5. Photographic Methods 260
    9.6. Diffractometers 284

10. **Data Processing** 310

    10.1. Introduction 310
    10.2. Primary Data Processing: Photographic Data . . 310
    10.3. Primary Data Processing: Diffractometer Data . 321
    10.4. Comparison of Equivalent Reflections 331
    10.5. Scaling Different sets of Data 331
    10.6. Absolute Scale and the Scaling of Derivative to Native Data 333

11. The Determination of Heavy Atom Positions 337

- 11.1. Introduction 337
- 11.2. Estimation of F_H 338
- 11.3. Heavy Atom Vector Maps 343
- 11.4. Difference Fourier Techniques for finding Heavy Atom Positions 349
- 11.5. The Use of Direct Methods in finding Heavy Atom Positions 351
- 11.6. Refinement of Atomic Parameters for Heavy Atoms 352
- 11.7. Correlation of origins 360
- 11.8. General Approach to finding Heavy Atom Positions 361

12. The Calculation of Phases 363

- 12.1. Introduction 363
- 12.2. The Treatment of Errors in the Method of Isomorphous Replacement 364
- 12.3. Inclusion of Anomalous Scattering 371
- 12.4. The Determination of the Absolute Configuration of the Molecule 373
- 12.5. A Simplified Representation of the Phase Probability Distributions, The Hendrickson–Lattman Method 375

13. The Interpretation of the Electron Density Map 381

- 13.1. Introduction 381
- 13.2. Resolution 381
- 13.3. Low Resolution Studies, $dm > 4.5$ Å 387
- 13.4. Medium Resolution Electron Density Maps 3.5 Å $> dm > 2.5$ Å 389
- 13.5. High Resolution Electron Density Maps $dm < 2$ Å . 389
- 13.6. The Display of Electron Density Maps and Model Building for Medium and High Resolution Studies . 390
- 13.7. Interpretation of Electron Density Maps at Medium and High Resolution 399
- 13.8. Determination of the Amino Acid Sequence . . . 400

14. Difference Fouriers 404

- 14.1. Introduction 404
- 14.2. Peak Heights in Difference Fouriers 407

CONTENTS   xiii

    14.3. Errors in Difference Fouriers 409
    14.4. Conformational Changes 411
    14.5. Other Types of Difference Fouriers 416

15. **Refinement and Improvement of Resolution** 420

    15.1. Introduction 420
    15.2. Techniques Available for the Refinement of Protein Models. 421
    15.3. Refinement of Atomic Positions 422
    15.4. Examples of Refinement of Protein Structures . . 432
    15.5. Direct Methods of Refinement of Protein Phases. . 436

16. **Molecular Replacement** 443

    16.1. Introduction 443
    16.2. Orientation of Molecules 444
    16.3. Positioning the Oriented Molecule: the Translation Problem 456
    16.4. Molecular Replacement and the Calculation of Phases 462

17. **The Use of Neutrons and Gamma-rays in Protein Crystallography** 465

    17.1. Introduction 465
    17.2. Generation of Neutron Beams 465
    17.3. The Scattering of Neutrons by Atoms 467
    17.4. Crystal Damage 471
    17.5. Data Collection 472
    17.6. Neutron Fouriers 477
    17.7. Comparison of X-ray and Neutron Amplitudes . . 478
    17.8. Isomorphous Replacement with Neutron Diffraction 478
    17.9. Anomalous Scattering of Neutrons 479
    17.10. Gamma-ray Resonance 482

18. **Electron Microscopy** 483

    18.1. Introduction 485
    18.2. The Principle and Design of the Electron Microscope 487
    18.3. Preparation of Biological Specimens 496
    18.4. Radiation Damage 504
    18.5. High Voltage Electron Microscopy 507

| | 18.6. | Image Analysis . . . . 509 |
| | 18.7. | The High Resolution Transmission Scanning Electron Microscope (STEM) . . . . 513 |

19. Achievements of Protein Crystallography 518

REFERENCES 541

SUBJECT INDEX 559

ACKNOWLEDGEMENTS

We wish to acknowledge the debt we owe to our teachers, Professor D. C. Hodgkin and Professor D. C. Phillips. Their continued help and encouragement has been a great inspiration to us in our work.

We wish to thank the many colleagues who have generously helped us in the preparation of this book, in particular: Dr. R. Ashaffenberg, Dr. S. H. Banyard, Dr. J. Beaumont, Dr. C. W. Bunn, Dr. R. Diamond, Dr. G. G. Dodson, Dr. P. A. M. Eagles, Dr. P. R. Evans, Mr. J. A. Jenkins, Dr. K. R. Leonard, Dr. S. Mason, Dr. J. Mosley, Dr. S. Neidle, Professor, A. C. T. North, Dr. K. W. Snape, Dr. I. J. Tickle, Miss I. T. Weber, Dr. K. S. Wilson, Dr. S. Wood, Dr. D. G. R. Yeates and Professor M. Zeppezauer.

We are most grateful to those authors, publishers and learned societies who allowed us to reproduce published figures and photographs and due acknowledgement is made in each case in the text.

Finally we wish to thank Gail Gibbs and Reiko Sakuma for typing the manuscript.

1

INTRODUCTION

We ought to be neither like spiders which spin things out of their own insides nor like ants which merely collect but like bees which both collect and arrange.

Sir Francis Bacon

1.1 INTRODUCTION

Protein Crystallography is the application of the techniques of X-ray diffraction (and to a lesser extent, those of neutron, γ ray and electron diffraction) to crystals of one of the most important classes of biological molecules, the proteins. Proteins are ubiquitous molecules. They include the enzymes, the biological catalysts which control all chemical processes in living systems; the storage and carrier molecules such as haemoglobin which is responsible for the transport of oxygen in the blood; the hormones, which act as messengers between different parts of the organism; and the antibodies which provide immunity against infection. It is known that the diverse biological functions of these complex molecules are determined by and are dependent upon their three-dimensional structure and upon the ability of these structures to respond to other molecules by changes in shape. At the present time X-ray analysis of protein crystals forms the only method by which detailed structural information (in terms of the spatial co-ordinates of the atoms) may be obtained. The results of these analyses have provided firm structural evidence which, together with biochemical and chemical studies, immediately suggests proposals concerning the molecular basis of biological activity.

The success of these physical techniques in describing biological structure and function does not mean that biology has become a branch of physics. As Bernal (1969) has written: "The use of physical or chemical knowledge to explain mechanical, electrical or chemical aspects of living organisms has brought into greater relief their biological aspects. These phenomena however well they could be described in physical terms do not occur in mechanisms made by some divine craftsman from ideal models laid down from all eternity, but in self-regulating and self-producing entities whose present form is the result of an evolution stretching back over billions of years".

Protein Crystallography has advanced extremely rapidly in the last decade and in certain cases protein structures may now be determined relatively easily using standard methods. (Previously such analyses had proved difficult and time consuming.) Many of the techniques which were being tentatively tried out a few years ago are now well established; other approaches have still to prove fruitful and in certain areas there is a need for new ideas. This recent knowledge is scattered in many different journals and doctoral theses or is simply passed by word of mouth from one generation of research students to the next. It seemed to us that the subject has now reached a stage of maturity when a detailed review in the form of a book is needed. Our aim has been to draw together in a systematic and comprehensible manner those methods that are now well established and to indicate those areas where further expansion is needed.

The subject attracts workers from many different disciplines, such as chemistry, biochemistry, biophysics, zoology, physics and mathematics, and the different insights and skills contributed by these workers are vitally important in this multi-disciplinary subject. The book is primarily intended for those who wish to do research in the field. However, we have attempted to write it in a style which will be comprehensible to workers from these various disciplines.

The principles of protein structure and the implications of the X-ray analyses of biological molecules form important parts of many undergraduate courses in both biological and physical sciences. We hope that this book will also prove useful to undergraduates especially where a deeper insight into the scope and limitations of the technique than is usually provided in review articles is required.

In this introductory chapter we first survey the biochemistry of the proteins which have been and are being studied by protein crystallographers and which recur frequently in our discussions of techniques. We then review the history of our subject and provide a synopsis of the special problems encountered in protein crystallography and the means by which they have been overcome. Thus this chapter to some extent forms a resumé of the book.

1.2 THE NATURE AND FUNCTION OF GLOBULAR PROTEINS

We have already seen that globular proteins may have many roles in organisms including biological catalysis and regulation of cellular processes. It was in fact their wide range of biological functions and their consequent importance in living systems which led to the name "proteins" which is derived from a Greek word meaning first. We shall see in Chapter 2 that proteins are characterized by their hydrolysis to amino acids in dilute solutions of mineral acids at boiling point. In fact all proteins are built up from linear polymers of amino acids called

1.2 THE NATURE AND FUNCTION OF GLOBULAR PROTEINS

polypeptide chains. They may be divided into two classes, fibrous proteins and globular proteins, on the basis of their structure and function.

Fibrous proteins are relatively insoluble in aqueous solutions. These include the keratins of wool, hair and horn, the myosin of muscle, the fibroin of silk and the collagen of tendon. These fibrous proteins have a structural role. They are either protective materials or they are nature's ropes and girders. The fibrous nature and the insolubility of these materials make crystallization with good three dimensional order almost impossible. Consequently the information we can gain about their structure is much less precise and the techniques of their structure analysis are quite different from that of globular proteins. The techniques are reviewed by Holmes and Blow (1966); we do not intend to discuss them in this volume.

In globular proteins the polypeptide chain has a complex folded structure. Most globular proteins are soluble in aqueous solutions, but others which are normally located in or on membrane structures will be more soluble in nonpolar solvents. The difficulty of purifying and stabilizing the membrane proteins has meant that most proteins studied by X-ray analysis at high resolution are water soluble proteins and so our present knowledge of structure is really rather unrepresentative of globular proteins as a whole. Further, the crystallographer has tended to study the more stable proteins and these have often been extracellular proteins.

Let us consider in more detail the globular proteins which have been the subject of X-ray analysis. For the reader who has no biochemical training we recommend one of the standard texts such as "Biological Chemistry" by Mahler and Cordes (1968) or "Biochemistry" by Lehninger (1975). Here we will try to outline the biological roles of some of the most intensely studied proteins. We hope that this will give the reader some impression of the purpose of the structural study when we are discussing the techniques of protein crystallography.

Almost all of the proteins studied by X-ray analysis can be classified in one of the five following groups:

1. Enzymes
2. Redox proteins
3. Carrier and storage proteins
4. Hormones
5. Antibodies

Let us consider enzymes first.

Enzymes are catalysts; they alter the rate of biochemical reactions without altering the position of equilibrium. They are important in all biological processes. They are classified according to the kind of reactions catalysed; the classification is illustrated in Table 1.I with the enzymes which are the subject of discussion in this book.

Many of the enzymes are hydrolases and are involved in degrading biological macromolecules. Many also are extracellular, such as pepsin and chymotrypsin; these enzymes are synthesized in the gastric mucosa of the stomach and in the pancreas respectively and are secreted from these organs to facilitate digestion of protein foods.

These enzymes have been extensively studied biochemically. This fact together with their stability and the fact that they can be obtained in large quantities and are easily purified has made hydrolases favourite topics of study by the protein crystallographer. Many of these enzymes are synthesized as precursors called zymogens, which have a longer polypeptide chain than the enzymes and are catalytically inactive. Examples are pepsinogen and chymotrypsinogen. Cleavage of a part of the peptide of the zymogen by enzymes leads to production of the active enzyme. This mechanism allows the cell to control the expression of the enzyme activities. The structures of chymotrypsinogen and chymotrypsin have been determined.

Three of these enzymes, chymotrypsin, trypsin and elastase, have similar chemical sequences and protein crystallographers have shown that they have similar three dimensional structures. They have the same proteolytic mechanism but differ in their specificity for the peptides cleaved. They are called homologous enzymes and it has been concluded that they have diverged from a common ancestral protein. However not all proteins with similar catalytic activities are homologous. The enzyme, subtilisin, is closely related to the chymotrypsin-like enzymes in its function but its molecular architecture is quite different. Whereas homologous enzymes are examples of divergent evolution, the similarities in function between subtilisin and chymotrypsin appear to result from convergent evolution.

More recently intracellular enzymes active in the synthesis and degradation of glycogen to small molecules such as pyruvate and lactate have been widely studied. They include the *oxido-reductases* like glyceraldehyde phosphate dehydrogenase (GAPDH) and lactate dehydrogenase (LDH) which are involved in oxidation – reduction reactions. They also include *transferases* such as the phosphorylating enzymes which are usually called kinases, for example phosphoglycerate kinase (PGK) or hexokinase, and *isomerases* such as triose phosphate isomerase. Many of these proteins have sub-units; they are composed of 2 or more identical or nearly identical polypeptide chains.

The degradation of glycogen (catabolism) gives rise to energy in the form of rather unstable (high energy) compounds such as ATP and also to electrons. The electrons flow down a potential gradient as they are transferred from one redox protein to another. The *redox proteins*, the second group of proteins, are associated with cell organelles, the mitochondria and chloroplasts, and many are membrane bound. They include the cytochromes, flavodoxin, rubredoxin, ferredoxin and high potential iron protein, all of which have been studied by X-ray analysis.

1.2 THE NATURE AND FUNCTION OF GLOBULAR PROTEINS

Carrier and storage proteins also have an important role. The electrons from the redox proteins eventually reduce oxygen to water. The oxygen in higher animals is carried by haemoglobins which have four subunits. Haemoglobin has been studied by X-ray analysis since 1937. It shows very sensitive cooperative effects between subunits which enable the haemoglobin to quickly unload its oxygen at the correct place. This mechanism is also sensitive to protons (pH) which act by changing the protein conformation by binding to a site which is distinct from the oxygen binding site, i.e. it is an allosteric effect. Some primitive haemoglobins (from chironomus, glycera and lamprey) have only one subunit. Myoglobin, the first molecule successfully studied by X-ray analysis at high resolution is a related protein with only one subunit; its function is to store oxygen.

The intracellular metabolic processes like the glycolytic pathway are regulated by *hormones*. Many of these are protein molecules. Activity in the glycolytic pathway depends on the availability of glucose and this is regulated by two hormones — insulin and glucagon. Insulin increases the transport into the cell and therefore decreases the levels of glucose in circulation in the bloodstream. The second hormone, glucagon, increases the sugar levels in circulation. Both hormones are synthesized and stored in the pancreas and their release is controlled by circulating sugar concentrations. Both hormones have been studied successfully by X-ray analysis.

The recognition of foreign bodies in an organism is achieved by *antibodies*. There is a wide diversity of antibodies and they are difficult to prepare in a purified state. However, some antibody cells are cancerous — myeloma — and produce large quantities of the same antibody. The X-ray analysis of some of these proteins has been carried out. (Antibodies are composed of 4 polypeptide chains — 2 light chains and 2 heavy chains.) They can be cleaved to give a variable antibody binding fragment, Fab, and a fragment which is common to many antibodies and easily crystallizes, Fc. The light chains, Bence-Jones proteins, of myeloma patients are also easily purified. All antibody structures studied by X-ray analysis are from myeloma patients and for these molecules it is very difficult to be sure about the nature of the antigen to which the antibody binds specifically.

One further kind of protein studied by X-ray analysis is the *agglutinant*, concanavalin. The protein has four subunits each of which will recognize a cell surface and cause cells to stick together. This protein has one thing in common with all other proteins we have discussed: it is designed to bind specifically another molecule. Molecular recognition (specific binding) is a function which proteins carry out very efficiently. Thus an enzyme recognizes a substrate, a carrier protein a metabolite, a hormone a receptor, an antibody an antigen and an agglutinant a cell surface. Most globular proteins also give a response on recognition. Thus an enzyme catalyses a reaction of the substrate, the hormone activates or deactivates a metabolic process and the antibody activates a

TABLE 1.I

Classification for some enzymes which have been studied by X-ray analysis

Class	Systematic name	Trivial name	Reaction catalysed
1. Oxidoreductases			
1.1.1.1	Alcohol: NAD oxidoreductase	Alcohol dehydrogenase ADH	Alcohol + NAD^+ ⇌ aldehyde or ketone + NADH
1.1.1.27	L-Lactate: NAD oxidoreductase	Lactate dehydrogenase LDH	L-Lactate + NAD^+ ⇌ pyruvate + NADH
1.1.1.37	L-Malate: NAD oxidoreductase	Malate dehydrogenase MDH	L-Malate + NAD^+ ⇌ oxaloacetate + NADH
1.2.1.12	D-Glyceraldehyde-3-phosphate: NAD oxidoreductase (phosphorylating)	Glyceraldehyde phosphate dehydrogenase GAPDH Triosephosphate dehydrogenase	D-Glyceraldehyde-3-phosphate + orthophosphate + NAD^+ ⇌ 1,3-diphospho-D-glyceric acid + NADH
1.2.4.1	Pyruvate: lipoate oxidoreductase (accepter-acetylating)	Pyruvate dehydrogenase	Pyruvate + oxidised lipoate ⇌ 6-S-acetyl dihydrolipoate + CO_2
1.11.1.6	Hydrogen-peroxide: hydrogen peroxide oxidoreductase	Catalase	$H_2O_2 + H_2O_2$ ⇌ $O_2 + 2H_2O$

2. Transferases

EC	Name	Systematic name	Reaction
2.1.3.2	Aspartate transcarbamoylase	Carbamoylphosphate: L-aspartate carbamoyl transferase	Carbamoyl phosphate + L-aspartate ⇌ orthophosphate + N-carbamoyl-L-aspartate
2.7.1.1	Hexokinase Glucokinase	ATP: glucose-6-phosphotransferase	ATP + D-glucose ⇌ ADP + D-glucose-6-phosphate
2.7.2.3	Phosphoglycerate Kinase PGK	ATP: 1,3-Diphosphoglycerate phosphotransferase	ATP + 3-P-glycerate ⇌ ADP + 1,3-Diphosphoglycerate
2.7.1.4	Pyruvate Kinase	ATP: pyruvatephosphotransferase	ATP + pyruvate ⇌ ADP + phosphoenol pyruvate
2.7.4.3	Adenylate Kinase	ATP–AMP phosphotransferase	ATP + AMP ⇌ ADP + ADP
2.4.1.1	(α-1,4-Glucan) Phosphorylase	(α-1,4-Glucan) orthophosphate glucosyltransferase	$Glycogen_{n+1} + P_i \rightleftharpoons glycogen_n + glucose$-1-P
2.8.1.1	Rhodanese	Thiosulphate: cyanide sulphurtransferase	$Na_2S_2O_3 + HCN \rightleftharpoons HSCN + Na_2SO_3$
2.7.5.1	Phosphoglucomutase	Phosphoglucomutase	α-D-glucose 1,6-diphosphate + α-D-glucose-1-P ⇌ α-D-glucose-6-P + α-D-glucose; 1,6-diphosphate

TABLE 1.1 (continued)

Class	Systematic name	Trivial name	Reaction catalysed
3. Hydrolases			
3.1.3.1	Orthophosphoric mono-ester phosphohydrolase	Alkaline phosphatase	An orthophosphoric monoester + $H_2O \rightleftharpoons$ an alcohol + H_3PO_4
3.2.1.1	α-1,4-Glucan-4-glucano-hydrolase	α-Amylase	Hydrolyses α-1,4 glucan links in poly-saccharides containing three or more α-1,4-linked D-glucose units
3.2.1.17	N-acetylmuramide glycanohydrolase	Lysozyme (Muramidase)	Hydrolyses β-1,4 linkages in polysaccharides composed of either N-acetylglucosamine (GlcNAc) or alternating residues of (GlcNAc) and N-acetylmuramic acid (MurNac). In the latter case only hydrolyses link between MurNac and GlcNAc and not between GlcNAc and MurNac.
3.4.2.1	α-Carboxypeptide amino acid hydrolase	Carboxypeptidase A	Hydrolyses peptides, splitting off the C-terminal L-amino acid residue unless this is a basic residue or proline
3.4.4.1	Peptide peptidohydrolases	Pepsin	Hydrolyses peptides, including those with bonds adjacent to aromatic or decarboxylic L-amino acid residues
3.4.4.3		Rennin (Chymosin)	Hydrolyses peptides with bonds adjacent to aromatic groups or methionines
3.4.4.4		Trypsin	Hydrolyses peptides, amides, esters etc. at bonds involving the carboxyl groups of L-arginine or L-lysine
3.4.4.5		Chymotrypsin	Hydrolyses peptides, amides, esters etc. especially at bonds involving the carboxyl groups of aromatic L-amino acids

Code	Name	Description/Reaction
3.4.4.7	Elastase (pancreatopeptidase E)	Hydrolyses peptides, amides, esters etc. especially at bonds involving the carboxyl groups of L-amino acids with small uncharged side chains i.e. L-alanine
3.4.4.16	Subtilisin (subtilopeptidase A)	Hydrolyses peptides, amides, esters with low specificity
	Thermolysin	Hydrolyses peptide bonds on the imino side of hydrophobic residues such as L-isoleucine or L-phenylalanine
3.4.4.10	Papain	Hydrolyses peptides, amides and esters with a L-phenylalanine next but one on the carboxyl side of the bond cleaved
3.1.4.7	Nucleic acid hydrolases Nuclease (DNAase)	Hydrolyses deoxyribonucleic acids
3.2.2.1	N-ribosyl-purine ribohydrolase Ribonuclease (RNase)	Hydrolyses ribonucleic acids
4. Lyases		
4.1.2.7	Ketose-1-phosphate aldehyde-lyase Aldolase	A ketose-1-phosphate $\rightleftharpoons$ dihydroxyacetone phosphate + an aldehyde
4.2.1.1	Carbonate hydro-lyase Carbonic anhydrase	$CO_2 + H_2O \rightleftharpoons H^+ + HCO_3$
4.2.1.11	2-phospho-D-glycerate hydro-lyase Enolase	2-P-Glycerate $\rightleftharpoons$ phosphoenol pyruvate
5. Isomerases		
5.3.1.1	D-glyceraldehyde-3-phosphate-ketol-isomerase Triosephosphate isomerase (TIM)	Dihydroxyacetone-P $\rightleftharpoons$ glyceraldehyde-3-P
5.4.2.1	D-phosphoglycerate 2,3-phosphomutase Phosphoglycerate mutase (PGM)	3-P-Glycerate $\rightleftharpoons$ 2-P-glycerate
6. Ligases		
6.1.1.b	L-Phenylalanine: tRNA ligase (AMP) Phenyl-tRNA Synthetase	ATP + L-phenylalanine + tRNA $\rightleftharpoons$ AMP + pyrophosphate + L-phenylalanyl-t-RNA

complement system which leads to the degradation of the foreign body. The secrets of the mechanisms of recognition and response of globular proteins lie in their chemical and three dimensional structures. We must now turn again to the techniques used in determining that three dimensional structure.

1.3 THE DEVELOPMENT OF PROTEIN CRYSTALLOGRAPHY AND ITS LIMITATIONS

Protein crystallography is a relatively young science. The stories of the early days make fascinating reading. Stories of how pepsin crystals (2 mm long) were carried in a coat pocket from Sweden to Cambridge for J. D. Bernal who made the first X-ray pictures (Bernal and Crowfoot, 1934). Stories of Dorothy Crowfoot-Hodgkin cycling back from Somerville late at night to make sure her insulin crystals had indeed diffracted. In the early days each new picture caused great excitement and speculation. They showed that these large molecules were ordered in a crystal lattice and that their structures might be determined by X-ray diffraction techniques. However at that time little was understood of the nature of proteins and the methods by which their structures might be solved were unknown. In 1937 M. F. Perutz chose the determination of the crystal structure of haemoglobin as the subject for his thesis. Fortunately his examiners did not insist on a complete structural analysis. In those days the analysis of small molecules containing only a few atoms was proving a formidable problem. It must have required fantastic vision to believe that the complicated patterns of protein crystals could be transformed into equally complex structures.

The experiments of Bernal, Hodgkin and Perutz clearly showed that protein crystallography differed from conventional crystallography both quantitatively and qualitatively. In the next fifteen years the techniques of X-ray analysis of crystals of smaller molecules were developed, and many crystal and molecular structures were solved. But little progress was made in the study of protein crystals. There were several differences which made progress difficult.

First, protein crystals are composed, typically, of about 50% water and lose their crystallinity if they become dehydrated. This was recognized by Bernal and was easily overcome by keeping the crystals in a sealed tube with a drop of their mother liquor. However, the open lattice structure of the protein crystals has a more serious consequence. The protein molecules are less ordered than is usually found for small molecules. With crystals of small molecules the diffraction pattern often extends to the diffraction limit set by the wavelength of the X-radiation so that atomic resolution is possible. With protein crystals the experimenter is forced to work at very much worse than atomic resolution.

In all X-ray crystal structure analyses the problem is to find not only the amplitudes of all the diffracted X-rays (usually known as reflections) but also

1.3 THE DEVELOPMENT OF PROTEIN CRYSTALLOGRAPHY

their phases. Knowledge of both the amplitudes and phases allows the crystallographer to reconstitute the electron density of the crystal which gave rise to the diffracted beams. The amplitudes can be deduced from the intensities of the diffracted X-rays but the phases cannot be directly measured. This is known as the "phase problem". For protein crystals this is a particularly difficult problem, and even measurement of X-ray intensities is not straightforward.

The protein molecules contain a very large number of atoms and there are therefore a correspondingly large number of X-ray diffraction measurements to be made. Moreover, the intensities of the diffracted beams are relatively weak when compared to those of small crystal structures. These difficulties raise technical problems in data collection. Development of more sensitive detectors and automatic methods capable of coping with large numbers of measurements are required. A variety of analytical methods are available for the solution of the phase problem for small molecule X-ray analyses. These include the so-called direct methods, Patterson methods, trial and error methods and heavy atom methods which we describe in chapter 5. At the present time the success of these methods is limited to those unit cells having <100 atoms.

Over twenty years were to elapse from the time of the first X-ray photograph of a protein to the time when Perutz and his colleagues working in Sir Lawrence Bragg's Cambridge laboratory were able to show that the method of isomorphous replacement could be used to solve the phase problem for protein structures (Green et al., 1954). They showed that a heavy atom (an electron dense atom) like a mercury atom could be introduced into a protein crystal without disturbing the crystal packing to produce an isomorphous derivative crystal with observable intensity changes in the diffraction pattern. The method of isomorphous replacement had been used successfully with small structures previously. Perutz's achievement was to demonstrate that it could be useful in protein structure analysis. In 1959 the method was put on a sound systematic basis by D. M. Blow and F. H. Crick (Blow and Crick, 1959, See also Dickerson et al., 1961) and their proposals have been the starting point for all subsequent developments in the subject. In 1960 these methods led to the successful solution of the first protein crystal structure, that of myoglobin (Kendrew et al., 1960). This was followed by an enzyme, lysozyme, in 1965 (Blake et al., 1965) where the method of isomorphous replacement was greatly strengthened by the use of anomalous scattering data. (Anomalous scattering is caused by interaction of X-rays with strongly bound electrons.) Other enzyme structures, those of ribonuclease (Kartha et al., 1967; Wyckoff et al., 1967b), chymotrypsin (Matthews et al., 1967) and carboxypeptidase (Reeke et al., 1967) followed in quick succession in 1967.

During the last decade many structures have been reported at high resolution, see Chapter 19, Table 19.I, but only recently have the structures of any of these proteins been refined in a way which has been usual for small molecules since the 1950's. Only now are we able to report objective criteria for the correctness of our protein structures.

In protein crystallography the structure should not be the end point of the analysis. A successful solution of the structure forms the starting point for a whole range of experiments designed to elucidate its biological function. These experiments which have usually involved the study of interactions of substrate and metabolites with protein molecules have also led to the development of new techniques.

However, in its attempts to provide answers to biological questions, protein crystallography has suffered and will continue to suffer from severe limitations and it is well that these limitations should be explained at the outset. First the protein must be crystallized. Does this crystallization change its structure? Fortunately the answer appears to be "no" if we accept that the structure in the crystal respresents a stable equilibrium conformation of the molecule. [For a more detailed account see reviews by Mathews (1974) and Hess and Rupley (1971).] The forces that hold molecules together in a crystal lattice are very much weaker than those that hold the protein structure together so that gross conformational changes are unlikely. In some cases the same proteins, or the same proteins from different species, have been crystallized under very different conditions and yet the crystal structures of the molecules are found to be closely similar. Evidently salt concentration, solvent and pH conditions used in crystallization have very little effect on the gross structure (for example see Drenth *et al.*, 1971b,c). However, *small* conformational differences between the protein in the crystal and the protein in solution cannot be ruled out, although it is encouraging that some proteins do retain their biological activity in the crystal (for example see Quiocho *et al.*, 1972). In the absence of detailed structural information for the protein in solution, we must accept the crystal structure as a starting point for the interpretation of the behaviour of the molecule in solution and search for compatibility between properties predicted by the structure and those observed in solution.

Biological processes are essentially dynamic. A second limitation of the X-ray method is that it is essentially static. A set of X-ray measurements takes several days to record and the resulting image of the molecule which is obtained is time-averaged over this period. We cannot hope to define the different transient structural states adopted by a protein as it carries out its biological function by X-ray techniques alone (at least with present day methods). Close collaboration with other disciplines is necessary in order to test hypotheses within the framework of the known static structure. The continuing development of nuclear magnetic resonance techniques appears most promising for the detection of the transient structures not easily accessible to crystallography and may provide important structural information that is at present beyond the limit of resolution of diffraction methods.

1.4 STEPS IN THE X-RAY ANALYSIS OF CRYSTALLINE PROTEINS

We have tried to organize this book in a way which reflects the sequence of steps usually followed in a successful protein crystal structure analysis. However, these steps require two kinds of background knowledge. The first of these is biochemical. It concerns the principles of protein structure and we review these in Chapter 2 for the reader who is a physicist or physical chemist. The second area of knowledge concerns crystallography and X-ray analysis and these are described in chapters 4 and 5 mainly for the benefit of the reader with a biological training. These chapters provide the background knowledge for the rest of the book. They are necessarily concisely written but we give references to other simple as well as more extensive texts to which the reader may wish to refer. The method of *isomorphous replacement* and the use of *anomalous scattering* are central to the X-ray structure analyses of proteins. The principles of these methods are outlined in chapters 6 and 7. Each of the steps in the methods is then discussed in the chapters following.

The X-ray analysis of a protein structure usually proceeds in eight distinct steps: crystallization, preparation of isomorphous heavy atom derivatives, collection of data, data processing, determination of heavy atom positions, calculation of phases, interpretation of the electron density maps, and refinement. The straight-forward systematic appearance of this list is misleading because in practice there is a continuous feedback between the various stages. Moreover the steps are not independent and the success of each step depends on the success of the preceding steps.

Step (i) Crystallization (Chapter 3)

The first step in any structural analysis is the crystallization of the molecule (Chapter 3). Many proteins are prepared in a "micro-crystalline" form by biochemists and new techniques of growing and keeping protein crystals of up to 1 mm in length have been developed. Advances have come from an understanding of the factors which affect protein crystallization such as solubility, pH, temperature, the effects of different solvents and additives and knowledge of variations in proteins among different species. Many proteins are available in very small quantities (the structure of the redox protein rubredoxin was solved with as little as 15 mg of protein) and in recent years there has been careful study of micromethods. In the past proteins were often chosen for study because they happened to crystallize well. Nowadays a protein can be chosen because of its intrinsic interest to biology and the whole battery of new crystallization methods can be applied to produce suitable crystals.

Step (ii) Preparation of heavy atom derivatives (Chapter 8)

In the method of isomorphous replacement a heavy atom must be introduced into the crystal in such a way as to cause minimum conformational changes in the protein molecule. In theory only one heavy atom derivative is required provided that the anomalous scattering is significant. In practice, usually more than one derivative is used. The introduction of the heavy atom, either by co-crystallization or diffusion into preformed crystals, has been mostly a "trial and error" process. The complexity of protein structures and the intricate nature of the molecular packing in the crystals has meant that the rationalization of the preparation of isomorphous derivatives has been difficult and rather unrewarding. Nevertheless the accumulated experience from many successful and unsuccessful experiments has allowed some generalizations which are proving helpful in new work. These are discussed in chapter 8.

Step (iii) Data collection (Chapter 9)

Having overcome the two most important rate-limiting steps in the analysis (i.e. crystallization and preparation of derivative crystals), the protein crystallographer is next faced with the measurement of the intensities of the X-ray diffraction pattern. This is the subject of Chapter 9. We have already seen that it is here that the quantitative difference between protein crystallography and small structure crystallography is most apparent. Not only are there a very large number of diffracted beams (or reflections) to be measured (typically some 10,000 for a high resolution study of a medium sized protein) but also these intensities are relatively weak. Rapid and efficient methods for the measurements are required. The sudden increase in the number of protein structures in the late nineteen-sixties was due in no small way to the availability of fully automatic diffractometers. Modern proportional and scintillation counters are able to count individual X-ray quanta and to record directly the intensity of each diffracted beam in terms of the number of quanta received per second. The combination of these detectors with an instrument capable of moving the crystal and detector from one diffracted beam to the next without the need for manual intervention provided an accurate and labour-free method of data collection. Modification of diffractometers to measure up to five reflections at a time has further improved efficiency. Nevertheless diffractometers remain relatively slow especially for large proteins. The recent development of fully automatic densitometers capable of measuring the integrated intensity of every spot on a photographic film has led to the return to photographic methods in many laboratories. For the future it is possible that we may have an instrument which combines the merits of the camera geometries previously used with photographic techniques and counter efficiency. Position-sensitive proportional counters of

1.4.6 CALCULATION OF PHASES (CHAPTER 12)

the type widely used in high energy physics are beginning to be used in protein crystallography and they may allow a reduction in recording times by a factor of sixty compared with photographic film.

All protein structures have been solved using copper K_α radiation (wavelength 1.5418 Å) from conventional X-ray tubes either of the sealed-off fine focus design or the rotating anode variety. The rotating anode provides an increase in brilliance of, approximately, a factor of four over the stationary target tube but for many experiments, such as those with very small crystals or with very large molecules, an even more intense beam is desired. Recently attention has turned towards the X-radiation emitted by an electron synchrotron. This radiation is most intense in the appropriate wavelength range (1–2 Å) and after monochromatization and collimation may be one hundred times more brilliant than that produced by a rotating anode tube. However the use of more intense sources is likely to exacerbate the problem of radiation damage, a problem which is always with the protein crystallographer. Very little work has been done on the fundamental nature of radiation damage or its prevention. Empirical radiation damage correction curves are frequently applied to data but these have little theoretical basis.

Step (iv) Data processing (Chapter 10)

Data processing has been greatly facilitated by the rapid advances in computer technology. Systematic corrections, comparison of equivalent reflections and the scaling of data sets from different crystals are readily carried out. Absorption probably forms the most important and most difficult systematic correction. Several semi-empirical methods have been developed especially for protein crystallography, and are discussed in chapter 10.

Step (v) Determination of heavy atom positions (Chapter 11)

In order to determine the phases of the protein reflections by the heavy atom isomorphous replacement method it is necessary to determine the positions of the additional heavy atoms within each of the derivative crystals. The reliability of the final result rests on the accuracy with which these positions are determined. There have been many useful advances in this field, especially with respect to the combination of isomorphous and anomalous scattering data to give a better estimate of the observed contribution of the heavy atom to the derivative factor amplitudes; these are reviewed in Chapter 11.

Step (vi) Calculation of phases (Chapter 12)

The heavy atom contribution of several isomorphous derivatives can be combined with the measured structure amplitudes of the native protein and derivatives and the anomalous scattering measurements of the derivatives to give information about the phases of the protein reflections. This stage in the analysis has led to many ideas about the use of isomorphous replacement and anomalous scattering data and the proper treatment of the errors in these data. The phases are combined with the structure factor amplitudes of the native protein in the calculation of the Fourier summation to give the electron density map.

Step (vii) Interpretation of electron density maps (Chapter 13)

The manner of interpretation of the map will depend on the resolution of the study. Little information is available at low resolution ($d_m > 5$ Å) beyond the overall shape and dimensions of the molecule. Such maps are frequently calculated, however, in order to check the progress of the structure analysis. At medium to high resolution ($d_m < 3$ Å) the course of the polypeptide chain may be traced and side chains identified. Since the resolution is usually less than atomic resolution, individual atoms cannot be identified, but bonded-groups of atoms separated from their non-bonded neighbours by Van der Waals distances are resolved and can be distinguished. The interpretation is usually made with the aid of mechanical components (on a scale of 2 cm/Å) which have been accurately manufactured so that they represent the standard bond lengths and angles established by X-ray studies on individual amino acids. The model and map may be viewed simultaneously by means of a half silvered mirror. The interpretation of the amino acid residues in the protein almost always depends on knowledge of the amino acid sequence from chemical studies. Indeed for large proteins it has proved impossible to give a full account of the structure unless the structural information is complemented by the chemical studies. Recently very high resolution maps ($d_m \sim 1.5$ Å) have been achieved for several of the smaller proteins, including the redox protein rubredoxin and the hormone insulin, and these have proved most promising for they show that many atoms can be located very precisely and that identifications of the great majority of the side chains can be made without the need for sequence information.

Mechanical models suffer from obvious disadvantages associated with their bulkiness, lack of flexibility and difficulty in recording co-ordinates. The development of computer methods, where the map and the model are displayed simultaneously on a cathode ray tube screen, overcomes these disadvantages to some extent and allows the construction of a precise and stereochemically reasonable model for the polypeptide chain. However, computer methods suffer from the disadvantages associated with the attempt to represent a three dimensional object in two dimensions.

Step (viii) Refinement of the structure (Chapter 15)

For many structures the analysis has terminated with the subjective interpretation of the electron density map. It is only very recently that objective methods of refinement have been developed, which provide the precise structural information on which the answers to many important biological questions depend. Routine application of conventional least square minimization of the differences between observed and calculated structure factor amplitudes is not possible, largely because the number of observables scarcely exceeds the number of unknown parameters, if all atoms are considered as independent. Several new techniques have been specially developed for protein crystallography. These are considered in Chapter 15. Analysis of the results of the refinements by these methods indicate that very much more detailed information on protein structures is likely to be available in future. The recent refinement of the small protein rubredoxin at 1.2 Å resolution suggests that protein crystallography is now beginning to approach the limits of accuracy placed upon it by the quality and order of the protein crystals themselves rather than the techniques used in the analysis.

1.5 ALTERNATIVE TECHNIQUES

One method other than isomorphous replacement and the use of an anomalous scattering seems to have a particular potential in protein crystallography. This is the method of *molecular replacement*. This method depends on the presence of related structures either in the same crystal asymmetric unit or alternatively in different crystals. It is particularly relevant to proteins with many identical subunits or to proteins which are homologous (evolved from the ancestral protein) and which have closely similar structures. The near identity of the structures implies relations between different structure amplitudes and phases which are helpful in solving the phase problem. These techniques are discussed in Chapter 16.

We include a discussion of the applications to protein crystallography of *neutron diffraction* and *gamma ray resonance* (Chapter 17) and *electron microscopy* (Chapter 18), methods which are relevant and complementary to the X-ray method. Through most of the book we shall be concerned with the techniques of protein crystallography, but it is implicit that the ultimate goal of protein crystallography lies not in the techniques but in their application to carefully chosen biological problems. Therefore the last chapter is concerned with a short description of some of the achievements of protein crystallography (Chapter 19).

2

THE PRINCIPLES OF PROTEIN STRUCTURE

2.1 INTRODUCTION

Proteins have many different functions in living systems. They act as catalysts, hormones, molecular carriers, antibodies and structural proteins. This impressively wide range of biological roles derives from many different structures. However, the structures themselves are built on simple principles which are common to all proteins.

Proteins are made from twenty different amino acids. These are the building blocks. They are linked together as linear polymers and then folded to give globular or fibrous proteins. Here we are concerned only with globular proteins, and in this chapter we consider the principles of their structure in some detail. For crystallization, preparation of heavy atom derivatives and, of course, for interpretation of the electron density maps, it is very important for the protein crystallographer to have a proper understanding of these structural principles.

2.2 AMINO ACIDS – THE BUILDING BLOCKS

The hydrolysis of proteins with aqueous solutions of mineral acids at boiling point yields a mixture of up to twenty amino acids. They have the general formula of $NH_2-CHR-COOH$. The central carbon atom is known as the α-carbon, the groups bonded to this are the α-amino and α-carboxyl groups, and the atoms in the sidegroup, R, are labelled with the Greek letters β, γ, δ and so on. Each amino acid has a tetrahedral arrangement of bonds around the α-carbon, and this leads to the possibility of two isomers, which are mirror images, as shown in Fig. 2.1. All amino acids found in proteins (except glycine) have one configuration which is known as the L-configuration, and is most easily remembered by using the "CORN law". The amino acid is viewed along the H–C bond. Remember the α-carboxyl as 'CO', the sidegroup as 'R' and the α-amino group as 'N'. These should occur in a clockwise arrangement CO, R, N for the L-amino acid.

2.2 AMINO ACIDS – THE BUILDING BLOCKS

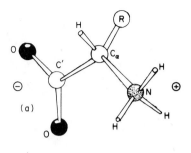

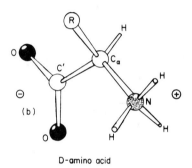

FIG. 2.1 The structures of (a) an L-amino acid and (b) a D-amino acid.

The wide range of amino acid side groups gives rise to the great variation of properties shown by proteins. Many of the side chains are polar (see Fig. 2.2). These can enter into weak interactions involving partial or complete transfer of protons.

The ability to bind a proton is measured by the pK.* (see Table 2.I)

Partial transfer gives rise to a hydrogen bond of the type, M...H...X, but complete transfer gives rise to the formation of an ion. For example the carboxylic groups of glutamic and aspartic acids will easily give up a proton to form an ion:

$$-C\begin{matrix}\nearrow O \\ \searrow OH\end{matrix} \longrightarrow -C\begin{matrix}\nearrow O^\ominus \\ \searrow O\end{matrix} \quad H^\oplus$$

Chemical groups such as these, which easily give up protons, are known as acid groups. On the other hand a number of other amino acid sidegroups, such as the

---

*For acid dissociation: $AH \rightleftharpoons H^+ + A^-$, $K = [H^+][A^-]/[AH]$ and $pK = -\log K$.

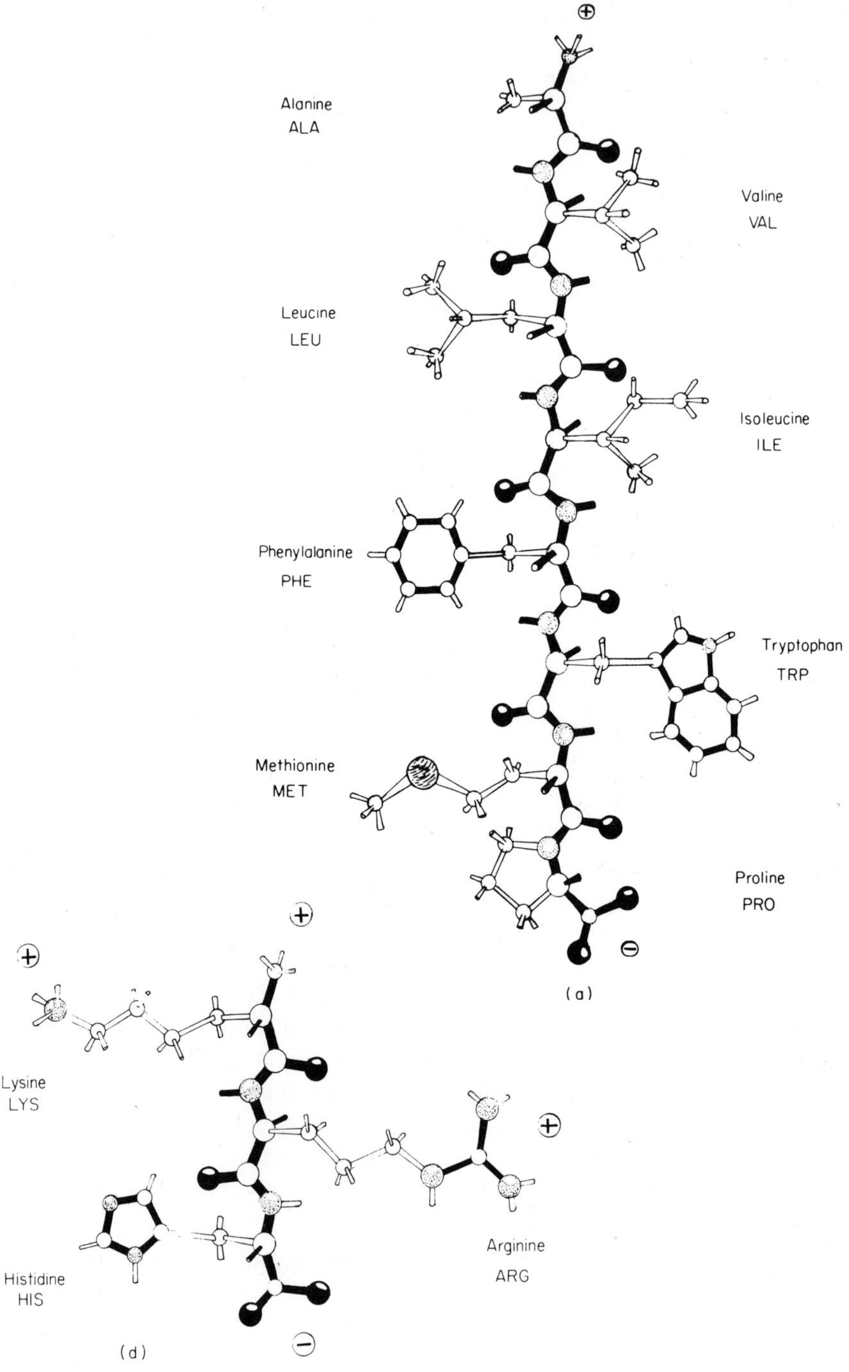

20   2. THE PRINCIPLES OF PROTEIN STRUCTURE

Alanine
ALA

Valine
VAL

Leucine
LEU

Isoleucine
ILE

Phenylalanine
PHE

Tryptophan
TRP

Methionine
MET

Proline
PRO

(a)

Lysine
LYS

Arginine
ARG

Histidine
HIS

(d)

2.2 AMINO ACIDS – THE BUILDING BLOCKS

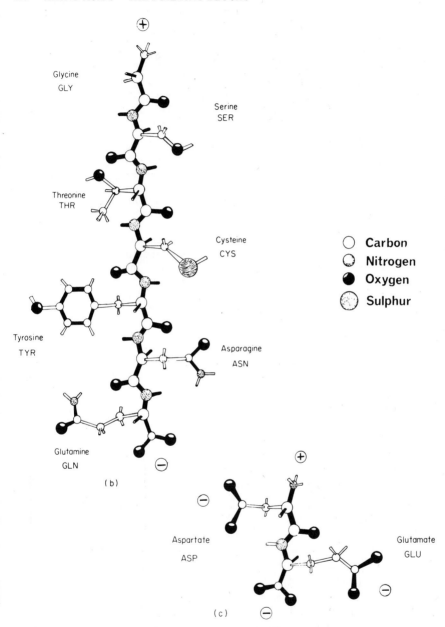

FIG. 2.2 The structures of different amino acid residues in polypeptide chains. The backbone is shaded black. Single bonds of the sidechains are shown in open lines, double bonds are shown shaded black. Non-polar (a), polar uncharged (b), acidic (c) and basic (d) amino acid side chains are illustrated in the forms predominating at pH 7.

TABLE 2.I

pK_1'	α-COOH	2.81–3.58
pK_2'	α-$NH_3^\oplus$	7.91–8.6
pK_R'	aspartic acid	~3.86
pK_R'	glutamic acid	~4.25
pK_R'	histidine	~6.0
pK_R'	cysteine	~8.33
pK_R'	tyrosine	~10.07
pK_R'	lysine	~10.53
pK_R'	arginine	~12.48

The values given of pK_1' and pK_2' are for α-COOH and α-$NH_3^\oplus$ groups in peptides or proteins. These values are rather different from those for free amino acids where the second charged group of the zwitterion is bound to C(α). The pK_R' values are those for free amino acids. They are close to those found in peptides and proteins but there will be some variation depending on the environment of the amino acid.

ε-amino and guanidinium groups of lysine and arginine, can accept a proton and become positively charged: i.e.

$$-NH_2 + H^\oplus \rightarrow -NH_3^\oplus$$

These are known as basic groups. When positively and negatively charged groups are close together they give a favourable electrostatic interaction and are known as an ion pair. In reality, interactions vary continuously between the extreme situations of an ion pair and hydrogen bond, as shown in Fig. 2.3. All groups which take part in interactions of these kinds can also bind molecules of water and for this reason are known as hydrophilic (water-loving) groups. Under physiological conditions in neutral solutions, the α-carboxyl and α-amino groups of an amino acid will also be charged; an amino acid acts as a double ion or "zwitterion" of the form $^\oplus H_3N-CHR-CO_2^-$. These charged groups may also participate in polar or hydrophilic interactions.

In contrast, many amino acid sidechains are non-polar. These sidechains are oily and tend to collect together rather than disperse in an aqueous environment; they are water hating or hydrophobic. The amino acids with such non-polar aliphatic or aromatic sidechains include alanine, leucine, isoleucine, valine and phenylalanine, and are shown in Fig. 2.2a. Hydrophobic interactions between non-polar amino acid sidegroups are an important aspect of the stability and structure of proteins.

2.2 AMINO ACIDS – THE BUILDING BLOCKS

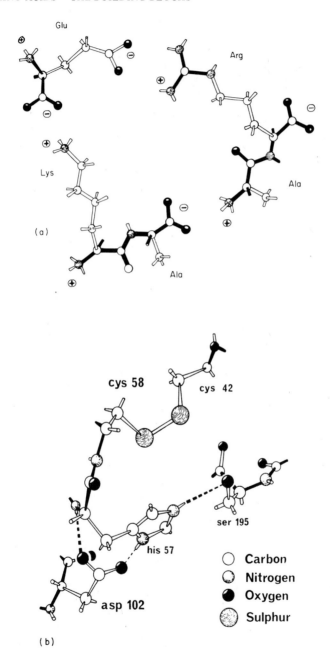

FIG. 2.3 Some examples of weak interactions found in protein structures. (a) ion pairs formed between cationic and anionic groups and (b) hydrogen bonds formed between polar groups at the active site of chymotrypsin.

```
                S―――――――――S
                |           |
A  H-Gly. Ile. Val. Glu. Gln. Cys. Cys. Thr. Ser. Ile. Cys. Ser. Leu. Tyr. Gln. Leu. Glu. Asn. Tyr. Cys. Asn-OH
                         S―S
                         |
                         S
                         |
B  H-Phe. Val. Asn. Gln. His. Leu. Cys. Gly. Ser. His. Leu. Val. Glu. Ala. Leu. Tyr. Leu. Val. Cys. Gly. Glu. Arg. Gly. Phe. Phe. Tyr. Thr. Pro. Lys. Ala-OH
   1    2   3    4    5   6    7    8    9   10  11   12   13  14   15  16   17  18   19  20   21  22   23  24   25  26   27  28   29  30
```

FIG. 2.4 The primary structure of insulin. The symbols for the amino acid residues are defined in Fig. 2.2. The amino acids comprise two chains, called the A and B chains. Each is numbered from the N-terminus.

2.3 THE PRIMARY STRUCTURE

The α-amino and α-carboxyl of the amino acids link together to give a protein molecule. They join end to end, the carboxyl group of one combining with the amino group of the next to form a peptide bond (−CO−NH−) and a molecule of water (HOH). The building blocks are, therefore, called "residues", and amino acid residues linked together in this way are known as peptides. Thus a dipeptide contains two, a tripeptide three, and a polypeptide many amino acid residues. The order of these amino acid residues in polypeptide chains is known as the *primary structure*.

The determination of primary structure has been one of the most exciting developments of chemistry in the last twenty years. The first success was obtained by du Vigneaud, who analysed, and subsequently synthesized, the peptide hormone, oxytocin, which has a cyclic structure of eight amino acid residues. The real breakthrough came with Sanger's determination of the amino acid sequence of a small protein, insulin (Sanger and Thompson, 1953; Sanger and Tuppy, 1951). The techniques used by Sanger's group in England, and by Stein and Moore in America for the sequence study of the larger protein, ribonuclease, form the basis of the techniques which have been used successfully to determine many protein structures. The reader should consult Hirs (1967) or Bailey (1967) for discussions of these methods. As an example the primary structure of insulin is given in Fig. 2.4.

Insulin contains two polypeptide chains which are known as the A and B chains containing 21 and 30 amino acid residues respectively. The amino acid residues are numbered from the amino terminus. The polypeptide chains are held together by the covalent bonds of the disulphide bridges, which link A7 and B7, and A20 and B19. There is also an intrachain disulphide bridge linking A6 and A11.

Studies of the primary structure have considerably increased our knowledge of the nature of proteins. We now know that proteins are always linear chains of amino acid residues; they never branch. However, the significance of the order of the amino acid residues in the chains has become clear only since the three dimensional structure of several proteins has been worked out by means of X-ray diffraction studies.

2.4 THE NATURE OF THE PEPTIDE LINK

To understand the three dimensional structure of proteins we must first consider the nature of the peptide link. From a survey of the information available in 1953 from X-ray structure analyses of various amino acids, peptides and related compounds, Pauling and Corey (Pauling *et al.*, 1951) arrived at a set of

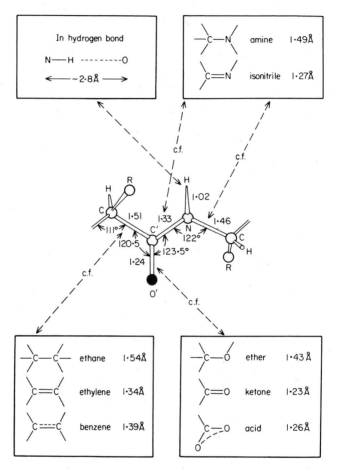

FIG. 2.5 The bondlengths (in Ångstroms) and angles (in degrees) of the peptide link. Lengths of similar bonds found in other molecules are shown for comparison.

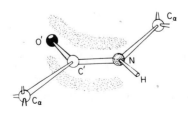

FIG. 2.6 Delocalization of π-electron density in the peptide unit which results in stabilization of a planar structure.

2.4 THE NATURE OF THE PEPTIDE LINK

dimensions for the peptide link. These are very close to those generally accepted today, which are given in Fig. 2.5. They showed that the most important features of the peptide link are its planarity and the related short C'—N and C'—O bond lengths. These tell us much about the bonding of this group which is known as the peptide unit. The shortness of a bond is indicative of its strength. Very short bonds are often double bonds, which have clear directional properties; rotation around the bond is restricted. If the double bond is delocalized over more than two atoms this gives rise to a planar structure. The peptide unit is such a group. The C'—O carboxyl bond of 1.24 Å is very short by comparison with the length of the carbon oxygen single bond of 1.43 Å. Also the C'—N bond, 1.33 Å, of the peptide unit is shorter than the C—N single bondlength, 1.47 Å. These facts and the planarity of the group are generally explained by delocalization of the double bond of the carbonyl group into the C'—N bond. The shape of the delocalized orbital is shown in Fig. 2.6. It will be

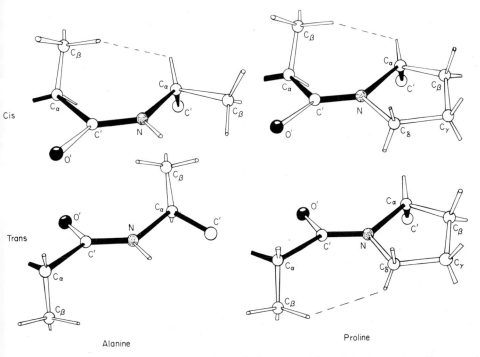

FIG. 2.7 Cis and trans arrangements of planar peptides of alanine and proline. For alanine, an unfavourable contact (indicated by – – – –) occurs between hydrogens bound to Cβ of one amino acid residue and Cα of the next in the cis arrangement; this does not occur in the more stable trans peptide. However, in proline unfavourable contacts occur in both cis and trans peptides, and they are equally stable arrangements. Thus cis arrangements are rarely found in polypeptide chains except in the case of peptides involving proline.

noted that neither of the bonds are short enough to be pure double bonds. However, the partial double bond character strongly favours a planar arrangement.

The arrangement shown in Fig. 2.6 has the C–N and C–C' bonds "trans" to each other with respect to the peptide bond C'–N. There is, of course, another arrangement where these bonds are "cis" to each other which still allows a planar structure and delocalization of the electron density. The "cis" peptide is less stable than the "trans" one because it brings the two adjacent residues closer to each other. The energy difference between the two is 2 Kcal/mole which makes the occurrence of the "cis" form uncommon in most free polypeptide chains. However, it may occur in a proline residue because the delta carbon atom of the ring occurs at a position equivalent to the alpha carbon in the "cis" peptide. This relationship is shown in Fig. 2.7. Therefore, we may expect to find a "cis" peptide only in a proline-containing peptide or in part of the structure where there is considerable strain induced in the polypeptide chain.

Although there is a very small difference of energy between the two planar configurations, the energy required for rotation about the peptide C'–N bond by 90° so that the C–C' and C–N bonds are in planes perpendicular to each other is much higher. It has been variously estimated to lie between 15 and 30 Kcal/mole (Ramachandran and Sasisekharan, 1968; Levitt and Lifson, 1969), and this conformation has not been observed. However, a twist of 10° from planarity would involve less than a Kcal/peptide, and it is now recognized that distortions from planarity of this order probably do occur in the polypeptide chains of native proteins (Winkler and Dunitz, 1971; Ramachandran and Kolaskar, 1973).

2.5 PEPTIDE STRUCTURE

We have seen that the peptide units in a protein probably have a planar trans configuration, and that rotation around the C–N bond is very difficult as it is a partial double bond. However, the lengths of C–C' and C–N bonds (1.51 Å and 1.46 Å respectively) are those expected for single bonds, and they will allow unrestricted rotation. In fact, rotation of the peptide around these bonds gives the polypeptide chain its flexibility. The different three dimensional structures obtained are known as conformations, and so the conformation of the polypeptide can be defined by rotation around these two bonds at each of the α-carbon atoms.

Note that we use the term "conformation" to refer to different structures which are generated by rotation about single bonds (e.g. we talk of the conformation of a polypeptide chain). We restrict the term "configuration" to describe arrangements of atoms about a single atom which are not superimposable by rotation (e.g. the D or L configuration of an amino acid). These two terms are often confused in old biochemistry textbooks.

2.5 PEPTIDE STRUCTURE

The rotations around the bonds C–N and C–C' are given by the dihedral angles ϕ (PHI) and ψ (PSI) respectively as shown in Fig. 2.8a. These angles are taken at 180° when the two units are coplanar in the arrangement shown in Fig. 2.8a. In the $\phi = \psi = 0$ conformation the two bonds adjacent to N_1–C (i.e. C'_1–N_1 and C–C'_2) and the two bonds adjacent to C–C'_2 (i.e. N_1–C and C'_2–N_2) are eclipsed. (Fig. 2.8b)

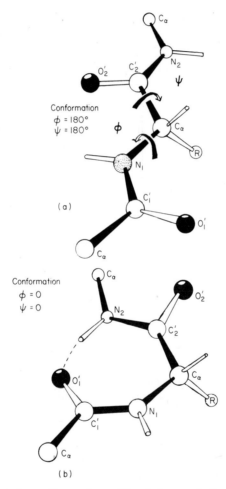

FIG. 2.8 Conformations of a polypeptide chain are defined by the dihedral angles (PHI) and (PSI) as shown in (a). Positive increases in the angles are accomplished by rotating the peptide unit clockwise (when viewed from Cα) round the C–N and C–C' bonds while the Cα is held in the same position. (a) shows an extended chain with $\phi = 180°$, $\psi = 180°$ and (b) shows an eclipsed conformation with $\phi = 0$, $\psi = 0$ which involves an unfavourable contact between O' of one peptide and the amino hydrogen of the next.

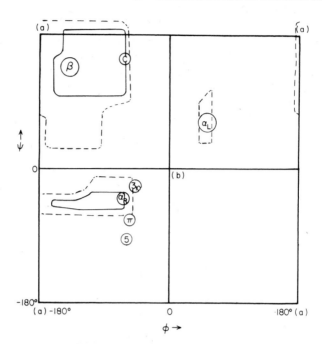

FIG. 2.9 A Ramachandran plot for a polypeptide chain of L-alanines. Continuous lines enclose regions with allowed conformations, and dashed lines define partially allowed conformations. The conformation (a) shown in Fig. 2.8(a) is close to the conformation of β-pleated sheet, (β). The eclipsed conformation (b) (shown in Fig. 2.8b) is in the centre of the plot and is disallowed. The 3_{10} helix, (3_{10}), the right-handed 3.6_{13} or alpha helix (α_R) and the π or 4.4_{16} helix (π) are in or close to allowed regions. The left-handed α-helix (α_L) is less stable than (α_R) for L-amino acids. (5) is the conformation in a five-membered planar ring.

Certain combinations of ϕ and ψ will have higher energy than others as a result of unfavourable interactions between non-bonded atoms. For example the eclipsed arrangement in Figure 2.8b has very high energy. The energy of each conformation may in fact be roughly predicted (Ramachandran and Sasisekharan, 1968; Scott and Scheraga, 1966; Leach et al., 1966). Conformations are classified in three groups. First, there are "allowed" conformations in which all the non-bonded interactions are acceptable. Secondly, there are "partially allowed" conformations which give rise to unfavourable repulsions between non-bonded atoms which might be overcome by attractive effects such as hydrogen bonds. Thirdly, there are "disallowed" interactions which give quite unreasonable nonbonded inter-atomic distances.

The allowed, partially allowed and disallowed conformations can be displayed on a two-dimensional plot — generally known as a Ramachandran plot after its

2.6 SECONDARY STRUCTURE OF POLYPEPTIDES

inventor (Ramakrishnan and Ramachandran, 1965). This is shown in Fig. 2.9. Allowed conformations have ϕ values less than 180°. This is because side groups give rise to unfavourable interactions for positive values. However, glycine has no sidechain and this amino acid alone can easily attain these unfavoured conformations. Glycines can then introduce added flexibility into a polypeptide chain.

2.6 SECONDARY STRUCTURE OF POLYPEPTIDES

Having considered the possible allowed arrangements of two linked peptides, we can now consider the conformation of a polypeptide chain containing many linked peptide units. There are two questions which need consideration. These are:—

(1) What structure will a polypeptide chain assume if subsequent amino acid residues have the same ϕ and ψ angles?
(2) Will any of these arrangements be further stabilized by the possibility of hydrogen bond formation between different peptide units along the polypeptide chain?

The conformation of a polypeptide chain and the intrachain hydrogen bonding scheme is known as the secondary structure of the protein.

Let us first consider the effect of linking together a number of peptide units using the same ϕ and ψ angles at each alpha carbon atom. In general, this results in a helix. Such helices are shown in Figs. 2.10, 2.11, 2.12 and 2.13 and they can be characterized by the number of units per turn of helix, n, and by the distance d, traversed parallel to the helix axis by each unit. The distance traversed by the n units parallel to the axis is given by $n \times d$ and this is called the helix pitch, p. Helices with 2 units per turn may be placed close to one diagonal of the Ramachandran plot. A special case of this type of helix is an extended chain shown in Fig. 2.10, and this is "allowed". The extended chain configuration is close to that found in silk fibrin (β). The second large allowed region corresponds to right-handed helices with n values 3 to about 4.5. Within this group is the alpha helix (α_R) with $n = 3.6$. The areas including the extended chain and the α-helix comprise the two large "allowed" regions of the Ramachandran plot. As the number of residues per turn of helix is increased, the

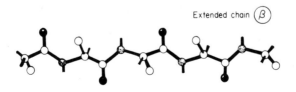

FIG. 2.10   The conformation of β-pleated sheet.

32 2. THE PRINCIPLES OF PROTEIN STRUCTURE

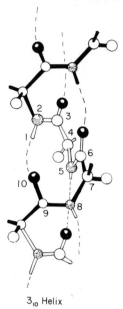

FIG. 2.11 The right-handed 3_{10}-helix (3_{10}); hydrogen bonds complete rings of 10 atoms.

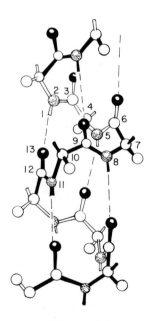

FIG. 2.12 The right-handed 3.6_{13} of α-helix (α_R); hydrogen bonds complete rings of 13 atoms.

2.6 SECONDARY STRUCTURE OF POLYPEPTIDES

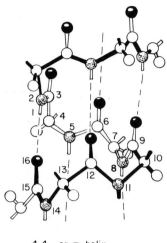

4.4$_{16}$ or π-helix

FIG. 2.13 The right-handed 4.4$_{16}$ or π-helix (π); hydrogen bonds complete rings of 16 atoms.

distance traversed along the axis of the helix, d, is decreased for any constant value of ϕ. Thus for a helix with $\phi = -60°$, "d" is maximum when $n = 2$ and it falls to zero when $n = 5$. In other words when $n = 5$ the pitch is zero ($p = 0$) and the polypeptide forms a closed five-membered ring 5. This value of $\phi = -60°$, $\psi = -90°$ is clearly a limiting case — it is also disallowed. With ψ less than this value the helix changes to a left-handed one.

Before considering the relative stabilities of the various possible allowed helices, we must answer the second question concerning the possibility of formation of hydrogen bonds. In a right-handed helix, the carbonyl group is most easily accommodated if the oxygen is pointing towards the carboxyl end of the polypeptide. The possibility of hydrogen bonds is then between the carbonyl group of an amino acid residue and the NH group of one of the succeeding residues. Some possibilities are shown in Figs. 2.11, 2.12 and 2.13. The different helices formed become looser as the number of residues between hydrogen bonded CO and NH groups increases. An alternative way of defining helices is by the number of residues per turn, n, and the number of atoms in the ring closed by the hydrogen bonds (written as a subscript). The helices are then known as follows: 3_{10}, 3.6_{13}, 4.4_{16}. The number of atoms is increased by three as the ring is increased by one peptide. The 3.6_{13} helix is the α-helix and the 4.4_{16} is sometimes known as the π helix. Of these helices, only the α-helix is fully allowed as can be seen from Fig. 2.9, and this no doubt explains its occurrence in both globular and fibrous proteins. The left-handed α-helix is of higher energy (see Figure 2.9). For a further description of secondary structure the reader is referred to Dickerson and Geiss (1969).

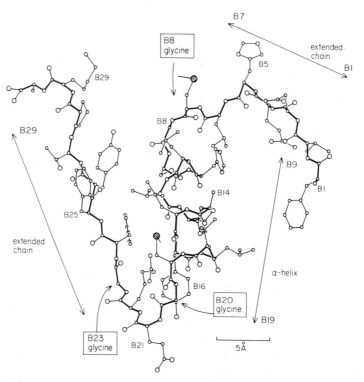

FIG. 2.14 The secondary structure of the insulin B-chain, the sequence of which is shown in Fig. 2.4. Lengths of extended chain and helix are indicated; these hinge at glycine residues. (After Blundell *et al.*, 1972.)

We must now ask whether the conformations which we have predicted are actually observed in proteins. We begin by discussing one protein, insulin, in some detail before describing structural features which are shared by limited numbers of protein molecules. The chemical structure of the insulin molecule has been described in Section 2.3. It contains two polypeptide chains, the A and B chains, bonded together by disulphide bridges. The arrangement of atoms in the B chain of insulin is shown in Fig. 2.14 (Adams *et al.*, 1969; Blundell *et al.*, 1971a,c, 1972). The ϕ and ψ values are given in the Ramachandran plot, Fig. 2.15. The chain can be described as follows. The first seven residues, B1 to B7 have an extended conformation. At residue B8 the chain takes a sharp turn and residues B9 to B19 form an α-helix. Residues B20 to B23 are in a "U" shaped configuration but the remaining residues again form part of an extended chain. We can describe the secondary structure in four parts separated by sharp turns. These turns hinge on residues B8, B20 and B23 which are all glycines. We have considered in Section 2.5 how glycine residues which have no sidechain allow more flexibility in the polypeptide chain. This seems to be the case in the insulin B chain where the ϕ angles are positive only for these three glycines. On the other hand, the residues B1 to B7 and B24 to B30 have angles lying in the allowed region around $\phi = -100°$ and $\psi = 120°$ (see Fig. 2.15). These are extended configurations in which the CO and NH groups point in directions perpendicular to that of the polypeptide chain. They are therefore unavailable for hydrogen bonding within the polypeptide chain but they can form hydrogen bonds to neighbouring chains. In fact two such lengths of B chain lie antiparallel

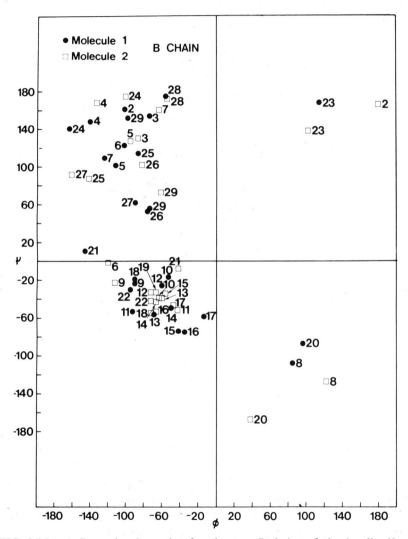

FIG. 2.15 A Ramachandran plot for the two B-chains of the insulin dimer. The numbers indicate the amino acid residue in the sequence (Fig. 2.4) the conformation of which is shown in Fig. 2.16. Note that glycine residues B8, B20 and B23, all have positive ϕ angles which are disallowed for L-amino acids (see Fig. 2.9). (Blundell *et al.*, 1972.)

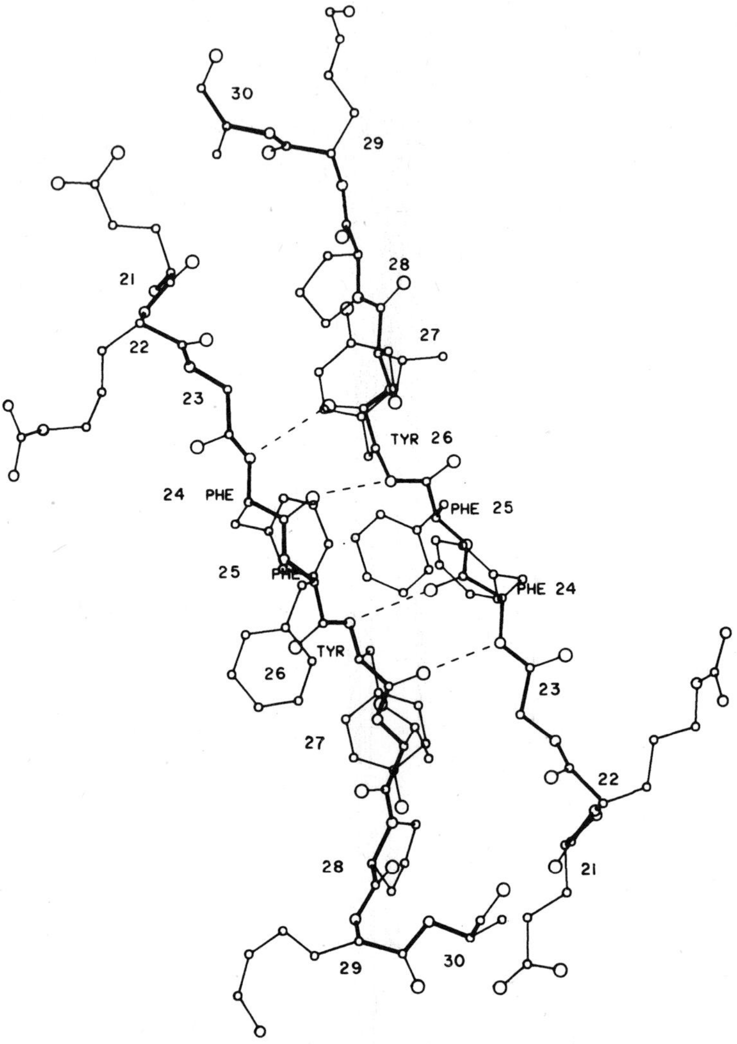

FIG. 2.16 Antiparallel pleated sheet structure formed by two equivalent extended chains [B21–B30] of molecules in the insulin dimer. (Blundell et al., 1972.)

in the dimer and hydrogen bond as shown in Fig. 2.16 and such a secondary structure resembles closely the hydrogen bonding scheme found in silk. The central part of the B chain B9 to B19 has ϕ and ψ angles in the right-handed helical region and corresponds very closely to the α-helix of $n = 3.6$. The helix is shown perpendicular to its axis in Fig. 2.17b where the hydrogen bonding scheme is also indicated. Figure 2.17a shows the same helix as viewed down the helix axis. The average angle of twist is $100°$ which is expected for an α-helix.

2.6 SECONDARY STRUCTURE OF POLYPEPTIDES

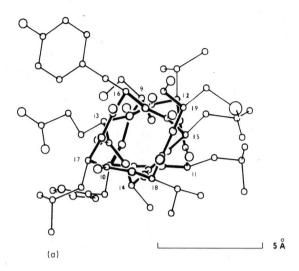

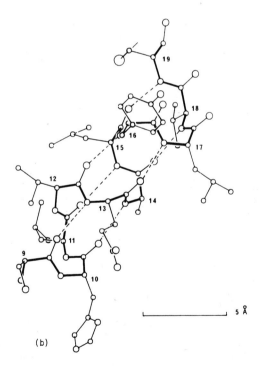

FIG. 2.17 The α-helix of residues B9—B19 of insulin viewed (a) down the helix axis and (b) at right angles to the helix axis. The sequence is given in Fig. 2.4). (Blundell *et al.*, 1972.)

38 2. THE PRINCIPLES OF PROTEIN STRUCTURE

The A chain of insulin is shown in Fig. 2.18 and the corresponding Ramachandran plot is given in Fig. 2.19. The A chain has 21 amino acid residues — shorter than the B chain — and they have a less extended structure. The first seven residues, A1 to A7, have a right-handed helical structure. The right-handed sense is retained for residues A8 to A10 so that the two half cystines at A6 and A11 can form a disulphide bridge (see Fig. 2.18). The

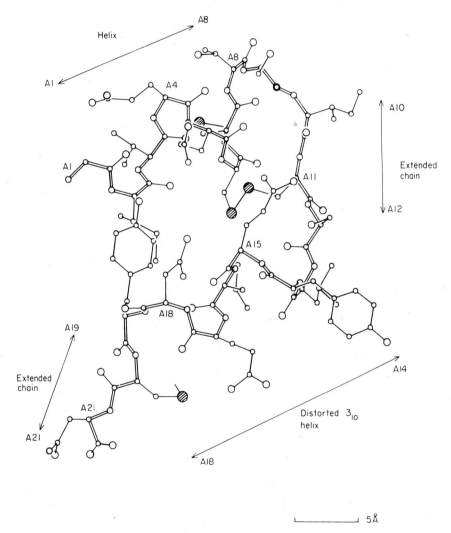

FIG. 2.18 The secondary structure of the insulin A chain, the sequence of which is shown in Fig. 2.4. Lengths of irregular helix and extended chain are indicated. (After Blundell *et al.*, 1972.)

2.6 SECONDARY STRUCTURE OF POLYPEPTIDES

progressive change in n can be seen from the Ramachandran plot. The ϕ and ψ angles change from values close to that of an α-helix, for residues A2 to A7, to an extended chain value for A11. The conformation of residues A13 to A19 is difficult to classify; the polypeptide forms part of an irregular right-handed helix in which few hydrogen bonds are formed. The remainder of the chain — residues A19 to A21 — has an extended structure.

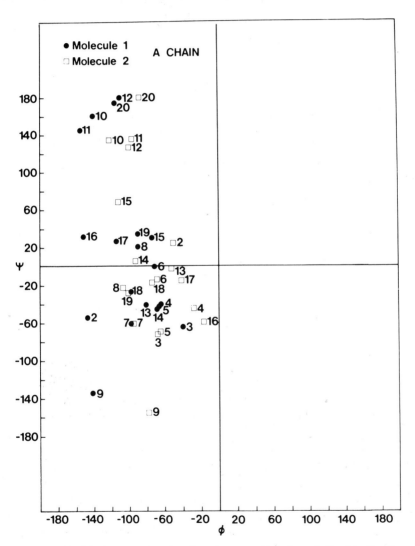

FIG. 2.19 A Ramachandran plot for the two A-chains of the insulin dimer. The numbers indicate the position of the amino acid residue in the sequence (Fig. 2.4). (Blundell *et al.*, 1972.)

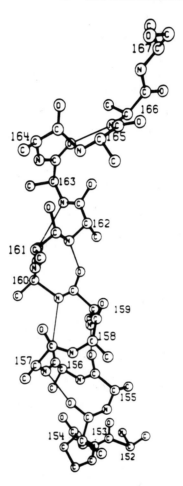

FIG. 2.20 A length of distorted α-helix found in the structure of carbonic anhydrase (Kannan et al., 1971).

The variety of different secondary structures found in insulin is characteristic of most proteins. However, the first protein studied at high resolution by X-ray diffraction techniques was myoglobin which contains eight segments of α-helix as shown in Fig. 19.5 (Kendrew et al., 1960, 1961; Kendrew, 1962; Watson, 1969). This structure perhaps led early workers to overestimate the importance of α-helix. The regularity of the α-helix also provides an attractive simplification in the conceptualization of protein structure. However, the subsequent successful X-ray studies of the enzymes, lysozyme (Blake et al., 1965; Phillips, 1967) and ribonuclease (Kartha et al., 1967; Wyckoff et al., 1967; 1970; Richards and Wyckoff, 1971), showed that these proteins contain little α-helix. In fact more

2.6 SECONDARY STRUCTURE OF POLYPEPTIDES

recent studies of enzymes have shown that the pleated sheet structures may be important features of globular proteins. It is now clear that proteins may have variable percentages of α-helix, β-pleated sheet and more irregular, non-repeating conformations.

Careful study of the secondary structure of a number of proteins has shown that the helices are distorted from the idealized arrangements of the α-helix postulated by Pauling and Corey (1951). Often the plane of the peptide group is rotated so that the CO group is pointing slightly outwards from the helix axis. The helices lie somewhere between the α-helix and the 3.0_{10} helix. This type of behaviour was first observed in the structure of lysozyme (Blake et al., 1965) and similar distortions have since been discovered in many proteins. The residues A13 to A19 of the insulin structure (see page 38) form such a helix. The eight helices of the enzyme carboxypeptidase (Reeke et al., 1967; Lipscomb et al., 1968, 1969, 1970; Hartsuck and Lipscomb, 1971) have unit rotations of between 90° and 107°, and there are between 3.4 and 3.9 units per turn. The idealized α-helix would have a unit rotation of 100° with 3.6 units per turn. In one length of helix of eleven residues, only one hydrogen bond exists. Similar distortions are found in the enzyme carbonic anhydrase (Kannan et al., 1971; Liljas, 1971) and an example of a distorted helix is shown in Fig. 2.20. In many helices the terminal residues are most distorted forming hydrogen bonds characteristic of a 3.0_{10} helix, or sometimes the more open structure of a 4.4_{16} or π helix as found at the carboxyl end of helix F of myoglobin (Kendrew, 1962; Watson, 1969).

The pleated sheet structures found in proteins also tend to be distorted from the idealized geometries of Pauling and Corey. For instance, in the enzyme, chymotrypsin, antiparallel chains are wrapped in very distorted cylinders so that there are only very short regions where the full, classical pleated sheet structure is found in α-chymotrypsin (Matthews et al., 1967; Birktoft et al., 1970; Blow, 1971; Birktoft and Blow, 1973). Both carbonic anhydrase (Kannan et al., 1971) (Fig. 2.21) and carboxypeptidase (Lipscomb et al., 1970) (Fig. 2.22) have parallel and antiparallel structure in the same sheet structure. The eight strands of sheet structure in carboxypeptidase shown in Fig. 2.22 give rise to four parallel and three antiparallel pairs in which there are 33 hydrogen bonds between 45 residues. Other enzymes such as lactate dehydrogenase (Adams et al., 1970) malate dehydrogenase (Hill et al., 1972) and subtilisin (Wright et al., 1969; Alden et al., 1970) have parallel pleated sheet alone. The secondary structure of malate dehydrogenase is indicated in Fig. 2.23. In all cases the sheet structures are not only distorted locally in irregular ways but also tend to have left-handed twists (Chothia, 1973). Each strand lies at a small angle to the previous one. In the eight strands of carboxypeptidase there is a twist of 120° from top to bottom of the sheet structure (Lipscomb et al., 1970).

As in the insulin structure, glycines are often found at sharp bends in the path of the polypeptide chain. Thus there are hairpin bends in the related enzymes

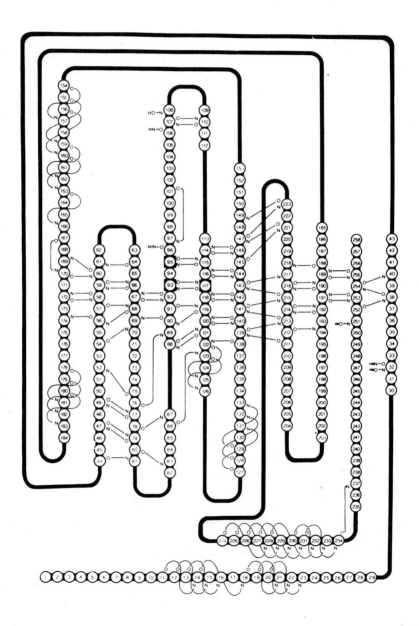

FIG. 2.21 The secondary structure of carbonic anhydrase indicating the extensive sheet structure; (a) a diagrammatic representation of the hydrogen

2.6 SECONDARY STRUCTURE OF POLYPEPTIDES

— chymotrypsin (Birktoft *et al.*, 1970), elastase (Shotton and Watson, 1970a,b) and trypsin (Stroud *et al.*, 1971, 1974) — and in each case there is a glycine which makes the hairpin bend possible. Similar sharp bends involving glycine also occur in cytochrome *c* (Dickerson *et al.*, 1971).

In summary, we can say that globular proteins have rather complex secondary structures, parts of which are very closely related but often not identical to the idealized geometries proposed by Pauling and Corey. Generally, conformations observed are close to those predicted from theoretical considerations; they are energetically allowed arrangements. Hydrogen bonds are a common feature of secondary structure.

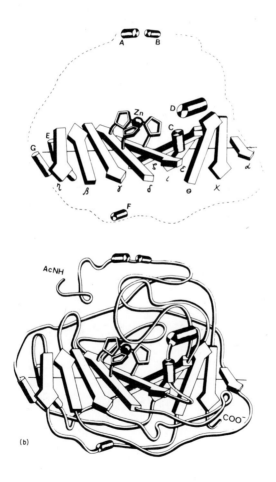

bonding; (b) the three-dimensional arrangement of the sheet and helices. (Kannan *et al.*, 1971; Liljas, 1971).

FIG. 2.22 The secondary structure of carboxypeptidase A; (a) a diagrammatic representation of the hydrogen bonding; (b) the three-dimensional arrangement (Lipscomb et al., 1970).

2.7 PROTEIN TERTIARY STRUCTURE

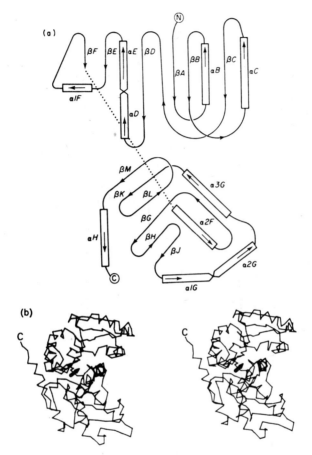

FIG. 2.23 The secondary and tertiary structure of malate dehydrogenase; (a) indicates the arrangement of sheet (β) and helix structures (α). Note the extensive parallel pleated sheet structure. The molecule is organized in two halves which form globular domains in the three dimensional structure shown in (b). (Hill et al., 1972.)

2.7 PROTEIN TERTIARY STRUCTURE

We now turn to the folding of the polypeptide which results in a three-dimensional structure. This is known as the tertiary structure. As an example we will continue our description of the structure of insulin (Blundell et al., 1971).

Figure 2.24 shows the arrangement of the two polypeptide chains of insulin. This figure can be directly related to Figs. 2.14 and 2.18 which show the A and B chains separately, viewed from the same direction.

The B chain is folded on itself to give a V shape bringing together widely separated nonpolar groups. The helix from B9 to B19 lies approximately

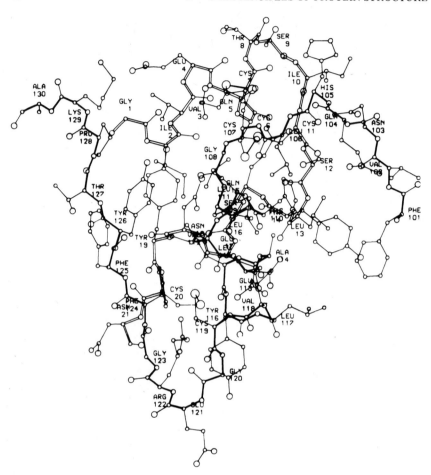

FIG. 2.24 The three-dimensional structure of the insulin monomer. The sequence is given in Fig. 2.4 and the amino acids are labelled 1−21 in the A chain and 101−130 in the B chain. The view is the same as that in Figs. 2.14 and 2.18 of the individual chains. (Blundell *et al.*, 1972.)

antiparallel to the extended chain of residues B23 to B27. Most of the contacts appear to be non-polar.

The A chain is also folded on itself; the helices A2−A7 and A13−A19 lie antiparallel to each other. However, in this case there are both polar and non-polar interactions.

Cystine disulphide bridges are very important in the general three-dimensional relationship between the A and B chains. The two half-cystines at B7 and B19 are at either end of the α-helix in the B chain, and these are covalently bonded to the A7 and A20 half-cystines. The two lengths of extended B chain − B1 to

2.7 PROTEIN TERTIARY STRUCTURE

B7 and B23 to B30 — are folded so that they pack against the more compact A chain to give the protein its globular nature. This tertiary structure is stabilized by both polar and non-polar interactions.

The centre of the three-dimensional structure comprises non-polar groups only. Leucines, isoleucines and cystines are packed in van der Waals contact so that water is excluded from the core of the insulin molecule. All of the polar groups are on the outside of the molecule; some form hydrogen bonds or ion-pairs with each other. There are also hydrogen bonds connecting different parts to the extended chain backbone. The result of these different interactions is a compact three dimensional structure.

Although the structures of the larger proteins must be more complex than the relatively small protein, insulin, the general structural principles are similar. Just as the A and B chains of insulin are folded back on themselves so the structure of longer polypeptides involves a similar folding resulting in a series of loops. For example in chymotrypsin (Birktoft *et al.*, 1970; Birktoft and Blow, 1973) there are a number of such loops which are stabilized either by hydrogen bonds or disulphide bridges. In many proteins these loops are built up into the twisted sheet structures already described. However these chains are often not in sequential order. In the case of chymotrypsin for instance, either the so called "serine" or "histidine" loops must be threaded through prefolded sections of chain depending on whether the protein is considered to fold from the N or C — terminal ends. The assembled pleated sheet structures in the enzymes subtilisin (Wright *et al.*, 1969; Alden *et al.*, 1970; Kraut, 1971) carbonic anhydrase (Kannan *et al.*, 1971) Fig 2.21 and carboxypeptidase (Lipscomb *et al.*, 1970) (Fig. 2.22) form a wall through the centre of the molecule, which may be important in the stability of the globular structure.

Helix may also be important in stabilizing the protein structure. We have seen how the helix of the B-chain of insulin appears to act as a spacer between two disulphide bridges. In subtilisin (Wright *et al.*, 1969) the eight helical segments are packed approximately parallel to each other, giving a compact structure. In the enzyme carboxypeptidase (Lipscomb *et al.*, 1970), six of the nine helices are on one side of the extensive sheet structure. This has the appearance of greater structural stability in comparison to the other side which contains many residues in a more irregular structure. It is interesting that it is the less structured side and not the side containing most of the helix which undergoes a large change of structure on adding a substrate molecule.

Many of the larger proteins seem to be organized in several distinct pieces. The enzymes lactate dehydrogenase (Adams *et al.*, 1970b) and malate dehydrogenase (Hill *et al.*, 1972) are apparently binuclear as shown in Fig. 2.23. Papain (Drenth *et al.*, 1968, 1971a,b,c) thermolysin (Colman *et al.*, 1972a; Matthews *et al.*, 1972a,c) (Fig. 2.25) and chymotrypsin (Birktoft *et al.*, 1970) are also binuclear. Subtilisin (Wright *et al.*, 1969) is even more complex; in this case there are three locally organized regions. The local organization of large proteins may be important in the folding of the protein. Each region may fold separately, and then come together to give the complete globular structure.

It is exciting that a common structural domain has been found in several enzymes which bind adenine nucleotides as part of their co-factors. This domain was first recognized in lactate dehydrogenase (Rossmann *et al.*, 1971) and has since been observed in other dehydrogenases such as soluble malate dehydro-

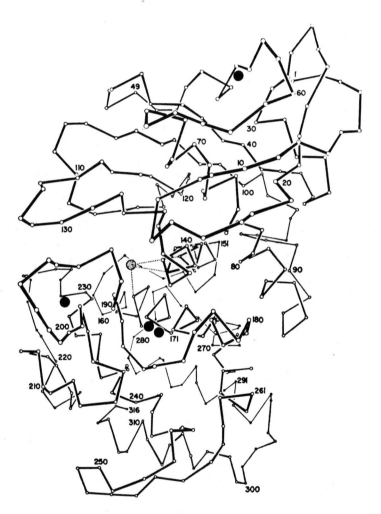

FIG. 2.25 The three-dimensional structure of thermolysin. The structure is clearly organized in two globular domains each comprising about 150 amino acid residues. Only the α-carbon positions are shown for each residue. There are four calcium atoms, ●, and one zinc atom, ○, which is at the active site in the cleft between the two globular domains. (Matthews *et al.*, 1972c.)

2.7 PROTEIN TERTIARY STRUCTURE

genase (Hill *et al.*, 1972), liver alcohol dehydrogenase (Branden *et al.*, 1974a,b; Ohlsson *et al.*, 1974) and glyceraldehyde-3-phosphate dehydrogenase (Buehner *et al.*, 1973), in horse and yeast phosphoglycerate kinase (Blake *et al.*, 1973; Bryant *et al.*, 1974) and in flavodoxin (Ludwig *et al.*, 1971; Anderson *et al.*, 1972; Watenpaugh *et al.*, 1972). In these enzymes about 150 consecutive residues form the common binding site while the remainder of the polypeptide chain has a different structure and is involved in other functions. The domain in common as perhaps best described as a "supersecondary structure" as most of the interactions stabilizing it are hydrogen bonds between peptide units. Thus the structure comprises an extensive area of beta-pleated sheet with helices lying on either side (Figs. 2.23 and 2.26). However, Nagano (1974) has observed that pleated sheet surfaces are frequently covered by helices. Indeed in a *parallel* pleated sheet the chain must traverse the surface. Alternatively a beta bend will give an *antiparallel* sheet.

Most proteins have a surprisingly compact globular structure. The shape varies from roughly spherical in subtilisin (diameter 42 Å) to ellipsoidal in papain (dimensions: 50 x 37 x 37 Å). Most of the charged sidechains are on the surface. When they occur away from the surface, they tend to be in association with other polar groups. Most hydrogen bond donors away from the surface form hydrogen bonds. Solvent molecules may be trapped in cavities and these appear to be involved in such hydrogen bonds. Most of the sidechains of the protein core are non-polar or hydrophobic — valines, isoleucines, leucines, phenylalanines — as found in insulin. The nature of this so called hydrophobic bonding is not really fully understood. However, an important feature appears to be the water structure, which becomes more random — or has more entropy — when it is not in association with non-polar groups. A similar mechanism appears to operate when oil is mixed with water; it collects together in drops. In fact, the folding of protein molecules to give a non-polar core is most easily understood by analogy with oil drop formation. Both phenomena are actually consequences of the Second Law of Thermodynamics, which says that spontaneous processes always give rise to an increase of entropy or randomness — randomness of solvent in this case.

The term "hydrophobic bonding" therefore is useful in emphasizing the fact that protein tertiary structure depends on the aqueous environment of the protein. The hydrophobic "bonds" keep the residues of the core away from the surface, while the polar residues which occur mainly on the surface, hydrogen, bond with water molecules and allow the protein to be soluble. If a protein is dried or if it is in the presence of organic solvents or urea, it will tend to lose its well-defined tertiary structure: it is then said to be denatured. Denaturation may also occur in strongly acid or alkaline solutions. Under these conditions, the protein may be so highly charged, that repulsions between the charges tend to denature the protein. Although the polar groups on the surface must not become predominantly either positively or negatively charged, there are also undesirable consequences of having the charge exactly balanced. At this point the protein will be least soluble, and it is known as the "isoelectric point". Knowledge of the solubility and stability of proteins is of great importance to the biochemist; he has to be very careful not to denature the protein but he must work under conditions where it is soluble. Of course the generalizations made here concern water soluble proteins. Lipid soluble proteins such as those found in membranes may have completely different properties: they probably have hydrophilic cores and certainly have hydrophobic surfaces.

The ease of protein denaturation underlines the fact that proteins unlike

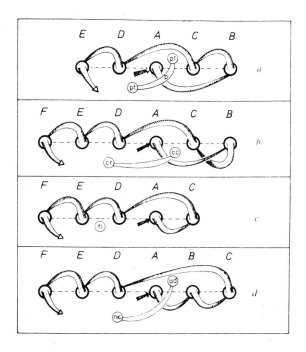

FIG. 2.26 Topologies of proteins containing a parallel sheet a, adenylate kinase; b, subtilisin; c, flavodoxin; d, nucleotide binding domains of lactate dehydrogenase, s-malate dehydrogenase, liver alcohol dehydrogenase, D-glyceraldehyde-3-phosphate dehydrogenase and phosphoglycerate kinase; pt, hydrophobic pockets, presumed binding sites for adenines; p, phosphate position in the catalytic centre of adenylate kinase; cr. hydrophobic crevice, that is, presumed binding site for aromatic or apolar side chain of subtilisin; cc, catalytic centre; fl. FMN; nic-ad. nicotinamide and adenine moieties of NAD. The carbonyl ends of the sheets are towards the viewer and the arrows indicate the direction of the polypeptide chains, that is strands ABCDEF in the sheet are ordered alphabetically along the course of the chain. Reproduced with permission from Schulz, G. and Schirmer, R. Nature (1974) 250, 143.

many chemical substances — steel or diamond for instance — are not very stable. In solution there is good evidence that the conformation found by X-ray analysis is in dynamic equilibrium with other slightly different structures involving localized unfolding. Nevertheless, the equilibrium conformation can be attained from the unfolded polypeptide — in the case of single chain polypeptides only; it is a self assembling system. The lack of structural rigidity is a consequence of the weak interactions — the hydrogen bonds and hydrophobic forces — which are important to the tertiary structure. However, the flexibility in protein structure is used to advantage in a sophisticated way in biochemical processes including the catalysis and control of chemical reactions in living organisms.

2.8 QUATERNARY STRUCTURE

Many proteins are oligomeric; they are aggregated in nature. They comprise several identical or closely related subunits. Aggregates of two subunits are called dimers, aggregates of three subunits are trimers, and higher aggregates tetramers, pentamers and hexamers. The arrangement and nature of the binding of the subunits together is known as the quaternary structure.

We must consider several aspects of quaternary structure. Are dimers, trimers, tetramers, pentamers etc. all found in nature? Are the subunits randomly arranged or are they symmetry related? What is the nature of protein—protein interactions? In order to answer these questions we will consider the quaternary structure of malate dehydrogenase (dimers), lactate dehydrogenase (tetramers) haemoglobin (tetramers) and insulin (dimers and hexamers).

However, first we will describe some principles of symmetry. The position of any two identical proteins can be related by a rotation and a translation. When the rotation is 180°, the proteins are said to be related by a two fold axis. When the rotation is 120°, the proteins are said to be related by a three fold axis. Generally when the rotation is 360/x, the proteins are related by an x fold axis. These symmetry axes are discussed further in chapter 4 and illustrated in Fig. 4.3. If the two molecules are related only by a rotation, the axis is a *proper rotation axis*. If there is a translation in the direction of the rotation axis, the axis is termed a *screw axis*.

In oligomers with more than two subunits there may be more than one rotation axis. Two rotation axes perpendicular to each other generate a third at right angles to the first two. The aggregate has 222 symmetry and the four subunits are arranged in an approximately tetrahedral arrangement. A two fold axis perpendicular to a three fold axis generates two extra two fold axes perpendicular to the three fold. There must be at least six subunits. The aggregate has 32 symmetry.

Malate dehydrogenase has two subunits which have identical chemical

FIG. 2.27 The structure of the dimer of soluble malate dehydrogenase. The two molecules are related by an approximate two-fold axis which lies perpendicular to the plane of the paper. (Hill *et al.*, 1972).

structures. The two molecules are related by an approximate two fold axis of symmetry (Hill *et al.*, 1972) as shown in Fig. 2.27. The homologous enzyme lactate dehydrogenase has four identical subunits. Two of these subunits are related by a two fold axis in the same way as the subunits of malate dehydrogenase. (Adams *et al.*, 1970b; Rossmann *et al.*, 1973) The other two subunits are related to these by two fold axes which are mutually perpendicular to each other and perpendicular to the first. The tetramer then has 222 symmetry; the arrangement is tetrahedral with one subunit at each vertex of the tetrahedron.

Lactate dehydrogenase has an arm of residues at the N-terminus which does not occur in malate dehydrogenase (see Fig. 2.28). The arm is extended around the adjacent subunit. This undoubtedly explains the further aggregation of lactate dehydrogenase to tetramers. All contacts are predominantly hydrophobic in character.

Haemoglobin also contains four subunits. However, these are of two different kinds called α and β. Each has a three dimensional structure which closely resembles that of myoglobin. (Perutz *et al.* 1968; Bolton and Perutz, 1970) (Fig. 2.29) As with lactate dehydrogenase, the subunits are arranged in an

2.8 QUATERNARY STRUCTURE

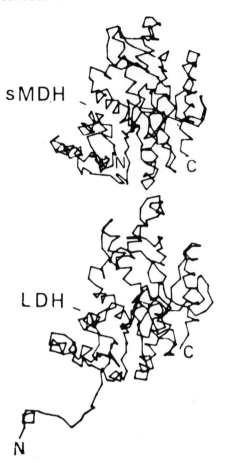

FIG. 2.28 A comparison of the structures of soluble malate dehydrogenase (sMDH) and lactate dehydrogenase (LDH). The two monomers have very similar structure and both form dimers of the kind shown in Fig. 2.27 for sMDH. LDH has a long arm at the N-terminus which can extend around the adjacent subunit so forming a tetramer. sMDH does not have this arm and does not form a tetramer (Hill *et al.*, 1972).

approximately tetrahedral arrangement. There is an exact two fold symmetry which relates the parts of identical α and β subunits together. The α_1 and β_1 are related by an approximate two fold axis perpendicular to this, and α_1 and β_2 subunits are related by a further approximate two fold axis perpendicular to both of the others. The tetramer as a whole has exact two fold symmetry and approximate 222 symmetry. Most of the contacts are between unlike chains, and are predominantly non-polar. There are very few contacts between like chains. An internal cavity extends all the way along the molecular two fold axis, and

54 2. THE PRINCIPLES OF PROTEIN STRUCTURE

this is lined with many polar groups. The oxyhaemoglobin tetramer resembles a spheroid with length 64 Å, and width 55 Å and height 50 Å.

Insulin in concentrations greater than 10^{-5} molar exists mainly as dimers. The dimers arise from interactions between the B-chains. (See Blundell *et al.*, 1972) There are many van der Waals interactions between non-polar groups such as valines, and phenylalanines. The molecules are related by an approximate two

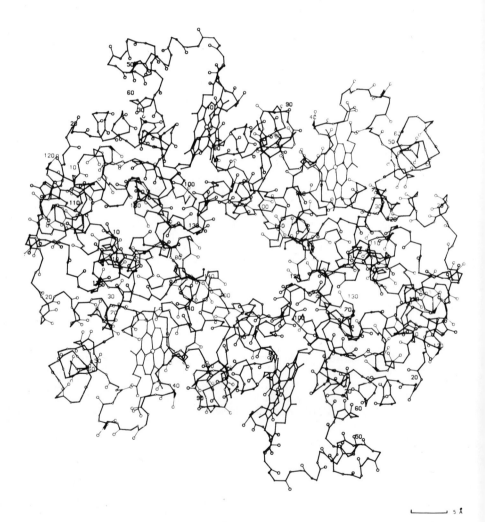

FIG. 2.29 The haemoglobin tetramer has four subunits related by perfect two-fold symmetry and approximate 222 symmetry. The tetramer is viewed down the two-fold axis. Only the backbones are indicated. (Drawn by Prof. A. C. T. North from coordinates supplied by Dr. M. Perutz.)

2.8 QUATERNARY STRUCTURE

fold axis so that the carboxyl terminal residues of the chains lie antiparallel and form hydrogen bonds characteristic of a pleated sheet. This is shown in Figs. 2.16 and 2.30. Distortions from exact two fold symmetry are widely distributed in the molecule and very obvious deviations occur in the vicinity of the two fold axis. For instance, one of the two B25 phenylalanines is right on the approximate two fold axis.

At neutral pH and in the presence of zinc ions the dimers aggregate further to hexamers. Three insulin dimers are related by a three fold axis perpendicular to the dimer two fold axes. The hexamer has 32 symmetry (Fig. 2.31). The contacts involve many non-polar contacts between A chain as well as B chain residues, but the dimers are less tightly bonded to each other than monomers. A cylindrical cavity lined with polar groups surrounds the three fold axis. Two zinc atoms lie in this cavity, each bound to three histidines, one from each insulin dimer. The insulin hexamer is an oblate spheroid.

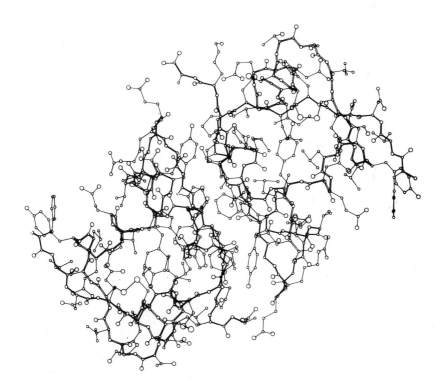

FIG. 2.30 The insulin dimer viewed down the approximate two-fold axis. The two molecules are held together by hydrophobic forces and an antiparallel pleated sheet shown in Fig. 2.16. (Blundell et al., 1972.)

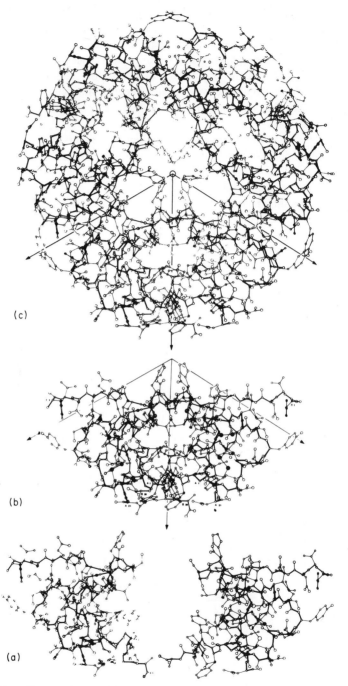

FIG. 2.31 The insulin hexamer. The molecules are viewed along the three-fold axis, and therefore at right angles to the approximate two-fold axes and Fig. 2.30. The two monomers (a) form dimers (b) which aggregate to give a hexamer (c) in the presence of zinc. (From Blundell et al., 1972.)

2.9 PROTEIN INTERACTIONS WITH SMALL MOLECULES AND IONS

Although it is rather unwise to base generalizations on such a few structures some patterns in quaternary structure emerge from these proteins. First, most proteins are built of dimers which aggregate to tetramers, hexamers and so on. Trimers have recently been found in glucagon (Blundell, 1975; Sasaki et al., 1975) but they are much less common. The dimer seems to have special stability.

Secondly, the dimeric basis of the aggregates is associated with exact or approximate two fold axes of symmetry. This type of aggregation no doubt arises because any interaction of the type $A_1 B_2$ (where the subscripts refer to the subunits) must also imply a potential $A_2 B_1$ interaction. Successful formation of both such interactions gives twice the bonding energy and implies at least approximate two fold symmetry. Attainment of exact symmetry in the aggregate may be inhibited by bulky residues close to the two fold axis or by the environment (especially in the crystal).

Thirdly, an antiparallel pleated sheet between subunits leads neatly to a two fold axis relationship. This type of interaction first found in insulin has now been found in the structures of concanavalin, (Hardman and Ainsworth, 1972; Edelman et al., 1972) prealbumin (Blake et al., 1974) and liver alcohol dehydrogenase (Bränden et al., 1973).

Finally, hydrophobic or non-polar interactions between the subunits seem to be important. In fact proteins which aggregate tend to have a larger percentage of their residues with a non-polar character than homologous monomeric proteins. For a discussion of protein–protein interactions see Chothia and Janin (1975).

2.9 PROTEIN INTERACTIONS WITH SMALL MOLECULES AND IONS

Many proteins can only express their biological activity when they are associated with ions or small molecules. These nonprotein structures are called *cofactors*. They bind to proteins with varying degrees of affinity. Many involve weak ionic interactions while some are covalently bound.

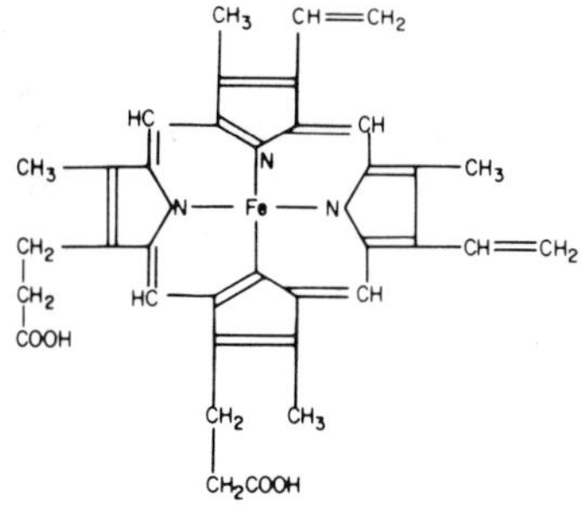

FIG. 2.32 The haem group.

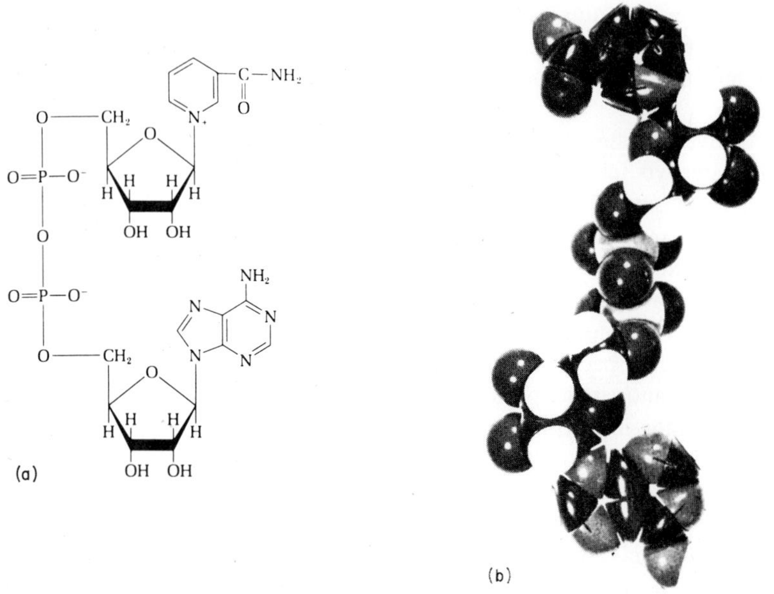

FIG. 2.33   The structure of NAD.

Metal ions are often cofactors and many enzymes such as carboxypeptidase (Zn^{2+}) or nuclease (Ca^{2+}) require ions for activity. These ions can be removed from the enzyme by addition of a strong chelating agent. Groups which are strongly bound are usually called *prosthetic groups*. An example is the haem group (shown in Fig. 2.32) and in myoglobin and haemoglobin this is covalently bound through a histidine. Many enzymes require cofactors and these are often complex organic molecules known as *coenzymes*. An example is nicotinamide adenine dinucleotide, NAD (shown in Fig. 2.33), which is required by lactate and malate dehydrogenase. The intact enzyme–coenzyme complex is called a *holoenzyme*. The inactive enzyme alone is the *apoenzyme*. Enzymes also bind their substrates and related molecules which inhibit their catalytic activity. The latter are known as *inhibitors*. The complex of enzyme–coenzyme–substrate or inhibitor is known as the *ternary complex*.

3

CRYSTALLIZATION OF PROTEINS

3.1 INTRODUCTION

The origins of protein crystallography may be traced to the first X-ray diffraction photographs of pepsin taken by Bernal and Crowfoot (1934). The crystals had been obtained a few months earlier in Uppsala, by chance, from a preparation which had been left undisturbed for several weeks (Hodgkin and Riley, 1968). Although detailed studies on this particular enzyme were not undertaken until several years later, the episode illustrates the manner in which almost all of the early protein projects were started. A somewhat opportunist approach was adopted. The proteins selected for study were chosen not on the basis of their biological significance, although this generally turned out to be interesting, but on their ready availability and the ease with which they could be crystallized. Recent experience suggests that most purified globular protein molecules can be crystallized,* provided sufficient time and expertise are invested in the project, and the current trend is to select enzymes for detailed study on a rather more rational basis than has hitherto been the case. The study of crystallization techniques must now be an important part of protein crystallography.

Protein crystallization is a biochemical technique much older than X-ray analysis itself. Crystallization has been used for many years by the biochemist for isolation and purification of proteins. Microcrystallinity of the preparation is often one of the first indications of an improvement of purity. In recent years the use of crystallization in this way has to some extent been superseded: crystallization is rarely quantitative and selective at the same time and anyway crystals can contain 10% or more of alien proteins. Nevertheless, the biochemist bequeaths the protein crystallographer a substantial body of information concerning crystallization.

For many biochemists, protein crystals have the appearance of a silky sheen, often termed "schlieren"; the size of the crystals is not their concern. The protein crystallographer has not only to crystallize the protein but also to grow

---

*It may be possible that the best packed regular or crystalline arrangement is more open (or, more correctly, has a higher free energy) than irregular arrangements. In this case the protein would never crystallize.

crystals of a substantial size. This is necessary because the intensity of the X-ray diffraction pattern of a crystal is roughly proportional to the volume of the crystal and inversely proportional to the crystal unit cell volume which cannot be smaller than one protein subunit. For proteins of molecular weight less than 50,000 a crystal of 0.1 mm in all dimensions will probably give a diffraction pattern allowing the basic crystal cell data to be deduced. However, a crystal of 0.3 mm in each dimension or about 30 times the volume will usually be needed for a high resolution crystal structure analysis. For proteins of even higher molecular weight, the size required will be larger. For neutron diffraction, crystals of 3 mm in each dimensional are needed for even the smallest proteins. Only in electron microscopy (chapter 18) are very small microcrystals required.

There are two problems: (1) to reach a point of supersaturation at which crystals are formed and (2) then to grow the crystals big enough for diffraction studies. The point of saturation depends on the variation of solubility with the protein concentration, ionic strength, the temperature, the presence of organic solvent, the pH and the binding of counter ions to the protein. The rate of nucleation and thus the number and size of the crystals formed under any set of conditions defined by these variables depends on the presence of foreign particles in the solution, the introduction of crystal seeds, the size and material of container, the concentration and the degree of supersaturation of the protein solution. We will discuss these factors before describing some of the techniques which are used by protein crystallographers.

3.2 FACTORS AFFECTING THE SOLUBILITY OF PROTEINS

The solubility in water of a molecule of any type depends on interactions of the molecule with water. A layer of water surrounds the solute interacting with the surface charges and dipoles, sometimes forming hydrogen bonds. The high dielectric constant of water also contributes to the solubility as it allows separation of charged ions. Fortunately, proteins have well defined structures with a surface containing charged and polar groups, and their solubility properties can be characterized and are reproduceable. We can consider a protein as a large polyvalent ion and discuss its solubility in terms of the Debye–Huckel theory which was developed to describe the properties of smaller ions. [See Physical Chemistry by Glasstone (1974).]

In most of the following discussion we consider the protein molecules in solution. This ignores a very important aspect; the solubility is dependent on a thermodynamic equilibrium between solid and liquid states and consequently the solubility will depend on the nature of the solid state. C. W. Bunn has been one of the few to emphasize the important point that the solubility of the amorphous material is generally greater than that of the crystals. Unfortunately solubilities quoted in the literature with the exception of those for rennin (Foltmann 1959, Bunn et al., 1971) do not distinguish values for crystalline and amorphous material. It is to be hoped that protein crystallographers will be more precise in their definition of solubilities in the future.

Factors affecting the solubility of proteins are discussed more fully by Green and Hughes (1955) and Dixon and Webb (1961).

3.2.1 Ionic strength

The Debye–Huckel theory recognized that the simple ionic model was unsatisfactory in describing the properties of electrolytes except in very dilute solution. As the concentration of ions is increased so an "atmosphere" of ions of opposite charge tends to form around each ion. This ionic atmosphere changes the interaction of the ions with the water molecules and therefore modifies the solubility.

At low ion concentrations the effect of this ionic atmosphere is to increase the solubility as it increases the possibilities for favourable interactions with the water molecules. This situation can be described by

$$\log S - \log S_0 = \frac{A z_+ z_- \sqrt{\mu}}{1 + aB\sqrt{\mu}} \tag{3.1}$$

The equation relates the solubility of the salt, S, at a particular ionic strength, μ, to the solubility in the absence of electrolyte, S_0. The valences of the ions of the salt are z_+ and z_- and the ionic strength of the medium is defined by

$$\mu = \tfrac{1}{2}\Sigma c_j z_j^2$$

This shows that ions of higher charge will be more effective in modifying the solubility. The constants A and B depend on the temperature and the dielectric constant. The denominator, $1 + aB\sqrt{\mu}$, allows for the fact that the ions have a finite average diameter, a, and therefore polarize the surrounding medium less than a point charge. An increase of solubility at low ionic strengths is characteristic of most ions and is commonly found with proteins, which are more soluble in the presence of a small amount of electrolyte than in pure water. The phenomenon is known as "salting in".

As the ionic strength is increased the ions added begin to compete with each other and with the solute for the surrounding water. The resulting removal of water molecules from the solute begins to decrease the solubility. Equation (3.1) must be extended to the following form

$$\log S - \log S_0 = \frac{A z_+ z_- \sqrt{\mu}}{1 + aB\sqrt{\mu}} - K_s \mu \tag{3.2}$$

The decrease in solubility at high ionic strengths is almost completely determined by the term, K_s, and this "salting out" process is therefore proportional to

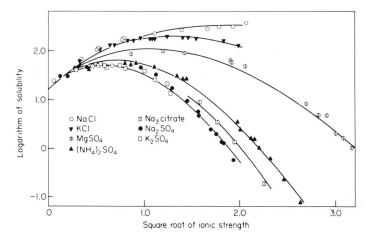

FIG. 3.1 The solubility of carbonmonoxyhaemoglobin in various electrolytes at 25°C. The curves are drawn according to equation 3.2 (from Green, 1932).

the ionic strength. Salting out is found with all water soluble species whether they be ions, organic molecules, gases or proteins.

The solubility dependence on ionic strength for a protein is shown in Fig. 3.1. It is well described by the equation (3.2). Of course the Debye-Huckel theory does not completely explain the solubility of proteins. Proteins are much larger than most ions and have surface charge patterns which are unrelated to their net charge. Furthermore the Debye-Huckel theory does not adequately consider the effect of ordering of solvent molecules around the solute which varies according to the nature of the group. For instance this leads to an unfavourable entropy term in the dissolution of non-polar groups. Finally the dissolution of a protein must take into account the free energy of the insoluble form. It is the difference of free energies of the solid and dissolved forms which is important. Thus crystalline material is generally much less soluble than amorphous material owing to the more favourable electrostatic interactions in the crystals. More extreme conditions are generally needed to dissolve protein crystals than are required for the amorphous protein.

Despite these limitations the Debye-Huckel theory does correctly indicate "salting in" and "salting out" phenomena. It indicates that the protein crystallographer may decrease the solubility of a protein dissolved in a medium of low ionic strength by increasing or decreasing the concentration of ions. Either technique may be used in the attempt to induce crystallization.

Different ions will affect the solubility of the protein in different ways. The effect will tend to follow the Hofmeister series: small highly charged ions are more effective than large low charged ions in salting out. This is generally true;

3.2.1 INONIC STRENGTH

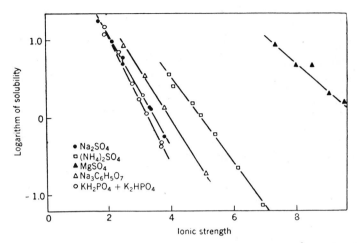

FIG. 3.2 The solubility of carbonmonoxyhaemoglobin at 25°C and pH6.6 in concentrated solutions of various electrolytes (from Green, 1931).

univalent salts such as KCl are relatively ineffective in salting out proteins. Fig. 3.2 shows the values of K_s for different salts where K_s is the slope of the salting out line. It decreases in the order: potassium phosphate, sodium sulphate, ammonium sulphate, sodium citrate, magnesium sulphate. For chlorides it is much lower. It can be seen that ammonium sulphate is more effective at a particular ionic strength in salting out than magnesium sulphate, although the magnesium ion is smaller. This is not in agreement with the Hofmeister series, but the explanation probably lies in the fact that according to the Debye–Huckel theory magnesium will have a rather large ionic atmosphere at high ionic strengths which decreases its effectiveness. It can be seen that ammonium sulphate has the expected effect on the protein solubility at high ionic strength. Its effects are not extreme. Its general use as a protein precipitant is related more to its great solubility than to its effectiveness at any particular ionic strength. A saturated solution of ammonium sulphate at 25°C contains 767 g of salt for each litre of water. However, use of an excessively high salt concentration should be avoided as it will tend to remove solvent from the crystal lattice and may eventually disorder the crystals.

Quite often ammonium sulphate proves to be troublesome as it tends to liberate ammonia which reacts with heavy atom reagents such as platinochloride used in making isomorphous derivatives (see page 221). In this situation alkali phosphates which are very soluble or alternatively quaternary ammonium salts can be used. For less soluble proteins a smaller ionic strength will be needed to bring the protein solution to saturation, and this can often be achieved by using magnesium sulphate, alkali sulphates or chlorides.

3.2.2 pH and counter ions

So far we have considered the protein to be a charged species with a fixed net valence. This is clearly incorrect. We can modify the net charge by adding or withdrawing protons i.e. by changing the pH or by specifically binding ions (so called counter ions) to the polar groups of the protein. Generally the protein is more soluble the more net charge it contains. The point of least solubility is when it has a net charge of zero. This is because the ions can pack in a solid form and interact electrostatically without accumulating a net charge of high energy. The point of zero net charge is the isoelectric point.

The effect of modifying the pH is illustrated in Fig. 3.3. The value of K_s is not changed, i.e. all the salting out lines are parallel. It is the solubility in the absence of electrolyte, S_0 which changes with pH and shifts the conditions for salting out. The variation of S_0 with respect to the pH is shown in Fig. 3.3b.

It reaches a minimum in the region of the isoelectric point and increases on the addition of either acid or base at this point. Occasionally other minima are shown. This could result from a conformational change in the protein from aggregation, or from the specific interaction with an ion or molecule in the mother liquor at a particular pH. Sometimes the isoelectric point is different at high and at low ionic strengths. This is thought to arise from the action of counter-ions at higher ionic strength which give rise to a net charge at the pH of the isoelectric point.

In summary, the pH and the presence of counter-ions are further factors which can be changed in order to modify the solubility. Adequate control of pH

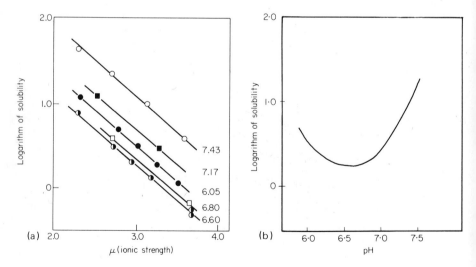

FIG. 3.3 The solubility of haemoglobin in concentrated phosphate buffers at various pH's. Figure 3.3b is derived from Fig. 3.3a (from Green, 1931).

3.2.3 Temperature

during crystallization necessitates the presence of suitable buffers. The theory of buffers is not presented here but the reader should refer to Glasstone (1974) for a theoretical discussion and to Hodgman (1949) for tables of buffer pH.

3.2.3 Temperature

Many of the factors which govern solubility have a marked temperature dependence. The dielectric constant decreases with increase of temperature and the entropy terms in the free energy of solution tend to dominate the enthalpy terms. It is therefore not surprising that the temperature coefficient of solubility varies from protein to protein, and with the conditions such as ionic strength and presence of organic solvents.

Variation in temperature at high ionic strength and pH does not usually affect K_s, but it does affect S_0. Thus we obtain a series of parallel lines representing salting out at different temperatures. (Fig. 3.4) At *high* ionic strength most proteins are less soluble at 25°C than at 4°C. The solubility sometimes decreases ten-fold on warming up a protein in ammonium sulphate solution. Occasionally proteins show a minimum solubility at about 25°C.

This negative temperature coefficient of solubility is characteristic only of high ionic strength media. Quite often the temperature coefficient is positive at *low* ionic strength: the solubility increases with increasing temperature. In pure water the temperature coefficient is positive for most proteins.

Generalizations concerning the temperature coefficient of solubility for proteins are difficult. Either increase or decrease of temperature may bring the protein solution to saturation: it depends on the protein and the experimental conditions. Furthermore the solubility of the crystals rather than the amorphous

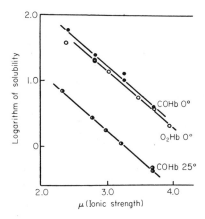

FIG. 3.4 The solubility of haemoglobin in concentrated phosphate buffers at various temperatures (from Green, 1931).

material is here especially important. The temperature coefficient of solubility may be positive and negative, respectively, for different crystal modifications of the same protein (Czok and Bücher 1960).

3.2.4 Organic solvents

The addition of organic solvents also produces marked changes in the solubility of proteins in aqueous solutions. This is partly the result of the lower dielectric constant of mixtures of water and organic solvents compared to water. Thus water has a dielectric constant of 80, but a mixture of about 34% by weight ethanol in water has a dielectric constant of 60. This decrease in the dielectric constant increases the coulombic attraction between unlike charges on the protein molecules and lowers the solubility. The organic solvents may also decrease the solubility through specific solvation of the protein and displacement of bound water. Generally the protein solubility decreases with decrease of temperature when substantial amounts of organic solvent are present. Organic solvents often denature proteins and this should be minimized by working at low temperature.

Ethanol is the most commonly used precipitant, but acetone is often advantageous as it is somewhat milder. However, the high volatility of these solvents makes them difficult to handle, and less volatile solvents which are miscible with water are preferable. Of these 2-methyl-2, 4-pentanediol (MPD) has proved useful. (The commercial MPD must be purified by use of mixed-bed ion exchangers and fractional distillation *in vacuo* over KBH_4). Other possible precipitating agents include dimethyl sulphoxide, dimethyl formamide and acetonitrile.

3.3 FACTORS AFFECTING CRYSTAL NUCLEATION AND GROWTH

The variation of parameters such as ionic strength, pH, type of counter ion, temperature and concentration of organic solvent can be used to decrease the solubility of the protein, so that the solution becomes saturated or supersaturated. Supersaturation is a metastable state; its conversion to a more stable state is kinetically controlled. A super-saturated solution eventually gives rise to nuclei some of which will grow into larger crystals. We must now consider the factors which lead to nucleation and favour the growth of few, large crystals.

Little research has been undertaken into the nucleation and growth of protein crystals. This is quite understandable for an enormous amount of effort has been put into the study of simpler systems such as crystals of metals and salts but this has shed little light on the theory or on the practical problems of growing crystals. With proteins the situation is far more complex; there are many ionic and molecular species interacting in equilibrium with each other.

For smaller molecules it seems that the formation of crystals is conditional

3.3 FACTORS AFFECTING CRYSTAL NUCLEATION AND GROWTH

upon the appearance of nuclei of a critical size. Aggregates of molecules less than the critical size will have a positive or unfavourable free energy of formation and will therefore tend to dissolve. The critical nucleus is sufficiently large that the interactions between each molecule and its neighbours is on the average conducive to an overall negative free energy of formation. The critical nucleus may contain between 10 and 200 molecules and the time for its formation varies with the conditions. However, it is clear that the rate of nucleation will increase considerably with the degree of supersaturation. This is almost certainly the case with protein solutions also. In fact the rate of nucleation increases sharply at a certain degree of supersaturation. In order to decrease the number of nuclei and hence the number of crystals the degree of supersaturation must be as low as possible. (This may not always be so. In some cases crystals have been observed to grow from amorphous precipitates.)

Supersaturation must be slowly approached to minimize nucleation and when a low degree of supersaturation has been obtained it must be carefully controlled. Addition of salt should be diffusion controlled and convection currents should be avoided. These give rise to local increases of temperature which often have the effect of decreasing the solubility and initiating unwanted nucleation.

The nucleation process is made thermodynamically more favourable if there are some foreign particles in the solution which act as nucleation centres. Therefore dust particles must be excluded from the solution by centrifugation, and small air bubbles should be avoided. Usually the protein should be very pure. In particular it should be free of other protein impurities. However, occasionally an important cofactor is lost on purification. This was the case with insulin where the loss of zinc on purification prevented crystallization at pH6 and ignorance of the importance of zinc held up attempts to repeat the original crystallization of insulin. Thus it is impossible to be dogmatic about the requirement of purity for crystallization. Denatured protein will tend to precipitate out more easily and may form undesirable centres for nucleation. Thus one should be careful to avoid organic solvent precipitation, or high temperatures if they are likely to denature the protein. Lyophilization should also be avoided as this often leads to denaturation. Freshly prepared protein solutions should be used as most proteins become slowly denatured in solution.

Even when the chance of heterogenous nucleation has been reduced by removal of dust particles, air bubbles and denatured protein, nucleation can still be initiated at the surface of the vessel. In many cases protein crystals grow on the vessel walls. A near spherical container would seem to be advantageous as this gives the maximum volume-to-surface ratio. It is often found that fewer, larger crystals are obtained from large scale crystallizations than from microtechniques.

Controlled nucleation is often achieved by minimizing heterogeneous nucleation and introducing a few crystal seeds if they are available. Seeding is best carried out by using a few, very small crystals. A seeding solution can be prepared by crushing a single crystal in a stabilizing solvent in which the crystal fragments will not redissolve. The solution should be centrifuged at a low speed to remove the larger fragments and diluted to obtain the optimal concentration of nucleation centres (dilutions of the order of 10^5 are sometimes necessary). The seeding solution should be of a very low degree of supersaturation that does not give crystals quickly. It can be introduced into the solution by touching the sample with a glass rod which has been wetted with the seeding solution. This technique has been found to be quite successful in producing large crystals.

Once the critical nuclei have formed, we have to encourage growth to a large crystal size. Crystals grow by forming molecular layers parallel to crystal planes which have high molecular density. The addition of a further molecule to an incomplete plane is often energetically favourable but the initiation of a new plane may not be so. It is thought that crystals grow by formation of a series of two dimensional critical nuclei on the crystal faces. Surface imperfections decrease the energy required for nucleation. Screw dislocations permit addition continuously to an incomplete plane. For growth in a spiral the growing plane is always incomplete and there is no need to wait for the formation of new surface nuclei. It is a strange conclusion that a thing as near to perfection as a crystal should depend on imperfections for its growth! Addition of small amounts of organic solvents can have favourable effects on crystal growth. Addition of 2% dioxane facilitates growth of horse phosphoglycerate kinase crystals and eliminates twinning in crystals of α-chymotrypsin. Addition of 4% dimethylformamide partly overcomes twinning in the case of aldolase. Thus it is possible that impurities in certain situations may catalyse the growth of good crystals, although similar molecules may initiate a too high rate of nucleation. These contradictory indications about the optimum conditions for nucleation and growth of large crystals merely emphasize the need for a rather empirical approach to the subject.

Saturation of the protein solution may be achieved at a variety of concentrations. For example it will occur at a lower concentration and lower ion strength closer to the isoelectric point than far from it. Larger protein crystals can often be obtained using a higher concentration of protein (but a low supersaturation). Thus large crystals are often best obtained far from the isoelectric point especially when the protein has a low solubility. The higher concentration of protein makes more protein available for crystal growth.

3.4 TECHNIQUES FOR CRYSTALLIZATION

So far we have described the parameters which may be varied to bring a protein to a state of low supersaturation which favours limited nucleation but growth of large crystals. We must now consider the experimental techniques which have

3.4.1   BATCH CRYSTALLIZATION   69

found favour with the protein crystallographer. In our previous discussion we have not been concerned with the most important controlling factor: the availability of the protein. Some protein crystallographic projects have consumed many hundreds of milligrams of protein. Some years ago, most crystallographers would have conservatively estimated a requirement of 100 mgs for crystallization trials and 1 g for a complete structure determination. However, more recently rubredoxin was solved with as little as 15 mg of protein. Now with the development of microtechniques many projects are initiated with this amount of material. For each general kind of technique we will consider the microtechniques which have been developed for use with small amounts of protein. These are especially useful for initial screening to find suitable conditions.

3.4.1 Batch crystallization

This is the classical technique described for early enzyme crystallization. The protein is dissolved at low ionic strength to give a solution of high concentration. The precipitating agent (salt or organic solvent) is then added to bring the solution to a state of low supersaturation. After standing for a few hours or perhaps a few months the crystals appear. Often the precipitant is added day by day in small amounts if the conditions for supersaturation are not known. A turbidity or opalescence in the solution probably indicates that the point of low supersaturation has been passed. Nevertheless quite often an amorphous precipitate will redisolve and crystals will slowly form in its place. This technique was used very successfully for the production of large crystals of lysozyme, ribonuclease and enzymes of the trypsin family, but it offers very few possibilities of control of crystal growth. It is only useful for proteins whose nucleation and growth velocities are extremely low even at high supersaturation or for those proteins which give well ordered crystals even at high growth rates.

Lysozyme is one of the easiest enzymes to crystallize. Large tetragonal crystals may be obtained from the following method: 400 mg lysozyme are dissolved in 5 ml 0.04 M acetate buffer at pH 4.7. The solution is gently stirred for at least 5 mins to avoid frothing and to ensure that all the protein is dissolved. 5 ml of 10% (w/v) sodium chloride solution is added over a period of 5 mins and stirring continued for another 5 mins. The solution is filtered into a plastic container and placed in a quiet corner at room temperature. Crystals appear after two days (see frontispiece).

Batch crystallization is not easy to scale down directly as it depends critically on the addition of a small but precise amount of precipitant. One modification which enables screening of conditions has been found useful for phycocyanin (Dobler et al., 1972). A small amount (300 μl) of protein at a high concentration is prepared at low ionic strength. One tenth of the volume is removed and placed in a small test tube. The volume is then made up again to 300 μl by addition of solvent. The protein is then at a lower concentration. 30 μl are again removed and stored, and the volume made up again to 300 μl with solvent. In this way a series of solutions with decreasing protein concentration are prepared. From each sample of 30 μl so prepared, 3 μl of solution is removed and stored. Each sample is then made up to 30 μl by adding 3 μl of a solution of precipitant. By repeating this process a series of solutions with increasing concentration of precipitant is prepared.

These solutions may be left to stand and the emergence of amorphous precipitate or crystals recorded as a function of protein and precipitant concentration. This is certainly a useful technique for preparing solubility curves even if it has the limitations characteristic of large scale batch crystallization.

As an alternative to directly adding and mixing the precipitating agent with protein solution, it can sometimes be arranged that the mixing is diffusion controlled. Thus the protein solution may be carefully placed in a capillary either above or below the precipitating agent so that the most dense solution is underneath. A further refinement of this technique is to freeze the protein solution, so that the diffusion is further slowed. Drenth and Hol have used this method to produce large crystals of papain, phospholipase and rhodanese from organic solvents. The protein solution is frozen to $-20°C$ in a test tube and the less dense solvent is placed above it. The test tube is then slowly warmed.

3.4.2 The hot box technique — use of the temperature gradient of solubility

A modification of the batch crystallization technique to exploit the temperature gradient of solubility has been useful in producing crystals of certain proteins. We have seen in section 3.2.3 that the temperature coefficient of solubility varies in sign and magnitude from protein to protein. Let us first consider the situation of a protein which is more soluble at higher temperatures than at room temperature.

Glucagon and insulin are both stable and quite soluble at $50°C$ in buffers of low ionic strength. The problem is to control the approach to supersaturation which is reached at lower temperatures. For this a "hot box" has proved useful. This is simply a thermos flask buried in the centre of a tea chest packed with pieces of polystyrene (Fig. 3.5). The protein is dissolved by warming it in a test tube to the required temperature.

The test tube is then suspended in a thermos flask of water at the same temperature and the thermos is placed in the hot box and left for a period of time. The final temperature reached by the hot box is most important. For instance larger crystals of insulin can be grown at $25°C$ than $20°C$ where the supersaturation is higher. For insulin the "hot box" technique can be used very sensitively by dissolving the protein at about pH 8 and than bringing it slowly towards the isoelectric point. At about pH 6.3 a slight turbidity occurs. At this point the solution is warmed up to clear the turbidity. The test tube containing the solution is then placed in the hot box and allowed to cool.

3.4.3 EQUILIBRIUM DIALYSIS

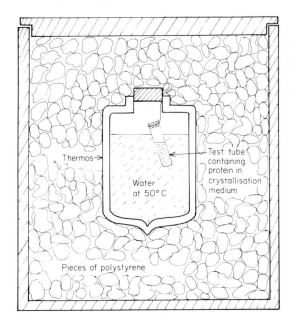

FIG. 3.5 Apparatus for slowly cooling the crystallization medium — the hot box. (As used by Dr. G. Dodson for insulin.)

Jakoby (1971) has developed a technique for microcrystallization of proteins which is most useful for producing seed crystals. The method depends on the decrease in solubility of proteins in ammonium sulphate solution with increase in temperature. The protein is precipitated with salt and is then extracted with ammonium sulphate solutions of decreasing concentration at or near 0°. Thus contaminating proteins soluble at high concentrations of ammonium sulphate are extracted first and removed. Extracts obtained in this manner are allowed to warm to room temperature during which period crystallization results.

3.4.3 Equilibrium dialysis

A suitable dialysis membrane is a semipermeable diaphragm which allows solvent and small ions to equilibrate but is not penetrated by protein molecules. Such membranes are prepared from cellophane and other cellulose derivatives. The ionic strength and pH of a protein solution enclosed in such a membrane is adjusted by equilibration against a protein free solution.

Such a technique was first described by Theorell (1932) and used in the crystallization of haemoglobin (Boyes-Watson et al., 1947). A volume of about 1 ml of protein solution can be enclosed in a bag made from dialysis tubing and submerged in a suitable buffer. The supersaturated state is approached slowly by gradually modifying the conditions in the buffer outside. On the appearance of slight turbidity or opalescence in the bag no further changes are made in the external buffer until crystals have grown to their maximum size. When growth is complete at one salt concentration or pH, sometimes further growth may be achieved by increasing salt concentration or moving the pH nearer to the isoelectric point, so reducing protein solubility. Dialysis membranes must be carefully cleaned before use as most commercial products contain impurities. They should be boiled with EDTA (0.01 M, Na salt) or a 1% solution of sodium bicarbonate for 10 minutes. Acetic anhydride can be used to decrease the pore size for proteins of molecular weight less than 10,000.

Equilibrium dialysis has many advantages over other techniques. The change of conditions are diffusion controlled so the point of nucleation can be slowly approached and growth subsequently regulated. The precipitate or microcrystals can be redissolved by reversing the conditions in the outside container. Thus several attempts can be made using the same dialysis cell. The method can be used for crystallization against low ionic strength buffer as well as conventional salting out procedures. It is also suitable for variation of pH, use of organic solvent and for studying the effect of small amounts of ions.

Equilibrium dialysis can also be modified for use with very small quantities of proteins. The development of various kinds of microdiffusion cells has been largely the work of Zeppezauer and his coworkers. (Zeppezauer et al., 1967; Zeppezauer, 1971).

A microdiffusion cell made from a Pyrex or Plexiglass capillary tube is illustrated in Fig. 3.6. The ends of the capillary are carefully rounded (see Fig. 3.6b) to avoid damage to the membrane which is secured at one end by a ring of transparent PVC tubing. The PVC ring has two or three "feet" which extend beyond the dialysis membrane. When the cell rests vertically on these feet, the membrane is held away from the bottom of the container. The protein solution is injected into the cell using thin tubing taking care to avoid formation of air bubbles or damage to the dialysis membrane. The upper end is then sealed with paraffin foil, again secured with a ring of PVC tubing and the cell is introduced into a relevant buffer so that the buffer covers the dialysis membrane. The conditions in the buffer are then modified in order to induce crystallization in the cell. The cells have been used successfully to crystallize lysozyme, phosphorylase and rabbit muscle aldolase. A typical cell of 1 mm inner diameter and 20 mm in length contains only 16 microlitres of protein solution, so the method is very economical in protein. An interesting modification to this type of cell is to add a second capillary in which a concentration gradient is established. The technique reduces the number of manipulations required to bring the solvent to the right concentration and has been useful in crystallizing glutamine synthetase (Weber and Goodkin, 1970; Eisenberg et al., 1971).

The disadvantages of the microdiffusion cells made from capillaries are two fold. Crystals forming on the membrane are difficult to observe without removing the cell which disturbs the concentration gradient within the cell.

3.4.3 EQUILIBRIUM DIALYSIS

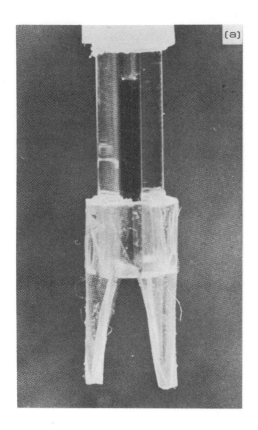

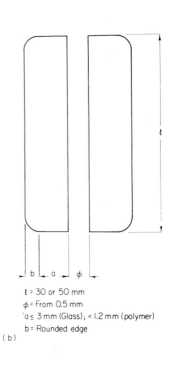

l = 30 or 50 mm
ϕ = From 0.5 mm
$a \leq$ 3 mm (Glass); < 1.2 mm (polymer)
b = Rounded edge

(b)

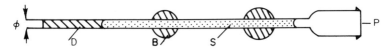

D = Acrylamide polymer diaphragm
S = Protein solution
B = Picein balls
P = Paraffin membrane
(c) ϕ = Inner diameter 0.7 or 1.0 mm

FIG. 3.6 Microdialysis cells, made from capillaries, developed by Zeppezauer et al. (1968). (a) Cell ready for use; (b) dimensions of suitable cells. Material: Pyrex glass or transparent polymer (Perspex, Plexiglas, Lucite or similar); (c) cell made from on X-ray capillary.

74 3. CRYSTALLIZATION OF PROTEINS

x (mm)	Volume of cell
1·4 mm	5 μl
2·0 mm	10 μl
3·0 mm	20 μl
4·0 mm	40 μl
4·5 mm	50 μl

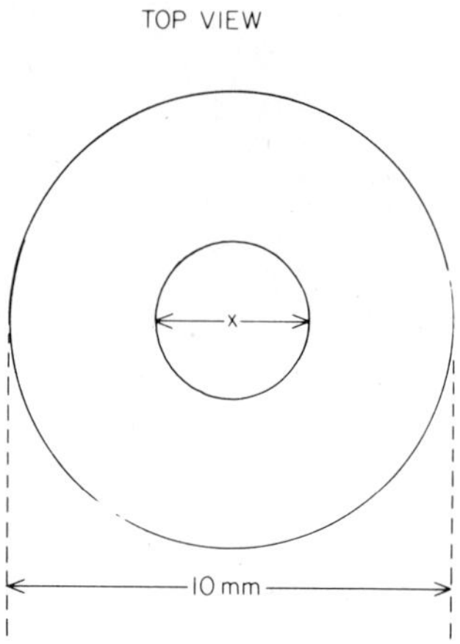

(a)

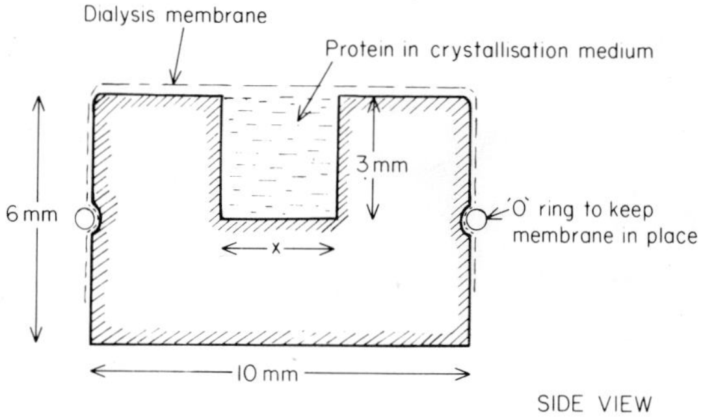

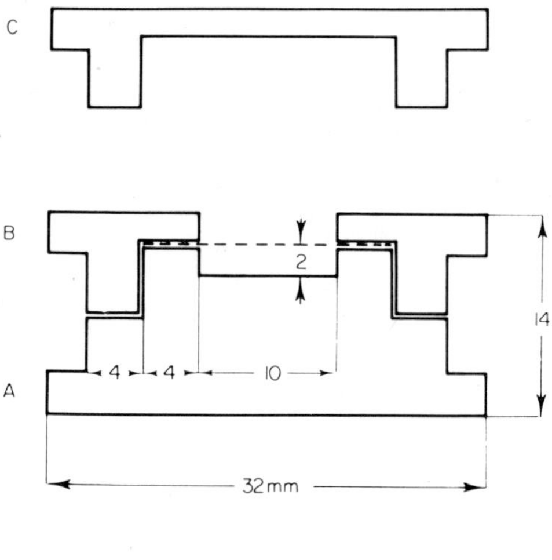

TOP VIEW

SIDE VIEW

(b)

Further, it is often difficult to remove the crystals from the capillary especially if it has a small diameter and the crystals are formed on the wall of the cell. This can be overcome by using a microdialysis cell of the type illustrated in Fig. 3.7. These cells can be made from optically treated glass, perspex or teflon. They avoid the use of PVC from which certain organic solvents occasionally extract plasticizers. The growth of crystals can be observed through a microscope without disturbing the cell. Despite the attractive features of these microdialysis cells, they have not been useful in growing large crystals but are very useful for finding the correct conditions for crystallization. They are also very difficult to clean.

Occasionally when only minute amounts of protein are available, crystallization can be carried out in an X-ray capillary tube (Zeppezauer, 1971) (Fig. 3.6c). Capillaries for this purpose have very thin walls, and they cannot be closed with a dialysis membrane. However, they can be converted to microdiffusion cells by sealing them with a semipermeable diaphragm of polyacrylamide.

3.4.4 Evaporation and vapour diffusion techniques

A traditional method for crystallization of smaller molecules is to effect the increase of concentration of the solution by evaporation of solvent. This technique is not generally useful with proteins as it is difficult to control and it is conducive to the crystallization of salts in the mother liquor. A more sensitive technique is to control the evaporation by equilibration with a more concentrated salt solution.

Thus a solution of the protein containing a salt solution 10% below the concentration needed for precipitation is equilibrated by vapour diffusion with a large volume of more concentrated salt solution (slightly below the concentration needed for precipitation of the protein). Both solutions are in open beakers within a larger carefully sealed beaker which is placed in a desiccator. The solvent is gradually transferred through the vapour phase from the protein solution to the more concentrated salt solution until the two are in equilibrium. Crystals tend to form as the protein solution becomes more concentrated. The method is reviewed by Davies and Segal (1971).

FIG. 3.7 Microdialysis cells manufactured for simple observation of crystallization. (a) A cell developed by Butler and Taylor at Cambridge and manufactured by Cambridge Repetition Engineers, Green's Road, Cambridge, CB43EQ; (b) a cell developed by Zeppezauer (1971). In (b) the cell is machined from Teflon rod. On both the cell, A, and the membrane holder, B, a simple ring (two edges) is carved out in order to simplify the gentle removal of the membrane holder when the experiment is completed. For storage of contents, a lid, C, is used instead of B. Similar cells made from optically treated glass are supplied on request by HELLMA, Optische Werkstätten, 7840 Mühlheim, Germany.

The vapour diffusion method can be easily modified to a micro technique. For instance the methods developed for t_{RNA} may be useful for proteins also. Neidle (private communication) suggests the following approach. The basic component of the system is a plastic Petri dish. The bottom half is divided into two as supplied by the manufacturers (Steralin Ltd., 12–14, Hill Rise, Richmond, Surrey). Additional sealing with chloroform is necessary near the

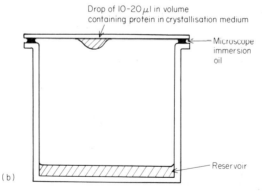

FIG. 3.8 Apparatus for crystallization by vapour diffusion on a microscale. (a) As used by Spencer and coworkers, King's College. Sixteen conditions can be investigated in equilibrium with a single reservoir. (b) The hanging drop method.

ends of the middle section, in order to prevent loss of solutions through hair-line cracks which apparently form during manufacture. One half of the base is used as a reservoir for the precipitant, and the other half is marked out as a 4 x 4 matrix for sample drops. Alternatively the solution can be placed on a microculture slide which has been siliconed to prevent spreading.* Motion of the solution in the reservoir during handling of the dish, is minimized by inserting four or five sections of filter paper in this compartment. The top of the dish is drilled to create a port through which solutions can be admitted to the reservoir. A perspex block, sealed with vacuum grease, closes this hole. The two halves of the dish are also sealed with the same grease. Sixteen trials per dish can be carried out. Each droplet is $10/\mu l$ in volume, and contains buffer, counter-ion and protein. The latter is adjusted to a concentration of about 1 mg/ml which is sufficient in many cases to give at least one really good-sized crystal. The dishes are, of course, kept in a controlled environment. Figure 3.8a shows one of the dishes.

Alternatively vapour diffusion methods can be used with a "hanging drop". This is illustrated in Fig. 3.8b and has been used very successfully by Wonnacott for the crystallization of glyceraldehyde phosphate dehydrogenase. The protein is dissolved in $10-20~\mu l$ of solvent at a salt concentration slightly below where the protein crystallizes. The method has the advantage of simplicity and is inexpensive to set up. It also allows the progress of crystallization to be followed easily. This method is particularly effective for growing large crystals, especially if the drop is seeded.

3.4.5 The crystallization of a new protein

These days most proteins which are available in large quantities have already been crystallized and are being actively studied by protein crystallographers. A new protein will probably be available in only small quantities. How should one then approach the preparation of large crystals?

The first step is to make a thorough study of the biochemical literature and discuss the project with biochemists who have prepared the protein in a purified state. Of particular importance is the range of conditions under which the protein is biochemically active: there is an advantage in studying an active

---

*Zeppezauer has observed that horse liver alcohol dehydrogenase wets siliconed glass surface which means that the protein is absorbed (and probably denatured) at such a surface. This may be a rare exception. The surface of acryl polymers (plexi glass, lucite) are slightly hydrated in contact with water, and these polymers might therefore be superior to siliconed glass or polystyrene. A coating of glass which prevents electroendoosmosis (probably by preventing access of ions to charged groups of the glass) but not wetting is made by polymerization of methylcellulose with formaldehyde (Hjerten and Mosbach, 1962).

protein! Often data are also available on conformational changes and aggregation which might be pH dependent.

The second step is to study the solubility dependence on pH, ionic strength, organic solvent and temperature within the limitations defined by previous biochemical study to produce a five dimensional matrix of variables. This can be done by using any of the microtechniques described in the previous section. It is probably best to study the pH dependence by performing experiments at pH intervals of 0.3 using one salt and one organic solvent. (Usually crystals are produced only over a very narrow pH range.) This will allow a qualitative picture of the solubility. The picture can then be refined by using smaller pH and temperature intervals and different salts and organic solvents where saturation occurs.

Such a study was carried out by Bunn et al. (1971) for the crystallization of rennin. The solubility dependence at pH 6.0 is illustrated in Fig. 3.9. The salt used was sodium chloride. Amorphous rennin shows the usual maximum solubility at intermediate ionic strengths but also shows an unusual minimum at very low salt concentrations. The temperature coefficient of solubility of amorphous rennin is positive at low ionic strength but negative at higher ionic strength as is very often found. Except at very low ionic strength crystalline rennin is much less soluble than amorphous rennin (although this result may not be entirely a thermodynamic phenomenon but rather kinetically controlled).

With such a solubility curve the rate of nucleation and growth can be studied. For instance Bunn takes amorphous precipitate of rennin, wet with strong salt solution (at point A in Fig. 3.9) and dilutes it with the minimum quantity of solvent. This brings the protein to point B in Fig. 3.9.

At this point the solution is highly supersaturated with respect to crystalline rennin (at point C) and very small crystals are obtained. Dilution with an equal

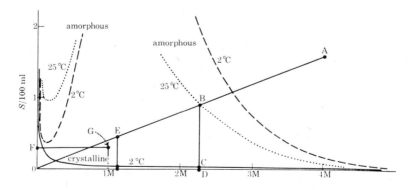

FIG. 3.9 The solubilities of amorphous and crystalline rennin (from Bunn et al., 1971).

volume of solvent brings the solution to point E where the supersaturation is less and larger crystals (0.2 mm) are formed.

As is often the case, preparation of even larger crystals depended on the identification of a factor peculiar to rennin. The secret appeared to rest with the addition of $AlCl_3$. This was added during purification and the precipitate was filtered off and discarded. Presumably $Al(OH)_3$ carries away some unknown impurity which otherwise prevented crystallization. Addition of a small amount was needed to initiate crystallization, but addition of more than the optimum amount usually produced no crystals unless seeds were introduced. Seeding often plays an important role in the production of larger crystals. This is well illustrated by the work of Lynen and his colleagues (Oesterhelt et al., 1969) on fatty acid synthetase, a multi enzyme complex comprising seven proteins and a total molecular weight 2.3×10^6. Microcrystals were first observed in a preparation which had stood at $4°C$ for 15 months. Crystallization conditions were then varied in a systematic manner with other preparations using these first crystals as seeds. Optimum conditions were established which produced hexagonal prisms of 0.1 mm in dimension in 2 days.

Even after crystals have been obtained, it is often worthwhile to continue to explore other conditions which may produce a crystal form or polymorph more suitable for X-ray study. The work on chymotrypsinogen was held up for years as a result of an unsuitable crystal form (Freer et al., 1970). Ten different crystalline forms of the allosteric enzyme, aspartate transcarbamylase (M-W, 300,000) have been produced. The form most suitable for study (which has one subunit per a symmetric unit) was found by accident (Wiley et al., 1971). Insulin also exhibits polymorphism (Schlichtkrull, 1958). The orthorhombic crystal form produced at acid pH is unstable at room temperature causing great problems for the X-ray analysis (Low and Berger, 1961). However, the rhombohedral form produced at pH 6.3 is quite stable (Crowfoot, 1935).

The species of animal or organism from which the protein is isolated forms yet another variable in the search for protein crystals (Dickerson et al., 1969). The same enzyme from a different species might crystallize better or provide a crystalline form which expresses the symmetry of the protein molecule (Wiley et al., 1971). Some examples of crystals are shown in the frontispiece.

3.5 PROTEIN CRYSTALS: THE SOLVENT OF CRYSTALLIZATION

Protein crystals differ from crystals of smaller organic molecules in that they contain a considerable quantity of liquid solvent. An example of a protein crystal packing is shown in Fig. 3.10. The crystal really contains two phases, a solid phase which comprises the proteins touching at a few points and forming an open lattice which is filled by the liquid phase. The solvent close to the protein molecules is often quite organized and strongly hydrogen bonded to the surface polar groups of the protein but the solvent in the centre of the channel which can be 20 Å wide is usually relatively disorganized in the same way as bulk liquid. The environment of a protein in a crystal lattice is not too different from that encountered in solution.

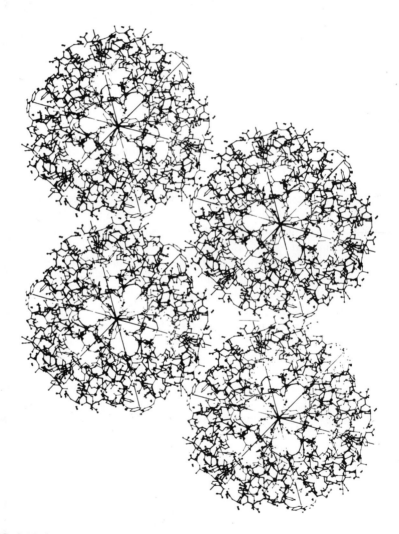

FIG. 3.10 The packing of insulin hexamers (see Section 2.8) into the rhombohedral 2-zinc crystals. Note the large channels between the hexamers and the relatively few parts of the surface which are not in contact with solvent (from Blundell et al., 1972).

3.6 CRYSTAL MOUNTING

As a protein crystal contains a liquid phase as well as a solid phase, it is necessary to keep the crystal in an environment of controlled high humidity. If the crystal is exposed to the air under normal laboratory conditions, the liquid in the crystal will quickly distil out. This will usually cause the protein molecules to pack more closely together but, at the same time, to loose the regularity of the crystal packing. This manifests itself eventually by the appearance of cracks in the crystals but often the drying out of crystals is difficult to identify except by taking an X-ray photo.

Protein crystals are usually mounted in thin-walled (0.001 mm thick) glass or quartz capillaries as described by King (1954) and by Holmes and Blow (1966). The glass used must not contain heavy atoms which will absorb and scatter the X-radiation. The crystals can be introduced into the capillaries by drawing them up by capillary action or gentle suction, an operation which is best achieved when viewed through a binocular microscope. Alternatively the crystal and a small volume of liquid may be drawn into a Pasteur pipette and then

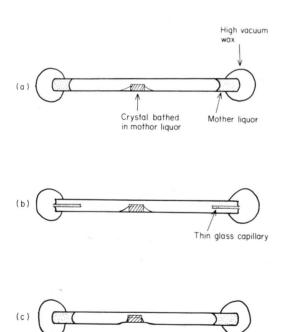

FIG. 3.11 Different ways of mounting protein crystals in thin-walled tubes. In (a) the solvent reservoir is held by capillary action of the mounting tube itself. In (b) the reservoir comprises two thin capillaries inside the mounting tube. In (c) the mounting tube has been flattened to prevent slippage.

pipetted directly into the capillary tube. The mother liquor is removed by drawing it away using a syringe, very thin glass capillaries or thin strips of filter paper.

A very small quantity of mother liquor should remain around the crystal so that the crystal is kept moist. The crystal adheres to the capillary wall by surface tension arising from this mother liquor. The capillary inner surface is carefully but quickly dried, and a few drops of liquid are placed at each end of the capillary which is then sealed with a high vacuum wax. The mounting of a crystal in a capillary in this way is illustrated in Fig. 3.11a. Sometimes it is more convenient to place the mother liquor in two small tubes as shown in Fig. 3.11b. Prior washing of the glass mounting capillaries with nitric acid and with buffer solution improves the stability of the mounted crystal of certain proteins.

An interesting combination of crystal growing and crystal mounting (Akervall and Strandberg, 1971) has been useful for crystals of tobacco necrosis satellite virus which are too fragile to be handled. A slight depression is introduced into the wall of a capillary tube as shown in Fig. 3.11c. A seed crystal is placed there and surrounded with crystallizing solvent containing more viruses. The tube is sealed until the crystal grows to an appropriate size and X-ray photographs are then taken without disturbing the crystal. The shape of the capillary is particularly effective in reducing crystal slippage.

The tools needed for protein crystal mounting are: thin walled glass capillaries (obtained from Pantak Ltd., Vale Road, Windsor, Berks); vacuum wax (melting temperature $78°C$) (obtained from H. J. Everett, Park Gate, Nr Southampton); Pasteur pipettes; filter paper cut into very thin strips; syringe and needle; very fine glass capillaries for drawing solvent from mounting capillaries; glass depression slide.

4

SYMMETRY, SPACE GROUPS AND OPTICAL PROPERTIES OF CRYSTALS

4.1 INTRODUCTION

An understanding of crystal symmetry forms the first essential step in the determination of any molecular structure by X-ray diffraction techniques. In certain favourable cases it may even be possible to determine the subunit structure of an oligomeric protein from an analysis of the symmetry of the crystal, and its density and water content.

Crystal symmetry itself forms a fascinating subject. The mathematical derivation of the finite number of symmetry operations which are possible for a crystal represents a remarkable achievement of nineteenth century mathematicians. The achievement is even more remarkable when we recall that these laws were put forward many years before they could be tested by X-ray diffraction. In this chapter we can do no more than to provide a synopsis of the laws which govern crystal symmetry. The reader is referred to Buerger (1963) for a geometrical derivation of these laws, and to Phillips (1963) for an introduction to crystal morphology and space groups. A full description of the 230 space groups is given in International Tables for Crystallography, Vol. I.

The study of the external shapes of crystals provides the logical starting point for understanding crystal symmetry. For any crystallographer an appreciation of the external appearance of a crystal forms an aesthetically pleasing introduction to the subject. A cursory glance at any crystal under the microscope will show that crystals exhibit clear cut faces and edges which are arranged with some degree of symmetry (see Frontispiece). These naturally formed plane faces are related to the underlying periodic arrangements of molecules in the crystal. We now wish to investigate the possible types of periodic arrangements. First we require a few definitions.

A *space lattice* is defined as an arrangement of points such that each point is in exactly the same environment and in the same orientation as every other point.

The *unit cell* of a crystal is defined as the basic parallelipiped shaped block from which the whole volume of the crystal may be built by regular assembly of

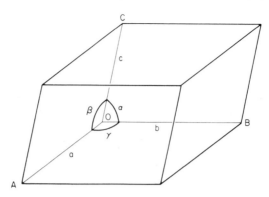

FIG. 4.1   The unit cell.

these blocks. Each cell contains a complete representation of the unit of pattern, the repetition of which builds up the crystal. In general it is convenient to choose a unit cell which has the shortest edges possible and at the same time is in accord with the highest possible symmetry of the crystal. The edges are defined by three vectors which, by convention, are given the symbols **a**, **b** and **c** and the angle between these defined as α, β and γ as shown in Fig. 4.1. The axial lengths are a, b, c. This system of axes then provides a framework within which any point in the unit cell may be specified with respect to a given origin by parameters x, y and z measured parallel to the axes and expressed as fractions of the cell edges.

The asymmetric unit. If the crystal contains symmetry elements, (indeed crystals which have no symmetry are rare) then the unit cell will contain more than one object (see Fig. 4.3). The asymmetric unit is the basic repeating object which is related to all the other identical objects in the unit cell by the operation of the symmetry elements and to the contents of the other unit cells by the translations **a**, **b**, and **c**.

4.1.1 The law of rational indices

In an optical examination of a crystal, it was natural that crystallographers should wish to label the observed faces of a crystal and relate these planes to some unit cell. For a particular choice of a unit cell it was found possible to define a plane, parallel to the crystal face, which made intercepts a/h, b/k and c/l of the unit cell edges where h, k and l are integers. This gave rise to the law of rational indices which states:

> The faces of a crystal are parallel to planes making intercepts a/h, b/k and c/l on the three axes, where h, k, and l are small integers.

4.1.1 THE LAW OF RATIONAL INDICES

The law is a consequence of the foundation of the crystal on a space lattice. The indices h, k, l are termed the Miller indices of the face and these are usually enclosed in brackets. In X-ray diffraction we will be concerned with the reflections from a set of planes. By convention the unbracketed symbol is used to denote the indices of a reflection. Hence the 111 reflection arises from planes parallel to the (111) face of the crystal. Braces are used to signify a set of equivalent faces. Thus for a cubic crystal we may describe the form by {100} which would mean the set of faces 100, $\bar{1}$00, 010, 0$\bar{1}$0, 001 and 00$\bar{1}$ of a cube. Square brackets (e.g. [100]) are used to enclose the indices of a direction in the real lattice (i.e. a zone axis) where the numbers are expressed in terms of a,b,c as units.

In Fig. 4.2 we show a two dimensional lattice and illustrate the indices for some sets of planes. For example the 100 planes form infinite intercepts on the b and c axes and divide the a axis by unity. They are therefore normal to the a axis. Likewise the 010 planes are normal to the b axis. The diagram shows a number of other types of planes which are inclined to both the a and b axes. The extension of these ideas to three dimensions is straight forward.

Although we may define planes which may take any values of the indices h, k, l, it turns out that crystal faces are only parallel to planes which have low indices, usually less than 3. Numbers greater than 5 are very rare. This is because crystal faces represent planes of high (although not necessarily the highest) density of lattice points. It can be seen from the diagram that the 320 planes, for example, contain a lower density of lattice points than the 010 planes.

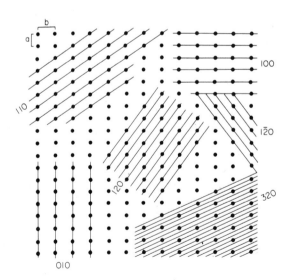

FIG. 4.2 A two dimensional lattice, illustrating the indices of certain sets of planes. (After Bunn, 1961.)

With these few definitions, we now consider the possible symmetry operations which are possible for a crystal lattice. We follow the historical approach and consider first the symmetry elements which are detectable in the external appearance of well-formed crystals and then consider the additional internal symmetry operations which are only revealed by X-ray diffraction.

4.2 THE THIRTY TWO CLASSES OF SYMMETRY

In an examination of several crystals of the same substance (for example, those shown in the frontispiece) it is observed that although they all have roughly similar shapes, some of the faces may be developed to different extents in different crystals. The problem of finding the fundamental similarities between these crystals of the same substance was solved experimentally with the aid of the optical goniometer. This enabled the direction of the normal to a crystal face to be defined from observations of the reflection of light from the mirror-like facets.

In an idealized crystal it should be possible to take a single point within the crystal and from this point draw lines which are normal to the crystal faces. This group of lines will represent the external crystal symmetry. From observations on a number of crystals and the determination of the directions of the normals to their crystal faces it was possible to build up such an idealized crystal and hence to determine the external symmetry.

We now ask what type of symmetry elements are possible for the external symmetry of a crystal. In mathematical terms we seek the set of operations which ensure that each point in the structure is brought back to its original position by the operation of the symmetry elements. This concept of a point group symmetry operation leads to the thirty-two classes of symmetry. There are three basic types of symmetry operators:

(i) Mirror plane (m)   Reflection across a plane of symmetry. Since biological molecules are composed of enantiomorphic molecules (e.g. D-sugars and L-amino acids), this type of symmetry operation, which converts a right-handed molecule into a left-handed one and vice versa, cannot occur in crystals of biological molecules.

(ii) Rotation axis (x)   An x-fold rotation axis rotates an object $360°/x$ about the axis. Hence a 2-fold rotation corresponds to a rotation of $180°$, a 3-fold rotation to $120°$ and so on. When rotational symmetry occurs in a crystal there are severe limitations on the type of rotation axes possible. This is because of the requirement that the crystal must be able to repeat indefinitely in three dimensions. It has been shown (e.g. Buerger, 1963, p. 33) that the angle of rotation α is restricted to values for which $\cos \alpha = M/2$ where M is a positive or negative integer and $\alpha = 360°/x$. Hence x may only equal 1 (the identity operator), 2, 3, 4 or 6 and may not take any other value.

Examples of these type of rotation axes are shown in Fig. 4.3. This diagram

4.2 THE THIRTY TWO CLASSES OF SYMMETRY

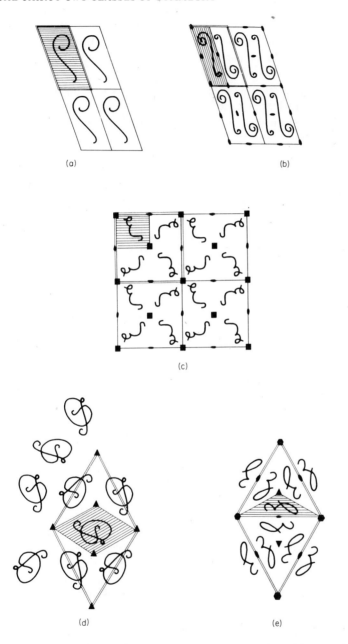

FIG. 4.3 The rotational symmetry elements (a)1 (b)2 (c)4 (d)3 and (e)62. A single unit cell is outlined by a double set of lines. Other unit cells are shown with single lines. The asymmetric unit is shown shaded. (After Kestleman, 1968.)

also illustrates the relationship of the asymmetric unit to the contents of the unit cell. Examination of the crystals shown in the frontispiece will demonstrate how some of these rotational symmetry elements are exhibited in the external crystal form.

Although the rotation axes in a crystal are restricted to those of degree 1, 2, 3, 4 or 6, this does not mean that no other rotation axes are possible for natural objects which do not have to repeat in a crystal lattice. For example many flowers contain a five-fold arrangement of their petals, spherical viruses and the skeletons of several radiolaria exhibit icosohedral symmetry and the base pairs in DNA are arranged with a ten-fold screw axis of symmetry. The restriction of the rotation axis simply means that none of these objects could crystallize in a lattice in which the petals, the virus subunit or the base pairs could form the asymmetric unit. On the other hand if an object contains rotational symmetry elements which are permissible in a crystal (such as for example 2-fold or 3-fold rotation axes), then it is possible for the object to crystallize in such a way that the subunit forms the asymmetric unit. There are many examples of this in protein crystallography when the protein itself is oligomeric. Such crystals are most favourable for detailed study since the size of the asymmetric unit is smaller than the size of the whole protein. This aspect of protein symmetry and crystal symmetry was discussed in Chapter 2.

(iii) Inversion about a point   symbolized in terms of degree of the rotation axis $\bar{1}, \bar{2}, \bar{3}, \bar{4}$, and $\bar{6}$. This operation carries out a rotation through the angle indicated followed by an inversion through a centre. Hence $\bar{1}$ is equivalent to a centre of symmetry and $\bar{2}$ is equivalent to a mirror plane. As biological molecules are enantiomorphic, this type of symmetry operation does not occur in crystals of biological molecules.

In addition to the symmetry elements described, it is also possible to have certain combinations and still fulfil the requirement of a point group. The complete set of operations is:

x	Rotation axis alone
$\bar{x}$	Inversion axis alone ($\bar{2}$ = m)
x/m	Rotation axis normal to a plane of symmetry
xm	Rotation axis with a vertical plane of symmetry
$\bar{x}m$	Inversion axis with a vertical plane of symmetry
$x2$	Rotation axis with a dyad axis normal to it
x/mm	Rotation axis with a vertical plane of symmetry and normal to a plane of symmetry.

When these operations are expanded in terms of the possible value for the rotation axes, they give rise to the thirty-two classes of symmetry. These are shown as stereograms in Fig. 4.4. Note that the only operations which need concern us in protein crystallography are x and $x2$, that is cyclic symmetry and dihedral symmetry.

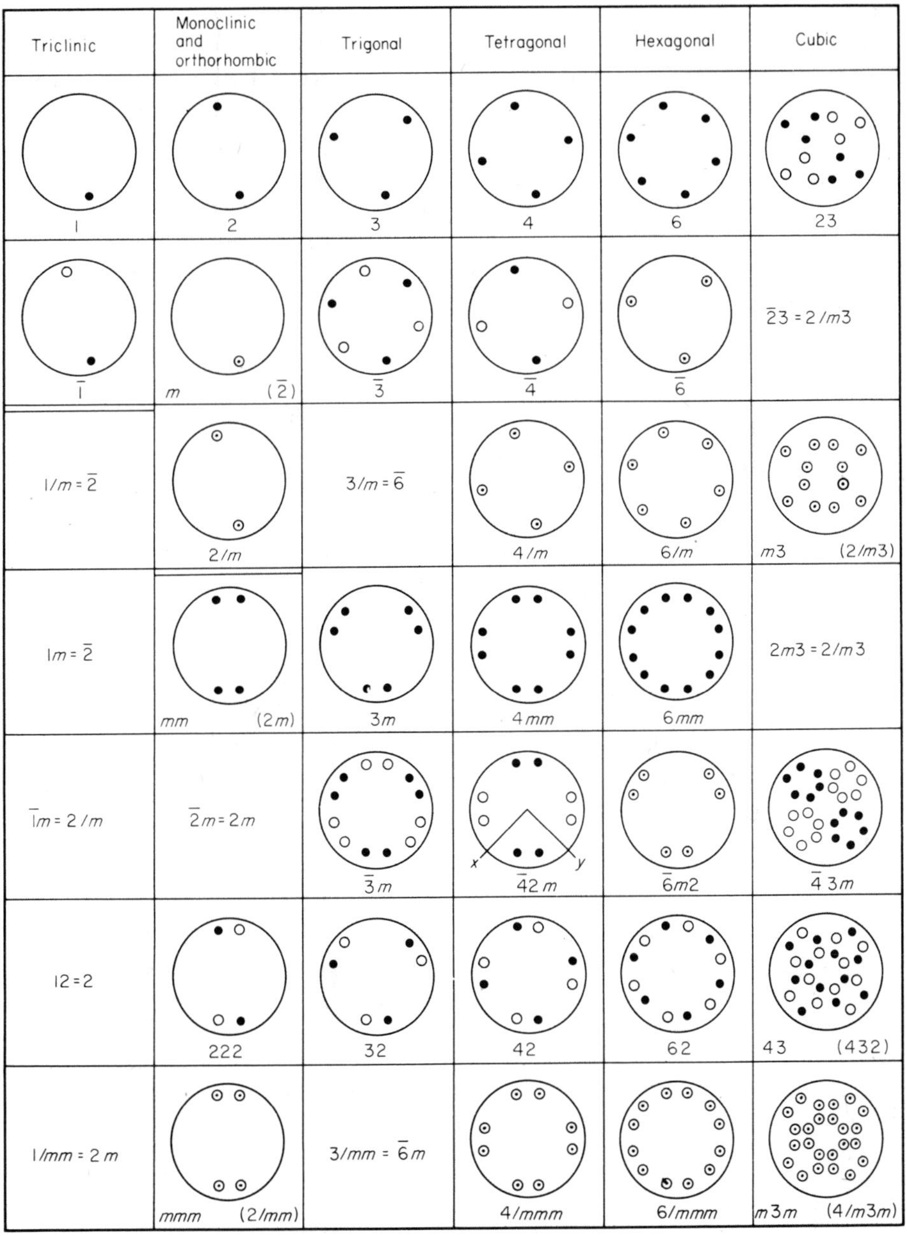

4.2 THE THIRTY TWO CLASSES OF SYMMETRY

Triclinic	Monoclinic and orthorhombic	Trigonal	Tetragonal	Hexagonal	Cubic
1	2	3	4	6	23
$\bar{1}$	m ($\bar{2}$)	$\bar{3}$	$\bar{4}$	$\bar{6}$	$\bar{2}3 = 2/m3$
$1/m = \bar{2}$	$2/m$	$3/m = \bar{6}$	$4/m$	$6/m$	$m3$ ($2/m3$)
$1m = \bar{2}$	mm ($2m$)	$3m$	$4mm$	$6mm$	$2m3 = 2/m3$
$\bar{1}m = 2/m$	$\bar{2}m = 2m$	$\bar{3}m$	$\bar{4}2m$	$\bar{6}m2$	$\bar{4}3m$
$12 = 2$	222	32	42	62	43 (432)
$1/mm = 2m$	mmm ($2/mm$)	$3/mm = \bar{6}m$	$4/mmm$	$6/mmm$	$m3m$ ($4/m3m$)

FIG. 4.4 The 32 crystal classes.

4.3 THE SEVEN CRYSTAL SYSTEMS

Crystals may be grouped into seven crystal systems on the basis of the highest rotational symmetry present. The seven systems are shown in Table 4.I. The presence of certain types of symmetry element require definite relationships between the unit cell axes. Thus a two-fold or higher rotation axis must always be normal to the other two axes. A three-fold or higher rotation axis requires the other two axes normal to it to be equal.

4.4 THE FOURTEEN BRAVAIS LATTICES

In the discussion so far we have implicitly assumed that the lattice is primitive (P), i.e. that it contains a point at each corner of the unit cell. Since each point at the eight corners is shared with eight nearest neighbours there is effectively one point per unit cell. However it is possible to construct other types of lattices which also meet the requirement of a space lattice. For example in the cubic system, if a point is placed at the centre of the cube, then a body centred lattice I (German Innenzentrierte), is obtained which still retains all the symmetry elements of a cube (Fig. 4.5). The environment of each point is the same as that of every other point and so the new arrangement meets the requirement of a space lattice. Of course, it is also possible to choose a primitive cubic lattice from the body centred lattice but in this case the angle between each pair of axes would be $109°28'$. This non-standard cell would not fit the requirements defined in Table 4.I. The body centred lattice contains two equivalent points per cell, that is it is doubly primitive. A third type of lattice is also possible for a cube. If a point is placed at the centre of each of the six faces, then a face centred cell (F) results which is quadruply primitive (Fig. 4.5).

The other crystal systems may be treated in a similar way. In the triclinic system with no restriction on cell size or shape a primitive cell can always be chosen. In the monoclinic system, centring on the C(001) face produces a new type of lattice (C). An A-centred lattice is equivalent to a C-centred lattice with a change of orientation. F and I lattices in the monoclinic system can be reduced to a C space lattice. The fourteen types of lattices which can be built up in this way are shown in Fig. 4.5. They were first described by Bravais (1848).

In rare cases research workers may prefer to work with a non-standard lattice because it facilitates certain aspects of the work. Thus for example in the case of lactate dehydrogenase (Rossmann *et al.*, 1967), the authors preferred to index their diffraction pattern on the basis of a larger face centred tetragonal unit cell (space group F422) than the smaller, standard body centred cell (space group I422) because the external morphology of the system made this assignment of axes easier. In general however there are compelling reasons for using the standard lattices described in International Tables Vol. I. All the relevant

TABLE 4.1
The seven crystal systems

Name	Possible Bravais Lattices	Axes of symmetry	Lattice	
Triclinic	P	No axes of symmetry	$a \neq b \neq c$	$\alpha \neq \beta \neq \gamma$
Monoclinic	P,C	1 dyad axis (parallel to b)	$a \neq b \neq c$	$\alpha = \gamma = 90° \neq \beta$
Orthorhombic	P,C,I,F	3 dyad axes mutually orthogonal	$a \neq b \neq c$	$\alpha = \beta = \gamma = 90°$
Tetragonal	P,I	1 tetrad axis (parallel to c)	$a = b \neq c$	$\alpha = \beta = \gamma = 90°$
Trigonal	P (or R)	1 triad axis (parallel to c)	$a = b \neq c$	$\alpha = \beta = 90°, \gamma = 120°$
			$a = b = c$	$\alpha = \beta = \gamma < 120°, \neq 90°$
Hexagonal	P	1 hexad axis (parallel to c)	$a = b \neq c$	$\alpha = \beta = 90°, \gamma = 120°$
Cubic	P,I,F	4 triad axes (along the diagonals of the cube)	$a = b = c$	$\alpha = \beta = \gamma = 90°$

92   4. SYMMETRY, SPACE GROUPS AND OPTICAL PROPERTIES

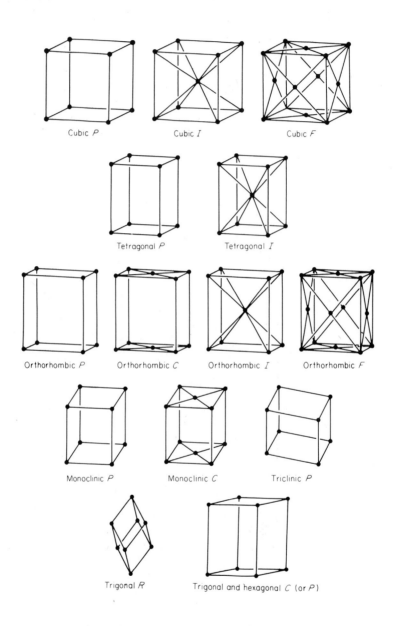

FIG. 4.5   The 14 space lattices.

4.5 INTERNAL SYMMETRY ELEMENTS

equations have been worked out and listed and any other worker in the field can understand the terminology. Crystallography is a systematic science.

4.5 INTERNAL SYMMETRY ELEMENTS

The set of symmetry elements which form a point group describe the symmetry of the crystal as a whole and may be recognized from the external shape of the crystal. The question we now ask is: is there more than one kind of internal symmetry arrangement which will give rise to the same external symmetry? For a point group operation, all possible combinations of symmetry operations were considered which brought each point of the structure back to its original position. In the internal symmetry of a crystal we wish to consider all the self consistent set of operations such that any symmetry operation or lattice translation brings all the remaining operations into coincidence. The fourteen Bravais lattices may be considered as internal symmetry operations of this kind. There are, in addition, two translational symmetry operations. These are:

(i) Glide plane   The structure is reflected across a plane and translated parallel to the plane. The translation must be for a distance which is one half the unit cell edge so that two successive reflections and translations are equivalent to one unit cell translation. Glide planes do not occur in enantiomorphic protein crystals.

(ii) Screw axis   The structure is rotated through an angle corresponding to the degree of the rotation axis and translated parallel to the rotation axis. For a two-fold rotation axis the translation must be one-half the unit cell edge parallel to the rotation axis so that again two successive operations will bring the structure into coincidence in the next unit cell. A two-fold screw axis is represented as 2_1.

For a three-fold screw axis, translations corresponding to one third (3_1) or two thirds (3_2) of the unit cell edge are possible. The structure is brought into coincidence in the next unit cell by three successive operations. These operations are shown in Fig. 4.6. They give rise to a clockwise and to an anticlockwise arrangement of the repeat motif about the rotation axis. The possibilities for a four-fold rotation axis are 4_1, 4_2, and 4_3 with translations corresponding to one quarter, two quarters and three quarters of the unit cell edge respectively and the possibilities for a six-fold rotation are 6_1, 6_2, 6_3, 6_4, and 6_5 with translations corresponding to one sixth, two sixths, three sixths, four sixths and five sixths respectively.

It can be shown that these translational symmetry elements still fill the requirement of a space lattice, that each point is in exactly the same environment as all other points, but they do not represent a point group. The result of these operations is a structure which continues indefinitely in space and results in a crystal terminated differently (at the molecular level) at opposite ends of

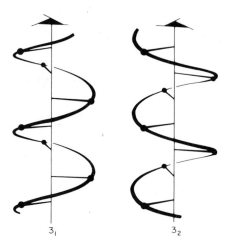

FIG. 4.6   The rotational screw operations 3_1 and 3_2.

the axis. But since the translation is of molecular dimensions it is not apparent in the external appearance of the crystal.

4.6 THE TWO-HUNDRED-AND-THIRTY SPACE GROUPS

The mathematical theories of Federov (1885), Schoenflies (1891) and Barlow (1894) at the end of the last century showed that the combination of the

TABLE 4.II
Enantiomorphic space groups

System	Class	Space group symbols
Triclinic	1	$P1$
Monoclinic	2	$P2, P2_1, C2$
Orthorhombic	222	$C222, P222, P2_12_12_1, P2_12_12, P222_1, C222_1, F222, I222, I2_12_12_1$
Tetragonal	4	$P4, P4_1, P4_2, P4_3, I4, I4_1$
	422	$P422, P42_12, P4_122, P4_12_12, P4_222, P4_22_12, P4_32_12, P4_322, I422, I4_122$
Trigonal	3	$P3, P3_1, P3_2, R3$
	32	$P312, P321, P3_121, P3_112, P3_212, P3_221, R32$
Hexagonal	6	$P6, P6_5, P6_4, P6_3, P6_2, P6_1$
	622	$P622, P6_122, P6_222, P6_322, P6_422, P6_522$
Cubic	23	$P23, F23, I23, P2_13, I2_13$
	432	$P432, P4_132, P4_232, P4_332, F432, F4_132, I432, I4_132$

4.6.2 ORTHORHOMBIC P2₁2₁2₁

external symmetry elements, represented by the thirty-two classes, with the internal symmetry elements, represented by the fourteen Bravais lattices and the translational symmetry operations, gave rise to a finite number of space groups. The total number is 230. In this book we are concerned only with crystals of biological molecules. Since these molecules are enantiomorphic we need only consider the enantiomorphic space groups. The total number of these is 65 and they are listed in Table 4.II.

We give a brief description of the arrangements of the symmetry elements for three commonly occurring space groups.

4.6.1 Monoclinic $P2_1$

This is space group number 4 in International Tables Vol. I and is based on the point group 2 (Fig. 4.7). The lattice is primitive with the two-fold screw axis parallel to the **b** axis of the unit cell and placed at the origin and at each lattice point. The upper part of the diagram shows a projection of the structure down the two-fold screw axis **b**, and the lower part shows the projection down **c**, where the two-fold screw axis is represented by the singly barbed arrows. An object located at a position x, y, z in the unit cell is brought to a position $\bar{x}, y + \frac{1}{2}, \bar{z}$ by operation of the symmetry elements. This generates a set of two-fold screw axes at the half-way positions which are also shown in the diagram. The unit cell contains two molecules. Although the structure itself is not centrosymmetric, the projection of the structure onto the xz plane is centrosymmetric with respect to the origin.

4.6.2 Orthorhombic $P2_1 2_1 2_1$

This is space group number 19 in International Tables and is derived from the dihedral point group 22 (Fig. 4.8). The presence of two mutually perpendicular two-fold axes generates a third two-fold perpendicular to the first two. The point group is therefore designated 222. In the space group P222 all three two-fold axes intersect at the origin. If 2 two-fold screw axes intersect then the space group which is generated is $P2_1 2_1 2$. If 3 two-fold screw axes intersect, then space group I222 is generated. For space group $P2_1 2_1 2_1$ the three screw axes must be non-intersecting and the origin is chosen half-way between the three pairs of non-intersecting screw axes.

The co-ordinates of the equivalent positions are shown in the diagram. There are four molecules per unit cell.

The projections down all three of the axes are centrosymmetric but the centre of symmetry is not at the origin. For example in the projection down the **c** axis, the centre is at $x = \frac{1}{4}, y = 0$.

Monoclinic 2   P 1 2₁ 1   No. 4 $P2_1$ / $C\frac{2}{2}$

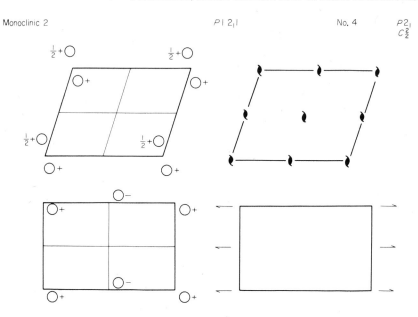

Origin on 2_1 ; unique axis b

Co-ordinates of equivalent positions

2nd SETTING

Number of positions,
Wyckoff notation,
and point symmetry

Conditions limiting
possible reflections

2   a   1 $x,y,z;\ \bar{x},\frac{1}{2}+y,\bar{z}$.

hkl: No conditions
$h0l$: No conditions
$0k0$: $k = 2n$

FIG. 4.7   Space group P2₁ (from International Tables for X-ray Crystallography Vol. I, International Union for Crystallography).

4.6.3 Tetragonal $P4_1 2_1 2$

This is space group number 92 in International Tables and is derived from the point group 42 in which the tetragonal axis is parallel to the z direction and the dyad axes are parallel to x and y directions (Fig. 4.9). In space group $P4_1 2_1 2$, these rotational axes become non-intersecting screw axes and the combination of these symmetry operations generates a dyad axes along the xy diagonal directions of the cell.

The space group contains eight equivalent positions per unit cell and is enantiomorphic to space group $P4_3 2_1 2$. There are three centrosymmetric

4.7   OPTICAL PROPERTIES OF CRYSTALS

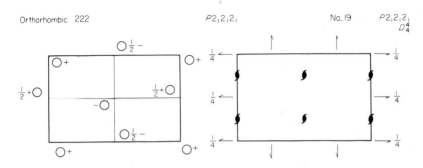

FIG. 4.8   Space group $P2_12_12_1$ (from International Tables for X-ray Crystallography Vol. I, International Union for Crystallography).

projections; the projection down **c**, the projection down **a** (which is of course equivalent to a projection down **b**) and the projection down the diagonal.

4.7 OPTICAL PROPERTIES OF CRYSTALS

Nineteenth-century crystallography was almost exclusively concerned with observations on the appearance and optical properties of crystals and such studies still form a useful preliminary characterization of a crystal prior to X-ray studies. With protein crystals, however, detailed observations on interfacial angles and refractive indices are difficult to make with any degree of precision because the protein crystal is usually observed either in a drop of mother liquor on a microscope slide or sealed in a thin walled glass capillary. Although protein crystallographers seldom carry out the detailed type of examination performed in classical crystallography, a preliminary study with the polarizing microscope can yield much useful information especially in connection with the quality of the crystal and relative orientation of the unit cell axes with respect to its external form.

$P4_12_12$ No. 92 $P4_12_12$ 4 2 2   Tetragonal
D_4^4

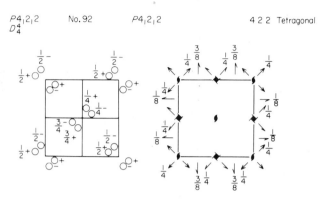

Origin at $2_1 2$ ([001] [110])

Co-ordinates of equivalent positions

Number of positions, Wyckoff notation, and point symmetry				Conditions limiting possible reflections
				General:
8	b	1	x,y,z; $\bar{x},\bar{y},\frac{1}{2}+z$; $\frac{1}{2}-y,\frac{1}{2}+x,\frac{1}{4}+z$; $\frac{1}{2}+y,\frac{1}{2}-x,\frac{3}{4}+z$; $y,x,\bar{z}$; $\bar{y},\bar{x},\frac{1}{2}-z$; $\frac{1}{2}-x,\frac{1}{2}+y,\frac{1}{4}-z$; $\frac{1}{2}+x,\frac{1}{2}-y,\frac{3}{4}-z$.	hkl: No conditions $00l$: $l=4n$ $h00$: $h=2n$

FIG. 4.9   Space group $P4_12_12$ (from International Tables for X-ray Crystallography Vol. I, International Union for Crystallography).

The reader is referred to Bunn (1961) and to Hartshorne and Stuart (1960) for more detailed descriptions of the optical properties of crystals.

4.7.1 The polarizing microscope

Light is an electromagnetic transverse wave motion which may be represented by two vectors, the magnetic vector, **H** and the electric vector, **E** which are always perpendicular to one another and to the direction of propagation of the light. Since it is the electric vector which stimulates the eye, we may neglect the magnetic vector in a discussion of visual optics. In ordinary light the vector **E** may adopt all possible orientations in the plane perpendicular to the direction of motion. In plane polarized light the vector **E** is constrained to lie in one direction only. In the propagation of plane polarized light the variations of **E** are therefore confined to a plane containing the direction of propagation.

When unpolarized light passes through an isotropic solid, where the distribution of matter is the same in all directions, the light is free to vibrate equally in all directions. Therefore there is no change in the character of the

4.7.1 THE POLARIZING MICROSCOPE

light. Examples of isotropic solids are the highly ordered and symmetrical cubic crystals and any amorphous material such as glass. If the object is anisotropic, that is the distribution of matter is different in different directions, then the electric vector is not free to vibrate in the same way in all directions. The dielectric constant, which is related to the refractive index, is different in different directions. All non-cubic crystals are anisotropic as are also a variety of semi-ordered non-crystalline substances such as plastics and muscle and collagen fibres. The electromagnetic theory of light has shown that in the passage of light through an anisotropic medium, the light can be resolved into two components (the ordinary and extraordinary rays) which vibrate in planes at right angles to each other. The rays travel different paths with different velocities. All such anisotropic objects are said to be doubly refracting or birefringent.

The best known example of the production of plane polarized light is the Nicol prism which is based on two calcite crystals cut and cemented together in such a way so that the ordinary beam is totally internally reflected and only the extraordinary ray is transmitted. Nowadays the polarizers are made of polaroid, a substance in which the oriented polymeric chains of a polyvinyl alcohol dyed with iodine strongly absorb one of the plane polarized components of the light and transmit the other.

In the polarizing microscope, the polarizers are used in pairs. The polarizer, which is located below the specimen stage, produces plane polarized light and the analyser which is located above the specimen is used to investigate the nature of the light after it has passed through the specimen (Fig. 4.10). When the polarizer and analyser are "crossed", that is they are arranged so that the plane transmitted by the polarizer is at right angles to the plane transmitted by the analyser, then no light will emerge and the field will appear dark. If an isotropic substance is placed on the specimen stage, then the field will remain dark as the character of the light is not altered. If non-cubic crystal is placed on the stage, then the object may well appear bright and exhibit colours. As the crystal is rotated on the specimen stage, there will be four positions $90°$ apart at which extinction occurs.

The explanation for this effect is shown in Fig. 4.11. When viewed along the direction of propagation of the light, the polarizer P will transmit light which vibrates in the plane P (i.e. the plane containing the line P and the direction of propagation). The crystal will normally resolve light into two components which vibrate in planes X and Z. If neither of these two directions are parallel to P, then the plane polarized light may be resolved into the two components parallel to X and Z. When these reach the analyser A, they will be resolved into their components parallel to the analyser direction A and will emerge as a single component. The field will appear bright. As the crystal is rotated about the axis of the microscope, it will reach the situation shown in Fig. 4.11b. The X direction is parallel to the direction of the polarizer, and the Z direction is normal to it. Hence only a component parallel to P will emerge and, since this is

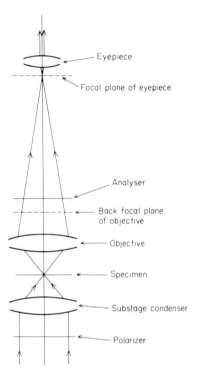

FIG. 4.10 The arrangement of lenses and polarizer and analyser in the polarizing microscope.

at right angles to the direction of A, no light will be transmitted and extinction results. A similar situation will occur when the crystal is rotated through 90°, when Z is parallel to P. Hence there will be four positions, separated by 90°, at which the crystal will show extinction and four positions 45° away from the extinction positions when the transmitted light will be brightest.

The colours seen with a birefringent specimen are due to the resolution of the light into the two components parallel to X and Z. The birefringent crystal will exhibit different refractive indices in these directions and hence the two rays will travel with different velocities through the crystal. When these two rays are recombined by the analyser, one ray will be slightly ahead of the other. The phase difference between the two rays will result in interference for a particular wavelength and this colour will be eliminated from the spectrum. The optical path difference between the two rays is given by:

$$R = (n_1 - n_2)d$$

where n_1 and n_2 are the refractive indices for the two rays and d is the thickness

4.7.2 THE QUARTZ PLATE

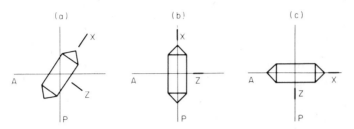

FIG. 4.11 The transmission of light through a crystal between crossed polarizer (P) and analyser (A). (After Bunn, 1961)

of the crystal. If R is zero (i.e. no difference in refractive indices or the crystal is very thin), then darkness results. There is no birefringence. $R = 230-260$ mμ, the crystal appears grey. $R = 400$ mμ then violet is extinguished and the crystal appears yellow. $R = 500-560$ mμ the blue-green colours are extinguished and the crystal appears red.

For increasing thickness of the crystal the colours will follow the order known as "Newton's scale", which is the same order as that of the interference colours given by an oil film on water. A chart of these colours is shown in Hartshorne and Stuart (1961). Some protein crystals, especially if they are large, give splendid colours but small crystals or those with small differences in refractive index will only give a dull grey colour. It is useful for the protein crystallographer to be able to recognize the bright colours given by salt crystals, such as ammonium sulphate!

4.7.2 The quartz plate

Sometimes with very small crystals it is difficult to detect whether or not there is any birefringence (even though the eye is very good at distinguishing between shades of grey). In such cases a quartz plate may be of use. The quartz plate is placed in the microscope so that its slowest direction (i.e. the direction of highest refractive index) is at 45° to the crossed polarizers. The colour of the background field will depend on the thickness of the quartz but in many microscopes it is dark violet red. When the crystal is oriented so that its slow (highest refractive index) is parallel to that of the quartz (and hence the fast directions of both the quartz and the crystal are also aligned), then the relative phase difference between the two components will be increased. A shift up the retardation scale will be observed which may take the colour of the crystal into the greenish-blue range. At 90° to this position the fast and slow directions of the quartz and crystal will be parallel and a shift down the scale will be observed. The crystal may appear orange. Often it is easier to distinguish these changes in colour, and to establish the birefringence of the crystal on this basis, than to distinguish between shades of grey and black as the crystal is rotated between crossed polars.

4.7.3 The indicatrix

The different refractive indices of a crystal are conveniently represented by a figure known as the indicatrix. This is constructed from lines drawn outwards from an imaginary point within the crystal such that the length of each line is proportional to the refractive index of light vibrating along that line. It is found that for all crystals the ends of these lines fall on the surface on an ellipsoid, which is termed the indicatrix. In general all sections passing through the centre of the indicatrix are ellipses and the semi-axes of any such section represent the vibration directions and the refractive indices for the two rays which travel in a direction normal to the section.

4.7.3.1 Uniaxial crystals
Crystals which belong to the tetragonal, trigonal or hexagonal systems have one direction, the direction of the unique axis c, in which they are effectively isotropic. This direction is called the optic axis. The section of the indicatrix normal to the optic axis is a circle in which the two refractive indices, which represent the semi-axes of the section, are equal. Uniaxial crystals when viewed down the optic axis do not exhibit birefringence.

4.7.3.2 Biaxial crystals
All other crystals, that is those which belong to the orthorhombic, monoclinic and triclinic crystal classes, are termed biaxial since there are two directions

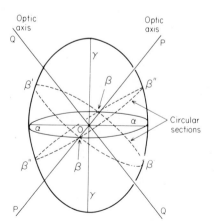

FIG. 4.12   The biaxial indicatrix. (After Bunn, 1961)

along which these crystals appear isotropic. The indicatrix is a triaxial ellipsoid in which the principal semi-axes have refractive indices α (the smallest refractive index), γ (the greatest refractive index) and β. The indicatrix has two circular sections and the radius of these sections is β (Fig. 4.12). The normals to these two circular sections represent the optic axes of the crystal, and the angle between them is termed the optic axial angle. In an orthorhombic crystal, the three principal axes of the ellipsoid α, β and γ are parallel to the unit cell edges. In a monoclinic crystal one of the axes of the indicatrix corresponds to the unique **b** axis and the other two axes may lie anywhere in the *ac* plane. In triclinic crystals there is no necessary relationships between the axes of the indicatrix and the crystallographic axes.

4.8 THE USE OF THE POLARIZING MICROSCOPE BY THE PROTEIN CRYSTALLOGRAPHER

In an initial examination of a crystal prior to X-ray studies, the polarizing microscope can be used to assess the following properties of the crystal.

4.8.1 Birefringence

If the object, supposed crystalline, shows no birefringence then it may belong to the cubic system in which case there may be some hint from its external morphology: it may be oriented so that the optic axis is parallel to the light beam, it may be too thin, in which case examination with the quartz plate in place may provide some information, or it may not be crystalline at all. If the object is birefringent then it must be based on some degree of order and may well be crystalline.

4.8.2 Quality of the crystal

For birefringent crystals some measure of the quality of the crystal may be obtained by observing the way in which the crystal extinguishes when rotated through 360° on the stage. The sharper the extinction, the more promising the crystal. Twinned crystals, in which two or more parts of the crystal are differently oriented, may sometimes be detected if different parts of the crystal extinguish at different positions. However, not all twinned crystals can be detected in this way.

4.8.3 Relative orientation of the crystal

For uniaxial crystals, the orientation of the unique axis may be found by noting the orientation of the crystal for which there is no birefringence. With the aid of a goniometer crystals can be aligned quite precisely for further study in the X-ray beam.

For biaxial crystals, the relative orientations of the optic axes bear no such direct relation to the directions of the unit cell edges and unless the three principal axes of the ellipsoid have been defined such information will not prove useful for the first crystal. Nevertheless a note of the extinction directions and the optic axes (if these can be found) with respect to the external morphology of the crystal will facilitate the setting up of other crystals.

4.9 THE DETERMINATION OF MOLECULAR WEIGHT FROM PROTEIN CRYSTALS

Molecular weights of protein molecules may be determined from preliminary crystallographic data. Values are often accurate to 10% and in some cases may be accurate to 4%. Although such estimates are not as accurate as measurements made with the analytical tools of solution chemistry, the crystallographic method is not complicated by association/dissociation phenomena. Thus the X-ray method has proved most useful in the determination of the subunit structure and the symmetry of oligomeric proteins.

The density of a crystal (D_c weight/volume) is given by

$$D_c = \frac{nM_c m_H}{V}$$

where M_c is the molecular weight of the contents of the asymmetric unit of one unit cell; n is the number of asymmetric units per unit cell (which is known from the space group of the crystal); m_H is the mass of the hydrogen atom (1.673×10^{-24} g); and V is the volume of the unit cell (which is known from preliminary X-ray studies).

The density of the crystals may be determined by the density gradient method (described below) and thus M_c may be calculated. However, the unit cell contains not only the protein molecule but also the solvent from which the protein is crystallized (Section 3.6). The solvent exists in two states: solvent firmly bound to the protein molecule and solvent free in the interstitial regions between molecules. The major uncertainty in the determination of molecular weights by the crystallographic method lies in estimates of the fraction of the unit cell that is occupied by solvent.

In a survey of 116 different crystalline proteins, whose molecular weights were known from other methods, it was found that the fraction of crystal volume occupied by solvent is usually near 43% but may vary between 27% and 65% (Matthews, 1968). These values correspond to an average crystal volume per unit molecular weight (V_m = volume of asymmetric unit/molecular weight) of

4.9 THE DETERMINATION OF MOLECULAR WEIGHT

2.4 $Å^3$/dalton with a variation of 1.68 $Å^3$/dalton to 3.53 $Å^3$/dalton. This range can be used to determine the approximate molecular weight of the asymmetric unit. If a rough value of the molecular weight of the protein is known, the state of aggregation of the protein in the crystal can be determined. For example crystals of phosphorylase b are tetragonal space group $P4_32_12$ (eight asymmetric units per unit cell) and have a unit cell volume of 1.94×10^6 $Å^3$. The molecular weight of the asymmetric unit using the average value of 2.4 $Å^3$ for (V_m is 100,000. Since phosphorylase is known to be in the dimeric state (total M.W. 200,000) under the conditions of crystallization, the X-ray results show that the asymmetric unit contains the monomer subunit and that the two subunits are related by a crystallographic dyad axis to form the dimer.

For a more precise determination of molecular weight it is necessary to measure the density and solvent content of the crystals.

The density of protein crystals may be measured by means of a density gradient column constructed from two miscible liquids (Low and Richards, 1952). A suitable tube is half filled with a liquid of fairly high specific gravity and a second, lighter liquid layered over the first. The linear gradient that develops near the interface is used for the measurement of density. Bromobenzene and xylene make a good combination of liquids and give a density range of approximately 1.15 to 1.35 g/cc for a 20 cm tube. The column is calibrated with drops of solution of known density. Crystals, blotted free of surplus mother liquor, are dropped into the column and their equilibrium positions noted. Alternatively the crystals may be added in a drop of their mother liquor and the excess liquid drawn off with a Hamilton syringe (Colman and Matthews, 1971). The second method avoids the danger of overdrying the crystal.

The solvent content of the unit cell is determined from measurements on the loss of weight of crystals on drying (North, 1959). Several large crystals are blotted free from surplus mother liquor and weighed wet with a micro-balance. The crystals are dried to constant weight and reweighed. The difference in weight is taken to represent the weight of water in the crystal. Protein crystals which have been air dried still retain 5–10% of their own weight in water (Haurowitz, 1950) and so an additional 7.5% is usually subtracted from the apparent weight of protein. The air dried crystals also contain a substantial amount of salt. Work on haemoglobin (Perutz, 1946) has shown that there appears to be a monolayer of bound water associated with each protein molecule which is not permeable to salt ions. The bound water corresponds to approximately 0.3 g per g of protein. In North's work on haemoglobin (North, 1959) it was assumed that of the estimated water content, 0.3 g per g of protein is salt-free water and the remainder has the same salt concentration as the mother liquor. The salt content of the mother liquor may be obtained by weighing a sample of mother liquor and then drying to constant weight. These results are used to derive the molecular weight of the protein in the following way (Matthews, 1974a,b):

Let d be the fractional loss of weight of a crystal on drying (g water/g wet crystal); s be the fractional mass of salt in the mother liquor (g salt/g mother liquor); u be the fractional mass of water retained in the crystal on drying (assumed to be in range 5–10%); and let w be the fractional mass of water which is firmly bound to the protein and is inaccessible to salt (approximately 0.3 g per g proteins). Then the weight of the crystal w_c is given by

$$w_c = w_p + dw_c + uw_p + s(w_c - w_p - ww_p)$$

where w_p is weight of protein

Hence the fractional weight of the protein is given by

$$\frac{w_p}{w_c} = \frac{1 - d - s}{1 + u - s - sw}$$

Hence the molecular weight of the protein (Mp) is given by

$$Mp = \frac{V D_c (1 - s - d)}{n m_H (1 + u - s - sw)}$$

Determination of molecular weights can be moderately accurate by this method especially in cases where there is a low salt content ($s \simeq 0$). The method requires the availability of several large crystals for air drying. Errors in measurements of d and estimates of u are likely to limit the accuracy.

The reader is referred to Matthews (1974b) for a description of other methods of determination of molecular weights from protein crystals. These include crystal density measurements coupled with knowledge of the partial specific volume of the protein measurements of mass of protein in a crystal of known volume and measurements of mass of protein in a unit mass of crystal.

5

THE PRINCIPLES OF X-RAY DIFFRACTION

5.1 INTRODUCTION – –THE WAVE NATURE OF X-RAYS

Almost all of the important properties of X-rays were reported by Röntgen in his paper published in 1896. Röntgen discussed the production of X-rays, their propagation in straight lines, their ability to penetrate matter and their interaction with photographic plates. He also reported the failure of X-rays to exhibit interference, reflection or refraction effects in *ordinary* optical apparatus. In spite of the failure of these experiments to demonstrate interference effects, the wave theory of X-rays had many adherents among the theoretical physicists of the time. The creation of electromagnetic waves seemed on the basis of the Lorentz-Maxwell equations, to be a necessary consequence of the sudden alteration in velocity of the electrons during the production of X-rays. Several years later, slight diffraction effects were observed after the passage of X-rays through a carefully machined wedge-shaped slit and Sommerfeld was able to calculate their wavelength as approximately 4×10^{-9} cm. With this information the conditions in Munich in the year 1912 were especially favourable for Laue's important experiment which was to show conclusively the wave nature of X-rays and to found the new science of X-ray crystallography. The impulse which led to this experiment came from a paper produced by Sommerfeld's student, Ewald, on the propagation of electromagnetic waves in a crystal. Laue was led to ask what would happen if the wave length of the rays was smaller than the repeat distance in the crystal. This was known to be of the order of 10^{-8} cm from measurements of the density of the crystal and the molecular weight of the compound. Laue estimated that the wavelength of X-rays should be just about right to produce interference effects in crystals. Two research students who had just received their doctorates under Röntgen, Friedrich and Knipping, agreed to try the first experiment. After many failures, they finally achieved the famous "Beerstein" photograph shown in Fig. 5.1 from a crystal of $CuSO_4.5H_2O$. Their results were published in 1912 and in the same year W. L. Bragg was able to provide a correct explanation for the formation of the spots. In 1913, Bragg used Laue's new photographs to work out the crystal structure of

108   5.   THE PRINCIPLES OF X-RAY DIFFRACTION

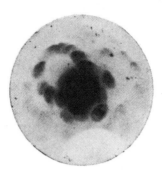

FIG. 5.1   The first photograph of X-ray diffraction by a crystal (Friedrich *et al.*, 1912).

NaCl, KCl, KBr and KI, the first crystals to be analysed by X-rays. Thus within a space of two years the final proof of the wave nature of X-rays had been established and the use of X-rays as a tool for crystal structure determination had been demonstrated. (See Bragg, 1962 for a more detailed account of the early history of X-ray diffraction.)

The electromagnetic wave nature of X-rays allows us to draw an analogy between the formation of an image from X-ray diffraction and the formation of an image with light rays. In the light microscope the object is illuminated by a beam of light. The object scatters the light and the scattered rays are collected by the objective lens and recombined to form an image of the object. For X-rays the first part of the image forming process is analogous. A crystal is illuminated by a beam of X-rays and the scattered rays may be recorded on a photographic film. However, for X-rays there is no analogy to the objective lens. Röntgen's observations on the inability of a wide range of substances to refract X-rays is still, more or less, correct. (In fact a small refraction of X-rays does occur: the refractive index of a calcite crystal is 0.999998.) This means there is no substance which can be used to focus X-rays. However, enough is known about the theory of diffraction to enable the scattered rays to be recombined analytically with the aid of a computer. There is only one problem. In order to reconstruct an image from its diffraction pattern both the phase and the intensity of each diffracted ray need to be known. But in a diffraction experiment only the intensities of the scattered rays can be measured. All information on their relative phase is lost. The computation of these phases forms the basic problem in all crystal structure analyses.

In this chapter we shall be concerned with the fundamental equations of X-ray diffraction. This will enable us to give a quantitative understanding of the phase problem and to show how X-ray diffraction data may be used to determine crystal structures. This chapter provides the groundwork for the more detailed discussion in later chapters of the techniques which are special to protein crystallography. For a more comprehensive treatment of the theory of X-ray diffraction, the reader is referred to the books by James (1957), Lipson and Cochran (1966), Lipson and Taylor (1958), Stout and Jensen (1968), Woolfson (1970).

5.2 BRAGG'S LAW

Bragg's achievement was to visualize the scattering of X-rays by a crystal in terms of reflections from planes of atoms (Bragg 1913). These are the same planes that were described in the previous chapter under the heading of Miller Indices but they are not necessarily restricted to those planes parallel to a crystal face.

The crystal planes are illuminated at a glancing angle θ and X-rays are scattered with an angle of reflection also equal to θ (Fig 5.2). The incident and the diffracted rays are in the same plane as the normal to the diffracting planes. Constructive interference between rays scattered from successive planes in the crystal will only take place if the path difference between the rays is equivalent to an integral number of wavelengths.

Let the separation between successive planes of atoms be d. The path difference between rays 1 and 2 is AB + BC = $2d \sin \theta$. Hence for constructive interference:

$$2d \sin \theta = n\lambda \quad \text{Bragg's Law} \tag{5.1}$$

where λ is the wavelength of X-rays (1.5418 Å for Cu Kα radiation) and n is an integer.

This fundamental equation predicts the position in space of any diffracted ray. Thus the first order spectrum from a particular set of planes will occur at an angle θ which satifies Bragg's law for $n = 1$; the second order spectrum will occur

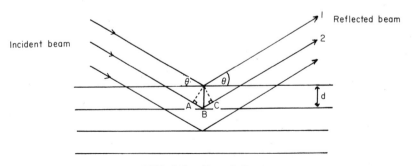

FIG. 5.2   Bragg's Law.

at $n = 2$ and so on. Note that the closer the separation of the planes (smaller d) the larger the angle of diffraction, θ. However the equation carries no information about the intensity of the diffracted ray. In order to understand how the variation in intensity of different diffracted rays is produced it is useful to draw on the analogy with light scattering.

5.3 THE ANALOGY WITH LIGHT SCATTERING

In Chapter 4 we discussed a crystal in terms of a three dimensional periodic arrangement of unit cells, each of which contained the same number of molecules arranged in such a way that each molecule is in the same environment as every other molecule. X-ray diffraction data from a crystal contain information about the structure of the molecule which forms the asymmetric unit. If the structure of the asymmetric unit is known, then the structure of all the other molecules in the crystal are known. In a sense, the existence of the crystal at this stage appears superfluous, although, as we shall show, its existence is entirely necessary in order for the diffraction data to be recorded. By making use of the analogy between X-ray diffraction and light diffraction, we can show the diffraction patterns expected for a single molecule and the modifications of these diffraction patterns which are produced when the molecules are placed in a crystal lattice. Although the problem of crystal structure determination is essentially a three dimensional problem, we shall restrict our examples to the two dimensional case. The extension to three dimensions is trivial once the simpler examples have been understood.

A schematic diagram of the apparatus which is used to record the optical diffraction patterns is shown in Fig. 5.3. A laser light source provides an intense monochromatic beam. The light emerges from a small circular aperture A placed in the front focal plane of the condenser lens C. The molecule or crystal is

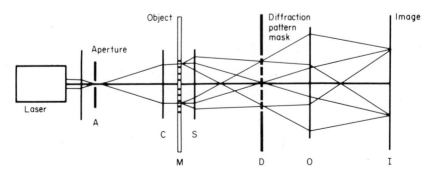

FIG. 5.3   Schematic diagram of the optical diffractometer.

5.3.1 THE DIFFRACTION PATTERN OF A TRANSPARENT DISC

represented by a set of holes punched in an opaque screen. This object is placed in plane M and is illuminated with parallel light. The scattered light rays are focused by a subsidiary lens S to form an image of the diffraction pattern in the plane D. The diffraction pattern may be photographed by placing a photographic film in this plane. Alternatively, the scattered rays may be allowed to continue and to be recombined by the objective lens system O to form an image of the mask in the image plane I.

In the case of X-rays, by using a finely collimated beam it is possible to record the diffraction pattern of a crystal on a photographic film. The optical set up is analogous to the X-ray case up to the stage of the diffraction plane D. However, since there is no lens capable of focusing X-rays, it is not possible to carry out a direct recombination of the scattered X-rays, unlike the situation in light diffraction.

5.3.1 The diffraction pattern of a transparent disc

First let us consider the diffraction pattern of a very simple object: a transparent disc in an opaque mask. The resulting optical diffraction pattern is shown in Fig. 5.4. It consists of a series of light and dark rings due to interference effects between light rays which traverse different parts of the disc. The smaller the diameter of the disc the larger will be the diameter of the first dark ring. (This is just one of the many examples of the reciprocal relationship between dimensions in real space and dimensions in diffraction space.) In X-ray diffraction the dimensions of the scattering objects (the atoms) are of the same order of magnitude as the wavelength of the X-rays. Hence in order to represent the

FIG. 5.4   The optical diffraction pattern of a transparent disc.

optical diffraction pattern of an atom, it would be necessary to punch a very small hole with dimensions of the order of the wavelength of light and whose transparency varied in the same way as the distribution of electron density in an atom. This is obviously rather difficult.

5.3.2 The optical diffraction pattern of a "molecule"

We now consider the diffraction pattern of a number of small holes arranged to represent a molecule. The six dots, arranged in the shape of a distorted hexagon, give rise to the optical diffraction pattern shown in Fig. 5.5b. (The size of the dots are small in comparison with their separation and we are effectively observing the low angle diffraction pattern within the radius of the first dark ring of the diffraction pattern of a hole.) The interference between rays scattered by the different dots gives rise to a continuous variation of weak and strong intensities in the diffraction pattern. The irregular distribution of intensity is centrosymmetric about the origin and this is a common feature of all diffraction patterns. From this example we may deduce that the variation of intensity in a diffraction pattern is due to the underlying structure of the molecule. The diffraction pattern of a single molecule is referred to as the *molecular transform*; it is the Fourier transform of the molecule.

In the case of light diffraction, it is possible to simulate the *molecular transform* because the interaction of light with matter is relatively strong. The interaction of X-rays with matter is weak and observations on the transforms of individual molecules are not possible. Hence it is necessary to crystallize the compound, so that all molecules are arranged in a regular fashion and the scattering from any one molecule is reinforced by the scattering of all the others. We now need to investigate the effect of the crystal lattice on the molecular transforms.

5.3.3 The diffraction pattern of a lattice

The diffraction pattern of a set of lines is shown in Fig. 5.6a. It is a row of dots perpendicular to the set of lines and whose separation is inversely proportional to the separation of the lines. In other words the Fourier transform of a set of lines is a row of dots. We may note in passing that the converse is also true: the Fourier transform of a row of dots is a set of lines. Again there is a demonstration of the reciprocal relationship between the object in real space and the diffraction pattern.

The diffraction pattern of a second set of lines arranged in a different orientation to the first set is shown in Fig. 5.6b. The two sets of lines can be built up to form a lattice by multiplying the two functions. We may now use a result of Fourier transform theory in order to predict the type of diffraction pattern to be expected for a lattice.

5.3.3 THE DIFFRACTION PATTERN OF A LATTICE

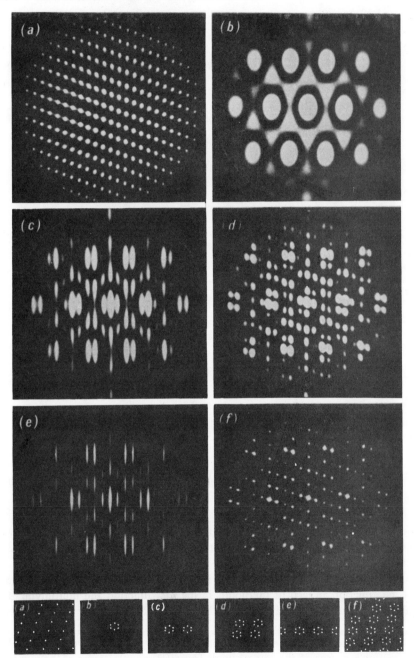

FIG. 5.5 Optical diffraction patterns illustrating scattering by (a) a lattice, (b) a single molecule, (c) two molecules, (d) four molecules, (e) a row of six molecules and (f) a small crystal. (From Taylor and Lipson, 1964.)

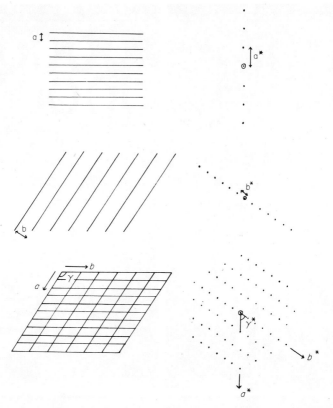

FIG. 5.6 The transforms of a set of lines and of a lattice. (After Holmes and Blow, 1966)

Theorem: The Fourier transform of the convolution of two functions is the product of their Fourier transforms.

The converse theorem states: The Fourier transform of the product of two functions is the convolution of the transform of one with the transform of the other.

The convolution of two functions is the result of placing the origin of the second function at each of the points in the first function and multiplying the value of the first function at that point by the second function.

Hence if a two dimensional lattice is viewed as the product of two sets of lines, then the diffraction pattern of the lattice is the convolution of the diffraction patterns of the sets of lines. The diffraction patterns of the two sets of lines are the two rows of dots. The convolution of these two rows of dots is a lattice (Fig. 5.6c). Hence we have the important result: *that the diffraction* pattern of a lattice is also a lattice but one whose dimensions are inversely proportional to the dimensions of the real lattice. (e.g. Fig. 5.5a.)

The separation between adjacent points of the diffraction pattern, the *reciprocal lattice*, are given by:

$$a^* = \frac{1}{a \sin \gamma}$$

$$b^* = \frac{1}{b \sin \gamma}$$

where γ is the angle between the two rows of the real lattice. If each point in the reciprocal lattice is indexed by two integers h and k, the distance of any reciprocal lattice point from the origin of the diffraction pattern is given by:

$$d^* = (h^2 a^{*2} + k^2 b^{*2} - 2hka^* b^* \cos \gamma)^{1/2}$$

5.3.4 The diffraction pattern of a molecule arranged in a lattice

The diffraction pattern of a molecule arranged in a two dimensional lattice may be deduced from the following observation. The crystal is the result of the convolution of the molecule with the two dimensional lattice (i.e. if we take the molecule and place it in turn at each of the lattice points we shall build up to a crystal). Hence the diffraction pattern of the molecular crystal is the product of the diffraction pattern of the molecule (the molecular transform) with the diffraction pattern of the lattice (the reciprocal lattice). The product of these two transforms results in a sampling of the molecular transform at each of the reciprocal lattice points. A reciprocal lattice spot will therefore appear strong in intensity if the underlying molecular transform is strong at that point or a spot will appear weak if the underlying molecular transform is weak (Fig. 5.5c,d,e and f).

Hence we see that the lattice nature of a diffraction pattern arises from the crystal lattice and the intensity of each spot is governed by the intensity at that point of the underlying molecular transform. In general, each part of the molecule contributes to every part of the diffraction pattern. Conversely in order to reconstruct the molecule from its diffraction pattern, it is necessary to measure the intensities of all of the diffraction spots.

In Fig. 5.7 we show the effects of the size and shape of the lattice on the appearance of the transform. Although the underlying molecular transform is the same in Fig. 5.7c,d,e and f, the appearance of the sampled transforms differ greatly. In Fig. 5.7e, for example, the large unit cell results in a fine sampling interval in the diffraction pattern so that the underlying transform is recognizable. The smaller unit cell of Fig 5.7d (which is a more realistic representation of the packing of molecules in a crystal) results in a large sampling interval and the underlying molecular transform is more difficult to detect.

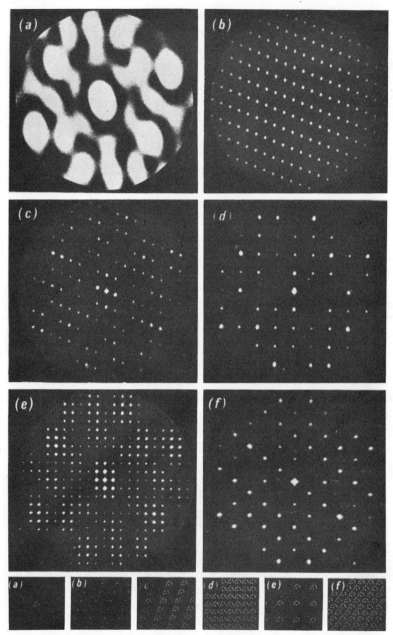

FIG. 5.7 Illustration of the effect of the size of the unit cell: (a) transform of a molecule; (b) transform of a lattice; (c) convolution of the two; (d), (e), and (f) are transforms of convolutions with lattices of different sizes. (Taylor and Lipson, 1964)

5.4 RESOLUTION

The concept of resolution in X-ray diffraction has the same meaning as the concept of resolution in image formation in the optical microscope. According to the Abbe theory of the microscope, a distance dm is resolvable if the objective lens can accept at least the corresponding first order diffraction spectra which arises from this spacing. This leads to the criteria for the performance of the microscope $dm = \lambda/2$ N.A. where λ is the wavelength of light and N.A. is the numerical aperture of the lens. The wider the cone of scattered rays accepted by the lens, the smaller the spacing that can be resolved. In X-ray diffraction the situation is slightly more complicated. We wish to know the effects to be expected in a three-dimensional Fourier representation of a crystal structure when the Fourier series is terminated while the coefficients are appreciable. Suppose diffraction data are collected to a minimum interplanar spacing dm where $dm = \lambda/2 \sin \theta (\max)$. It may be shown (James, 1957) that, in the resulting Fourier synthesis, point atoms are not imaged as points but as central maxima surrounded by negative and positive diffraction rings corresponding to the Fraunhofer diffraction pattern of a sphere radius $1/dm$ (see Fig. 5.4 for Fraunhofer diffraction pattern of a disc). In three-dimensions the position of the first zero surrounding the diffraction sphere of positive density is given by $r = 0.715 \, dm$. Thus it is assumed that the smallest detail that can be faithfully imaged in such electron density maps corresponds to features separated by more than $0.715 \, dm$. In practice in protein crystallography it is usual to quote the nominal resolution of a protein electron-density map in terms of dm, the minimum interplanar spacing for which F's are included in the series.

In protein crystallography the terms "6 Å (low) resolution" and "2.0 Å (high) resolution" mean that diffraction data have been collected to the limit of that particular interplanar spacing. The likelihood of detail of this resolution being apparent in the electron density map is dependent on many other factors such as the degree of isomorphism of the heavy atom derivatives, the degree of disorder in the crystal and the precision of the intensity measurements. The meaning of these terms in the context of protein crystallography is discussed in more detail in the chapter on electron density maps (Chapter 13).

In Fig. 5.8 we illustrate the effect of limiting the diffraction data on the detail observable in the final image.

The image in Fig. 5.8b has been formed from the most complete diffraction pattern as shown in Fig. 5.8a. The structure is readily recognizable and individual "atoms" are resolved. In Fig. 5.8d and f the images have been formed from progressively limited portions of the diffraction data shown in Fig. 5.8c and e respectively. In the low resolution image (Fig. 5.8f), the object is hardly recognizable: only its gross features are present.

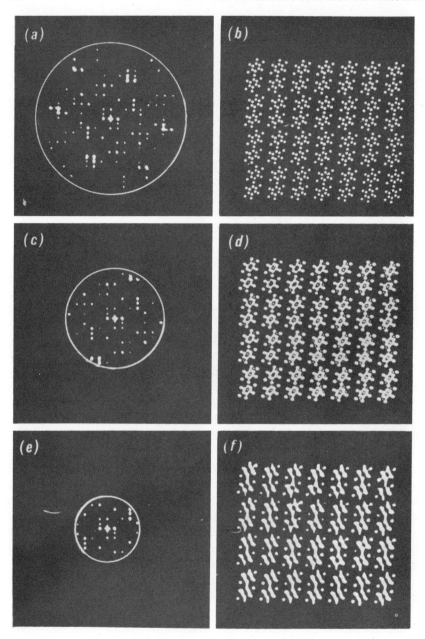

FIG. 5.8 Optical illustration of the effects of series termination in Fourier synthesis. The transforms of the portions of the diffraction patterns (a), (c), and (e) are shown in (b), (d), and (f) respectively. (Taylor and Lipson, 1964)

5.5 THE SCATTERING OF X-RAYS BY AN ATOM

When a beam of X-rays interacts with matter scattering occurs from two processes: (i) Thomson or coherent scattering; (ii) Compton or incoherent scattering.

(i) *Thomson scattering*. When X-radiation impinges on a free electron, the influence of the fluctuating electromagnetic field of the incident wave forces the electron into oscillations of the same frequency as the incident wave. The oscillating charge forms the source of a secondary scattered ray which is of the same wavelength as the incident radiation but differs in phase by 180°. All the scattered rays from a single electron have the same phase relationship to the incident beam. Hence the scattering is coherent.

The electromagnetic theory of this scattering process, developed by J. J. Thomson, shows that the scattered intensity $I_{2\theta}$, at an angle 2θ from a beam of unpolarized X-rays is given by:

$$I_{2\theta} = I_0 \frac{ne^4}{2r^2 m^2 c^4} (1 + \cos^2 2\theta)$$

where I_0 is the incident beam intensity, n is the effective number of independently scattering electrons, r is the distance from the scatterer, e is the charge of the electron m is the mass of the electron and c is the velocity of light. The $(1 + \cos^2 2\theta)$ terms represents the partial polarization of the scattered X-rays.

The inverse relationship between the scattered intensity and m, the mass of the electron, demonstrates why only the electrons in the atom contribute to the coherent scattering. The mass of the proton is approximately 2000 times that of the electron and hence the proton is far too heavy to become a secondary emitter.

The proportion of X-rays scattered by matter is weak. Woolfson (1970) has calculated that the fraction of incident X-radiation scattered by a "crystal" composed solely of free electrons and of dimensions 1 mm is less than 2%.

The total power of X-rays scattered at all angles is found by integration of the above equation through all space. This relation was used by Barkla in 1903 to determine the number of electrons per atom that are effective in scattering X-rays. He found this number to be equal to the atomic number of the scatterer within the accuracy of his experiments.

(ii) *Compton scattering*, or incoherent scattering, is essentially a billiard ball effect. Here the particle aspect of X-radiation is emphasized. The incident photon collides with a comparatively loosely bound electron and is deflected from its original path with some loss of energy. Because of the change in energy there is also a change in wavelength of the scattered ray. For a single atom, considered in isolation, Compton scattering can be significant in comparison to Thompson scattering especially at large scattering angles. However in the diffraction of X-rays by a crystal, the co-operative coherent scattering of many atoms becomes significantly greater than the sum of the incoherent contributions. Incoherent scattering is therefore usually ignored in X-ray crystallography.

5.5.1 The atomic scattering factor

The coherent scattering by a free electron is independent of scattering angle, apart from the $(1 + \cos^2 2\theta)$ term which represents the partial polarization of the scattered ray. In an atom the electrons occupy a finite volume and are bound in certain well-defined energy states. In order to obtain the expression for scattering by an atom it is necessary to take into account the spatial distribution of the electrons.

Let the electron density at a distance **r** from the centre of the atom be $\rho(\mathbf{r})$. Consider the wave scattered at position **r** in a direction **s**, relative to the wave scattered by a unit electron at the centre of the atom. The total wave scattered depends on the phase difference between the scattered waves.

Define the direction of the incident radiation by a vector $\mathbf{s}_0$ and the direction of the scattered radiation by a vector **s**. For simplicity in later equations let $|\mathbf{s}_0| = |\mathbf{s}| = 1/\lambda$ where λ is the wavelength of the incident radiation.

From Fig. 5.9 the path difference between ray 1 and ray 2 is $p - q$.

$$p = \lambda \mathbf{r} \cdot \mathbf{s}_0 \qquad q = \lambda \mathbf{r} \cdot \mathbf{s}$$

$$\text{phase difference} = \frac{2\pi}{\lambda} \times \text{path difference}$$

$$= 2\pi \mathbf{r} \cdot (\mathbf{s}_0 - \mathbf{s})$$

$$= 2\pi \mathbf{r} \cdot \mathbf{S}$$

where $\mathbf{S} = \mathbf{s}_0 - \mathbf{s}$.

The vector **S** is used to describe a position in diffraction space, in the same way that the vector **r** is used to describe a position in real space.

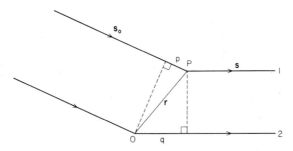

FIG. 5.9   Scattering at a point P relative to an origin O.

5.5.2 TEMPERATURE FACTOR

The total wave scattered by a small unit of volume dv at a position r relative to the wave scattered at the origin will therefore have an amplitude proportional to $\rho(\mathbf{r})dv$ and phase $2\pi\,\mathbf{r}\cdot\mathbf{S}$ i.e. wave scattered $= \rho(\mathbf{r})\exp(2\pi i\mathbf{r}\cdot\mathbf{S})dv$.

Hence the total wave scattered by the atom is calculated by summing the individual contributions over the volume of the atom

$$F(\mathbf{S}) = \int_{\substack{\text{vol. of}\\\text{atom}}} \rho(\mathbf{r})\exp(2\pi i\mathbf{r}\cdot\mathbf{S})dv. \tag{5.2}$$

The expression represents the *atomic scattering factor*. The variation in electron density over the entire volume of the atom is only known precisely, from wave mechanics, for the hydrogen atom. For larger atoms it is necessary to introduce various approximations in order to solve the wave equations. For light atoms the Hartree–Fock self-consistent field method has been used while for heavier atoms the Thomas–Fermi statistical approach has proved useful. Atomic scattering curves for most atoms have been calculated and are tabulated in International Tables for Crystallography, Volume III.

Some atomic scattering factor curves are shown in Fig. 5.10. The assumption is usually made that the electron density of the atom is spherically symmetric. Hence the atomic scattering factor is also spherically symmetric i.e. it is independent of the direction $\mathbf{S}$ and dependent only on the magnitude of $\mathbf{S}$. The assumption that $\rho(\mathbf{r})$ is spherically symmetric also implies that $\rho(\mathbf{r})$ is centrosymmetric, i.e. $\rho(\mathbf{r}) = \rho(-\mathbf{r})$. This means that in the summation the imaginary components cancel out and the summation is real. Hence $f(S)$ is real for all S.

We note that at $S = 0$, $\quad f(0) = \int_{\substack{\text{vol. of}\\\text{atom}}} \rho(r)dv = Z$

where Z is the total number of electrons in the atom.

5.5.2 Temperature Factor

The expression derived for the atomic scattering factor represents the scattering by an atom at rest. Changes in temperature affect the thermal motion of atoms and this in turn affects the scattered intensities. Shortly after the first X-ray experiments, Debye (1914) showed that the thermal motion of atoms causes a decrease in intensity by a factor $\exp[-B(\sin^2\theta/\lambda^2)]$ where $B = 8\pi^2\bar{u}^2/3 = 8\pi^2\bar{u}_x^2$ and $\bar{u}$ is the mean displacement of atoms along the normal to the reflecting planes and $\bar{u}_x$ its x-component. There is no change in the sharpness of the Bragg reflections.

A large temperature factor implies a rapid fall off of intensity with Bragg angle θ and hence limits the resolution of the structure. In small molecule crystals, values of B are typically in the range 2–6Å^2, which correspond to a displacement of the atoms about their mean positions $(\bar{u}_x^2)^{\frac{1}{2}}$ of between 0.12Å

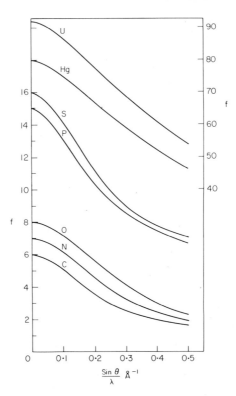

FIG. 5.10 Atomic scattering factors for uranium and mercury (top curves, scale on right hand side of figure) and sulphur, phosphorus, oxygen, nitrogen and carbon (lower curves, scale on left hand side of figure).

and 0.27Å. In protein crystals temperature factors may be of the order of $B = 12 - 20 Å^2$ corresponding to a mean displacement of between 0.15Å and 0.5Å. In almost all crystal structures, B is determined empirically and also takes into account a variety of other factors such as static disorder, wrong scaling of measurements, absorption, and incorrect atomic scattering curves. Hence values of the mean displacement may only be taken as a rough approximation, although undoubtedly the high values of the Debye factor encountered in protein crystals do represent the greater mobility of the atoms of the protein molecule. This is a consequence of high water content of protein crystals.

In the more general case, when atoms are not free to vibrate equally in all directions, the isotropic temperature factor B is replaced by an anisotropic temperature factor which represents an ellipsoid of vibration in reciprocal space, defined by six parameters $b_{11}, b_{12}, b_{13}, b_{22}, b_{23}, b_{33}$. The reduction in intensity is given by

$$\exp - (b_{11}h^2 + b_{12}hk + b_{13}hl + b_{22}k^2 + b_{23}kl + b_{33}l^2).$$

5.6 THE SCATTERING OF X-RAYS BY A MOLECULE

We now derive the expression for the scattering by an assembly of atoms placed at defined positions in a unit cell.

Let us consider atom 1 in Fig. 5.11 which is at a distance $\mathbf{r}_1$ from the origin. This shift in origin from the centre of the atom, means that the distance $\mathbf{r}$ in the equation for the scattering by an atom becomes $\mathbf{r} + \mathbf{r}_1$. Hence the scattering by atom 1 relative to the new origin is

$$\mathbf{f}_1 = \int_{\substack{\text{vol. of} \\ \text{atom}}} \rho(\mathbf{r}) \exp(2\pi i (\mathbf{r}_1 + \mathbf{r}) \cdot \mathbf{S}) dv$$

$$= f_1 \exp(2\pi i \mathbf{r}_1 \cdot \mathbf{S})$$

where

$$f_1 = \int_{\substack{\text{vol. of} \\ \text{atom}}} \rho(\mathbf{r}) \exp(2\pi i \mathbf{r} \cdot \mathbf{S}) dv$$

Similar expressions may be deduced for atoms 2, 3 and all the other atoms in the unit cell.

The total wave scattered by all the atoms is given by the vector sum of the individual contributions from each of the atoms (Fig. 5.12). Hence the total wave scattered is given by

$$G(\mathbf{S}) = \mathbf{f}_1 + \mathbf{f}_2 + \mathbf{f}_3 + \ldots + \mathbf{f}_N$$

i.e.

$$G(\mathbf{S}) = \sum_{j=1}^{N} f_j \exp(2\pi i \mathbf{r}_j \cdot \mathbf{S}) \tag{5.3}$$

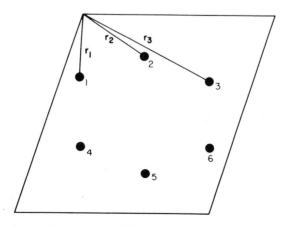

FIG. 5.11   The positions of atoms in a unit cell.

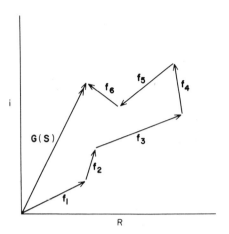

FIG. 5.12 Vector diagram illustrating the total wave scattered in a direction **S** in diffraction space by six atoms.

Equation 5.3 represents the *molecular transform*. The function is complex and varies continuously over all of diffraction space **S**.

5.7 SCATTERING OF X-RAYS BY A CRYSTAL

In order to work out the expression for the scattering by a crystal we first consider the case of a one-dimensional crystal which is composed of a linear array of unit cells with a repeat distance a. The total wave scattered by the crystal will be the sum of the waves scattered by each unit cell.

The wave scattered by the first unit cell relative to the origin is simply **G(S)**. The wave scattered by the second unit cell relative to the same origin is $G(S) \exp 2\pi i \mathbf{a} \cdot \mathbf{S}$ since all distances are shifted by the vector a. The wave scattered by the nth unit cell is therefore $G(S) \exp 2\pi i(n-1)\mathbf{a} \cdot \mathbf{S}$. Hence the total wave scattered is

$$F(S) = \sum_{n=1}^{T} G(S) \exp 2\pi i(n-1)\mathbf{a} \cdot \mathbf{S}$$

where T is the total number of unit cells.

The way in which these individual contributions add up may be seen from the vector diagram Fig. 5.13a. The wave from each unit cell is out of phase with its neighbour by an amount $2\pi \mathbf{a} \cdot \mathbf{S}$. Hence as the number of unit cells become large, the total wave scattered, **F(S)**, will be approximately of the same order of magnitude as **G(S)**, the molecular transform, which as we have already noticed

5.7 SCATTERING OF X-RAYS BY A CRYSTAL

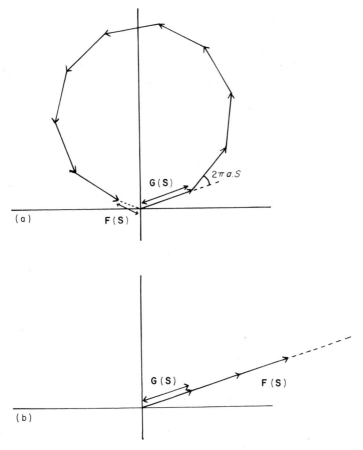

FIG. 5.13 Vector diagrams illustrating the total wave scattered by a molecule in a crystal lattice. (a) The phase difference between waves scattered by adjacent unit cells is $2\pi \mathbf{a}.\mathbf{S}$ (b) the phase difference is an integral multiple of 2π.

for X-rays is too small to be observed. How then is any scattering observed for a crystal?

Scattering will only be observed when the phase difference between the waves scattered by successive unit cells is equal to an integral multiple of 2π (Fig. 5.13b),

i.e. $2\pi \, \mathbf{a} \cdot \mathbf{S} = 2\pi h$

i.e. $\mathbf{a} \cdot \mathbf{S} = h$ where h is an integer.

Under these circumstance the waves add up constructively to form a significant scattered wave which is proportional in magnitude to $T \times |\mathbf{G}(\mathbf{S})|$. For

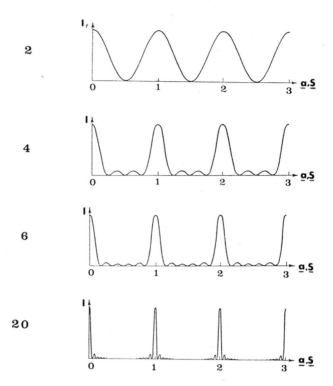

FIG. 5.14 The intensity distribution for the scattering from 2, 4, 6 and 20 unit cells. The maximum intensities have been scaled to give the same value for each case. In practice, the intensity scattered from 20 unit cells would be very much greater than that scattered from 2 unit cells.

a crystal 1 mm long and with unit cell dimensions 100 Å, $T = 10^5$ and hence the amplification factor is significant.

In summary, for a one dimensional lattice, scattering will only be observed in a particular direction S where S satisfies the equation $\mathbf{a} \cdot \mathbf{S} = h$. (Fig. 5.14.)

When the problem is extended to three dimensions with a unit cell defined by vectors $\mathbf{a}$, $\mathbf{b}$ and $\mathbf{c}$, the condition for diffraction becomes

$\mathbf{a} \cdot \mathbf{S} = h$

$\mathbf{b} \cdot \mathbf{S} = k$

$\mathbf{c} \cdot \mathbf{S} = l$ (5.4)

where h k and l are integers. These equations are known as the *Laue equations*. They are the mathematical expression of our previous statement (Section 5.3.3) that the diffraction pattern of a lattice is also a lattice. They enable us to rewrite the total wave scattered by a crystal in a direction S in the following way.

5.7 SCATTERING OF X-RAYS BY A CRYSTAL

$$F(S) = \sum_{j=1}^{N} f_j \exp 2\pi i \mathbf{r}_j \cdot \mathbf{S}$$

neglecting the proportionality constant T. Let the fractional co-ordinates of the jth atom be x_j, y_j, z_j.

i.e. $\mathbf{r}_j = \mathbf{a}x_j + \mathbf{b}y_j + \mathbf{c}z_j$

Hence

$$\mathbf{r}_j \cdot \mathbf{S} = x_j \mathbf{a} \cdot \mathbf{S} + y_j \mathbf{b} \cdot \mathbf{S} + z_j \mathbf{c} \cdot \mathbf{S}$$
$$= hx_j + ky_j + lz_j \quad \text{from Laue's equations}$$

$$F(hkl) = \sum_{j=1}^{N} f_j \exp 2\pi i (hx_j + ky_j + lz_j) \tag{5.5}$$

where $\mathbf{S}$ has been replaced by hkl on the left hand side of the equation.

Equation 5.5 is known as the *Structure Factor equation*. It represents the molecular transform sampled at the reciprocal lattice points hkl. If the positions of all the atoms in the unit cell are known then the corresponding diffraction pattern can be calculated.

The structure factor for a particular reflection from a crystal is a complex quantity which may be represented by its amplitude and phase, i.e.

$$F(hkl) = \sum_{j=1}^{N} f_j \exp 2\pi i (hx_j + ky_j + lz_j)$$
$$= F(hkl) \exp i\alpha(hkl)$$

where $F(hkl)$ is the amplitude and $\alpha(hkl)$ is the phase.

Alternatively the structure factor can be represented by its real and imaginary parts i.e.

$$F(hkl) = A + iB$$

where $A = F(hkl) \cos \alpha(hkl)$ and $B = F(hkl) \sin \alpha(hkl)$.

These real and imaginary components are indicated in the Argand diagram shown in Fig. 5.15.

If the structure is centrosymmetric (i.e. if for every atom at x, y, z, there is also an atom at $-x, -y, -z$) then it can easily be shown that the imaginary component B is 0. The phase $\alpha = 0$ or π and the structure factor is real. Although protein crystals themselves are never centrosymmetric, they may possess centrosymmetric projections because of crystal symmetry. Such projections are most useful in the heavy atom isomorphous replacement method of phase calculation.

When the scattered X-radiation is recorded on a photographic film or by a proportional or scintillation counter, all information on the phase is lost and

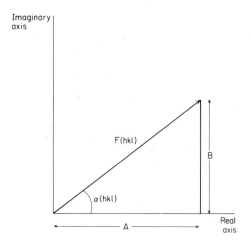

FIG. 5.15   Argand diagram.

only a measurement of the energy, that is the intensity of the diffracted ray, is recorded. The intensity is given by

$$I(hkl) = \mathbf{F}(hkl) \cdot \mathbf{F}^*(hkl)$$
$$= F(hkl)^2$$

where the asterisk indicates the complex conjugate.

In the recording of the optical diffraction patterns described in Section 5.3 only the intensity of the scattered light rays are monitored by the photographic film. Hence the transform shown in Fig. 5.5b represents $\mathbf{G(S)}^2$, and the optical analogy of a molecule in a crystal lattice (Fig. 5.5f) represents $F(hkl)^2$.

The calculated transform of the distorted hexagon is shown in Fig. 5.16 which may be compared with the optical transform of Fig. 5.5b. In Fig. 5.16 we show the phase map of the transform of the distorted hexagon. Since the structure is centrosymmetric the phases are either 0 or π, i.e. $\mathbf{G(S)}$ is real and has a positive or negative sign according to whether the phase is 0 or π. It is seen that the positions of the maxima (phase 0) and minima (phase π) succeed each other with a certain minimal distance of separation which depends inversely on the overall dimensions of the molecule in the corresponding direction. These observations form the basis of the minimum wavelength principle (Bragg and Perutz, 1952) which was used in the early days to provide phase information for haemoglobin diffraction data.

5.7 SCATTERING OF X-RAYS BY A CRYSTAL

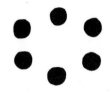

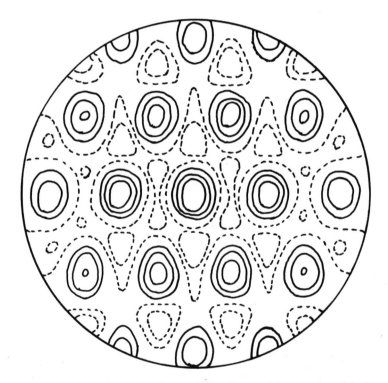

FIG. 5.16 The computed transform of the distorted hexagon used in Fig. 5.5. Since the structure is centrosymmetric, the phases are either 0 or π. Continuous contours represent peaks with phase 0 and the dashed contours represent peaks with phase π.

5.8 THE RELATIONSHIP BETWEEN THE LAUE CONDITIONS FOR DIFFRACTION AND BRAGG'S LAW. THE RECIPROCAL LATTICE

From our original definition of $S = s - s_0$ we can see that if s is to represent a diffracted ray, then, according to Bragg's law, it must be scattered through an angle 2θ where θ satisfies the equation

$$2d \sin \theta = n\lambda$$

Since

$$|s| = |s_0| = \frac{1}{\lambda}$$

$$S = \frac{2 \sin \theta}{\lambda} = \frac{1}{d} \quad \text{for } n = 1$$

The magnitude of the vector **S**, which defines a position in diffraction space, is inversely proportional to the interplanar spacing of the planes which give rise to that particular diffraction peak.

We have already noted that the diffraction pattern of a lattice is itself a lattice with dimensions inversely proportional to the dimensions of the real lattice. Let us define the diffraction lattice (*the reciprocal lattice*) by the vectors **a***, **b*** and **c***. The diffraction vector **S** may be specified by the distance from the origin to a particular point in the reciprocal lattice.

i.e. $S = h\mathbf{a}^* + k\mathbf{b}^* + l\mathbf{c}^*$

We now seek the relationship between the reciprocal lattice and the real lattice vectors.

From Laue's equations

$$\mathbf{a} \cdot \mathbf{S} = h, \quad \mathbf{b} \cdot \mathbf{S} = k, \quad \mathbf{c} \cdot \mathbf{S} = l$$

Hence

$\mathbf{a} \cdot \mathbf{a}^* = 1$	$\mathbf{a} \cdot \mathbf{b}^* = 0$	$\mathbf{a} \cdot \mathbf{c}^* = 0$
$\mathbf{b} \cdot \mathbf{a}^* = 0$	$\mathbf{b} \cdot \mathbf{b}^* = 1$	$\mathbf{b} \cdot \mathbf{c}^* = 0$
$\mathbf{c} \cdot \mathbf{a}^* = 0$	$\mathbf{c} \cdot \mathbf{b}^* = 0$	$\mathbf{c} \cdot \mathbf{c}^* = 1$

From these equations it is apparent that **a*** and **b*** are perpendicular to the real axis **c**. Conversely **c*** is perpendicular to **a** and **b**. This point is important, especially in connection with alignment of crystals in the X-ray beam on the precession camera and other instruments.

The relationships between the real and reciprocal unit cell dimensions are shown in Table 5.I for the different crystal systems.

TABLE 5.I
Relation between axes and angles of direct-lattice and reciprocal-lattice primitive unit cells for the various systems
(from International Tables for Crystallography, Vol. I)

Symbols	
a, b, c	Lengths of edges of direct-lattice unit cell.
α, β, γ	Inter-axial angles of direct-lattice unit cell.
a^*, b^*, c^*	Lengths of edges of reciprocal-lattice unit cell.
$\alpha^*, \beta^*, \gamma^*$	Inter-axial angles of reciprocal-lattice unit cell.
K	Reciprocal constant.
$V; V^*$	Volume of direct-lattice unit cell; of reciprocal-lattice unit cell.

Triclinic

$$a^* = \frac{Kbc \sin \alpha}{V}; \quad b^* = \frac{Kca \sin \beta}{V}; \quad c^* = \frac{Kab \sin \gamma}{V}$$

where $V = abc\{1 + 2\cos\alpha \cos\beta \cos\gamma - \cos^2\alpha - \cos^2\beta - \cos^2\gamma\}^{1/2}$

$$= 2abc\{\sin s \cdot \sin(s-\alpha) \cdot \sin(s-\beta) \cdot \sin(s-\gamma)\}^{1/2}; \quad V^* = \frac{1}{V}$$

$2s = \alpha + \beta + \gamma$

$$\cos\alpha^* = \frac{\cos\beta \cos\gamma - \cos\alpha}{\sin\beta \sin\gamma}; \quad \cos\beta^* = \frac{\cos\gamma \cos\alpha - \cos\beta}{\sin\gamma \sin\alpha};$$

$$\cos\gamma^* = \frac{\cos\alpha \cos\beta - \cos\gamma}{\sin\alpha \sin\beta}$$

Monoclinic
1st setting

$$a^* = \frac{K}{a \sin\gamma}; \quad b^* = \frac{K}{b \sin\gamma}; \quad c^* = \frac{K}{c}; \quad \alpha^* = \beta^* = 90°; \quad \gamma^* = 180° - \gamma$$

2nd setting

$$a^* = \frac{K}{a \sin\beta}; \quad b^* = \frac{K}{b}; \quad c^* = \frac{K}{c \sin\beta}; \quad \alpha^* = \gamma^* = 90°; \quad \beta^* = 180° - \beta$$

Orthorhombic

$$a^* = \frac{K}{a}; \quad b^* = \frac{K}{b}; \quad c^* = \frac{K}{c}$$

$\alpha^* = \beta^* = \gamma^* = 90°$

Tetragonal

$$a^* = b^* = \frac{K}{a}; \quad c^* = \frac{K}{c}$$

$\alpha^* = \beta^* = \gamma^* = 90°$

Cubic

$$a^* = b^* = c^* = \frac{K}{a}$$

$\alpha^* = \beta^* = \gamma^* = 90°$

Hexagonal

$$a^* = b^* = \frac{2K}{a\sqrt{3}}; \quad c^* = \frac{K}{c}$$

$\alpha^* = \beta^* = 90°; \quad \gamma^* = 60°$

Rhombohedral

$$a^* = b^* = c^* = \frac{K \cdot a^2 \sin\alpha}{V}$$

where $V = a^3[1 - 3\cos^2\alpha + 2\cos^3\alpha]^{1/2}$

$$\cos\alpha^* = \cos\beta^* = \cos\gamma^* = \frac{\cos^2\alpha - \cos\alpha}{\sin^2\alpha}$$

$$= -\frac{\cos\alpha}{(1 + \cos\alpha)}$$

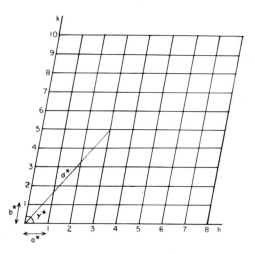

FIG. 5.17 The reciprocal lattice showing d^* ($=S$) for the 350 reflection.

Finally we may note (Fig. 5.17) that

$$|S| = \frac{2\sin\theta}{\lambda} = d^*$$

$$d^* = (h^2 a^{*2} + k^2 b^{*2} + l^2 c^{*2} + 2hk\, a^* \cdot b^* + 2hl\, a^* \cdot c^* + 2kl\, b^* \cdot c^*)^{1/2}$$

Notation: A particular point in diffraction space may be referred to by the vector S. If the point is a Bragg reflection it is more convenient to refer to it by the indices hkl. In later sections we shall abbreviate the indices hkl to h. Thus $F(hkl)$ is represented by F_h or $F(h)$.

5.9 THE EWALD CONSTRUCTION

The condition for diffraction of X-rays by a crystal may be expressed either in terms of Bragg's Law or in terms of the Laue equations. Ewald (1921) proposed a simple geometrical construction which encompasses both of these laws and which has proved a most useful aid in the description of diffraction.

A sphere is drawn with centre at the crystal (C) and radius $1/\lambda$. The origin of the reciprocal lattice is placed at a point 0 where the unreflected X-ray beam travelling in a direction AC meets the sphere (Fig. 5.18). The condition that a particular ray, CB, is a diffracted ray may then be expressed as: X-rays will be diffracted in the direction CB if the point B represents a reciprocal lattice point (h,k,l) i.e. The vector OB is a reciprocal lattice vector $S = h a^* + k b^* + l c^*$.

Hence in order to bring the reflection hkl into the diffraction position the

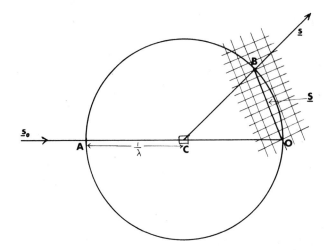

FIG. 5.18   The Ewald construction.

crystal must be rotated in such a way that this particular reciprocal lattice point cuts the sphere of reflection.

Note In order to be consistent with our earlier equations we have continued to define the magnitudes of the incident vector (s_0) and the diffracted vector (**s**) to be equal to $1/\lambda$. This convention results in dimensions of Å^{-1} for the reciprocal lattice vectors. In practice it is more convenient to work with dimensionless reciprocal lattice units. The scale factor λ is thus taken into the calculation of the reciprocal lattice dimensions and the radius of the Ewald sphere is taken to be unity.

5.10 FRIEDEL'S LAW

Friedel's law refers to the relationship between a reflection hkl and its centrosymmetrically related partner $\overline{hkl}$ where $\overline{h}, \overline{k}, \overline{l}$ refer to negative indices. The relationship may be readily deduced as follows:—

$$\mathbf{F}(hkl) = \sum_{j=1}^{N} f_j \exp 2\pi i(hx_j + ky_j + lz_j)$$

$$\mathbf{F}(\overline{h}\,\overline{k}\,\overline{l}) = \sum_{j=1}^{N} f_j \exp 2\pi i(-hx_j - ky_j - lz_j)$$

$$\therefore F(hkl) = F(\overline{h}\,\overline{k}\,\overline{l})$$

$$\alpha(hkl) = -\alpha(\overline{h}\,\overline{k}\,\overline{l})$$

Hence the recorded diffraction pattern will appear centrosymmetric (i.e. $I(hkl) = I(\bar{h}\,\bar{k}\,\bar{l})$) even though the structure itself may not possess a centre of symmetry. Deviations from Friedel's Law occur in the case of anomalous scatters and in such cases these small differences may be used to provide phase information (Chapter 7).

5.11 THE ELECTRON DENSITY EQUATION

We have seen from section 5.7 that the diffraction pattern of a crystal can be calculated from the structure factor equation, once the structure of the crystal is known. We now ask: if we know the diffraction pattern, how do we compute the structure? The answer is provided by Fourier transform theory. The diffraction pattern is the Fourier transform of the structure. It follows therefore that the structure is the Fourier transform of the diffraction pattern.

In order to show this, the expression for the structure factor is rewritten in terms of a continuous summation over the volume of the unit cell.

$$\mathbf{F(S)} = \sum_{j=1}^{N} f_j \exp 2\pi i \mathbf{r}_j \cdot \mathbf{S}$$

$$= \int_{\substack{\text{Volume of} \\ \text{unit cell}}} \rho(\mathbf{r}) \exp 2\pi i \mathbf{r} \cdot \mathbf{S}\, dv$$

where $\mathbf{S}$ is used to denote the position in diffraction space.

By multiplying both sides by $\exp -2\pi i \mathbf{r}' \cdot \mathbf{S}$ and integrating over the volume of diffraction space it may be shown that

$$\rho(\mathbf{r}) = \int_{\substack{\text{Volume of} \\ \text{diffraction} \\ \text{space}}} \mathbf{F(S)} \exp -2\pi i \mathbf{r} \cdot \mathbf{S}\, dv_s$$

where dv_s is a small unit of volume in diffraction space.

The integration can be replaced by a summation since $\mathbf{F(S)}$ is not continuous and is non-zero only at the reciprocal lattice points. Hence

$$\rho(xyz) = \frac{1}{V} \sum_{h=-\infty}^{\infty} \sum_{k=-\infty}^{\infty} \sum_{l=-\infty}^{\infty} \mathbf{F}(hkl) \exp -2\pi i(hx + ky + lz) \quad (5.6)$$

Thus if the structure factors, $\mathbf{F}(hkl)$, are known for all reflections, hkl, then the electron density may be calculated for each point xyz, in the unit cell. The electron density represents the structure of the crystal.

In a crystal structure determination, the electron density is computed at grid points whose intervals Δx, Δy, Δz are chosen to correspond to a third or a

5.11 THE ELECTRON DENSITY EQUATION

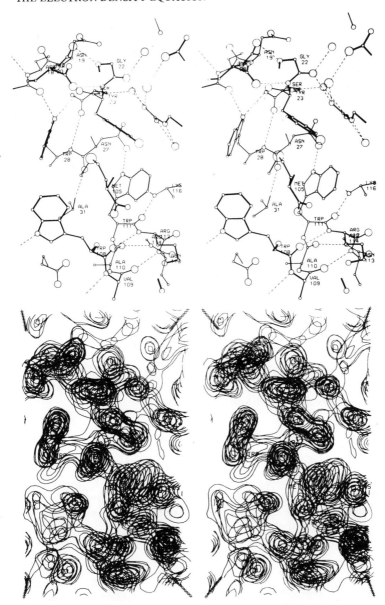

FIG. 5.19 A portion of the high resolution electron density map of hen egg white lysozyme and its interpretation. The region illustrates the so-called "hydrophobic box" where four aromatic residues (tryptophans 28, 108 and 111 and tyrosine 23) surround a methionine residue (met 105). The rings and the sulphur of the methionine are clearly visible in the map, although individual atoms are not resolved at this resolution. [By permission of Professor D. C. Phillips. From Snape (1974)]

quarter of the resolution of the diffraction data. It is convenient to carry out the calculation in sections perpendicular to one of the crystal unit cell axes. The result of the computer calculation is a set of figures where each number represents the value of the electron density at that particular grid point in the unit cell. These values will be high at or near the atomic positions and close to zero elsewhere. In order to emphasize the structure and to show the shapes of the atoms, the electron density map is contoured at an appropriate contour level (e.g. $0.25 e/Å^3$ for a protein structure). The contoured sections are transferred to transparent sheets and the sections stacked to give a three-dimensional representation of the molecule (Fig. 5.19). In Chapter 13 we discuss the special problems encountered in the calculation and interpretation of electron density maps of protein crystals.

Projections: The projection of a crystal structure down a crystallographic axis (c, say) is given by

$$\rho(xy) = \int_c \rho(xyz)dz$$

It may readily be shown that such a projection is given by the transform

$$\rho(xy) = \frac{1}{A} \sum_{h=-\infty}^{\infty} \sum_{k=-\infty}^{\infty} F(hk0)\exp - 2\pi i(hx + ky)$$

where A is the area of the xy projection of the crystal. This means that the projection of a structure may be determined from the data obtained from the corresponding zero level of the diffraction pattern.

5.12 THE PHASE PROBLEM

In order to calculate the electron density we need to know $F(hkl)$: that is we need to know both the amplitude $F(hkl)$ and the phase $\alpha(hkl)$ of the structure factor. This is emphasized by rewriting equation 5.6 as

$$\rho(xyz) = \frac{1}{V} \sum_{h=-\infty}^{\infty} \sum_{k=-\infty}^{\infty} \sum_{l=-\infty}^{\infty} F(hkl)\exp i\alpha(hkl)\exp - 2\pi i(hx + ky + lz)$$

In the recorded diffraction pattern, only the intensities, and hence the amplitudes, of the diffracted rays may be measured. All information on the phases is lost. Hence it is seen that it is impossible to determine a structure directly from measurements on the recorded diffraction pattern since part of the information is missing. *The problem of phase determination is the basic problem in any crystal structure determination.*

There are four methods by which the phase problem may be overcome.

(1) The Patterson summation. This is a Fourier summation based on the experimentally observable $F(hkl)^2$. It is essentially a vector map of the structure which, for molecules containing relatively few atoms (<25), can often be interpreted in terms of a rough approximation to the structure. This interpretation may subsequently be improved by refinement. (2) Direct methods in which mathematical relationships between the reflections can be used to provide phase information. (3) The method of heavy atom isomorphous replacement in which a heavy atom is introduced into a light atom structure and is used as a marker atom to provide phase information. (4) Anomalous scattering in which phase information is obtained from the information contained in the scattering by an atom whose natural absorption frequency is close to the wavelength of the incident radiation. A brief summary of the basic principles of the first two of these methods is given here. A detailed account of the last two methods and their special application to protein crystallography is given in Chapters 6 and 7.

5.13 THE PATTERSON FUNCTION

The Patterson function (Patterson, 1934) is a convolution function which may always be calculated from a set of X-ray diffraction data. It is defined as

$$P(uvw) = \int_{\substack{\text{Vol. of} \\ \text{unit cell}}} \rho(xyz)\rho(x+u, y+v, z+w)dv$$

The value of the function at a particular point $\mathbf{u}(u,v,w)$ is calculated from the product of the two values of the electron density at positions $\mathbf{x}$ and $\mathbf{x} + \mathbf{u}$ (i.e. separated by a vector $\mathbf{u}$) summed over the whole unit cell.

Writing

$$\rho(\mathbf{x}) = \frac{1}{V} \sum_{\mathbf{h}=-\infty}^{\infty} F_\mathbf{h} \exp - 2\pi i \mathbf{h} \cdot \mathbf{x}$$

$$\rho(\mathbf{x} + \mathbf{u}) = \frac{1}{V} \sum_{\mathbf{h}'=-\infty}^{\infty} F_{\mathbf{h}'} \exp - 2\pi i \mathbf{h}' \cdot (\mathbf{x} + \mathbf{u})$$

where $\mathbf{x}$ and $\mathbf{h}$ represent the vectors (x,y,z) and (h,k,l) respectively and the summation over $\mathbf{h}$ represents the triple summation over h,k,l.

Then

$$P(\mathbf{u}) = \frac{1}{V^2} \sum_\mathbf{h} \sum_{\mathbf{h}'} F_\mathbf{h} \cdot F_{\mathbf{h}'} \exp - 2\pi i \mathbf{h}' \cdot \mathbf{u} \int_{\substack{\text{Vol. of} \\ \text{unit cell}}} \exp - 2\pi i (\mathbf{h} + \mathbf{h}') \cdot \mathbf{x} dv$$

The integration is equal to zero, unless $\mathbf{h} = -\mathbf{h}'$ when it is equal to V

$$P(\mathbf{u}) = \frac{1}{V} \sum_{\mathbf{h}} F_{\mathbf{h}}^2 \exp{-2\pi i \mathbf{h} \cdot \mathbf{u}}$$

By Friedel's Law $F_{\mathbf{h}} = F_{\bar{\mathbf{h}}}$, hence the expression may be simplified to

$$P(uvw) = \frac{2}{V} \sum_{h=-\infty}^{\infty} \sum_{k=-\infty}^{\infty} \sum_{l=0}^{\infty} F(hkl)^2 \cos 2\pi(hu + kv + lw) \quad (5.7)$$

Because the Patterson function is the Fourier transform of $F(hkl)^2$, rather than $F(hkl)$, it may always be calculated from a set of recorded diffraction intensities.

It can be seen from Fig. 5.20 that $P(\mathbf{u})$ will have a large value if the positions $\mathbf{x}$ and $\mathbf{x} + \mathbf{u}$ both represent atomic positions: that is when $\mathbf{u}$ is an interatomic vector. The Patterson function therefore represents a map not of the individual atomic positions but of the interatomic vectors. It may be visualized as the convolution of the structure with its inverse since $F_{\mathbf{h}}^2 = F_{\mathbf{h}} \cdot F_{\mathbf{h}}^*$ (see p. 114).

Note that the Patterson summation always contains a large peak at the origin due to the superposition of all the self–self vectors. $P(000) \propto \Sigma_{j=1}^{N} Z_j^2$.

In general if an atom i contains Z_i electrons and an atom j contains Z_j electrons then the peak in the Patterson summations which represents the vector

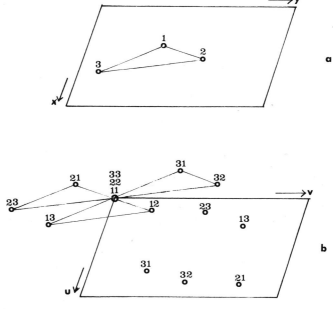

FIG. 5.20 A simple structure and its Patterson function.

5.13 THE PATTERSON FUNCTION

TABLE 5.II
The 11 Laue-symmetry Groups (from International Tables for crystallography Vol. I)

Triclinic	Monoclinic	Tetragonal	Trigonal	Hexagonal	Cubic
1 $\bar{1}$	2 m $2/m$	4 $\bar{4}$ $4/m$	3 $\bar{3}$	6 $\bar{6}$ $6/m$	23 $m3$

	Orthorhombic				
	222 $mm2$ mmm	422 $4mm$ $\bar{4}2m$ $4/mmm$	32 $3m$ $\bar{3}m$	622 $6mm$ $\bar{6}m2$ $6/mmm$	432 $\bar{4}3m$ $m3m$

Notes
1. Point groups belonging to the same Laue group are enclosed in a rectangle.
2. The symbol of the Laue group is that of the centrosymmetrical point group which is given last within each rectangle.

between atoms i and j will have a weight proportional to Z_iZ_j. Therefore the heavier atoms will give rise to heavier peaks in the Patterson map.

The Patterson summation also contains a centre of symmetry at the origin, although the structure itself may not be centrosymmetric. The value of the Patterson function at a position $\mathbf{u}_{12}$, say, which represents the vector from atom 1 to atom 2 is equal to the value of the function at $-\mathbf{u}_{12}$ which corresponds to the vector from atom 2 to atom 1.

The centrosymmetric nature of the Patterson synthesis implies that the space group of the synthesis is different from that of the real structure. The space group of the Patterson synthesis may be derived from the space group of the real structure by addition of a centre of symmetry and loss of the translational symmetry elements along screw axes or across glide planes. The symmetries of the corresponding Patterson synthesis for the 32 point groups are shown in Table 5.II.

It is often possible to solve the structure of a simple compound from an inspection of the Patterson function and even quite complicated structures have been solved using an almost intuitive feel for where atoms should lie. A good example of this type of interpretation was in the structure of rubidium benzyl penicillin (Crowfoot *et al.*, 1949), where it was possible to identify the stronger rubidium–rubidium vectors in the Patterson summation and to use these positions as a starting point in a structure factor calculation to solve for the other atoms (e.g. the sulphur atoms) by difference Fourier techniques. However the *ab initio* solution of a protein structure by Patterson methods is impossible, even with the modern automated peak searching programmes. This is because of problems of resolution, the complexity of the structure and because Patterson space is very much more crowded than Fourier space. If a structure contains N atoms per unit cell, then the numbers of vectors in the Patterson map is N^2, N of which are the self—self vectors ($u = 0$) which superimpose at the origin. The remaining $N(N - 1)$ vectors are distributed throughout the volume of the unit cell, which in a Fourier synthesis contains only N peaks. The possibility of recognizing any detailed structural features in a protein Patterson is therefore unlikely. In spite of this a thorough understanding of the Patterson synthesis and its interpretation is necessary for the protein crystallographer because this technique forms the basis for the location of heavy atoms in the isomorphous replacement method.

Although Patterson syntheses of protein molecules cannot be interpreted directly, they do contain structural information. Some of this information can be extracted by means of the rotation and translation functions which are used to determine the relative orientations of subunits in an oligomeric protein. These methods are discussed in Chapter 16.

5.14 DIRECT METHODS

In our study of molecular transforms we have seen that every atom of the molecule contributes to every part of the transform: different parts of the transform are related. It is therefore not surprising that certain arrangements of peaks and troughs reoccur in the molecular transform. This means that mathematical equations can be written relating different phases and amplitudes of the transform. When the molecular transform is sampled by the reciprocal lattice as in a diffraction pattern, we can write equations between the phases and amplitudes of different structure factors. Such equations may be used to predict unknown phases in a mathematical process known as a "direct" method; it is direct because it involves no intuitive or subjective steps.

To understand the nature of the equations, we must examine a molecular transform in more detail. Figure 5.21 is the Fourier transform of a centrosymmetric array of equal atoms (see Fig. 5.16). Note the vector AB relating peak B to the origin A. Similar vectors EF and CD relate other peaks in the transform. Furthermore, peaks A, C and E are positive while peaks B, D and F are negative. The product of the signs at the ends of the vectors is the same in all cases. We can write:

$$S_A S_B = S_C S_D = S_E S_F$$

where S_A stands for the sign of transform at A etc. If AB is a vector **h** in reciprocal space, AD is a vector **k̄** and AC is a vector **h − k**, then we can write a

5.14 DIRECT METHODS

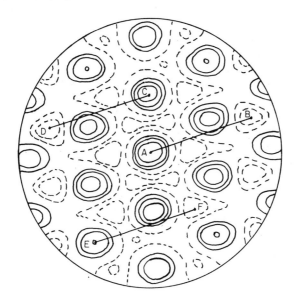

FIG. 5.21 The relation of the Sayre equation to the molecular transform. The transform is of the centrosymmetric hexagon shown in Fig 5.16. Continuous contours represent peaks with phase 0 (plus) while dashed contours are peaks with phase π (minus). The vector AB relates the negative peak B to the origin. Similar vectors EF and CD relate other peaks in the molecular transform. In all cases the product of the signs at the ends of vectors is negative, i.e. $s_A s_B = s_C s_D = s_E s_F$.

relation between structure factors,

$$s(F_0)s(\mathbf{F_h}) = s(\mathbf{F_{\bar{k}}})s(\mathbf{F_{h-k}})$$

but

$$s(\mathbf{F_{\bar{k}}}) = s(\mathbf{F_k}) \quad \text{and} \quad s(F_0) \text{ is positive}$$

Therefore

$$s(\mathbf{F_h}) = s(\mathbf{F_k})s(\mathbf{F_{h-k}}) \tag{5.8}$$

This kind of relationship can be written for many structure factors. From the Fourier transform in Fig. 5.21 it is evident that this mathematical relationship is more often correct when the vectors lie with their ends on peaks in the transform (large structure factors).

This example shows in a qualitative way a relationship between signs of different structure factors. We must now make it quantitative. For this it is useful to modify the structure amplitudes so that they correspond to a structure of point atoms at rest. These are the *normalized structure factors* defined by:

$$E_\mathbf{h} = \frac{F_\mathbf{h}}{[\overline{F^2}]^{1/2}} \qquad (5.9)$$

where $[\overline{F^2}]^{1/2}$ is the root mean square value of structure amplitudes with $\sin\theta$ values close to that of $F_\mathbf{h}$. Consequently the average value of E^2 is 1.

Let us now consider expanding the R.H.S. of equation 5.8.

$$E_\mathbf{k} E_{\mathbf{h}-\mathbf{k}} = [\overline{F^2}]^{-1} \left\{ \sum_{j=1}^{N} f_j \exp(2\pi i \mathbf{k} \cdot \mathbf{r}_j) \right\} \left\{ \sum_{j'=1}^{N} f_{j'} \exp(2\pi i(\mathbf{h}-\mathbf{k}) \cdot \mathbf{r}_j) \right\}$$

$$= [\overline{F^2}]^{-1} \sum_{j=1}^{N} f_j^2 \exp(2\pi i \mathbf{h} \cdot \mathbf{r}_j) \quad \text{(i)}$$

$$+ [\overline{F^2}]^{-1} \sum_{j \neq j'} \sum f_j f_{j'} \exp(2\pi i[\mathbf{k} \cdot \mathbf{r}_j + (\mathbf{h}-\mathbf{k}) \cdot \mathbf{r}_j]) \quad \text{(ii)}$$

Consider now $\langle E_\mathbf{k} E_{\mathbf{h}-\mathbf{k}} \rangle$. In this average the double summation (term ii) tends to zero, and therefore

$$\langle E_\mathbf{k} E_{\mathbf{h}-\mathbf{k}} \rangle = [\overline{F^2}]^{-1} \sum_{j=1}^{N} f_j^2 \exp(2\pi i \mathbf{h} \cdot \mathbf{r}_j)$$

This is proportional to the structure factor of the squared-structure $E'_\mathbf{h}$. Thus:

$$\langle E_\mathbf{k} E_{\mathbf{h}-\mathbf{k}} \rangle = [\overline{F^2}]^{-1} [\overline{F^4}]^{1/2} E'_\mathbf{h} \qquad (5.10)$$

The squared structure is illustrated in Fig. 5.22. It is apparent from this figure that if all the atoms are of equal, positive electron density, the squared-structure factor is proportional to the normal structure factor. We can then write

$$\langle E_\mathbf{k} E_{\mathbf{h}-\mathbf{k}} \rangle = N^{1/2} E_\mathbf{h} \qquad (5.11)$$

where N is the number if atoms in the unit cell.

This result was originally derived by Sayre (1952). If two normalized structure factors, $E_\mathbf{k}$ and $E_{\mathbf{h}-\mathbf{k}}$, have very large values then the following relationship is likely to be correct

$$s(E_\mathbf{h}) = s(E_\mathbf{k} E_{\mathbf{h}-\mathbf{k}}) \qquad (5.12)$$

This is expression 5.8 derived by inspection of the molecular transform.

Equation 5.12 can be used to relate together the signs of the very large structure factors. For a three dimensional centrosymmetric structure three signs can be chosen to define the origin for a primitive cell. Further signs can then be chosen arbitrarily and the equation 5.11 used to generate a complete set from these. Success depends to some extent on choosing structure factors for initial sign fixing which interact with many other large structure factors. If four signs

5.14 DIRECT METHODS

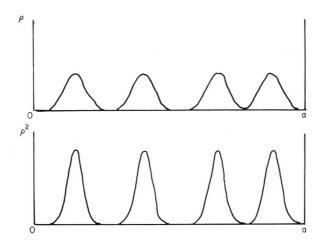

FIG. 5.22 A comparison of the functions ρ and ρ^2 for an equal-atom structure.

are chosen arbitrarily there will be sixteen possible sets of signs. These can be evaluated by calculating a consistency index

$$C = \frac{\langle | E_\mathbf{h} \sum_\mathbf{k} E_\mathbf{k} E_\mathbf{h-k} | \rangle}{\langle E_\mathbf{h} \sum_\mathbf{k} E_\mathbf{k} E_\mathbf{h-k} \rangle} \tag{5.13}$$

where the average is over all the reflections. The correct solution usually has the largest value of C. Alternatively the probability of the sign predictions can be used to choose between signs sets. The probability that a sign is positive is given by

$$P(E_\mathbf{h}) = \tfrac{1}{2} + \tfrac{1}{2}\tanh N^{-1/2} E_\mathbf{h} \sum_\mathbf{k} E_\mathbf{h-k} E_\mathbf{k} \tag{5.14}$$

This emphasizes that the sign relationships will give a better prediction if the magnitudes $E_\mathbf{h}$, $E_\mathbf{h\text{-}k}$ and $E_\mathbf{k}$ are large and N is small.

If the structure is non-centrosymmetric, we must derive relationships between phases rather than signs. For this purpose we can expand equation 5.11 in a different way.

Now

$$\mathbf{E_h} = E_\mathbf{h} \exp(i\psi_\mathbf{h})$$
$$= E_\mathbf{h} \cos \psi_\mathbf{h} + iE_\mathbf{h} \sin \psi_\mathbf{h}$$

and

$$E_k E_{h-k} = E_k E_{h-k} \exp\{i(\psi_k + \psi_{h-k})\}$$

$$E_h \cos \psi_h = \langle E_k E_{h-k} \cos(\psi_k + \psi_{h-k}) \rangle_k \qquad (5.15)$$

$$E_h \sin \psi_h = \langle E_k E_{h-k} \sin(\psi_k + \psi_{h-k}) \rangle_k \qquad (5.16)$$

$$\tan \psi_h = \frac{\langle E_k E_{h-k} \sin(\psi_k + \psi_{h-k}) \rangle_k}{\langle E_k E_{h-k} \cos(\psi_k + \psi_{h-k}) \rangle_k} \qquad (5.17)$$

This is the tangent formula of Karle and Hauptman (1956). It can be used to generate phases from a set of starting phases. For non-centrosymmetric crystals, three phases can be chosen to define the origin and the value of a further phase is limited by the choice of enantiomorph. Further phases can be chosen arbitrarily and used to generate phase sets. The correct set usually gives the lowest Karle R value defined by

$$R = \frac{\Sigma |E_{obs} - E_{calc}|}{\Sigma E_{obs}} \qquad (5.18)$$

and using equations 5.15 and 5.16

$$E_{calc}^2 = \langle E_k E_{h-k} \cos(\psi_k + \psi_{h-k}) \rangle^2 + \langle E_k E_{h-k} \sin(\psi_k + \psi_{h-k}) \rangle^2$$

There are many different variations in approach to the use of direct methods (see Woolfson, 1961; Karle and Karle, 1966) but they all depend on the same basic assumptions made in the derivation of the expressions described above. First they depend on the assumption that electron density is positive. Second, the expressions are correct only for molecular structures containing resolved, equal atoms. Thirdly, the relations are only good for structures with a few atoms (< 100).

5.15 REFINEMENT

The problem of refinement arises whenever a trial structure has been proposed from an interpretation of the Patterson synthesis, direct methods, or any other method. Improvements in the positional and thermal parameters are sought so that the structure corresponds as closely as possible to reality. Reality in this case is represented by the set of observable structure factor amplitudes. The observed and calculated structure factors must agree as closely as possible. The treatment of this type of problem was discussed by Legendre in 1806 in a paper describing the calculation of the orbits of comets from a number of observations (Legendre, 1806). Legendre suggested that the best and most plausible parameters were those for which the sum of the squares of the differences between the observed and calculated values was a minimum.

5.15 REFINEMENT

Let a set of n unknown parameters $x_1, x_2, \ldots x_n$ be related to a set of m observable values $g_1, g_2, \ldots g_m$ by a set of linear equations.

$$g_1 = a_{11}x_1 + a_{12}x_2 + \ldots + a_{1n}x_n$$
$$g_2 = a_{21}x_1 + a_{22}x_2 + \ldots + a_{2n}x_n$$
$$\ldots$$
$$g_m = a_{m1}x_1 + a_{m2}x_2 + \ldots + a_{mn}x_n$$

where a_{11}, a_{12} etc. are known constants.

These equations are known as the *observational equations*. There are n unknown parameters and m observations. If $n > m$ the problem cannot be solved; if $n = m$, the problem is uniquely determined; if $n < m$ the problem is overdetermined and the difficulty lies in finding the best values of the parameters. Suppose approximate values for $x_1, x_2, \ldots x_n$ are known, then error terms E_1, E_2 etc. are defined as:

$$E_1 = a_{11}x_1 + a_{12}x_2 + \ldots + a_{1n}x_n - g_1$$
$$E_2 = a_{21}x_1 + a_{22}x_2 + \ldots + a_{2n}x_n - g_2$$
$$\ldots$$
$$E_m = a_{m1}x_{m1} + a_{m2}x_2 + \ldots + a_{mn}x_n - g_m$$

Legendre's principle suggests that the "best" values of the parameters are those for which $\sum_{r=1}^{m} E_r^2$ is a minimum.

i.e.

$$\frac{\partial \left(\sum_{r=1}^{m} E_r^2 \right)}{\partial x_1} = \frac{\partial \left(\sum_{r=1}^{m} E_r^2 \right)}{\partial x_2} = \ldots = \frac{\partial \left(\sum_{r=1}^{m} E_r^2 \right)}{\partial x_n} = 0$$

Hence there are now n equations for the n unknowns. The first of these is given by:

$$\frac{\partial \left(\sum_{r=1}^{m} E_r^2 \right)}{\partial x_1} = \sum_{r=1}^{m} a_{r1}(a_{r1}x_1 + a_{r2}x_2 + \ldots + a_{rn}x_n - g_r) = 0$$

These equations are known as the *normal equations*. These may be solved directly for $x_1, x_2, \ldots x_n$ by matrix methods.

In matrix notation the observational equations are given by $\mathbf{Ax} = \mathbf{g}$, the normal equations by $\mathbf{A}^T\mathbf{Ax} = \mathbf{A}^T\mathbf{g}$, and the solution for the parameters by $\mathbf{x} = (\mathbf{A}^T\mathbf{A})^{-1}\mathbf{A}^T\mathbf{g}$ where $\mathbf{A}^T$ is the transpose of matrix $\mathbf{A}$.

Often some observations will be judged more reliable than others. This can be taken into account by including suitable weighting terms in the equations as discussed in, for example, the *Calculus of Observations* by E. Wittaker and Robinson. The weighting factors have been omitted here for simplicity.

In a crystallographic problem, the unknown parameters are the positional and thermal parameters for each of the atoms and the observables are the structure factor amplitudes $F_0(\mathbf{h})$ for each reflection $\mathbf{h}$. The quantity to be minimized is

$$\sum_{\mathbf{h}} (F_0(\mathbf{h}) - F_c(\mathbf{h}))^2$$

where the summation is over all reflections and $F_c(\mathbf{h})$, the calculated structure factor amplitude, is given by

$$F_c(\mathbf{h}) = \left| \sum_{j=1}^{n} f_j \exp\left(-B_j \frac{\sin^2 \theta}{\lambda^2}\right) \exp 2\pi i \mathbf{h} \cdot \mathbf{x}_j \right|$$

where B_j may represent isotropic or anisotropic thermal vibrations.

However, $F_c(\mathbf{h})$ is not a linear function of the atomic parameters $\mathbf{x}_j$ and B_j and hence the equations of condition are non-linear. In order to overcome this difficulty, let us suppose that the trial values of the parameters are not too far from the truth and we wish to calculate small corrections to these. $F_c(\mathbf{h})$ is expanded by Taylor's theorem to give

$$F_c(\mathbf{x}^0 + \delta \mathbf{x}) \simeq F_c(\mathbf{x}^0) + \sum_{j=1}^{n} \delta x_j \left.\frac{\partial F_c}{\partial x_j}\right|_{\mathbf{x}^0}$$

where $F_c(\mathbf{x}^0)$ is the value of the calculated structure factor amplitude for a particular reflection at the trial values of the atomic parameters $\mathbf{x}^0$, and $\delta \mathbf{x}$ are the corrections sought for the n unknowns. (Note that the notation for the atomic parameters has been changed and is consistent with that given earlier for the unknown parameters i.e. atomic parameters $x_1, y_1, z_1, B_1, x_2, y_2, z_2, B_2 \ldots B_N$ are now represented by $x_1, x_2, x_3, \ldots x_n$.)

With this approximation for F_c, the observational equations become

$$F_0(\mathbf{h}) - F_c(\mathbf{h}) = \sum_{j=1}^{n} \delta x_j \frac{\partial F_c}{\partial x_j}$$

The equations are now linear in terms of the correction parameters δx_j and may be solved by the method of least squares. The correction terms are applied to the atomic parameters, new values of F_c calculated and the process iterated until convergence is reached.

It will be noted that after squaring and differentiating the error terms from the observational equation, the matrix $\mathbf{A}^T \mathbf{A}$ in the normal equations will contain off diagonal terms which include factors of the form $(\partial F_c/\partial x_i)(\partial F_c/\partial x_j)$. Since there may be many hundreds of observations, the computation and solution of the equations is a lengthy process. In order to simplify the calculations it is often

5.15 REFINEMENT

fair to assume that, in the first approximation, there is no correlation between the atomic parameters so that terms of the type $(\partial F_c/\partial x_i)(\partial F_c/\partial x_j)$ may be ignored for $i \neq j$. This is known as diagonal least squares (see the discussion in Rollett (1965) for a more detailed account). In diagonal least squares the normal equations are given by

$$\delta x_j \sum_{\mathbf{h}} \left(\frac{\partial F_c}{\partial x_j}\right)^2 = \sum_{\mathbf{h}} \left(\frac{\partial F_c}{\partial x_j}\right) (F_0(\mathbf{h}) - F_c(\mathbf{h}))$$

In X-ray crystallography the progress of the refinement is usually assessed by the reliability index R

$$R = \frac{\sum_{h} |F_0(\mathbf{h}) - F_c(\mathbf{h})|}{\sum_{h} F_0(\mathbf{h})}$$

In a small structure refinement, the final value of this residual is usually below 0.10 and for present day structures based on good intensity measurements may well be below 0.05. This implies that the standard deviations in the atomic parameters, which are also calculated during the least squares procedure, are of the order of 0.004 Å and bond lengths may be determined to a precision of 0.006 Å. In protein crystallography, the method of least squares is used to refine heavy atom positions but it is difficult to give any rules for the expected value of R as this will depend on the resolution of the data, the accuracy of the measurements, the number of heavy atom sites and many other factors. The refinement of heavy atoms is discussed in Chapter 11. The improvement of a protein structure is seldom attempted by straightforward least squares methods because the magnitude of the problem and the relatively low ratio of observables to unknown parameters make this impracticable. Various novel methods have been developed for the refinement of protein structures and these are discussed in Chapter 15.

The values of R expected for a totally random structure have been derived from the known distributions of X-ray intensities (Wilson, 1949, 1950) and have been shown to be $R = 0.83$ for a centric and $R = 0.59$ for an acentric distribution. The fact that R is inherently higher for reflections with a centric distribution, even when the structure is nearly correct, reflects the higher dispersion of the distribution function for these terms. The distribution function also implies that the effect of the average value of F on the denominator will be different for centric and acentric reflections since

$$\langle |F| \rangle = 0.798 \left(\sum_{j=1}^{n} f_j^2 \right)^{1/2} \quad \text{for centric data}$$

$$\langle |F| \rangle = 0.887 \left(\sum_{j=1}^{n} f_j^2 \right)^{1/2} \quad \text{for acentric data}$$

148   5. THE PRINCIPLES OF X-RAY DIFFRACTION

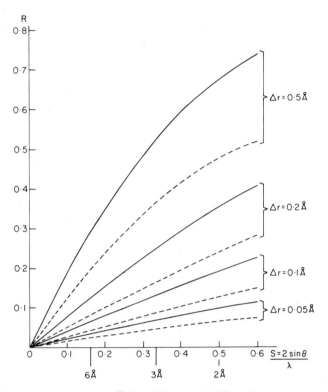

FIG. 5.23 Curves of $R = \Sigma\,|F_o(\mathbf{h}) - F_c(\mathbf{h})|/\Sigma\,F_o(\mathbf{h})$ versus $2\sin\theta/\lambda$ for different values of Δr, the errors in atomic positions. (From Luzzat, 1952.)

There will also be a larger proportion of weak reflections in the centric data. All other things being equal therefore, it is expected that R for centrosymmetric structures or projections will be about 10% of its value larger than for non-centrosymmetric structures.

An analysis of R as a function of the errors in atomic positions has been carried out by Luzzati (1952) and is shown in Fig. 5.23. The derivation assumed that the errors in position are normally distributed and that these errors are the sole cause of the differences between $F_0(\mathbf{h})$ and $F_c(\mathbf{h})$. This is rarely the case, expecially in protein crystallography, but the curves do provide an upper limit to the expected error Δr.

5.16 SUMMARY

The wave nature of X-rays, and the fact that their wavelength is of the same order of magnitude as the spacings in a crystal, provide the basis for the diffraction of X-rays by a crystal.

5.16. SUMMARY

A set of crystal planes, separation d, will scatter X-rays, wavelength λ, through an angle 2θ where the angle θ is defined by *Bragg's Law*

$$2d \sin \theta = n\lambda \qquad (5.1)$$

where n is an integer.

An alternative condition for diffraction by a crystal, in which the lattice is defined by unit cell vectors **a**, **b**, and **c**, is given by *Laue's equations*

$$\mathbf{a} \cdot \mathbf{S} = h \quad \text{where } h, k, l \text{ are integers.}$$
$$\mathbf{b} \cdot \mathbf{S} = k$$
$$\mathbf{c} \cdot \mathbf{S} = l \qquad (5.4)$$

S the diffraction vector, is defined by $\mathbf{s} - \mathbf{s}_0$ where $\mathbf{s}_0$ represents the direction of the incident beam and **s** the direction of the diffracted beam.

$$|\mathbf{S}| = \frac{2 \sin \theta}{\lambda} = \frac{1}{d}$$

The scattering of X-rays by an atom is described by the *atomic scattering factor*.

$$f(\mathbf{S}) = \int_{\substack{\text{Vol. of} \\ \text{atom}}} \rho(\mathbf{r}) \exp 2\pi i \, \mathbf{r} \cdot \mathbf{S} \, dv \qquad (5.2)$$

where approximations for $\rho(\mathbf{r})$, the electron density distribution for the atom, may be obtained from wave mechanics.

The scattering of X-rays by a molecule is described in terms of the *molecular transform*

$$G(\mathbf{S}) = \sum_{j=1}^{N} f_j \exp 2\pi i \, \mathbf{r}_j \cdot \mathbf{S} \qquad (5.3)$$

where f_j and $\mathbf{r}_j$ represent the atomic scattering factor and the position of the jth atom respectively, and the molecule contains a total of N atoms.

The diffraction by a molecule in a crystal lattice is given by the *structure factor equation*, which represents the sampling of the molecular transform at the reciprocal lattice points defined by the conditions on **S** expressed in the Laue equations.

$$F(hkl) = \sum_{j=1}^{N} f_j \exp 2\pi i (hx_j + ky_j + lz_j) \qquad (5.5)$$

Hence the diffraction pattern can always be calculated from a known crystal structure. The diffraction pattern is the Fourier transform of the structure. Conversely the crystal structure, described in terms of the *electron density* of the unit cell $\rho(xyz)$, can be calculated from the Fourier transform of the diffraction pattern

$$\rho(xyz) = \frac{1}{V} \sum_{h=-\infty}^{\infty} \sum_{k=-\infty}^{\infty} \sum_{l=-\infty}^{\infty} F(hkl)\exp-2\pi i(hx+ky+lz) \quad (5.6)$$

In order to calculate the electron density both the amplitude $F(hkl)$ and the phase $\alpha(hkl)$ of the structure factor need to be known. From the recorded diffraction pattern only measurements of the amplitude may be obtained. All information on the phase is lost.

The *phase problem* is the fundamental problem in crystal structure determination.

The *Patterson synthesis* defined by

$$P(uvw) = \frac{2}{V} \sum_{h=-\infty}^{\infty} \sum_{k=-\infty}^{\infty} \sum_{l=0}^{\infty} F(hkl)^2 \cos 2\pi(hu+kv+lw) \quad (5.7)$$

may always be calculated from experimental observations. This synthesis represents a vector map of the structure and may be used in structure determination of small molecules.

The only successful method of phase determination in protein crystallography is the method of *heavy atom isomorphous replacement*, coupled with *anomalous dispersion* data.

Direct methods, efficiently programmed on fast computers, have revolutionized small crystal structure determination. These methods are not applicable to *ab initio* phase determination for protein crystals but they appear most promising for the improvement of a set of isomorphous phases, for the extension of resolution beyond the limit defined by the isomorphous replacement method, and for the location of heavy atom derivatives.

6

ISOMORPHOUS REPLACEMENT

6.1 INTRODUCTION

The method of isomorphous replacement is central to the X-ray analysis of protein crystals. It was first exploited successfully by Perutz and his coworkers (Green et al., 1954) in their study of haemoglobin twenty years after the first X-ray photographs of proteins were recorded. Most successful protein structure analyses have since relied on this technique and the method of isomorphous replacement seems likely to retain a central role in the foreseeable future.

A perfectly isomorphous derivative is one in which the only change in electron density between it and the native crystal is a peak at the site of heavy atom substitution. With small molecules this is usually achieved by replacing an atom with a similar but heavier atom, for instance the replacement of selenium for sulphur or iodine for bromide. This method was first used by Robertson and Woodward (1937) for analysis of phthalocyanines. With proteins it is difficult to replace atoms in this way except in cases where a metal atom such as zinc is easily removed and replaced specifically by a heavier metal such as mercury or lead. However, protein crystals offer an alternative way of making an isomorphous derivative. They contain large channels of mother liquor. It is often possible to bind a heavy atom to the surface of the protein so that it does not disturb the molecular or crystal structures but rather occupies a position in one of these channels. The channels also provide routes along which heavy atoms can diffuse so that isomorphous derivatives can be made from the native crystals. Thus these derivatives are really prepared by isomorphous addition rather than replacement, for it is usually assumed that the heavy atom replaces disordered solvent molecules which give little contribution to the diffraction pattern except at low resolution.

If the heavy atom is sufficiently electron dense it will provide observable changes in the X-ray diffraction pattern (see transforms in Fig. 6.1). Typical changes in an X-ray photo due to the addition of a heavy atom are illustrated in Fig. 6.2a-d. The intensities of the native and derivative diffraction patterns can be used to solve the phase problem if the position of the heavy atom added is

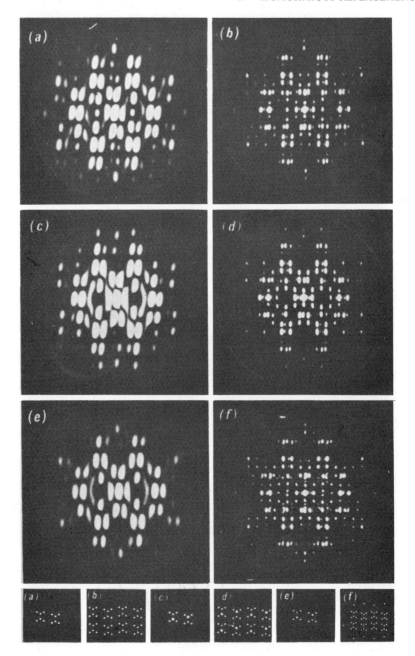

FIG. 6.1 Transforms of molecular arrangements showing the effect of replacement of an atom of the native structure (a, b) with a heavier atom (c, d) and addition of a further atom (e, f) (from Taylor and Lipson, 1964).

6.2. SINGLE ISOMORPHOUS REPLACEMENT

known. For centrosymmetric crystals, one heavy atom derivative will unambiguously determine the phase. For non-centrosymmetric protein crystals there is an ambiguity in the phase determination of most reflections but this can be overcome by using a further derivative. The use of several isomorphous derivatives is called multiple isomorphous replacement. (M.I.R.)

For the protein crystallographer the method of isomorphous replacement has five stages:

1. The preparation of heavy atom derivatives.
2. The measurement of intensities of X-ray diffraction patterns for native and derivative crystals.
3. The reduction and correction of the intensity data.
4. The determination of the heavy atom positions.
5. The determination of phases.

In the following chapters we will discuss these stages of the analysis in detail. Here we are concerned only with the general theory of the method of isomorphous replacement.

6.2 SINGLE ISOMORPHOUS REPLACEMENT

Let $\mathbf{F}_P$ be a structure factor for the native protein. $\mathbf{F}_P$ is a vector and can be described in terms of a structure factor magnitude, F_P, and a phase, α_P. The structure factor, $\mathbf{F}_{PH}$, of the heavy atom derivative is also a vector with magnitude, F_{PH}, and phase α_{PH}. $\mathbf{F}_{PH}$ can be derived from $\mathbf{F}_P$ by the vector addition of $\mathbf{F}_H$, which is the contribution of the heavy atoms to the structure factor of the derivative. This can be written:

$$\mathbf{F}_{PH} = \mathbf{F}_P + \mathbf{F}_H$$

or illustrated as a vector diagram (Fig. 6.3).

Let us assume that we have measured F_{PH} and F_P and that we know the arrangement of heavy atoms in the crystal unit cell, i.e. we can calculate the vector $\mathbf{F}_H$. What can we then derive about the phase? From Fig. 6.4 using the cosine law:

$$\alpha_P = \alpha_H + \cos^{-1}\left(\frac{F_{PH}^2 - F_P^2 - F_H^2}{2F_P F_H}\right) = \alpha_H \pm \alpha' \tag{6.1}$$

This equation shows that there are two possible values for α_P which cannot be distinguished with one isomorphous derivative. Only when the vectors $\mathbf{F}_P$ and $\mathbf{F}_H$ are colinear is there no ambiguity.

The ambiguity can be illustrated graphically in a construction devised by Harker (1956) as shown in Fig. 6.5. The vector $-\mathbf{F}_H$ is drawn from the centre,

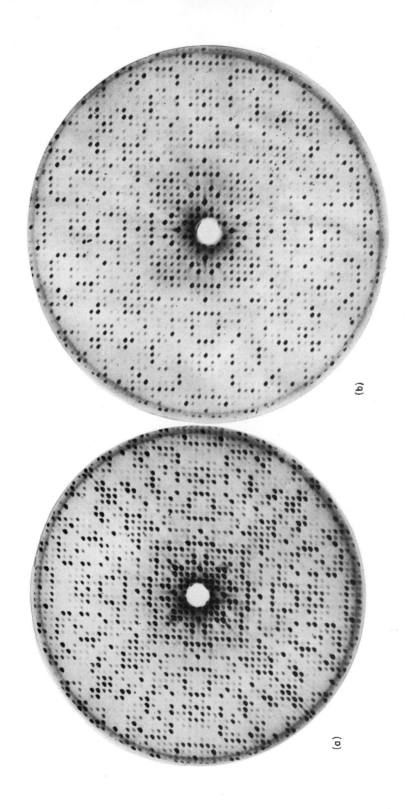

(a)

(b)

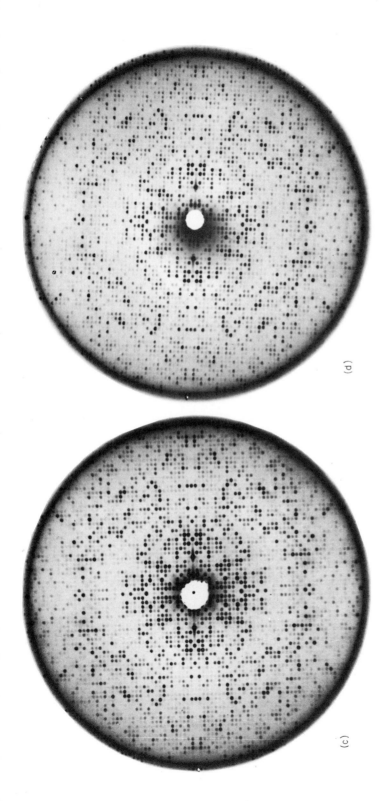

FIG. 6.2 X-ray precession photographs ($\mu = 15°$) of native protein and heavy atom derivative crystals. (a) Native lysozyme (hh0 zone); (b) lysozyme p-chloromercuribenzene sulphonate (PCMBS); (c) native phosphorylase (h01 zone); (d) phosphorylase ethyl mercuri thiosalicylate (EMTS).

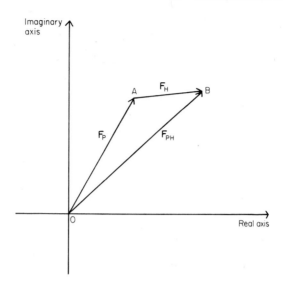

FIG. 6.3 A vector diagram (Argand diagram) illustrating the native protein (F_P) and heavy atom (F_H) contributions to the structure factor (F_{PH}) for the heavy atom derivative of the protein.

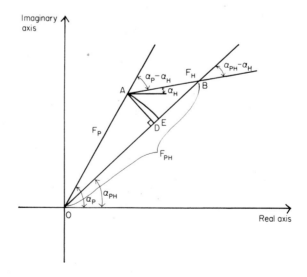

FIG. 6.4 A vector diagram defining the structure factor amplitudes and phases referred to in the text.

6.2. SINGLE ISOMORPHOUS REPLACEMENT

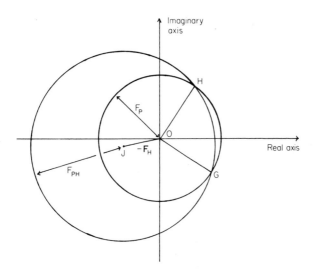

FIG. 6.5 The Harker construction for phase calculation by the method of single isomorphous replacement corresponding to the situation shown in Fig. 6.4. The vector J represents $-\mathbf{F}_H$. Circles of radii F_P and F_{PH} are drawn with their centres at O and J respectively. The vectors OH and OJ represent the two possibilities for $\mathbf{F}_P$.

O. A circle of radius F_{PH} is drawn centred on the end of the vector at J. A second circle of radius F_P is drawn with its centre at O. There are generally two points of intersection of the circle, i.e. at G and H in Fig. 6.5. The vectors OG and OH represent the two possibilities for $\mathbf{F}_P$. From equation 6.1 it can be seen that the vectors OG and OH are symmetrically disposed about $\mathbf{F}_H$. When $\mathbf{F}_H$ and $\mathbf{F}_P$ are colinear there is only one point where the circles touch and the solution is unambiguous.

For a single isomorphous derivative there are two closely similar ways of including the phase information from single isomorphous replacement in a Fourier synthesis. The mean of the two possible phases can be used with suitable weighting (Blow and Rossmann, 1961). The phase will always be either α_H or $\alpha_H + \pi$. Alternatively both phases can be used in the Fourier calculation. This is known as a "double-phased synthesis" (Bijvoet, 1949; Bokhoven et al., 1951) and is very close to the β-isomorphous synthesis of Ramachandran and Raman (1959). It can be shown that these syntheses using single isomorphous replacement are actually equivalent to the protein structure plus the inverse structure convolved with the phase-squared structure of the H atoms (Ramachandran and Srinivasen, 1970) (if the heavy atoms also have a noncentrosymmetric array). The latter is a convolution of structures, and gives rise to general background features. However, if the heavy atoms form a centrosymmetric array, then the electron density map will contain the inverse structure (with the centre of inversion at the heavy atom centre of symmetry). This means the map is quite uninterpretable. Thus the method of single isomorphous replacement is useful for non-centrosymmetric structures as long as the heavy atoms are not in special positions or related by a centre of symmetry. The method can then give a protein electron density map as was shown by Blow and Rossmann (1961) but the high level of background has meant that it has been rarely used by protein crystallographers.

Usually the positions of the heavy atoms added are not known and therefore the vector $\mathbf{F}_H$ cannot be calculated. However, the amplitudes F_P and F_{PH} can also be used to give an estimate of the amplitude F_H. This can be seen from the following.

From Fig. 6.4 we can write

$$BD = F_H \cos(\alpha_{PH} - \alpha_H) \tag{6.2}$$

$$OE = F_P \tag{6.3}$$

$$OD = F_P \cos(\alpha_P - \alpha_{PH}) \tag{6.4}$$

From (6.2), (6.3) and (6.4)

$$F_{PH} - F_P = F_H \cos(\alpha_{PH} - \alpha_H) - F_P\{1 - \cos(\alpha_P - \alpha_{PH})\}$$

$$= F_H \cos(\alpha_{PH} - \alpha_H) - 2F_P \sin^2\left\{\frac{\alpha_P - \alpha_{PH}}{2}\right\} \tag{6.5}$$

If F_H is small compared to F_P and F_{PH}, the sine term will be very small and

$$F_{PH} - F_P \simeq F_H \cos(\alpha_{PH} - \alpha_H) \tag{6.6}$$

When the vectors F_P and F_H are colinear, then

$$|F_{PH} - F_P| = F_H \tag{6.7}$$

These expressions show that we can use the isomorphous differences to derive some estimate of the heavy atom magnitudes. These in turn can be used to calculate the heavy atom positions and so the values of $\mathbf{F}_H$. Of course, these equations assume that F_H is small compared to F_P and F_{PH}. If F_H can be large we may get a "cross over" and

$$F_H = |F_{PH} + F_P|$$

The determination of heavy atom positions is discussed in detail in Chapter 11.

6.3 CENTRIC ZONES

We have seen that isomorphous replacement gives an unambiguous determination of the phase when F_P, F_{PH} and F_H are colinear. This is the situation for a centrosymmetric structure where the phases are always 0 or π. Proteins always crystallize in non-centrosymmetric structures because only one enantiomorph (that comprising L-amino acids) is present. Nevertheless, in certain space groups the structure will appear centrosymmetric when projected in a particular direction and so some two dimensional zones will be centric. For these centric zones F_P, F_{PH} and F_H are colinear† and the phase, α_P, is given unambiguously if F_P, F_{PH} and F_H are known.

A crystal structure will appear centrosymmetric when projected down any evenfold rotation axis. Thus space groups $P2_1$ and $C2$ have one centric zone, h 0 l, if **b** is the unique axis. In a similar way P222, $P2_12_12_1$, or $I2_12_12_1$ have three centric zones which are h 0 l, h k 0 and 0 k l. On the other hand space groups such as P1 or R3 have no centric zones.

Proteins which crystallize in space groups with centrosymmetric projections offer very special advantages in the method of single isomorphous replacement. Phases for these projections can easily be determined. This can be summarized in the following way:

$$\text{if } F_{PH} > F_P \quad s(\mathbf{F}_P) = s(\mathbf{F}_H) \tag{6.8}$$

where s indicates "sign of"

$$F_{PH} < F_P \quad s(\mathbf{F}_P) = -s(\mathbf{F}_H) \tag{6.9}$$

$$F_{PH} \simeq F_P \quad \text{No information}$$

But if F_P is very small and F_H large there is a possibility of "cross over" and in this case

$$s(\mathbf{F}_P) = -s(\mathbf{F}_H) \tag{6.10}$$

These situations are illustrated in Fig. 6.6. A complete phase determination for the centric zones does not give the three-dimensional structure of a complex protein, but it is very useful in the initial evaluation of heavy atom derivatives, the determination of heavy atom positions (see Chapter 11) and the estimation of errors (see Chapter 12).

---

†The phases of centric zones are not always 0 or π but may be $\pi/2$ and $-\pi/2$ for some reflections if the origin does not correspond to a centre of symmetry. Nevertheless, the vectors F_P, F_{PH} and F_H are still colinear. For example, this is true of space groups $P2_12_12_1$, $I2_12_12_1$ or $P2_13$.

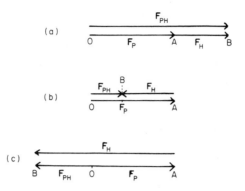

FIG. 6.6 Phase determination in a centric zone by the method of single isomorphous replacement. (a) Equation 6.8: $s(\mathbf{F}_P) = s(\mathbf{F}_H)$. (b) Equation 6.9: $s(\mathbf{F}_P) = -s(\mathbf{F}_H)$. (c) Equation 6.10: $s(\mathbf{F}_P) = -s(\mathbf{F}_H)$.

6.4 MULTIPLE ISOMORPHOUS REPLACEMENT

For general reflections the ambiguity in phase determination given by the method of single isomorphous replacement must be overcome by using at least two derivatives (Bokhoven et al., 1951; Harker, 1956; Blow, 1958; Bodo et al., 1959).

Double isomorphous replacement will give the situation shown in Fig. 6.7 (which is an extension of Fig. 6.5). The first heavy atom derivative indicated, that $\mathbf{F}_P$ may be one of two possible vectors given by OG and OH (see Section 6.2 on single isomorphous replacement). The vector for a heavy atom contribution of the second derivative is drawn as $-\mathbf{F}_{H2}$. Two possible values of $\mathbf{F}_P$ are given by the vectors OH and OL. The vector OH is indicated by both derivatives and this must be the correct choice for $\mathbf{F}_P$. From this graphical construction it is obvious that if $\mathbf{F}_{H2}$ is colinear with $\mathbf{F}_{H1}$ the second derivative will not help in sorting out the ambiguity in the phasing. Two derivatives with different heavy atoms or different occupancies (but the same relative occupancies for different sites) at the same atom positions will not be very useful.

Although for most reflections two isomorphous derivatives give an estimate of α_P, it is advisable to include phase information from more than two heavy atom derivatives. This is for several reasons. For some reflections $\mathbf{F}_H$ for one of the derivatives will be very small; for others F_{H1} and F_{H2} may be colinear by chance: but there is a more important consideration. The values used for F_{PH}, F_P, F_H and α_H will be estimates which may be somewhat in error. Consequently the circles will usually not intersect at one point, H, but rather they will give rise to several intersections which may be quite widely separated if there are large

6.5. THE SIZE OF THE HEAVY ATOM CONTRIBUTION

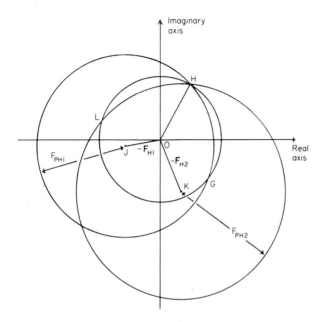

FIG. 6.7 The Harker construction for phase determination by the method of double isomorphous replacement. The construction is an extension of that shown in Fig. 6.5. The vector OK represents $-F_H$ for the heavy atom contribution to the structure factor of the second heavy atom derivative. A circle of radius F_{PH2} is drawn with its centre at K. F_P is given unequivocally by OH.

errors. This is illustrated in Fig. 12.2 on page 367. In fact we need to estimate the errors for each phase determination (Blow and Crick, 1959). This is a complicated subject and Chapter 12 is entirely devoted to its consideration.

6.5 THE SIZE OF THE HEAVY ATOM CONTRIBUTION

How heavy must the heavy atom be? This is a question which we have so far ignored. We must be able to measure the intensity differences between the derivative and native diffraction patterns. The differences due to the contribution of the heavy atom must be larger than the errors in their measurement.

The average change of intensity for acentric reflections due to the addition of heavy atoms has been estimated by Crick and Magdoff (1959) as

$$(2N_H/N_P)^{1/2} \cdot (f_H/f_P)$$

where there are N_H heavy atoms of scattering factor, f_H, and the protein is assumed to comprise N_P equal light atoms of scattering factor, f_P. This means

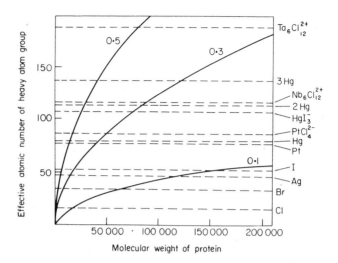

FIG. 6.8 A graph to illustrate the average change in intensity in the X-ray diffraction pattern of a protein by various "heavy atom" groups. The effective atomic number of the "heavy atom" group is plotted against the molecular weight of the protein in the asymmetric unit. The curves give values of the root mean square fractional change in intensity $\langle(\Delta I)^2\rangle^{1/2}/\langle I\rangle$ of 0.1, 0.3 and 0.5 calculated assuming that the scattering factor of a protein atom is 7 and its atomic weight is 14. (After Eisenberg, 1970).

that addition of a mercury atom of 80 electrons to a protein of molecular weight of about 24,000 gives an average change of intensity of about 40%. This may be ten times the standard error of measuring the intensities according to standards of the present time. In fact an interpretable electron density map of nuclease — inhibitor complex was obtained using a single isomorphous derivative made by the addition of an iodine atom of 46 electrons. Bigger proteins obviously need heavier atoms, and for very large proteins metal clusters such as $(Ta_6Cl_{12})^{2+}$, $(W_6Cl_8)^{4+}$ or the polytungstate ions may be needed. A graph showing the percentage intensity change caused by addition of certain heavy atoms to proteins of differing molecular weight is shown in Fig. 6.8.

6.6 TESTS FOR ISOMORPHISM

We have defined an isomorphous derivative as one which differs from the parent by a change of electron density only at the site of heavy atom substitution. This is neat, but not very useful. Rather we need an operational definition.

Unfortunately this is not really possible as lack of isomorphism can manifest itself in many different ways some of which are impossible to identify unless the

6.6. TESTS FOR ISOMORPHISM

structure of the native protein crystal is already known. For instance, movement of a sidechain or even part of a polypeptide chain in a way which did not disturb the rest of the protein structure or the crystal packing would modify the electron density at a site other than the position of the heavy atom substitution; but the distribution of intensity differences between the X-ray diffraction patterns would not be qualitatively different from that caused by a heavy atom.

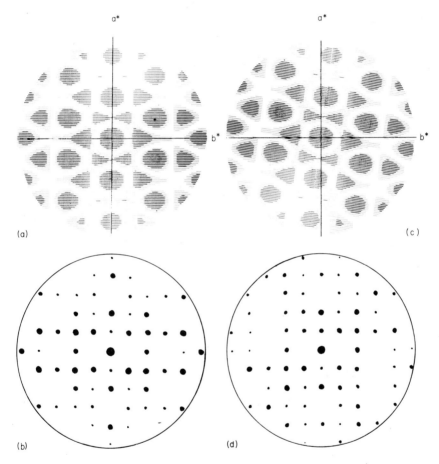

FIG. 6.9 Lack of isomorphism and its effect on diffraction patterns at different resolutions. (a) Shows the modulus of a model molecular transform and (b) is the diffraction pattern (weighted reciprocal lattice) given for the molecule in a cell; (c) and (d) show the equivalent transform and diffraction pattern for a situation in which the molecule has rotated by a small angle. The diffraction patterns shown in (b) and (d) are similar at low $\sin\theta/\lambda$ but are very different at high $\sin\theta/\lambda$. Thus lack of isomorphism has more serious consequences in high resolution studies and can be recognized by intensity differences which increase with $\sin\theta/\lambda$.

Sometimes lack of isomorphism may be identified by a physical method. For instance large changes in the diffraction pattern without a concomitant increase of density of the crystal would imply changes in the protein or crystal structure. In a similar way quantitative analysis by radioactivation analysis or X-ray fluorescence may be a sensitive indication of the presence or absence of the required heavy atom.

Often a lack of isomorphism will lead to easily observable changes in the cell dimensions. Crick and Magdoff (1956) calculated that a change of all cell dimensions by 0.5% will give changes of intensity averaging approximately 15% in the 3Å sphere. Therefore careful measurements of the cell dimensions of all derivatives must be carried out to make sure cell dimension changes are not larger than this.

Rotation of the protein in the unit cell will also give rise to changes in intensity; a rotation of $\frac{1}{2}°$ around the centre of gravity will give intensity changes of 15%. Such lack of isomorphism does not necessarily give rise to changes of cell dimensions. However, it does give rise to a recognizable change in the radial distribution of differences between the intensities of native and derivative diffraction patterns. Figure 6.9a shows the modulus of the molecular transform of a model protein. a* and b* indicate the directions of the reciprocal cell axes which give rise to a sampling of the transform in the diffraction pattern shown in Fig. 6.9b. If the protein rotates by a small angle in the unit cell, the orientation of the reciprocal cell axes changes with respect to the molecular transform and the transform is sampled at different points. This is shown in Fig. 6.9c,d. It is evident from this that the sampling of the transform differs little near the origin, i.e. at low $\sin \theta$, but becomes very obviously different at higher resolution. Thus while the diffraction patterns shown in Fig. 6.9b,d are rather similar at low $\sin \theta$, they are very different at high $\sin \theta$. This implies that the difference of intensity will become larger with increasing resolution. *Lack* of isomorphism can be identified by plotting $|\overline{F_{PH} - F_P}|/\overline{F_P}$ against the resolution. If the ratio increases on average for higher resolution, there is probably lack of isomorphism. Sometimes this trend can be identified from visual inspection of the X-ray photos. A similar distribution will arise from changes of cell dimensions.

Even if there is lack of isomorphism, the data may still be useful for a low resolution study. Conversely, strict isomorphism is essential in accurate high resolution study.

7

ANOMALOUS SCATTERING

7.1 INTRODUCTION

We have seen that small differences in the X-ray intensity pattern caused by addition of a heavy atom to an otherwise unchanged crystal structure can be used to determine the phases of the reflections. Here we are concerned with other small differences in the X-ray intensities caused by these heavy atoms as a result of so-called "anomalous scattering". These anomalous scattering differences are also useful in the calculation of phases and have played an important role in the structure analysis of several proteins.

7.2 THE DISPERSION AND ABSORPTION EFFECTS

So far we have assumed that the electrons of atoms scatter as if they were free electrons. This assumption must be incorrect for we know that electrons, particularly those in the inner K and L shells, are bound quite tightly. In terms of classical theory we might consider the atoms to scatter as if they contained electronic dipole oscillators having certain natural or resonance frequencies (Hönl, 1933). These resonance frequencies are the absorption frequencies of the atoms. The scattering factor for an electron may be approximated to

$$f = \frac{\omega^2}{\omega^2 - \omega_s^2 - ik\omega} \tag{7.1}$$

where ω is the frequency of the incident wave, ω_s is the resonance frequency and k is the damping factor.

The imaginary part is only large when ω is comparable to ω_s. For the moment we will ignore it and approximate the scattering factor

$$f' = \frac{\omega^2}{\omega^2 - \omega_s^2} \tag{7.2}$$

For the general case, there will be many electrons and also many frequencies which give rise to absorption of X-rays. We may therefore write

$$f' = \sum_s \frac{g(s)\omega^2}{\omega^2 - \omega_s^2} \tag{7.3}$$

which may be written in the form

$$f' = f_0 + \Delta f' = \sum_s g(s) + \sum_s \frac{g(s)\omega_s^2}{\omega^2 - \omega_s^2} \tag{7.4}$$

where $g(s)$ is the oscillator strength corresponding to the resonance frequency ω_s. The assumption of free electrons would correspond to the situation where the natural frequency ω_s is zero. For this situation we define the scattering factor

$$f_0 = \sum_s g(s)$$

The real part of the increment in the scattering factor due to the binding of the electrons is $\Delta f'$. When $\omega_s^2 \ll \omega^2$, the contribution to the increment will be positive for the oscillator of frequency ω_s. However, when $\omega_s^2 \gg \omega^2$ the increment will be negative and the scattering factor, f', will be less than f_0, that for free electrons. In the region of the natural frequency the increment becomes large and changes sign. The oscillator model is an oversimplification of the true situation. The simple model (C in Fig. 7.1) gives a reasonable approximation for incident frequencies which are not comparable to ω_s. The actual situation is shown as curve A in Fig. 7.1.

We must now ask how the real contribution, $\Delta f'$, to the scattering factor

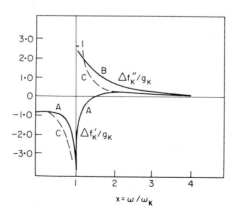

FIG. 7.1 Corrections to the scattering factor for anomalous scattering due to K electrons. Curve A shows $\Delta f'_K/g_K$, curve B is $\Delta f''_K/g_K$, and curve C is the curve corresponding to A for a simple harmonic oscillator (After Hönl, 1933).

7.2. THE DISPERSION AND ABSORPTION EFFECTS

affects the primary beam. Any individual oscillator scatters π out of phase with the primary beam. There are many similar oscillators in the scattering medium and these combine together to give a resultant wave which is $\pi/2$ out of phase with respect to the primary beam. This may be proved by considering the waves scattered from a plane parallel to the scattered wave front. The wavelets arriving from the first Fresnel zone combine together to give a resultant wave. This is $\pi/2$ out of phase with the wavelets scattered so that angles of incidence and reflection are equal with respect to the reflecting plane (see James, 1957, pages 35 and 36). This changes the phase velocity of the primary wave of X-rays in the medium, which manifests itself as a change of refractive index. The change with wavelength or frequency of the radiation is known as a *dispersion* effect. $\Delta f'$ is the *dispersion component* of the anomalous scattering. It does not itself involve absorption. Indeed at incident frequencies less than ω_s there will be no absorption although the dispersion components will be measurable.

If ω is comparable to ω_s but slightly greater we can no longer ignore the factor $ik\omega$ in the denominator of (7.1) and f will become complex. We must then write the scattering factor f, in terms of the real part, f', and an imaginary part $\Delta f''$. Thus

$$f = f' + i\Delta f'' = f_0 + \Delta f' + i\Delta f'' \tag{7.5}$$

The imaginary part lags $\pi/2$ behind the primary wave, i.e. it is always $\pi/2$ in front of the scattered wave. The resultant imaginary wave formed from all the equivalent oscillators will be a further $\pi/2$ out of phase and so the scattered wave due to the imaginary component scatters out of phase with the primary wave. It therefore results in an absorption of X-rays. $\Delta f''$ is known as the *absorption component* of the anomalous scattering.

The magnitude of $\Delta f''$ is proportional to the absorption coefficient. Thus it is only large when the incident frequency is close to an absorption edge. Suppose, for example, we are considering the electrons of the K shell. Then if the incident frequency ω is greater than the natural frequency ω_k, the K absorption frequency, ω must always coincide with the frequency of some oscillator of the K continuum. $\Delta f''$ will have a value which is not negligible. Since the absorption coefficient $\mu_k(\omega)$ is zero if $\omega < \omega_k$, $\Delta f''$ will differ from zero only if $\omega > \omega_k$. The dependence of $\Delta f''$ on the wavelength in the region of an absorption edge is illustrated in curve B of Fig. 7.1.

In our discussion so far we have considered a simple oscillator model and this is of course an over simplification of the true situation. We should consider a wave mechanical model with a continuum of absorption energies. Cromer and Liberman (1970) have used relativistic Slater–Dirac wave functions for computations of $\Delta f'$ and $\Delta f''$. These calculations are very elaborate, but they show good agreement with the experimental data for dispersion corrections either from measuring refractive indices or from measuring atomic scattering factors at frequencies close to those of the absorption edges of the atom. In

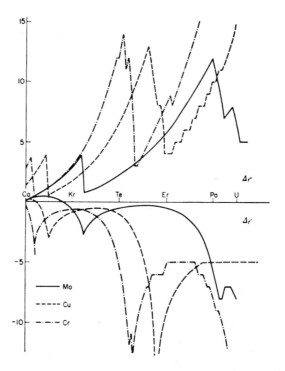

FIG. 7.2 The variation of the theoretical values of $\Delta f'$ and $\Delta f''$ (Dauben and Templeton, 1955) with atomic number for MoKα, CuKα and CrKα.

general these experimental results, except for interferometric results (Hart, 1974), cannot yet be used *in place* of theoretical values. In Fig. 7.2 we illustrate the variation of the theoretical values of $\Delta f'$ and $\Delta f''$ with atomic number and in Table 7.1 we give the values of $\Delta f'$ and $\Delta f''$ along with the absorption coefficient for a selection of light and heavy atoms which are of interest to the protein crystallographer. We must now consider how the anomalous scattering effects might be used by the crystallographer.

7.3 THE USE OF ANOMALOUS DISPERSION TO SIMULATE ISOMORPHISM

We have seen that both the real and the imaginary parts of the anomalous scattering vary with the wavelength of the X-radiation especially in the region of the absorption edge of the anomalous scatterer (see Fig. 7.2 and Table 7.II). The consequent differences in the X-ray diffraction patterns given by two different incident wavelengths can be used to simulate isomorphism and thus to calculate

7.3. THE USE OF ANOMALOUS DISPERSION TO SIMULATE ISOMORPHISM

TABLE 7.I
Values of the real (dispersion) $\Delta f'$ and imaginary (absorption) $\Delta f''$ components of anomalous scattering for CuKα X-rays. The atoms selected are those found in proteins and those used frequently for heavy atom derivatives. μ is the absorption coefficient (see Section 9.3.4, p. 249)

	Atomic number	Cu Kα radiation 1.542				
		$\Delta f'$		$\Delta f''$		
		$\sin\theta/\lambda = 0$	$= 0.6$	$\sin\theta/\lambda = 0$	$= 0.4$	μ
C	6	0	0	0	0	4.6
N	7	0	0	0	0	7.52
O	8	0	0	0.1	0.1	11.5
S	16	0.3	0.3	0.6	0.6	89.1
Fe	26	−1.1	−1.1	3.4	3.3	308
Zn	30	−1.7	−1.7	0.8	0.7	60.3
Pd	46	−0.5	−0.6	4.3	4.1	206
Ag	47	−0.5	−0.6	4.7	4.5	218
I	53	−1.1	−1.3	7.2	6.9	294
Sm	62	−6.6	−6.7	13.3	12.8	397
Gd	64	−12	−12	12.0	11.6	439
Lu	71	−7	−7	5	5	153
Pt	78	−5	−5	8	7	200
Au	79	−5	−5	8	8	208
Hg	80	−5	−5	9	8	216
Pb	82	−4	−5	10	9	232
U	92	−4	−5	16	16	

estimates of the phase of the protein by a method which is exactly equivalent to that of isomorphous replacement. (Pepinsky and Okaya, 1956; Ramaseshan and Venkatesan, 1957; Mitchell, 1957).

The variation of the atomic scattering factor with the wavelength can be calculated by a vector of $\Delta f''$ to $f_0 + \Delta f'$ for different values of $\sin\theta/\lambda$ at each wavelength.

TABLE 7.II
The variation of the atomic scattering factor with wavelength.

	Atomic number	CuKα radiation $\lambda = 1.542$		CrKα radiation $\lambda = 2.291$	
		$\Delta f'$	$\Delta f''$	$\Delta f'$	$\Delta f''$
Fe	26	−1.1	3.4	−1.6	0.9
Zn	30	−1.7	0.8	−1.0	1.5

The difference is most marked at high $\sin\theta/\lambda$ values. This is because the electrons giving rise to the anomalous scattering are tightly bound and close to the nucleus. The anomalous scattering by these electrons is much less dependent on the value of $\sin\theta/\lambda$ than the normal scattering of the electrons as a whole.

The changes in the diffraction pattern with change of wavelength are equivalent to the changes due to isomorphous replacement but in this case the crystals are perfectly isomorphous. Although the differences in intensities will be less than for isomorphous replacement the method is not subject to the same systematic errors due to non-isomorphism. However, the radiation has to be chosen carefully in order to maximize the dispersion effect. The absorption of the two wavelengths will be different and corrections for absorption must be carefully calculated (see Section 9.3.4). Also it is advantageous to measure the same reflection with two wavelengths at the same time or within a small interval of time. The variable wavelength of synchrotron X-radiation is potentially useful in this work.

If the position of the anomalous scatterer is known, the vector corresponding to the difference of anomalous scattering for the two wavelengths can be calculated. Assuming the other atoms do not scatter anomalously, information about the phases of the crystal structure can then be deduced in a way which is analogous to the method of isomorphous replacement. Unlike the methods using the $\Delta f''$ described in Sections 7.5 and 7.6 this method can be used in centric zones. Initial studies on erythrocruorin (chironomus haemoglobin of MW 16,000), using the haem iron atom as anomalous scatterer and NiKα and CuKα X-radiations, are being carried out using the two wavelength diffractometer (Hoppe, 1975). The measurements necessary to determine the phase of one reflection take about 1 hour and the average phase error is estimated to be about 50°. The method could be quite useful for metal containing proteins where isomorphous heavy atom derivatives cannot be prepared.

7.4 THE BREAKDOWN OF FRIEDEL'S LAW

Table 7.I indicates that for CuKα radiation the absorption coefficients for most light atoms are very small. This is true for carbon, nitrogen, oxygen and hydrogen, the atoms most abundant in proteins. For crystal structures containing only these light atoms we can then ignore the imaginary part of the scattering factor and consider only the real part.

Let us consider the structure amplitudes of the direct and inverse reflections namely hkl and $\bar{h}\bar{k}\bar{l}$.

We can write

$$\mathbf{F}_P(+) = \sum_{j=0}^{P} f_j' \exp(2\pi i \mathbf{h} \cdot \mathbf{r}_j) \quad \ldots \quad \text{for } hkl \tag{7.5}$$

7.4. THE BREAKDOWN OF FRIEDEL'S LAW

$$F_P(-) = \sum_{j=0}^{P} f_j' \exp(-2\pi i \mathbf{h} \cdot \mathbf{r}_j) \quad \ldots \quad \text{for } \bar{h}\bar{k}\bar{l} \tag{7.6}$$

where the summation is over all the atoms in the unit cell. Thus the magnitude of the structure factors for hkl and $\bar{h}\bar{k}\bar{l}$ are equal. This is known as Friedel's law.

Consider now a protein crystal structure to which we have added some heavy atoms such as a uranium atom. We can no longer consider the scattering factor as real and we write it as

$$f_j = f_j' + i\Delta f_j''$$

Thus the contribution of the heavy atom to the direct reflection of the derivative crystal can be expressed

$$F_H(+) = \sum_{j=0}^{H} f_j' \exp(2\pi i \mathbf{h} \cdot \mathbf{r}_j) \tag{7.7}$$

$$F_H''(+) = \sum_{j=0}^{H} f_j'' \exp(2\pi i \mathbf{h} \cdot \mathbf{r}_j) \tag{7.8}$$

for the real and imaginary parts respectively, where H is the number of heavy atoms in the unit cell. Corresponding expressions can be written for the inverse reflections. We can then write the direct and inverse reflections for the derivative as

$$F_{PH}(+) = F_H(+) + iF_H''(+) + F_P(+) \tag{7.9}$$

$$F_{PH}(-) = F_H(-) + iF_H''(-) + F_P(-) \tag{7.10}$$

These relationships are expressed on a vector diagram in Fig. 7.3. It can be seen that for the general case $F_{PH}(+)$ is no longer equal to $F_{PH}(-)$; Friedel's law is no longer obeyed. The difference, $F_{PH}(+)^2 - F_{PH}(-)^2$, is known as the Bijvoet difference or the anomalous scattering difference (Coster et al., 1930; Bijvoet, 1949; Peerdeman et al., 1951).

Anomalous scattering differences are only observed if there are anomalous scatterers amongst non-anomalous scatterers. A crystal composed entirely of the same anomalous scatterer obeys Friedel's Law.

In Fig. 7.3 the vectors $\mathbf{F}_H''$ and $\mathbf{F}_H$ are at right angles to each other. This is correct if the anomalous scattering atoms are all identical. It is not generally true if there is more than one type of anomalous scatterer. It will only be true if the ratio $f'/\Delta f''$ is the same for all the anomalous scatterers. If this is not the case our analysis of the anomalous scattering differences must be more complex. The protein crystallographer is usually concerned with adding only one type of anomalous scatterer to the protein and it is a useful assumption to say that $\mathbf{F}_H''$ and $\mathbf{F}_H$ are at right angles to each other.

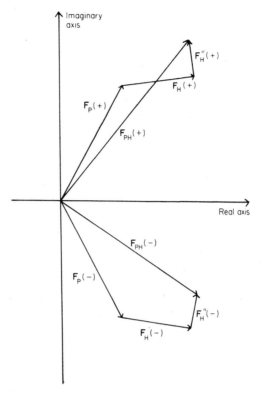

FIG. 7.3 Vector diagrams for the structure factors for the direct and inverse reflections. 'P' are atoms which scatter normally. 'H' are atoms which scatter anomalously.

The difference in $F_{PH}(+)$ and $F_{PH}(-)$ depends on the non-colinearity of $\mathbf{F}_P(+)$ and $\mathbf{F}_H(+)$. If these vectors are colinear then $F_{PH}(+) = F_{PH}(-)$ and Friedel's law is obeyed. This will happen often by chance. However, the vectors will always be colinear when the structure is centrosymmetric, and so there are no anomalous scattering differences for a centrosymmetric structure. Centrosymmetric structures obey Friedel's law. Likewise all reflections in centric zones of non-centrosymmetric structures obey Friedel's law.

The anomalous scattering differences are largest when $\mathbf{F}_P$ and $\mathbf{F}_H$ are approximately at right angles. This is illustrated in Fig. 7.4a. This has important consequences in the uses of anomalous scattering differences. We note that when $\mathbf{F}_P$ and $\mathbf{F}_H$ are colinear, the isomorphous difference is maximum and the anomalous scattering difference zero (Fig. 7.4b). Conversely the anomalous scattering difference is largest when the isomorphous difference is small. We

7.4. THE BREAKDOWN OF FRIEDEL'S LAW 173

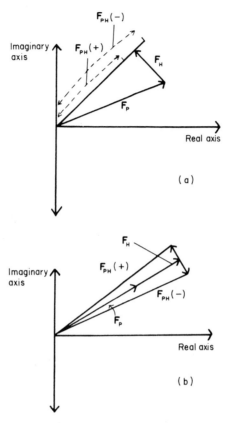

FIG. 7.4 The maximum differences of a Bijvoet pair is when F_H and F_P are almost mutually perpendicular (a). There is no anomalous difference when they are colinear (b). The vectors for the structure factor of the inverse reflections are shown reflected across the real axis.

might expect to find the biggest anomalous scattering differences in reflections which show little isomorphous change on addition of the heavy atom. In fact isomorphous and anomalous scattering differences generally give complementary information, and taken together provide a very powerful tool in structure analysis.

If the crystal has rotation axes, pairs of reflections other than hkl and $\bar{h}\bar{k}\bar{l}$ may be compared to determine anomalous scattering differences. In general for normal scattering the equivalence of reflections is determined by the Laue group symmetry. However, for anomalous scattering, the equivalent reflections are in accordance with the point group symmetry (Ramaseshan, 1963). Let us take for example the space group $P2_1$. The Laue group is $2/m$ so that

$$F_{hkl} = F_{\bar{h}k\bar{l}} = F_{h\bar{k}l} = F_{\bar{h}\bar{k}\bar{l}}$$

so that for normal scattering these are equivalent reflections. However, the point group is 2 so that: $F_{hkl} = F_{\bar{h}k\bar{l}} \neq F_{h\bar{k}l} = F_{\bar{h}\bar{k}\bar{l}}$ for anomalous scattering. This is apparent also from consideratation of the phases which are:

Group I   Group II

$\alpha_{hkl} = \alpha$ $\alpha_{\bar{h}\bar{k}\bar{l}} = -\alpha$

$\alpha_{\bar{h}k\bar{l}} = \pi + \alpha$ $\alpha_{h\bar{k}l} = \pi - \alpha$

Thus we compare any reflection in Group I with any in Group II to determine the anomalous scattering difference. Any such pair is known as a Bijvoet pair.

We must now consider the measurement of anomalous scattering differences caused by the addition of one anomalous scatterer to a protein. Clearly the anomalous scatterer must be near to an absorption edge. From Table 7.I and Fig. 7.2 we can see that there are a number of good candidates using CuK_α radiation. The scattering factors of heavy atoms from platinum to uranium in the periodic table have increasingly large imaginary parts. $\Delta f''$ for lead is 10 electrons while for uranium it is 16 electrons. Similarly the scattering factors of the rare earth atoms up to samarium where $\Delta f'' = 13$ electrons have large imaginary components, but rare earths with a larger atomic number than samarium such as lutetium have smaller values of $\Delta f''$.

Anomalous scattering differences in diffraction patterns of protein derivatives are usually quite difficult to find visually but occasionally they are more obvious. Fig. 7.5 shows a lead–insulin $hk0$ photo which has point group symmetry 3, instead of Laue group symmetry $\bar{3}$ expected for normal scattering and found in the native protein X-ray photo. In particular the differences in the Bijvoet pairs marked 'o' in Fig. 7.5 are large although the intensities of the reflections are small. But this is a fortunate situation. Usually the data must be measured on a diffractometer before convincing evidence of anomalous scattering differences are obtained.

The small size of most anomalous scattering differences underlines the importance of precision in measurement. Friedel pairs should be measured close together in time so that radiation damage, movement of fluid in the mounting tube or changes in voltage of the X-ray generator do not affect the differences. Pairs of reflections with the same absorption should be compared. It is often possible to measure two equivalent reflections which are a Bijvoet pair but which are not a true Friedel pair. In this case it is possible that they may have different absorption; this situation should be avoided.

7.5. FOURIER SYNTHESES FOR STRUCTURES

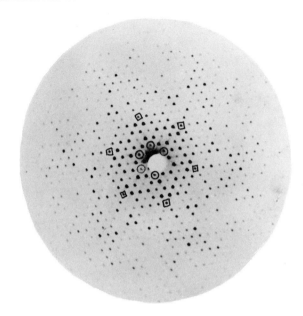

FIG. 7.5 The breakdown of Friedel's law due to anomalous scattering. The X-ray photograph of the hk0 projection of lead-insulin. The space group is R3. The photo shows point group symmetry 3, and not $\bar{3}$. Reflections indicated with ○ or □ should be equivalent in $\bar{3}$; but in fact form two sets of three equivalent reflections. (By permission of Dr. G. Dodson.)

7.5 FOURIER SYNTHESES FOR STRUCTURES WITH AN ANOMALOUS SCATTERER

Before discussing anomalous scattering methods favoured by protein crystallographers to calculate phases we will first quickly describe Fourier techniques which have been useful for small molecules. These have been well reviewed by Ramachandran and Srinivasen (1971).

Most of these methods are really equivalent and are based on the Patterson function in the presence of anomalous scatterers. This approach was largely pioneered by Pepinsky and Okaya (1956) and Ramachandran and Raman (1956). We can write a normal Patterson

$$P(\mathbf{r}) = \frac{1}{V} \sum_{\mathbf{h}} F_{PH}^2 \exp(-2\pi i \mathbf{h} \cdot \mathbf{r})$$

If Friedel's law holds then

$$F_{PH}(+)^2 = F_{PH}(-)^2$$

and we can write the Patterson as a summation over one half of the reciprocal space

$$P(\mathbf{r}) = \frac{2}{V} \sum_{\mathbf{h}}' F_{PH}^2 \cos(-2\pi i \mathbf{h} \cdot \mathbf{r}) \qquad (7.11)$$

However, in the presence of an anomalous scatterer Friedel's law breaks down and we must write

$$P(\mathbf{r}) = \frac{1}{V} \sum_{\mathbf{h}^+}' F_{PH}(+)^2 \exp(-2\pi i \mathbf{h} \cdot \mathbf{r})$$

$$+ \frac{1}{V} \sum_{\mathbf{h}^-}' F_{PH}(-)^2 \exp(+2\pi i \mathbf{h} \cdot \mathbf{r})$$

$$= \frac{1}{V} \sum_{\mathbf{h}^+}' [F_{PH}(+)^2 + F_{PH}(-)^2] \cos(2\pi \mathbf{h} \cdot \mathbf{r})$$

$$- \frac{i}{V} \sum_{\mathbf{h}^+}' [F_{PH}(+)^2 - F_{PH}(-)^2] \sin(2\pi \mathbf{h} \cdot \mathbf{r})$$

$$= P^e(\mathbf{r}) - iP^o(\mathbf{r}) \qquad (7.12)$$

where $P^e(\mathbf{r})$ and $P^o(\mathbf{r})$ are Pepinsky and Okaya's "even" and "odd" Patterson functions.

Now the odd function has peaks of heights $f_j' f_k'' - f_j'' f_k'$ at positions $\mathbf{r} = \mathbf{r}_i - \mathbf{r}_j$. There is no origin peak, but there is a centre of anti-symmetry. However, coincidence of some of the positive peaks with the equal number of negative peaks due to the anomalous scatterers limits the use of this synthesis.

However, the even function will have positive peaks at $\mathbf{r} = \mathbf{r}_i - \mathbf{r}_j$ and $\mathbf{r} = \mathbf{r}_j - \mathbf{r}_i$. A weighted sum of the two functions can be used to compensate for the negative peaks of the odd function.

Raman (1959) has suggested related synthesis called the α_{ano} and β_{ano} syntheses. Of these the β_{ano} synthesis seems to be more useful. Let us assume that we know the positions of the heavy atoms which scatter anomalously so that we can compute $\mathbf{F}_H$ and $\mathbf{F}_H''$ for the direct and inverse reflections by expressions 7.7 and 7.8. The β_{ano} synthesis is defined by

$$\beta_{ano} = \frac{[\frac{1}{2}\{F_{PH}(+)^2 - F_{PH}(-)^2\} - \{\mathbf{F}_H(+)\mathbf{F}_H''(-) + \mathbf{F}_H(-)\mathbf{F}_H''(+)\}] \exp(i\alpha_H'')}{F_H''}$$

This simplifies to the form

$$\beta_{ano} = \mathbf{F}_P(+) + \mathbf{F}_P(-)\exp(2i\alpha_H'') \qquad (7.13)$$

Thus the β_{ano} synthesis contains the correct structure, i.e. $\mathbf{F}_P$, but it also contains the inverse structure convolved with the phase-squared structure

7.6. THE USE OF ANOMALOUS SCATTERING DIFFERENCES

$\exp(2i\alpha_H'')$ of the H remaining atoms. The latter term should mainly result in a small, negative background. The arrangement of the P atoms should be evident above this.

These two functions have been quite useful for X-ray analysis of small molecules but suffer from the difficulty that the mathematical expressions do not easily allow the incorporation of further phase information from isomorphous replacement and heavy atom techniques. They have therefore found little use in protein crystallography. Rather protein crystallographers have preferred to use the anomalous scattering differences to give a direct estimate of the phase (Bijvoet, 1949) along with some estimate of error. We must now turn to this kind of analysis which expresses the phase information from anomalous scattering differences in a different form.

7.6 THE USE OF ANOMALOUS SCATTERING DIFFERENCES IN THE DETERMINATION OF PHASES

We first consider the relation of the anomalous difference $F_{PH}(+) - F_{PH}(-)$ to the phases α_{PH} and α_P (Bijvoet, 1949). From Fig 7.6 using the cosine rule:

$$F_{PH}(+)^2 = F_{PH}^2 + F_H''^2 - 2F_{PH}F_H'' \cos\left(\alpha_{PH} - \alpha_H + \frac{\pi}{2}\right)$$

$$F_{PH}(-)^2 = F_{PH}^2 + F_H''^2 - 2F_{PH}F_H'' \cos\left(\alpha_{PH} - \alpha_H - \frac{\pi}{2}\right)$$

Hence

$$F_{PH}(+)^2 - F_{PH}(-)^2 = 4F_H''F_{PH} \cos\left(\alpha_{PH} - \alpha_H - \frac{\pi}{2}\right)$$

but

$$F_{PH}^2(+) - F_{PH}(-)^2 = [F_{PH}(+) + F_{PH}(-)][F_{PH}(+) - F_{PH}(-)]$$
$$\approx 2F_{PH}[F_{PH}(+) - F_{PH}(-)]$$

Therefore combining these results

$$F_{PH}(+) - F_{PH}(-) \approx \frac{2}{k} F_H \cos\left(\alpha_{PH} - \alpha_H - \frac{\pi}{2}\right) \quad (7.14)$$

$$= \frac{2F_H}{k} \sin(\alpha_{PH} - \alpha_H) \quad (7.15)$$

where $k = F_H/F_H''$.

Hence $\alpha_{PH} = \alpha_H + \pi/2 + \theta$ where

$$\theta = \pm\cos^{-1}\left\{\frac{k[F_{PH}(+) - F_{PH}(-)]}{2F_H}\right\} \quad (7.16)$$

178   7. ANOMALOUS SCATTERING

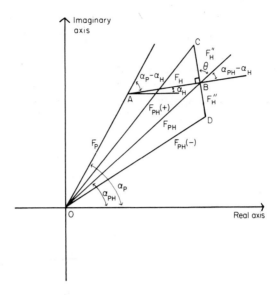

FIG. 7.6   Structure factors for the real and inverse reflections (shown reflected across the real axis).

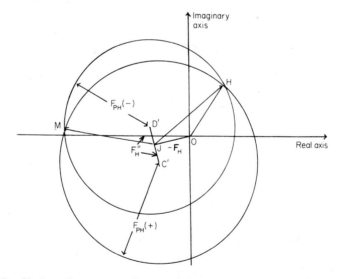

FIG. 7.7   Harker diagram to illustrate phase determination using anomalous scattering. The diagram is constructed in the same way as that for isomorphous replacement (Fig. 6.5). Circles of radii F_{PH} (+) and F_{PH} (−) are drawn with their centres at $-(\mathbf{F}_H + \mathbf{F}_H'')$ and $-(\mathbf{F}_H - \mathbf{F}_H'')$. The two points of intersection indicate the two possible vectors for $\mathbf{F}_{PH}$, i.e. JH and JM. Two possibilities for $\mathbf{F}_P$ are given as OH and OM.

7.6. THE USE OF ANOMALOUS SCATTERING DIFFERENCES

Thus we cannot determine the phase of the heavy atom derivative α_{PH}, unambiguously. There are two possible phases which are symmetrically disposed about $\alpha_H + \pi/2$. These two possibilities are illustrated in Fig 7.7. This diagram is constructed in the same way as that for isomorphous replacement by drawing circles of radii $F_{PH}(+)$ and $F_{PH}(-)$ with their centres at D' and C'. The two points of intersection indicate the two possible vectors for $\mathbf{F}_{PH}$, i.e. *JH* and *JM*.

In the absence of other phase information the crystallographer has no alternative but to use both possibilities in a double phased Fourier synthesis (Ramachandran and Raman 1956) or to take the mean of the two possibilities and use this along with a weight (possibly $\cos\theta$). These techniques are equivalent to the β_{ano} synthesis described in Section 7.5. Argos and Mathews (1973) have shown that the methods can give useful information about protein phases even when the anomalous scatterer is only an iron atom in a protein of molecular weight ~10,000.

For small molecules and possibly very small proteins such as glucagon (MW 3,500) the two possibilities for the phase of the heavy atom derivative may be distinguished if the heavy atom, i.e. the anomalous scatterer is sufficiently large. In thise case the correct phase will in most cases be that which is near to the heavy atom phase. This method of choosing the phase is often called the Bijvoet–Ramachandran–Raman method (Ramachandran and Raman, 1956) and for examples of its use with small structures the reader should refer to Raman 1959; Dale *et al*., 1963; Moncrief and Lipscomb, 1966.

For isomorphous replacement with protein crystals F_H will often be small compared to F_{PH} and F_P, and α_{PH} will be close to α_P. When this is so the two possibilities for α_P given by anomalous scattering are also disposed about $\alpha_H + \pi/2$. In the case of isomorphous replacement the two possibilities for α_P are symmetrically placed about α_H. Thus we have information which is complementary in the two situations and which when combined gives an unambiguous determination of α_P (Peerdeman *et al*., 1951). The complementary nature of the information given by isomorphous replacement and anomalous scattering is also evident from equations 6.6 and 7.15. From isomorphous replacement we determine $\cos(\alpha_{PH} - \alpha_H)$ while from anomalous scattering we determine $\sin(\alpha_{PH} - \alpha_H)$. In Fig. 7.8 we have combined the information from isomorphous replacement (Fig. 6.5) and anomalous scattering (Fig. 7.7) to give a construction which allows the unambiguous determination of the protein phase, α_P. In this diagram the three circles drawn with radii F_P, $F_{PH}(+)$ and $F_{PH}(-)$ intersect at one point indicating that $\mathbf{F}_P$ is given by the vector *OH*. Theoretically this method can be used to estimate α_P for reflections where F_H is known and $F_H > 0$. The estimates of α_P can then be used in a Fourier synthesis to reveal the structure of the protein.

In practice this is not the case. In our discussion we have ignored the fact that there will be errors in the measurements of F_P, $F_{PH}(+)$ and $F_{PH}(-)$ and in the

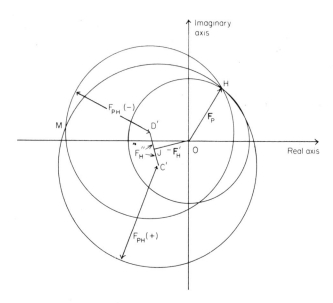

Fig. 7.8 Harker diagram to illustrate phase determination using anomalous scattering (Fig. 7.7) combined with isomorphous replacement for situation shown in Fig. 7.6. The intersection of the three circles of radii F_P, F_{PH} (+) and F_{PH} (−) at H indicates that F_P is given by the vector OH.

calculation of F_H. There may also be a lack of isomorphism. These factors will introduce errors into the phase calculations so that the circles do not necessarily all intersect at one point. We therefore need to carefully consider the errors in the phase calculations (Blow and Rossmann, 1961, North 1965 and Matthews, 1966), and this is discussed in Chapter 12.

7.7 CHOOSING THE ABSOLUTE CONFIGURATION USING ANOMALOUS SCATTERING

So far we have ignored the fact that only L-amino acids occur in globular proteins; but the protein enantiomorph must be chosen correctly. With normal scattering alone the two enantiomorphs cannot be distinguished; they give the same diffraction pattern. However if there is anomalous scattering and $F_{PH}(+) \neq F_{PH}(-)$, then the correct enantiomorph can be identified (Bijvoet, 1949, 1954).

For small molecules this is normally achieved in the following way. As the structure is known in atomic detail the values of $F_{PH}(+)$ and $F_{PH}(-)$ can be calculated and compared to the observed values. Usually $[F_{PH}(+)/F_{PH}(-)]$ obs and $[F_{PH}(+)/F_{PH}(-)]$ calc. are compared for each reflection. If the enantio-

7.7. CHOOSING THE ABSOLUTE CONFIGURATION

morph chosen is incorrect the calculated ratios will tend to be >1 when the observed ratios are <1 and vice versa. If the enantiomorph is correct, the ratios will be in better agreement. Of course this assumes that we have distinguished $F_{PH}(+)$ and $F_{PH}(-)$ correctly when measuring the data, i.e. we have chosen a right-handed set of axes. If this is achieved we can specify the enantiomorph. With proteins we will not usually know the atomic positions until we have reached a rather advanced stage in the calculations, but we still need to know the correct enantiomorph in the initial low resolution study. For isomorphous replacement alone the choice depends critically on the choice of the hand of the heavy atom arrangement in the derivative. If this is chosen incorrectly the enantiomorph will be wrong and we would find D-amino acids and right-handed helices in our electron density if it were at sufficiently high resolution to distinguish these features. However, if there are anomalous scattering measurements we can recognize the correct enantiomorph without the need to distinguish these features at the molecular level.

Let us suppose that we have calculated α_P from isomorphous replacement and we are unsure whether it corresponds to the correct hand. Let us further suppose that the protein contains an anomalous scatterer such as an iron atom. We can then calculate the imaginary part of the difference Fourier (Kraut, 1968) using equation 7.15.

$$Im\{\Delta e(r)\} = \frac{1}{V} \sum_{\mathbf{h}}{}' [F_P(+) - F_P(-)] \sin(\alpha_P(\mathbf{h}) - 2\pi \mathbf{h} \cdot \mathbf{r}) \qquad (7.17)$$

$$= \frac{2}{kV} \sum_{\mathbf{h}}{}' F_H \sin(\alpha_{PH} - \alpha_H) \sin(\alpha_P(\mathbf{h}) - 2\pi \mathbf{h} \cdot \mathbf{r}) \qquad (7.18)$$

The imaginary part is smaller than the real part by the factor, $2/k$, but it should be everywhere positive. If the enantiomorph is incorrectly chosen this will not be the case and the Fourier gives rise to negative holes at loci which are related by inversion through the origin to the anomalous scatterers. This has been demonstrated for high potential iron protein (Strahs and Kraut, 1968). If the native protein contains no anomalous scatterer the same Fourier technique can be used with the heavy atom derivative data. It was exploited in this way in the low resolution study of chymotrypsinogen (Freer et al., 1970), where the incorrect enantiomorph had been previously chosen. By these procedures the enantiomorph of a protein can be chosen correctly.

An alternative procedure for finding the absolute configuration (when the native protein contains an anomalous scatterer) is to calculate the phases, α_p, using expression 7.16. These phases can then be used in difference maps which give the absolute configuration of the heavy atoms in a derivative (Argos and Mathews 1973). The difference maps tend to be confused by the images of the inverse structure (discussed on page 176) but the positions of these images can be calculated. With the correct enantiomorph for the heavy atoms known, the protein enantiomorph is easily fixed.

In most cases protein crystallographers have used the anomalous scattering data in a slightly different way to find the absolute configuration. If only one derivative is available, the phases are calculated twice by combining isomorphous and anomalous scattering data once for each heavy atom configuration. The two phase sets are distinguished by examining the protein electron density maps for recognizable features. Alternatively if more than one heavy atom derivative is available the two phase sets are distinguished by using them in difference Fourier syntheses to find the second heavy atom position. The correct heavy atom enantiomorph should give phases leading to the largest peak at the heavy atom position of the second derivative. These techniques are discussed in greater detail in Chapter 12.

8

PREPARATION OF HEAVY ATOM DERIVATIVES

8.1 INTRODUCTION

In the two previous chapters we have discussed the use of isomorphous replacement and anomalous scattering in phase determination. We have seen the need to introduce a heavy atom into the crystal without disturbing either the crystal packing or the structure of the protein. In most cases the addition of an iodine, mercury or platinum atom gives suitable isomorphous differences in the X-ray diffraction pattern. Anomalous scattering differences caused by samarium, lead, uranium and other heavy atoms can also be used to help solve the phase problem. In this chapter we are concerned with the preparation of the heavy atom derivatives for use with these techniques.

Protein crystals exist as two phases: a solid phase of protein molecules packed in an open lattice and a liquid phase of solvent occupying the channels and spaces in this lattice (Section 3.6). Addition of a heavy atom to a specific position on the protein surface can be accomplished by displacement of solvent without distortion of the protein or the crystal structure. Nevertheless, the preparation of heavy atom derivatives is usually a "trial and error" process. We would need to understand the crystal structure in detail to make a completely rational approach. Knowledge of the chemical sequence may give a few clues about suitable heavy atom reagents. However, some amino acids are buried in the protein core or involved in contacts between proteins in the lattice and will be inaccessible. Others will bind the heavy atom because the geometrical arrangement of the protein lattice chelates the reagent. Chelation tends to be entropy driven rather than determined by the internal energy of the bonds and so bonds may form with unusual ligands. In fact chelation of heavy atom reagents in the protein lattice is a major factor causing lack of specificity.

There are some situations where prior knowledge of the geometry of part of the protein is available. If the protein contains a metal ion cofactor such as zinc or calcium this may sometimes be removed and replaced by a different metal. From the size of these ions, we can infer that they may be replaced with only a little distortion by some heavier atoms. The specificity of the active site of an enzyme or carrier protein site may also be known. Heavy atom analogues of the substrate often act as inhibitors and can be used to prepare heavy atom derivatives. Prior knowledge of the geometry of individual amino acids in the chain is also available and their replacement by a synthetic analogue of similar geometry is an attractive idea. The aminoterminal residue offers advantages in this approach as it can be removed specifically by the Edman technique and replaced by the usual methods of protein synthesis.

Apart from these possibilities the lack of knowledge about the available surface side chains and the possibility of non-specific reaction of heavy atoms in the protein crystal lattice make a rational approach difficult. In this chapter the reader will not find any certain method for making an isomorphous derivative: rather he is presented with a discussion of the approaches which on the basis of previous structure analyses give the best chance of success. We begin by discussing the more rational approaches: the replacement of a metal cofactor, the use of a labelled inhibitor and the replacement of an amino acid residue. We then consider reaction with surface amino acid residues, and discuss the importance of pH, buffer, salting out agent and temperature which may determine the nature and extent of the reaction.

8.2 METAL ION REPLACEMENT IN METALLOPROTEINS

A metal ion cofactor may often be removed and replaced by a heavy atom to give an isomorphous derivative. This will not be possible for some covalently bound metals, removal of which may destroy the tertiary structure of the protein, but where ionic or partially ionic bonds are concerned the metals may usually be removed more easily. This is often true when zinc or calcium cofactors are bound by carboxylate or imidazole ligands; for instance metal ion replacement has given useful heavy atom derivatives of carbonic anhydrase, carboxypeptidase, thermolysin and insulin, which all contain zinc, and in staphylococcal nuclease, which contains calcium.

Initial attempts at removing zinc cofactors in solution were unsuccessful. For instance, addition of 1,10 phenanthroline to carbonic anhydrase gave a zinc free enzyme which did not crystallize (Tilander *et al.*, 1965): and insulin crystallized in a different space group in the absence of zinc (Low and Berger, 1961). More success was achieved by soaking the crystals themselves in a solution of a suitable chelating agent. Thus dialysis of carbonic anhydrase crystals against 2,3-dimercaptopropanol in an hydrogen atmosphere produced crystals of the zinc free enzyme (Tilander *et al.*, 1965). Use of 5-hydroxyquinoline-8-sulphate and ethylenediaminetetramine (EDTA) with carboxypeptidase (Lipscomb *et al.*, 1966) and rhombohedral 2-Zn insulin (Adams *et al.*, 1966) respectively gave zinc free crystals. It appears that the crystal packing stabilizes the zinc free structure

8.2. METAL ION REPLACEMENT IN METALLOPROTEINS

and the conformational change or disaggregation which occurs in solution. Electron density maps of insulin later showed that the sidechains in the regions vacated by the zinc atoms were rather disordered and were not reordered on addition of heavy atoms with the exception of cadmium.

Imidazole groups of histidines are frequently involved in coordination of replaceable zinc atoms: there are three histidine ligands to zinc in carbonic anhydrase and in insulin and two histidines and a glutamate in carboxypeptidase and thermolysin. Zinc is smaller than calcium (see Table 8.II). It is more polarizing and prefers soft ligands such as imidazole but will also bind water molecules or carboxylates. It is most easily replaced by metal ions such as Fe^{2+}, Co^{2+}, Cd^{2+}, and Hg^{2+}, but only the latter two are heavier than zinc and give a useful change of scattering matter. Dialysis of the apoenzyme against 0.003M mercuric acetate for 10 days leads to substitution of mercury at the zinc site in carbonic anhydrase (Tilander et al., 1965). Dialysis against 0.005M mercuric chloride gives a similar mercury substitution for carboxypeptidase (Lipscomb et al., 1966).

Cadmium gives a useful insulin derivative which can also be prepared by cocrystallization (Adams et al., 1966). Plumbous ions (Pb^{2+}) give low substitution at the zinc sites in insulin while in zinc free carboxypeptidase two sites 4 Å from the zinc site were given. These results are not unexpected. The non-group valence of a plumbous ion means that it is larger than mercuric and less polarizing. It tends to bind electronegative groups such as carboxylate rather than imidazole and will not easily replace zinc.

Calcium is more electropositive and larger than zinc (although the atomic number of calcium is smaller than that of zinc). It tends to bind oxygen ligands such as carboxylate sidechains. It is best replaced by other alkaline earth ions such as Sr^{2+} or Ba^{2+} or with trivalent lanthanide ions. The calcium ion of nuclease was exchanged for barium by soaking the crystals in a solution of barium chloride (Arnone et al., 1971). In solution the lanthanide, neodymium, will replace calcium in trypsin and trypsinogen (Darnall and Birnbourn, 1970) and in α-amylase (Smocka et al., 1971). Colman et al. (1972b) have demonstrated that three of the four calcium ions of thermolysin may be replaced by either lanthanide ions or by strontium or barium. The crystals were first equilibrated with calcium free tris-acetate buffer at pH 5.5 for one hour and then a solution of metal ions in the same buffer were added. The four calcium sites in thermolysin can be characterized in the following way:

Ca 1, the inner double site
Ca 2, the outer double site
Ca 3, the single site at Asp 57
Ca 4, the single site at Asp 200

as shown in Fig. 8.1 (Matthews and Weaver, 1974).

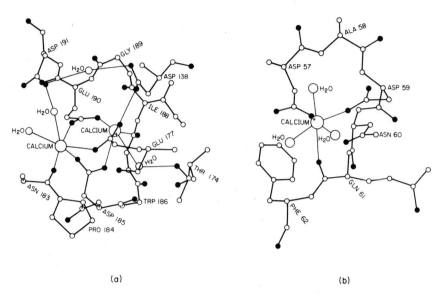

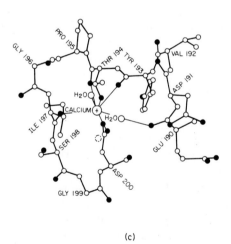

FIG. 8.1 The calcium and europium binding sites in thermolysin. Oxygen atoms are represented by solid circles. Water molecules and calcium ions, indicated by large circles are labelled. The calcium ligands are indicated by thin lines, and the positions of the bound europium ions are shown by crosses: (a) sites 1 and 2; (b) site 3; (c) site 4. An additional water molecule bound when europium substitutes for calcium is shown as a broken circle. (Reproduced with permission from Matthews and Weaver, 1974)

8.2. METAL ION REPLACEMENT IN METALLOPROTEINS

TABLE 8.I
Refinement statistics for calcium substitutions in thermolysin. The sites 1–4 refer to the four calcium sites described in the text. R_c is defined in Section 11.6.2.

Ion occupancies	Ca^{2+}	La^{3+}	Nd^{3+}	Eu^{3+}	Er^{3+}	Lu^{3+}	Sr^{2+}	Ba^{2+}
Site 1	–	20.8	20.8	20.8	23.2	26.6	10.2	3.9
Site 2	–	−16.6	−16.8	−16.5	−16.6	−16.2	+8.6	+12.1
Site 3	–	22.5	23.0	23.9	24.4	23.8	4.3	12.9
Site 4	–	21.6	20.6	22.5	22.1	17.3	8.3	6.4
R_c	–	49.3	49.6	48.2	44.4	44.3	65.4	62.4
r(Å)	0.99	1.15	1.08	1.03	0.96	0.93	1.13	1.35

The lanthanides bind at sites 1, 3 and 4 and in so doing calcium is ejected from site 2 also. This presumably is the result of the higher charge of the lanthanides; barium and strontium replace all four calciums. The occupancies of the replacement atoms at the calcium sites are given in Table 8.I.

A suitable heavy atom often has a larger radius than the atom it replaces. Table 8.II gives the ionic radii of some metal ions which are of interest in this context. In the thermolysin study, it is apparent from the occupancies and the crystallographic R values (see Section 11.6.2) given in Table 8.I that a significantly more isomorphous derivative is given when the metal ion has a smaller radius than calcium. Replacement of calcium by the smaller lanthanides such as Lu^{3+} gives the most isomorphous derivatives (Matthews and Weaver, 1974).

TABLE 8.II
The ionic radii of some metal ions found in proteins and of heavy atoms which might be used to replace them in making a heavy atom derivative

Zn^{2+}	0.74 Å	Co^{2+}	0.56 Å
Cd^{2+}	0.97	Fe^{2+}	0.628
Tl^+	1.44	La^{3+}	1.15
Pb^{2+}	1.21	Nd^{3+}	1.08
Ca^{2+}	0.99	Eu^{3+}	1.03
Sr^{2+}	1.13	Sm^{3+}	0.964
Ba^{2+}	1.35	Er^{3+}	0.96
Hg^{2+}	1.04*	Lu^{3+}	0.93

Crystal radii after Pauling.
*Derived from HgF_2. Note that mercuric rarely forms a regular ionic coordination sphere. Often there are short covalent bonds as well as longer ionic interactions.

positions of europium and calcium atoms in thermolysin are illustrated in Fig. 8.I. Replacement of calcium by barium involves the introduction of a larger metal ion and it is not surprising to discover that the barium atom in nuclease is 0.75Å from the calcium ion position (Arnone et al., 1971). Similarly the mercury positions in carbonic anhydrase (Liljas, 1971) and carboxypeptidase (Lipscomb et al., 1970) differ by 0.7Å and 0.25Å respectively from the zinc positions. The calculations used in the method of isomorphous replacement should take account of this. The position of the metal ion cofactor must be found precisely by initial Fourier studies or by use of anomalous scattering; it should be included as a negative atom in the structure factor calculation. If this is done correctly, replacement of a metal ion cofactor can produce a very useful heavy atom derivative.

8.3 HEAVY ATOM LABELLED SPECIFIC INHIBITORS OF ENZYMES AND CARRIER PROTEINS

The binding sites of many enzymes and carrier proteins have been the subject of detailed biochemical studies. Often specific inhibitors have been designed and these have been modified by the protein crystallographer to provide useful heavy atom reagents. Indeed such heavy atom-labelled inhibitors have played an important part in the preparation of isomorphous derivatives of several enzymes.

The chymotrypsin studies are a good example of this kind of approach. Chymotrypsin has a serine in the active site which reacts to give a covalent intermediate in proteolysis. It was shown in 1949 that the active site serine could be reacted with phosphoryl fluorides (Jansen et al., 1949). Pollit and Blow (Sigler et al., 1966) exploited this observation to prepare crystals of α-chymotrypsin inhibited by heavy atom derivatives of alkyl phosphoryl fluorides, but these proved unstable. The difficulty was overcome by using the stable aromatic sulphonyl fluorides, which could be easily substituted with iodine or mercury atoms. The reagents used are shown in Table 8.III. The X-ray maps showed that all these gave substitution at the serine in the active site. Unfortunately the PCMBSF also gave substitution at a further site and neither site was fully occupied. The pipsyl derivative was completely isomorphous with the BSF and tosyl derivatives and showed a peak of electron density in the expected position and of the expected weight for the iodine atom. Unfortunately the inhibitor-enzyme complexes were not completely isomorphous with native enzyme crystals. This made it difficult to combine the information from these derivatives with other heavy atom derivatives, and the tosyl derivative had to be used as the parent protein in the method of isomorphous replacement.

In fact the addition of a substrate or an inhibitor often causes a conformational change. In lysozyme there are small (1Å) but significant movements of several amino acids on binding the inhibitor tri-N-

8.3. HEAVY ATOM LABELLED SPECIFIC INHIBITORS OF ENZYMES

TABLE 8.III
Heavy atom labelled sulphonyl fluoride inhibitors of α- and γ-chymotrypsin

Trivial name	Aromatic substituent	Chemical structure
BSF	Benzene	C$_6$H$_5$–SO$_2$F
tosyl	p-toluyl	CH$_3$–C$_6$H$_4$–SO$_2$F
pipsyl	p-iodophenyl	I–C$_6$H$_4$–SO$_2$F
PMBSF	p-methoxyphenyl	CH$_3$–O–C$_6$H$_4$–SO$_2$F
PCMBSF	p-chloromercuriphenyl	ClHg–C$_6$H$_4$–SO$_2$F

acetylchitotriose (Blake et al., 1967a), while the binding of glycyl-tyrosine to carboxypeptidase produces a shift of a tyrosyl residue of 14Å (Lipscomb et al., 1970). Changes of this kind make heavy atom-labelled inhibitors of limited use. Certainly errors will occur in the phase determination if the data are combined with the native protein data. However, even if another enzyme-inhibitor is used as the parent protein there are still difficulties in the interpretation of the electron density map for knowledge of the arrangement of the protein sidechains at the active site is one of the main objectives of the study of an enzyme. The fact that inhibitor molecules may interfere with this part of the protein by reacting with the sidechains of the active site thus inducing a conformational change may restrict their use for making derivatives.

Nevertheless this approach to making heavy atom derivatives has been widely exploited. Sulphonyl fluorides have been helpful in the study of other serine proteases, elastase and subtilisin. A similar series of substituted sulphonamides has been used by Strandberg and his colleagues (Tilander et al., 1965) to prepare crystalline heavy atom-labelled inhibitor complexes of human carbonic anhydrase. The high resolution study has shown that these sulphonamides bind directly to the zinc ion in the active site, as shown in Fig. 8.2 and replace the water molecule bound in the native crystals (Liljas, 1971).

Labelled competitive inhibitors have also been used. In the study of nuclease, Cotton and his colleagues replaced the 5 methyl groups of the inhibitors thymidine $3^1,5^1$-diphosphate with iodine (Arnone et al., 1971). This gave a

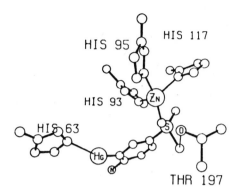

FIG. 8.2 The binding of a sulphonamide inhibitor to carbonic anhydrase. (Lilias. 1971)

useful derivative which was perfectly isomorphous with the nuclease inhibitor complex. However, the inhibitor 5-iodouridine 2′,3′-phosphate was useful in the study of ribonuclease-S only to calculate initial protein phases, with which the more complex binding of the uranyl ion was studied by means of difference Fouriers (Wyckoff *et al.*, 1967b). In the study of ribonuclease-A, an arsenate ion bound in the active site was used as a derivative. The heavy atom anions $Pt(CN)_4^{2-}$ and $Au(CN)_2^-$ often act as inhibitors. In carbonic anhydrase they bind close to the zinc atom (Liljas, 1971; Kannan *et al.*, 1971). In liver alcohol

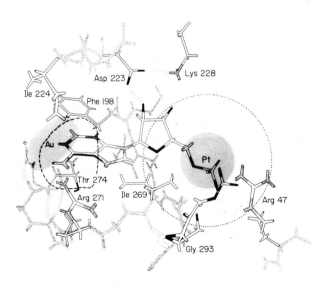

FIG. 8.3 The binding of $Pt(CN)_4^{2-}$ and $Au(CN)_2^-$ in the coenzyme binding sites of liver alcohol dehydrogenase (LADH) (Branden *et al.*, 1975).

8.4. REPLACEMENT OF AN AMINO ACID

dehydrogenase they bind in strict competition with the coenzyme, NAD (Gunnarsson et al., 1974). The auricyanide ion binds at two sites. One site is normally occupied by the phosphate groups of the coenzyme and the other by the adenosine part. The platicyanide ion binds only to the phosphate site (Eklund et al., 1975). These sites are shown in Fig. 8.3 and the ligands are listed in Table 8.IX on p. 226.

8.4 REPLACEMENT OF AN AMINO ACID WITH A HEAVY ATOM LABELLED ANALOGUE

In the search for general methods of making heavy atom derivatives, a systematic approach would be to replace specific amino acids with synthetic analogues containing heavy atoms. The heavy atom analogues must closely resemble the amino acid residue replaced both in shape as well as charge. The number of heavy atoms which can be used is rather restricted by the fact that few form stable covalent bonds with carbon, nitrogen or oxygen atoms unless they are highly coordinated and thus rather bulky. The exceptions seem to be mercury, iodine or selenium.

Mercury is very suitable as it forms covalent, linear two coordinated compounds and may be substituted on aromatic rings. Iodine is also quite useful in forming aromatic substituted derivatives but aliphatic iodine compounds tend to be rather unstable especially in an X-ray beam. Selenium is similar to sulphur in its chemistry and could be introduced as a cysteine or methionine analogue. Although these substitutions seem an attractive idea on paper, they are in fact extremely difficult to achieve in the laboratory.

The most straightforward approach appears to be replacement of the N-terminal amino acid after removing it by the Edman procedure. This has been attempted for insulin by Borras—Cuesta (1972) and by Brandenburg (1969). The N-terminal amino acid in the B chain of insulin is phenylalanine. As this was known to be unimportant for either tertiary structure or biological activity, these workers attempted to replace it with a p-iodophenylthiocarbonyl group and with p-iodophenylalanine. The method was made technically difficult by the need for gentle chemical reactions which do not destroy other bonds or denature this multichain protein. Extensive masking of other amino groups is also required. Unfortunately, after this very careful organic chemistry, the modified insulin failed to crystallize isomorphously.

To the present time there seem to be no examples of the successful use of this method, but it may have advantages for small proteins and peptides where other methods of making heavy atom derivatives are unsuccessful and where, unlike insulin, there is a single polypeptide chain.

8.5 DIRECT BINDING OF THE HEAVY ATOM SALTS

We have seen that the reaction of heavy atom salts with amino acid side chains on the surface of the protein is likely to be rather unspecific. This lack of specificity precludes generalizations but nevertheless some general guidelines can be given which allow the protein crystallographer to maximize the chances of preparing a series of derivatives with different heavy atom sites. For a background to this discussion the reader should refer to Phillips and Williams, (1966); some very good general discussion is also included in Thomson *et al.* (1972).

8.5.1 Thermodynamic stability of metal complexes

Let us first consider the potential ligands. These include not only the functional groups of the protein but also the ions of the buffer salting out agent and heavy atom reagent. Ligands can be classified in terms of whether they are hard or soft (Pearson, 1963). "Hard" ligands are electronegative and form electrostatic interactions; there is little tendency to delocalize electron density and form covalent bonds with cations. They include fluoride ions, water molecules, glutamate, aspartate and terminal carboxylates, and hydroxyls of serine and threonine as well as acetate, citrate and phosphate buffer ions. In contrast soft ligands are polarizable and form covalent bonds. They include chloride, bromide, iodide, sulphur ligands, cyanide and imidazole. Methionine, cysteine, cystine and histidine are soft ligands. In fact the distinction between soft and hard ligands is rather arbitrary in some cases such as ammonia and it is better to write the ligands in series of increasing hardness.

$$I^- > Br^- > Cl^- \gg F^- \gg H_2O$$

$$RS^- > R_2S \gg NH_3 > H_2O$$

$$CN^- > R-NH_2, \quad N > Cl^- > CO_2^-, \text{alcohol}-OH$$

We can now classify the metal ions according to their preference for hard ligands (on the right) or soft ligands (on the left). Class (a) metal ions tend to bind preferentially to the hard ligands. They include the cations of the A metals such as alkali metals, the alkaline earths, the lanthanides, some actinides and group III A, IV A and V A transition metals. In contrast class (b) metal ions are fairly soft and polarizable so that they can form covalent bonds with the soft ligands. They include the heavy metals at the end of the transition series such as Pt, Au, Hg. Their complexes are covalent and often anionic for example $Pt(CN)_4^{2-}$ $Au(CN)_2^-$, $PtCl_4^{2-}$ and HgI_4^{2-}. The most stable ions are formed with the softest ligands so that methionine, cysteine and imidazole will displace chloride from $PtCl_4^{2-}$ but not from $Pt(CN)_4^{2-}$. The positions of class (a) and class (b) cations in the periodic table are shown in Fig. 8.4. From this it can be seen that class (b)

8.5.1. THERMODYNAMIC STABILITY OF METAL COMPLEXES

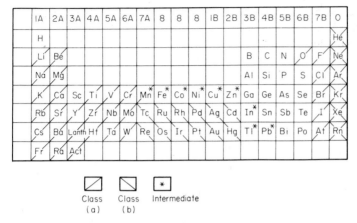

FIG. 8.4 The classification of elements according to their class (a) or class (b) character.

metal ions are at the end of the transition series and therefore have outer shells of 'd' electrons which are quite polarizable or soft. In contrast class (a) metal ions have the inert gas configuration and the electrons are held very strongly and are not polarizable: the ions are hard. In the middle and towards the end of the first transition series there are a number of ions which have intermediate properties between class (a) and class (b). These can be placed in order of increasing class (b) character:

$$Fe^{2+} < Co^{2+} < Ni^{2+} < Cu^{2+} < Zn^{2+}$$

All these ions can be found in nature with either soft or hard ligands. For instance, Zn^{2+} the ion with the most class (b) character is usually found with soft sulphur or imidazole ligands but carboxylate or water ligands are also found. The metal ions Tl^+ and Pb^{2+} are also intermediate in their characters as they have the so called "inert pair" of electrons in their outer shell. They tend to be more class (a) than class (b) and prefer coordination by carboxylate rather than imidazole or sulphur ligands. The classification of metal ions as class (a), class (b) or intermediate gives us a useful way of rationalizing the thermodynamic stability of metal ion complexes. Given that a protein has a series of different side chains and main chain groups we can use the classification as an initial guide to the kind of complexes which we would expect to find when metal ions bind to proteins. Class (a) metals will tend to bind to hydroxyl and carboxylate groups, water molecules and acetate, citrate or phosphate of the buffer medium. Class (b) metals, on the other hand, will tend to bind methionine, cysteine, cystine, imidazole or amino groups and will be very susceptible to chloride and ammonia in the buffer solution. The intermediate type ions will tend to be less specific in their interactions.

These generalizations only apply to equilibrium involving free ligands. Where a number of ligands can form a multidentate or chelating system, the specificity is lower. Thus histidine forms a bidentate complex:

$$\begin{array}{c}\text{histidine-Pt bidentate chelate complex}\end{array}$$

10

through the soft ligands imidazole and amino groups and as expected leaves the carboxylate group free. However, glycine also forms a bidentate complex but with no sidechain group available it chelates through the amino and the carboxylate groups. Thus although the carboxylate group is quite hard it will bind platinum in a chelating system as this is then entropically favourable. In fact most interactions of metal ions with proteins are through chelation and this major reason why the reactions used in making heavy atom derivatives lack specificity.

A second reason for the lack of specificity is the interaction of a metal anion such as $Pt(CN)_4^{2-}$ with a cationic side chain. With the thermo-dynamically stable cyanide complexes in particular this must be the major mode of interaction and it is not surprising to find that the binding of cyanide and halogen complexes of class (b) metals often differ.

8.5.2 Lability of metal complexes

The thermodynamic stability of a metal complex does not tell us how fast ligands enter or leave the complex. The rate is very important in the preparation of heavy atom derivatives, and will be of special significance in the case of the covalent complexes of platinum, gold and mercury.

The reaction may involve the loss of a ligand (which is rate determining) followed immediately by the addition of a further ligand. This reaction would be first order, $S_N 1$. Some complexes will undergo attack in a second order $S_N 2$ reaction, and this is the case for many square planar complexes such as $PtCl_4^{2-}$. For $S_N 1$ reactions the rate depends only on the leaving group whereas $S_N 2$ reactions depend on the attacking nucleophile as well as the leaving group. Let us postulate an $S_N 2$ reaction mechanism for a complex PtL_4^{2-}, i.e.

$$PtL_4^{2-} \xrightarrow{X'} PtL_4 X^{3-} \longrightarrow PtL_3 X^{2-} + L'$$

8.5.4. EFFECT OF pH

The relative rates of attack are $RS^- > I^- > Br^- > -N > Cl^- > RO^-$. The rate of leaving follows the order

$$H_2O > Cl^- > NO_2^- > CN^-$$

Thus the protein crystallographer could slow down nucleophilic attack by changing from $PtCl_4^{2-}$ to $Pt(NO_2)_4^{2-}$. The rate of leaving is also determined by the group trans in the square planar complex. Polarizable ligands such as I' increase the rate of leaving of the group trans to them whereas NH_3 does not have this affect. This means that once an ammonia complex is formed it is difficult to displace the groups trans to the ammonia.

Sulphur ligands like methionine are not only poor leaving groups but also good nucleophiles. They form thermodynamically stable complexes.

The rate also depends on the metal ion. For instance $PtCl_4^{2-}$, $AuCl_4^-$ and $PdCl_4^{2-}$ are all square planar complexes, and their relative reactivities with different reagents are similar. However the rates are

$$PdCl_4^{2-} > PtCl_4^{2-} > AuCl_4^-$$

Thus if the reaction is too slow with $PtCl_4^{2-}$, it may be useful to try $PdCl_2^{2-}$ instead. Also $AuCl_4^-$ may react more slowly if the reaction is disadvantageously fast, but this anion can easily be reduced and is somewhat less stable than platinochloride. Its reactions will depend on the amount of chloride ion in solution.

8.5.3 Oxidation states of metal ions in protein crystals

Williams and his coworkers have drawn attention to the importance of the relative stabilities of oxidation state in the biological media (Thomson et al., 1972). In particular the complexes of transition metals easily undergo rapid oxidation-reduction. This is also true of protein solutions and crystals and generally the following oxidation states will tend to dominate:

Os(II), Ru(II), Ir(III), Rh(III), Pt(II), Pd(II), Au(I)

The relative stabilities of the oxidation states will depend critically on the ligand but addition of heavy atoms in higher oxidation states can lead to reduction.

8.5.4 Effect of pH

The properties of the protein ligands available are also dependent on the pH. The proton will compete with the metals for binding to the ligand. Thus it is important to consider the effect of pH on both the thermodynamic and kinetic stability of different potential ligands.

Table 2.I (p. 22) gives the pK values of amino acids. A wider range of values is likely to be found in real proteins. Of the hard ligands, carboxylate groups of glutamate or aspartate will be protonated at a higher pH than the aliphatic hydroxyls of serine or threonine. Thus although carboxylate groups will generally bind cations such as Sm^{3+} or Pb^{2+} more quickly and strongly above pH 3.5, the strength of binding will be strongly decreased below this pH unless the carboxylate has an unusually depressed pK. Thus the number and occupancy of sites can be decreased by lowering the pH. Sites involving serine or threonine oxygens then may be relatively strengthened.

A similar effect on the rates of reaction would be expected at higher pH for the softer ligands. For instance the nucleophilicity of histidine increases in the region pH 6.5 when it loses its proton and positive charge; the nucleophilicity of cysteine also increases sharply when it becomes ionized at about pH 8.3. The attacking groups have the order

$$RS^- > R_2S > RSH$$

Methionine is a better nucleophile at pH 6.0 than cysteine; but at pH 8.5 the cysteine would have lost its proton and become much more effective.

8.5.5 The effect of salting-out agents and buffers

Salting-out agents and buffers are a potential source of alternative ligands for the heavy atom reagent. They may complex or precipitate the metal and consequently interfere with the reaction of the protein. This was first realized by Bluhm et al., (1958) and discussed for $PtCl_4^{2-}$ by Sigler and Blow (1965). Of the salting-out agents, ammonium sulphate has most commonly been observed to interfere with the reaction of heavy atoms. The ammonium ion can dissociate at higher pH increasing the amount of ammonia available. The nucleophilic strength is in the order,

$$NH_3 > Cl^- > H_2O$$

If there is an excess of ammonia, it will react:

$$PtCl_4^{2-} \longrightarrow cis\ PtCl_2(NH_3)_2 \longrightarrow Pt(NH_3)_4^{2+}$$

The cationic platinum ammonia complex, $Pt(NH_3)_4^{2+}$, is less susceptible to reaction due to the low trans effect of ammonia. Reactions also occur with palladium, silver, gold and mercury complexes. The occurrence of this reaction may be decreased by lowering the pH (preferably to about pH 4) when NH_4^+ will be formed but this may also decrease the nucleophilicity of the protein ligands.

In order to avoid the effect of substitution by ammonia, many protein crystallographers transfer their crystals to sodium/potassium phosphate or magnesium sulphate (Sigler and Blow, 1965). But this is also a nuisance in some respects. An excess of phosphate will also tend to displace chloride from $PtCl_4^{2-}$,

TABLE 8.IV
The solubility of metal ions

	Acetate	Cl$^-$	Citrate	OH$^-$	NO$_3^-$	O^{2-}	PO$_4^{3-}$	SO$_4^{2-}$
Zn^{2+}	30	262	V.S.	i	327	0.00016	i	86.5
Ba^{2+}	76.4	31		5.6	S		i	i
Cd^{2+}	V.S.	140		0.0026	S		i	75.5
Ce^{3+}	26.5	100	i	i		i	i	
Au$^+$				S		i		
Pb^{2+}	45.6	0.67		0.0155		0.0017	0.000014	0.00425
Hg^{2+}	25	3.6				0.0052	i	d
Hg$_2^{2+}$	0.75	0.00021				0.0051	i	0.06
Ag$^+$	0.72	0.000089	0.028		122	0.0013	0.00065	0.57
Tl$^+$	V.S.	0.32		25.9	9.55	1.48	0.5	4.87
UO$_2^{2+}$	7.69	320			170		i	

The solubilities are in grams/100 ml. Where previous figures are unavailable: i = insoluble, S = soluble, V.S. = very soluble, d = decomposes.

which changes charge of the anion (see Section 8.6.2). With most cations [especially class (a)], phosphate (and sulphate) forms insoluble complexes (see Table 8.IV) Acetate will also complex class (a) metal ions such as uranyl or the lanthanide ions and decrease their reactivity. Citrate, a chelating anion, binds even more strongly and can totally inhibit the binding of these cations in some cases. Tris appears to be the buffer ion which has the least effect on the reactivity of cations; its complexes with cations are not particularly stable.

As the reactions of buffer and salting-out agent are rather complicated and depend critically on the nature of the heavy atom reagent, we consider them further in section 8.6.

8.5.6 The hydrophobicity of the ion

Many complexes are hydrophilic, and will not penetrate into the interior of the protein. They will most probably bind to the surface residues of the protein. Some groups may be more hydrophobic. For instance the mercury reagents PhHgCl and EtHgCl or the platinum complex $Pt(NH_3)_2Cl_2$ can penetrate the core of the protein and thus bind groups which are inaccessible to reagents such as Hg^{2+} or $PtCl_4^{2-}$. Even charged groups might penetrate into hydrophobic pockets especially if they contain polarizable iodine atoms.

8.5.7 Effect of concentration of reagent and temperature in heavy atom binding

The highest concentration of the metal ion reagent that can be used will depend on the stability of the crystal structure with respect to the ionic strength. As many proteins are crystallized by salting out, usually this will not present problems. However, for proteins which are crystallized by decreasing the ionic strength it may give rise to difficulties which can only be overcome by cross-linking the protein molecules (see Section 8.5.8). Most heavy atom derivatives have been prepared by using dilute solutions of metal reagents at between 0.0001 M and 0.01 M. Dilute solutions will favour the more tightly bound heavy atom sites and decrease non-specific interactions. They minimize the amount of heavy atom reagent in the lattice which may otherwise give rise to absorption of X-rays without contributing to the diffraction pattern except at low sin θ values.

Sometimes increasing the concentration will give rise to specific binding at a further site. This was found to be the case in the analysis of rhombohedral insulin crystals. A solution of lead acetate at 0.01M concentration gave a derivative with one major lead site. Increase of the concentration to 0.1M led to increase of the occupation of a minor site from 10% to about 60%; and this was utilized to provide a further derivative (Adams et al., 1969).

Often an increase of the heavy atom reagent concentration will decrease the time of soaking required without changing the nature of the substitution. If the heavy atom reagent causes denaturation of the protein it may be advantageous to soak the protein for a short time at a higher concentration.

The temperature also changes the rate of reaction and sometimes the degree of binding. For rhombohedral insulin, uranyl acetate reacted 20 times more slowly at 4°C than at room temperature and this allowed the more effective control of the extent of substitution (Blundell, unpublished results). Conversely in study of cytochrome c_2 heavy atom reagents did not bind at 4°C, but at 37°C several heavy derivatives were easily prepared (Salemme *et al.*, 1973)

8.5.8 Crosslinking protein crystals

Although it is desirable to change the pH, salt concentration and buffer of crystals in order to facilitate reaction of various heavy atom salts, this may not always be possible.

If the conditions are changed the crystals will often become fragile and cracked, and may even finally dissolve. In other cases the cell dimensions may change. Quiocho and Richards (1964) found that treatment of carboxypeptidase crystals with glutaraldehyde

$$OCH.CH_2.CH_2.CH_2.CHO$$

resulted in the crystals having greater mechanical strength and total insolubility in deionized water. The bifunctional agent appears to crosslink the protein molecules in the crystal lattice by reacting with adjacent lysines. The reagent in solution forms a cyclic hemiacetal of its hydrate which polymerizes. The exact chemical nature of the reaction is unclear (Richards and Knowles, 1968), although it probably proceeds through the free reagent, the concentration of which increases with temperature (Korn *et al.*, 1972).

Crosslinking studies have been useful in the studies of carboxypeptidase A which is crystallized at low ionic strength (Lipscomb *et al.*, 1966). Calf rennin, which is crystallized from 2M NaCl, can also be crosslinked and transferred into salt free solution. Under these conditions reaction with heavy atom reagents is much more extensive than at high salt concentration (Bunn *et al.*, 1971). Many protein crystals are slowly disordered by crosslinking. This is true of insulin crystals (Blundell and Dodson, unpublished results). However, it appears that the crosslinking reaction occurs mainly at the surface of the insulin crystal in a short time and this is sufficient to make them stable under a wide range of conditions. Other larger bifunctional aldehydes such as adipaldehyde give less disorder.

8.6 RESULT OF DIRECT BINDING OF HEAVY ATOM REAGENTS

We will now consider the binding of heavy atom reagents to protein crystals. These are summarized in Tables 8.V–8.X. Where references are given in the

TABLE 8.V
The binding of uranyl, lanthanide, plumbous and

Protein	Reagent	Conc. of reagent	Buffer	pH
Ribonuclease s	UO_2Ac_2	3 mM	3.2 M AS 0.1 M acetate	5.5
Insulin	UO_2Ac_2	1 mM	0.05 M acetate 0.01 M $ZnAc_2$	6.3
Cytochrome b_5	UO_2Ac_2	100 mM	4 M AS 0.1 M tris	7.5
Ferricytochrome c_2 (Rhodospirullum rubrum)	$UO_2(NO_3)_2$ at 37°C	10 × protein conc.	3 M AS	5.8
Bacterial ferredoxin	$UO_2(NO_3)_2$	100 mM	3.3 MAS 0.7 M Tris/HCl	7.5
Rubredoxin	$UO_2(NO_3)_2$	~100 mM	3.5 M AS	4
Tosyl elastase	$UO_2(NO_3)_2$	5 mM	1–2 M Na_2SO_4 0.01 NaAc	5.0
Human lysozyme	$UO_2(NO_3)_2$	10 mM	3 M NaCl 0.02 M NaAc	
Prealbumin	$UO_2(NO_3)_2$	10 mM	3.1 M AS	5.0
Lysozyme (hen egg white)	$UO_2(NO_3)_2$		0.85 M NaCl	4.7
α-Chymotrypsin	$UO_2(NO_3)_2$			3.6
Concanavalin A	$UO_2(NO_3)_2$	0.1 mM		
Ribonuclease A	$K_3UO_2F_5$		40% ethanol	5.2
Insulin	$K_3UO_2F_5$	1 mM	0.05 M citrate 0.01 M $ZnAc_2$	6.3
Tosyl elastase	$K_3UO_2F_5$	30 mM	1.2 M Na_2SO_4 2 0.01 M NaAc	5.0
Human lysozyme	$K_3UO_2F_5$	3 mM	3 M NaCl 0.02 M NaAc	4.5

8.6. RESULT OF DIRECT BINDING OF HEAVY ATOM REAGENTS

thallous cations and their complexes to protein crystals

Time of soak	Site number	Z	Binding site	Authors' reference
	1	58		Wyckoff et al. 1967b
	2	20		
	+5 minor		Clusters of sites	
3 or 4 hours 18°C	1		Bl3 Glu, Bl3 Glu'	Adams et al. 1969
3 or 4 days 4°C	2			
	3		Cluster of sites	
	4			
	1	57	Asp 66; Glu 48	Mathews et al. 1971
	2	50	Glu 78	
	3	26	Asn 11	
	4	30	Asp 83, Lys 86; Glu 38, Glu 30	
	5	32	Glu 43, Glu 37	
	6	38	Glu 44	
7–14 days	1	49	Glu 64	Salemme et al. 1973
	2	52	Thr 63	
	3	27	Lys, 97	
	4	13	Gly 37, Lys 112	
13 days	12 sites		Clusters of sites, some common with Yb^{3+} and Sm^{3+}	Adman et al. 1973
14 days	6 sites		Most highly occupied close to Asp 47. Cluster of 4 sites	Herriott et al. 1970
14 days	1	92	Glu-70, Glu-80, Tyr 82, Val 67, Leu 73 (Mainchain CO)	Watson et al. 1970
	1	33		Blake and Swan 1971
1 day	6 major sites		Between 2 Glu one on each of two subunits	Geisow 1975
	5 sites		Asp and Glu carboxylates Thr 89	Blake, 1968
			Major site. Glu 21, Asp 153	Tulinsky, 1974
5 days	1	40	Asp 80, Asp 83	Reeke, 1974
	2	5	Not close to any sidechains	
	1	48		Carlisle et al. 1974
	2	35		
	3	23		
7 days	1		2 sites close together	Adams et al. 1969
	2		Glu B13, Glu B13' (similar to $UO_2 Ac_2$ but lower occupancy)	
14 days	1		Glu 70, Glu 80, Tyr 82, Val 67, Leu 73 (mainchain). (Same as $UO_2(NO_3)_2$ but lower occupancy)	Watson et al. 1970
	1	34		Blake and Swan, 1971
	2	19		

TABLE 8.V

Protein	Reagent	Conc. of reagent	Buffer	pH
Lysozyme (hen egg white)	$K_3UO_2F_5$		0.85 M NaCl	4.7
High potential Iron protein	$K_3UO_2F_5$	10 mM	3.2 M AS	6.5
Insulin	$SmAc_3$ $GdAc_3$ $DyCl_3$	1 mM	0.05 M NaAc 0.01 M $ZnAc_2$	6.2
Bacterial ferredoxin	$Sm(NO_3)_3$ $PrCl_3$	40 mM	3.3 M AS 0.7 M Tris/HCl	7.5
Flavodoxin clostridium MP	$SmCl_3$	20 mM	2.5 M AS 0.1 M Tris	6.8
Thermolysin	$LaCl_3$ $EuCl_3$ $ErCl_3$ $LuCl_3$		tris/acetate	5.5
Flavodoxin clostridium MP Oxidized	$SmCl_3$		2.5 M AS	6.8
Concanavalin A	$Sm(NO_3)_3$	10 mM		
Insulin (Apo protein)	$PbAc_2$	10 mM	0.05 M acetate	6.3
Insulin	$PbAc_2$	100 mM	0.05 M acetate	6.3
Carboxypeptidase	$PbCl_2$	3 mM	0.01 Na citrate 0.02 M tris	7.5
Concanavalin A	$Pb(NO_3)_2$		0.3 M $NaNO_3$ 0.01 M Sodium maleate	6.8
Subtilisin novo	TlF		0.05 M glycine/ NaOH 55% acetone	9.1
Insulin	TlAc	10 mM	0.05 M NaAc 0.01 M $ZnAc_2$	6.3

8.6. RESULT OF DIRECT BINDING OF HEAVY ATOM REAGENTS

(continued)

Time of soak	Site number	Z	Binding site	Authors' reference
	1		Asp and Glu groups; (same as two largest $UO_2(NO_3)_2$ sites)	Blake, 1968
	2			
7 days	1	6		Carter et al. 1974b
	2	1.9		
1 day	1		Glu B13, Glu B13' (same as $UO_2 Ac_2$ major site)	Blundell, 1970
3 days	~10		Many sites clustered and some common with $UO_2(NO_3)_2$	Adman et al. 1973
	1	46	Carboxylate ?	Anderson et al. 1972
	2	13		
	1	23	Ca^{2+} double site	Colman et al. 1972b
1 hour	2	23	Asp 57 See Table 8.I	
	3	22	Asp 200	
			3 sites close together	Burnett et al. 1974
1 day	1	30	Glu 87, Asp 136, Asn 131,	Reeke, 1974
	2	22	Gln 132, Gly 152, Asp 80,	
	3	9	Asn 82, Asp 83	
1 day	1	57	B13 Glu, B13 Glu	Adams et al. 1969
	2		His B10	
	3		His B10	
	4		N terminus B1, A17 Glu	
1 day	4 sites		As at 10 mM but site 4 is higher in occupation	Adams et al. 1969
40 days	1	58	Glu 270	Lipscomb et al. 1970
	2	53	Citrate, not protein	
	1	64	Gln 87, Asp 136, Asp 80,	Edelman et al. 1972
	2	63	Lys 82, Asp 83	
	1	11.7	Asp 197	Hol, 1971
7 days			B13 Glu, B13 Glu (same as uranyl but lower occupancy)	Blundell, 1970

tables, they are not given in the text. The chemical formula and structures of mercurials named in abbreviated form in the text are given in full in Table 8.VI.

8.6.1 Reaction of class (a) or hard cations

Let us first consider the binding of class (a) cations to proteins. We will be concerned mainly with lanthanide and actinide ions. (see Figure 8.4 and Table 8.V.)

The most commonly used cation is the uranyl ion. This is a hard cation with a linear geometry

$$(O=U=O)^{2+}$$

The oxygens are covalently bound but further coordination is predominantly ionic and occurs in the plane perpendicular to the uranyl-oxygen bonds. Oxygen or fluorine ligands are most common. Table 8.V shows that UO_2^{2+} binds carboxylate groups of glutamate or aspartate, and occasionally hydroxyl sidechains of threonine or serine as expected of a hard cation. Uranyl acetate and nitrate have been used successfully; nitrate coordinates less strongly to the uranyl and its salt tends to give more extreme effects. Ribonuclease-S, rubredoxin and insulin give multisite uranyl derivatives in which there is much variation in occupancy between sites. Quite often sites are clustered together having low occupancy. This sometimes presents problems in the determination of the positions, so that a high resolution study is required. The reaction of the uranyl cation can be modified by complexing it with a hard ligand such as fluoride in $(UO_2F_5)^{3-}$ or acetate in $(UO_2Ac_3)'$. In lysozyme, elastase and insulin the binding of the uranylfluoride anion is close to sites occupied by other uranyl salts, but the binding is less extensive. The carboxylate ions almost certainly coordinate by displacing the fluorine ligands.

Lanthanides also bind to carboxylate sidechains (Table 8.V). The lanthanides show a decreasing ionic radius from La^{3+} to Lu^{3+}: this is the lanthanide contraction (see Table 8.II). The most stable oxidation state for all the lanthanides is III. Samarium in the form of its acetate binds to two glutamate carboxylate groups in insulin. Similar substitution is given by gadolinium and dysprosium. In flavodoxin, samarium binds close to an aspartate and in lysozyme, europium and gadolinium bind between the active site glutamate and aspartate groups. The difference in specificity of the lanthanides is mainly due to difference in size; heavier and smaller lanthanides may often bind where others will not. This has been neatly illustrated in the work on thermolysin where lanthanides bind at calcium sites (see Section 8.2).

We have seen that the B metals, thallium and lead, are intermediate between class (a) and class (b) in character. Tl^+ and Pb^{2+} form many salts which are isomorphous with alkali metal and alkaline earth salts. In fact Tl^+ is often found

8.6.2. COVALENT REACTION OF CLASS (B) IONS

along with Rb^+ and Pb^{2+} is found with Ba^{2+} in minerals. In proteins these metal ions often bind as class (a) or hard cations (Table 8.V.) In insulin the major site for both Pb^{2+} and Tl^+ is at two glutamates in the same way as Sm^{3+} or UO_2^{2+} ions. In carboxypeptidase, Pb^{2+} binds to glutamate but in concanavalin it binds to glutamine and aspartic acid. Tl^+ can replace Na^+ and K^+ in subtilisin Novo at a site which has one aspartate residue and a number of other oxygen and nitrogen atoms.

The intermediate character of the Pb^{2+} ions compared to uranyl and lanthanide ions is shown by the fact that Pb^{2+} binds to a small extent with the imidazoles which bind zinc in zinc free insulin and to a terminal amino group when the lead is at high concentration.

In conclusion, of the class (a) cations the lanthanides appear to be more selective than uranyl ions. In contrast to uranyl which often gives multisite substitution, samarium (and lead) may give one major site. In insulin the extent and rate of lead binding was found to be very dependent on the concentration of ions and the temperature.

The use of uranyl and lanthanide ions is sometimes prevented by the insolubility of their phosphate and hydroxides (see Table 8.IV). However, they do not bind ammonia strongly: changing from ammonium sulphate to phosphate may be advantageous in the preparation of heavy atom derivatives of platinum and mercury but it represents a step backwards in the preparation of derivatives of lanthanide and uranyl ions. Chelating agents such as citrate will bind the metal ions and inhibit the binding; for instance uranyl acetate and samarium acetate show a very slow reaction with insulin in citrate buffer. The most suitable buffer is probably acetate or tris.

8.6.2 Covalent reaction of class (b) ions and their complexes

Class (b) ions of platinum, mercury and gold are polarizable and form covalent complexes. We have seen in Section 8.5.1, that they will tend to bind to soft ligands such as sulphydryl, imidazole, thioethers rather than to harder ligands such as carboxylate or hydroxyl groups. Their chemistry has been reviewed by Blake (1968), Dickerson et al., (1969) Thomson et al., (1972) and by Petsko (1973). Their chemistry is dependent on the structure and charge of the complexes and the nature of the buffer, salting-out agent and pH (Sections 8.5.4 and 8.5.5). Here we will review the result of binding studies to proteins and relate these to the simple theories already outlined.

8.6.2.1 Mercuric reagents

The high binding constants of mercury compounds to sulphydryl groups were the basis of the first use of isomorphous replacement by Perutz and his colleagues (Green et al., 1954). They used the different reactions of the sulphydryls in the α- and β-chains of haemoglobin to prepare a series of heavy

TABLE 8.VI

Mercurials used in preparing heavy atom derivatives

Name of mercurial	Chemical formula	References to use in protein crystallography
PCMB p-chloromercuribenzoate	Cl–Hg–C$_6$H$_4$–CO$_2'$	Haemoglobin (Perutz et al. 1968) Papain (Drenth et al. 1968) LDH (Adams et al. 1969)
PHMB p-hydroxymercuribenzoate	HO–Hg–C$_6$H$_4$–CO$_2'$	
PCMBS p-chloromercuri benzene sulphonate	Cl–Hg–C$_6$H$_4$–SO$_3'$	Papain (Drenth et al. 1968) Lysozyme (Blake et al. 1962) Myoglobin (Kendrew, 1962)
PHMA p-hydroxymercuri-aniline	HO–Hg–C$_6$H$_4$–NH$_2$	Papain (Drenth et al. 1968)
PCMA p-chloromercuri-aniline	Cl–Hg–C$_6$H$_4$–NH$_2$	Myoglobin (Bluhm et al. 1958; Kendrew, 1962; Banaszak et al. 1965)
PHMBS p-hydroxymercuri-benzene-sulphonate	HO–Hg–C$_6$H$_4$–SO$_3'$	Bence Jones Dimer (Schiffer et al. 1973) MDH (Hill et al. 1972)
MSSS 3-hydro-mercuri-5-sulpho-salicylic acid	HO–Hg–C$_6$H$_2$(OH)(COOH)–SO$_3$H	Basic trypsin inhibitor (Huber et al. 1970)

TABLE 8.VI (continued)

Name of mercurial	Chemical formula	References to use in protein crystallography
HMTS or MHTS 2-hydroxy-mercuri-toluol-4-sulphonic acid	HO—Hg—C₆H₃(CH₃)—SO₃H	Myoglobin (Bluhm et al. 1958) Lysozyme (Blake et al. 1962) Basic trypsin inhibitor (Huber et al. 1970)
HMSA 3- or 5-hydroxymercuri-salicylic acid	(HOHg)—C₆H₂(COOH)(OH)—(HgOH)	Basic trypsin inhibitor (Huber et al. 1970)
PMA Phenylmercuri acetate	C₆H₅—Hg—OOCCH₃	Chymotrypsin (Sigler et al. 1966) Fab NEW' (Poljak et al. 1973) Calcium binding protein (Kretsinger and Nockolds, 1973)
PAMA or APMA Acetamino phenyl-mercuri acetate p-acetoxymercuri-acetanilide	CH₃—C(O)—O—Hg—C₆H₄—NH—CO—CH₃	Basic trypsin inhibitor (Huber et al. 1970) Fab NEW' (Poljak et al. 1973)
EMTS Thiomersalate or thiomerosal or thiomersal or merthiolate sodium or ethyl mercuri thiosalicylate	CH₃CH₂Hg—S—C₆H₄—CO₂Na	Calcium binding protein (Kretsinger and Nockolds, 1973) LADH (Branden et al. 1973) Bence Jones Dimer (Schiffer et al. 1973)

TABLE 8.VI (*continued*)

Name of mercurial	Chemical formula	References to use in protein crystallography
Mersalate sodium mersalyl or salyrganic acid	(benzene ring)–O–CH$_2$–COONa; –CO–NH–CH$_2$–CH–CH$_2$–Hg–OH with OCH$_3$	Ferricytochrome *c* (Dickerson *et al.* 1971); Chironomus haemoglobin (Huber *et al.* 1969b)
Merethoxylline sodium	(benzene ring)–O–CH$_2$–COONa; CONHCH–CH$_2$–Hg–OH, O–CH$_2$–CH$_2$–O–CH$_3$	
Nitromersol	(benzene ring with CH$_3$, OH, HgOH, NO$_2$)	
Esidrone	(pyridine ring)–COONa; N–CO–NH–CH$_2$–CH–CH$_2$–Hg–OH with OCH$_3$	
Meroxyl Mercuhydrin	HO–Hg–CH$_2$–CH–CH$_2$–NH–CO–NH–CO–CH$_2$–CH$_2$–COOH with OCH$_3$	Fab' NEW (Poljak *et al.* 1971)
Chlormerodrin (Neohydrin) 3-chloromercuri-2-methoxy propyl urea	Cl–Hg–CH$_2$–CH–CH$_2$–NH–CO–NH$_2$ with OCH$_3$	Fab' NEW (Poljak *et al.* 1971)

TABLE 8.VI (*continued*)

Name of mercurial	Chemical formula	References to use in protein crystallography
Sodium mercaptomerin	NaOOC—⟨cyclopentyl⟩—C(=O)—NH—CH$_2$—CH—CH$_2$—Hg—S—CH$_2$—COONa, with OCH$_3$	
DMDX Diacetoxymercuri dipropylene dioxide	CH$_3$COO—Hg—⟨dioxane ring with CH$_3$, CH$_3$⟩—Hg—OOCCH$_3$	Fab' NEW (Poljak *et al.* 1971)
Mercurochrome	⟨anthracene-dione with OH, Hg, NaO, Br, Br, COONa, phenyl⟩	
DMA Dimercuriacetate	NO$_3$—Hg NO$_3$—Hg ⟩CHCOO'	LDH (Adams *et al.* 1969) Haemoglobin (Perutz *et al.* 1968) Thermolysin (Colman *et al.* 1972b)
Baker's dimercurial or 1,4-diacetoxymercuri-2,3-dimethoxybutane	CH$_3$COO—Hg—CH$_2$—CH(OCH$_3$) CH$_3$COO—Hg—CH$_2$CH(OCH$_3$)	Haemoglobin (Perutz *et al.* 1968) LDH (Adams *et al.* 1969) Fab' NEW (Poljak *et al.* 1973)

atom derivatives. The first sulphydryl reagent used was PCMB(p-chlormercuribenzoate; see Table 8.VI) which attached to the more reactive β-chain sulphydryl residues. Derivatives involving the α-chain binding sites were produced by first blocking the β-chain sulphydryls with iodoacetamide and then crystallizing in the presence of mercuric acetate. Further derivatives were prepared by the use of sulphydryl reagents such as dimercuriacetate (DMA) and 1,4-diacetoxy mercuri-2,3-dimethoxybutane (Baker's mercurial) which contain two mercury atoms, found crystallographically to have their mercury atoms separated by 1.7Å and 4.9Å respectively. An equally sophisticated approach was used by Rossmann and his coworkers in their study of lactate dehydrogenase. This enzyme has two sulphydryl groups reactive with PCMB, $HgCl_2$ and DMA, but only one reactive with the more bulky reagent Baker's mercurial. All the derivatives used for calcium binding protein were mercury derivatives — EtHgCl, $HgBr_2$ and CMMPU — which bind to one cysteine sulphydryl. Lists of possible sulphydryl reagents are given in Table 8.VI.

Most of these reagents have a covalent mercury—carbon bond which is not easily broken. The chloride, acetate or nitrate ligands are not bound strongly; the mercury will have a positive charge if these become dissociated in solution. The mercury atom will then be particularly reactive to the negatively charged and polarizable $-S^-$ groups. The cysteines will be less reactive at lower pH when the sulphydryl is protonated. Ammonia will complex the mercury but will not displace carbon substituents and of course will never change the charge. An excess of chloride (as precipitating agent) may slow the rate of reaction by complexing with the mercury and giving it a net negative charge; this has been observed with calf rennin which is crystallized from 2M NaCl.

The $-SMe$ of methionine is less nucleophilic than the sulphydryl group, $-S^-$, and any way has no negative charge. This probably explains the fact it rarely binds mercury reagents. One example is the binding of HgI_4^{2-} to rubredoxin.

Sulphur atoms are not the only possible ligands for mercurials. In fact histidines very often bind mercury reagents. Imidazole becomes a very good ligand above pH 6—7 when it looses its proton. For instance, Table 8.VII shows that thermolysin, which has no methionine or cysteine, binds DMA, mercuri-succinimide and $HgCl_2$ through histidines. The mercury-containing compound, mersalyl, has been called a histidine specific reagent as it binds to a histidine in cytochrome c and subtilisin BPN'. However, neither of these proteins have sulphydryl groups. In calcium binding protein, mersalyl binds to a sulphydryl group. The specificity, if any, of mersalyl is most likely due to its large size.

The specificity of mercurials is in fact very often due to their size, shape and substituent groups. We have seen that in lactate dehydrogenase, the bulky Baker's mercurial binds only one sulphydryl while smaller reagents bind two. The binding of mercurials to the immunogloblin fragment Fab' New was also studied by varying the nature of the reagent until ones were found which bound specifically. In alcohol dehydrogenase, which contains fourteen sulphydryls,

8.6.2. COVALENT REACTION OF CLASS (B) IONS

most mercurials denatured the protein but one, thiomersal, gave a more specific reaction leading to a good derivative. Variation of the nature and size of the mercurial seems to be a very useful way of producing a good heavy atom derivative.

The kind of interaction which may give rise to binding of mercurials to proteins was evident from the early studies on myoglobin and lysozyme. In myoglobin, PCMBS binds to a histidine. There is also an important interaction with the sulphonate group of PCMBS. Figure 8.5 shows that its negative charge interacts with the positive charge of a lysine on a neighbouring molecule in the lattice. PCMA does not bind at this site; this has no negative charge. In fact this kind of interaction may easily give rise to binding of PCMBS in a way which does not involve the mercury atom at all. In lysozyme, PCMBS binds only through the sulphonate!

Mercury can also be used in an anionic form by complexing it with halides to give $HgCl_4^{2-}$, $HgBr_4^{2-}$ or HgI_4^{2-}, or with thiocyanate to give $Hg(SCN)_4^{2-}$. Despite the fact that the anions have a negative charge the mercury can still become bound by the negatively charged cysteine as in lamprey haemoglobin. This is probably due to the dissociation of the complex. Thus HgI_4^{2-} dissociates to give rise to HgI_3^-, HgI_2 and I^- in solution, and the reaction may be through an uncharged species. On the other hand the anionic forms can interact electrostatically as in lysozyme or in the minor sites of ferrocytochrome and myoglobin or through hydrophobic interactions as in myoglobin or chironomus haemoglobin. This is discussed in Sections 8.6.3 and 8.6.4.

8.6.2.2 Mercuration of the disulphide bond

Steinberg and Sperling (1967) have suggested that heavy atom derivatives may be prepared by inserting a mercury atom in the disulphide bonds of cystine. The –S–Hg–S– system formed in this way is linear but is about 3Å longer than the disulphide bond, S–S.

Nevertheless substitution into one disulphide of ribonuclease (Sperling et al., 1969) or papain (Arnon and Shapira, 1969) appears to have little effect on the conformation or biological activity, and insertion of mercury atoms into immunoglobulin fragments also had little effect on antibody binding (Steiner and Blumberg, 1971). Mercurated ribonuclease crystallized but showed cell dimension differences of up to 1% in the cell length and 2° in the monoclinic cell angle (Sperling et al., 1969). Similarly the mercurated immunoglobulin fragment, Fab' New (Poljak quoted in Steiner and Blumberg, 1971) also crystallized non-isomorphously. However, mercurated Bence–Jones dimer gave crystals isomorphous with the native which were used in the crystallographic study (Ely et al., 1973).

The mercuration of cystine may be carried out by reduction of the disulphide followed by reaction with mercuric ions (Sperling et al., 1969).

TABLE 8.VII
The binding sites of mercuric

Protein	Reagent	Conc. of reagent	Buffer	pH
Carbonic anhydrase (apoenzyme)	$HgAc_2$		2.3 M AS	8.5
Calcium binding protein	$HgAc_2$ $HgBr_2$	0.8 mM	4 M phosphate	6.8
Thermolysin	Hg (succinamide)$_2$	5 mM	5% DMS 0.01 M $CaAc_2$ 0.01 M tris/acetate	4
Thermolysin	$HgCl_2$; LiCl	1 mM 20 mM	5% DMS 0.01 M $CaAc_2$ 0.01 M tris acetate	6
Carboxypeptidase A	$HgCl_2$	0.8 mM	0.2 M LiCl 0.02 M tris	7.5
Myoglobin	$Hg(NH_3)_2^{2+}$ (HgO in AS)	equimolar with protein	3 M AS	6.5
Glycera haemoglobin	$HgAc_2$			6.8
Concanavalin	$HgAc_2$		2.1 M phosphate	6.0
Hen egg white lysozyme chloride	$HgCl_2$		0.85 M NaCl	4.7
Prealbumin	$HgAc_2$	0.5 mM	3.1 M AS	5.0
Human lysozyme	$HgAc_2$	50 mM	3 M NaCl 0.02 M NaAc	4.5
Chironomus haemoglobin	$HgAc_2$		3.75 M phosphate	7
Lamprey haemoglobin	$Hg(CN)_2$	0.5 mM	2 M phosphate 10 μM NaCN	6.8
Lactate dehydrogenase	$HgCl_2$	10x conc. of protein		
Papain (HgCl Blocked-SH)	$HgCl_2$	mM		
Haemoglobin	$HgCl_2$		1.9 M AS	7.0

8.6.2. COVALENT REACTION OF CLASS (B) IONS

reagents in protein crystals

Time of soak	Site number	Z	Binding site	Authors' reference
	1		His 93, His 95, His 117, (Zinc site)	Liljas, 1971
10–30 days	1		Cys 18	Kretsinger and Nockolds, 1973
5 days	1		His 231	Colman et al. 1972b
2 days	1		His 231	Colman et al. 1972b
150 days	1	47	His 69, Glu 72, His 196, Zinc site	Lipscomb et al. 1970
	2	46	His 120	
	3	48	His 29, Lys 84	
	4	–	His 303	
	1		His GH-1 close to Ag^+ site Asn GH-4 same as Zn^{2+} Lys A-14	Bluhm et al. 1958
	1		Cys B30	Love, 1971 Padlan and Love, 1968
	1	36	His 127	Hardman, 1974
	2	47	Met 129, His 127	
	3	39	Asp 118	
	4	68	Asp 80, Asx 83, Tyr 100, His 205	
	5	36	Lys 135	
	6	31	Tyr 12, His 205	
	1 major		Arg 14, His 15, Asn 93, Lys 96, Arg 128 (Same as $PdCl_4^{2-}$, $PtCl_6^{2-}$)	Blake, 1968
1 day	4 sites	64	Cys 10 (one per monomer)	Geisow, 1975
	1	49		Blake and Swan, 1971
	2	29		Osserman et al. 1969
	1	70	His G2, Asn G7	Huber et al. 1971
	2	16	His G19	
7 days	1	47	Three sites close to SH of Cys 141 position. i.e. side-chain occupies different positions. ($HgCN_4^{2-}$ does not bind)	Love, 1971
	2	20		
	3	6		
	1		Cys (SH)	Adams et al. 1970a
	2		Cys (SH)	
	1	43	His 159	Drenth et al. 1962,
	2	18	Asn 194	1967, 1971a,b,c
	1	80	Cys 104α (SH)	Cullis et al. 1961
	2	80	Cys 93β (SH)	

8. PREPARATION OF HEAVY ATOM DERIVATIVES

TABLE 8.VII

Protein	Reagent	Conc. of reagent	Buffer	pH
Glycera haemoglobin	Hg(Ac)$_2$	2 × protein conc. prior to crystallisation	–	–
Insulin	EtHgCl	Sat.	0.05 M acetate 0.01 M ZnAc$_2$	6.3
Calcium binding protein	EtHgCl MeHgCl	0.8 mM	4 M phosphate	6.8
Lactate dehydrogenase	MeHgNO$_3$			
Adenyl kinase	MeHgNO$_3$	0.05 mM		
Concanavalin	MeHgCl		2.1 M phosphate	6.0
Ribonuclease A	PCMBS		40% ethanol	5.2
Papain	PCMBS	10 mM	Methanol Water	9.3
Myoglobin	PCMBS	equimolar with protein	3 M AS	6.5
Lysozyme (hen egg white)	PCMBS		0.85 M NaCl	4.7
Papain	PCMB PCMA	1.5 mM	Methanol Water	9.3
Bovine pancreas basic trypsin inhibitor	PCMB	Sat.	2.25 M phosphate	10
Calcium binding protein	PCMB PCMA	0.8 mM	4 M phosphate	6.8
Myoglobin	PCMA	equimolar with protein	3 M AS	6.5
Lactate dehydrogenase	PCMB	10 × protein conc.		
Haemoglobin	PCMB			
Bovine pancreas basic trypsin inhibitor	HMSA	4 mM	2.25 M phosphate	10

8.6.2. COVALENT REACTION OF CLASS (B) IONS

(*continued*)

Time of soak	Site number	Z	Binding site	Authors' reference
	1		Cys 30 (B39)	Padlan and Love, 1974
	1		His B5	
4 days	1		Cys 18-SH	Kretsinger and Nockolds, 1973
			Cys (SH)	
5 days	1	70	Cys 25	Schulz et al. 1974
	2	21	Cys 187	
	3	12	His 36	
	1	55	Met 129, His 127	Hardman, 1974
	2	72	His 205	
	3	53	Tyr 100	
	4	57	Asp 118	
	5	33	Tyr 100	
	6	22	His 180, Gln 87, Trp 88, Trp 182	
	1	26		Carlisle et al. 1974
	2	8		
	1		His 81	Drenth et al. 1962,
	2		His 159	1968, 1971a,b,c
	1		His G14 (Hg binding) Asn H8, Lys FG2, Ser F7 (SO_3 binding)	Bluhm et al. 1958
	1	39	Sulphonate of PCMBS binds Arg 68	Blake, 1968
	1	45	His 81 (Same as PCMBS)	Drenth et al. 1962,
	2	19	His 159	1967, 1968, 1971a,b,c
	1	51	C terminus	Huber et al. 1970
	2	14	Lys 41, Tyr 10	
	3	12	N terminus	
10–30 days	1	80	Cys 18 (SH)	Kretsinger and Nockolds, 1973
	1	80	His GH1 Asn GH-4 Lys A-14	Bluhm et al. 1958
	1	80	Cys (SH)	Adams et al. 1970a
	2	12	Cys (SH)	
	1	80	Cys 93β (SH)	Cullis et al. 1961
	1	24	Asn 24, Gln 31, Lys 15	Huber et al. 1970
	2	14	N terminus	
	3	8	N terminus	
	4	8	Asn 24, Gln 31, Lys 15	
	5	6	C-terminus	

8. PREPARATION OF HEAVY ATOM DERIVATIVES

TABLE 8.VII

Protein	Reagent	Conc. of reagent	Buffer	pH
Bovine pancreas basic trypsin inhibitor	MSSS	6 mM	2.25 M phosphate	10
Bovine pancreas basic trypsin inhibitor	HMTS	6 mM	2.25 M phosphate	10
Lysozyme	HMTS		0.85 M NaCl	4.7
Haemoglobin	Baker's Dimercurial		1.9 M AS	7
Lactate dehydrogenase	Baker's Dimercurial	10 × protein conc.		
Lactate dehydrogenase	PHMB	4 × protein conc.		
Malate dehydrogenase	PHMBS	0.1 mM	2.8 M AS 0.1 M NaAc	5
α-Chymotrypsin	PMA PhHgAc		3.5 M phosphate 2–4% dioxane	4.2
Carbonic anhydrase	MMTGA	Co-crystallize		
Carbonic anhydrase	CH_3COOHg — H_2N—⟨C$_6$H$_4$⟩—SO_2NH AM SULF			
Calcium binding protein	Thiomersal PhHgAc $PhHgNO_3$ $HgPh_2$ 3-chloro mercuri-2-methoxypropyl urea	0.8 mM	4 M phosphate	6.8
Thermolysin	DMA	0.001 M	5% DMS 0.01 M $CaAc_2$ 0.01 tris/acetate	7.5
Liver alcohol dehydrogenase	Thiomersal	10^{-2} mM	Tris/HCl 0.05 M	8.4
Lactate dehydrogenase	DMA	10 × protein conc.		
Haemoglobin	DMA			
Ferricytochrome c (horse)	Mersalyl		4.6 M phosphate	6.2
Cytochome b_5	Mersalyl	0.3 mM	3 M phosphate	7.5

8.6.2. COVALENT REACTION OF CLASS (B) IONS

(*continued*)

Time of soak	Site number	Z	Binding site	Authors' reference
	1	75	N terminus	Huber *et al.* 1970
	2	48	Tyr 21, Arg 19	
	3	17	Asn 24, Gln 31, Lys 15	
	4	15	N terminus	
	5	9	Lys 46	
	1		N terminus	Huber *et al.* 1970a
			Sulphonate binding to Arg 68	Blake, 1968
	1	80	Cys 93β (SH)	Cullis *et al.* 1961 Perutz *et al.* 1968
	1	80	Cys (SH)	Adams *et al.* 1970a
		80	Cys (SH)	Adams *et al.* 1970a
2 days	3 sites		Cys (SH)	Hill *et al.* 1972
	1, 2	55	N-terminus S–S Cys 1–27	Sigler *et al.* 1966
	1		His 63, Glx 66, Gln 91	Liljas, 1971
	2		Cys 210	
	1		Asp 9, His 13	Liljas, 1971
	2		His 63, Glx 66, Gln 91	
	3		Cys 210	
	4		Arg 25, His 35 (carbonyl)	
10–30 days	1		Cys 18 (SH)	Kretsinger and Nockolds, 1973
14 days			His 231	Colman *et al.* 1972b
1 day	1		Cys 240 Cys 9	Branden and Zeppezauer, 1974
	1	13.5	Cys (SH)	Adams *et al.* 1970a
			β-Chain SH	Cullis *et al.* 1961
	1	72	His. 33	Dickerson *et al.* 1971
	1	42	Glu 48	Mathews *et al.* 1971
	2	23	Tyr 27(O) Arg 84	

TABLE 8.VII

Protein	Reagent	Conc. of reagent	Buffer	pH
Calcium binding protein	Mersalyl	0.8 mM	4 M phosphate	6.8
Subtilisin BPN'	Mersalyl	0.9 mM	2.1 M AS 0.05 M NaAc	5.9
Chironomus haemoglobin	Mersalyl		3.75 M phosphate	7
Concanavalin A	Mersalyl	0.1 mM		
Ferricytochrome c (tuna)	K_2HgI_4	0.3 mM	4 M AS	6
Rubredoxin	K_2HgI_4	High conc.	3.3 M AS	4
Papain	K_2HgI_4	5 mM	Methanol water	
Myoglobin	$KHgI_3$	Same as protein conc.	3.0 M AS	6.5
Lysozyme	K_2HgI_4		0.85 M NaCl	4.7
Chironomus haemoglobin	$K_2Hg(SCN)_4$		3.75 M phosphate	7
Lamprey haemoglobin	$K_2Hg(SCN)_4$	0.5 mM	2 M phosphate 10 μM NaCN	6.8
Glycera haemoglobin	K_2HgI_4	1.5 mM	3.2 M AS phosphate	6.8

However, it now appears that the reaction may be achieved in one stage by using the reducing mercurous ions:

$$Hg_2^{2+} + RSSR = 2(RSHg)^+ = RSHgSR + Hg^{2+}$$

In insulin this leads to a selective mercuration of one disulphide (Sperling and Steinberg, 1974).

This method of making heavy atom derivatives leads to a modified protein which must have changes of position in the two polypeptide chains linked by the disulphide, and in this respect is not very suitable for the method of

8.6.2. COVALENT REACTION OF CLASS (B) IONS

(*continued*)

Time of soak	Site number	Z	Binding site	Authors' reference
10–30 days	1		Cys 18-SH	Kretsinger and Nockolds, 1973
10–40 days	1	60	His 64	
	2	33	N terminus	
	3	21	His 64	
	1	58	His G2, Asn 67, His G19.	Huber *et al.* 1970
	2	79	Same as HgAc$_2$	
5 days	1	25	His 127, Met 129	Reeke, 1974
	2	6	His 127	
	1	18	Gln 16	Takano *et al.* 1971
			Cys-17 thioether bridge to haem	
	2	15	Gly 37, Asn 60	
14	1	49	Gly 43, Met 1	Herriot *et al.* 1970 Jensen, 1971
	1	32	N-terminus. Same as His 159	Drenth *et al.* 1971a,b,c
	2	11	IrCl$_6^{3-}$ and PtCl$_6^{2-}$	
	1	56	Next to haem in hydrophobic pocket	Bluhm *et al.* 1958 Kretsinger *et al.* 1968
	2	28	Lys FG2, Gln F6, Asn H8, Gln H4	
	1 major		Arg 13, Arg 13'	Blake, 1968
	2 minor		(2 sites 5.6 Å apart) (same as IrCl$_6^{3-}$, AuCl$_4'$, PdI$_4^{2-}$)	
	1	27	haem	Huber *et al.* 1970
	2	68	His G19	
1 day	1	80	His 73	Love, 1971
	2	40	Cys 141	
	3	32	Cys 141	
25 days	3 sites		Cys B30 (major site)	Padlan and Love, 1974

isomorphous replacement. Nevertheless, in large proteins this may be a useful way of making a specifically modified protein, and for this prize a certain lack of isomorphism may be tolerable.

8.6.2.3 Silver ions

Table 8.VIII shows the reactions of protein crystals with silver nitrate either with cysteine as in haemoglobin or more often with histidine as found in myoglobin and carboxypeptidase. The ions react in a similar way to mercuric ions such as Hg(NH$_3$)$_2^{2+}$ and probably also react as an ammonia complex Ag(NH$_3$)$_4^+$ when

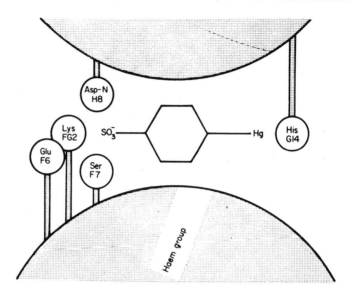

FIG. 8.5 The binding of PCMBS to myoglobin (Watson et al., 1964).

ammonium sulphate is present. Ag^+ is less polarizing and not as reactive as Hg^{2+}. This may explain why in acetate buffer Ag^+ ions give a good derivative of glucagon which is similar to the mercuric chloride derivative but shows less disorder (Blundell, 1975). In a reaction where the metal ion displaces a proton, Hg^{2+} will usually react at a lower pH than Ag^+.

TABLE 8.VIII
The binding sites of silver

Protein	Reagent	Conc. of reagent	Buffer	pH
Thermolysin	$AgNO_3$	5 mM	5% DMS 0.01 M $CaAc_2$ 0.01 M tris/Ac	5.5
Carboxypeptidase A	$AgNO_3$	5 mM	0.2 NaAc	8.0
Myoglobin	$AgNO_3$	equimolar with protein	3 M AS	6.5
Haemoglobin	$AgNO_3$		1.9 M AS	7.0
Trypsin	$AgNO_3$	12 mM	$MgSO_4$	7.5

8.6.2.4 Platinochloride, aurichloride and analogous reagents

One of the most successful reagents used to prepare heavy atom derivatives has been the platinochloride ion, $PtCl_4^{2-}$. The reaction conditions and binding sites of this and related heavy atom reagents are summarized in Table 8.IX.

We have seen in Section 8.5.1 that platinum, palladium and gold are class (b) metals: they are fairly soft, forming stable covalent complexes with soft ligands such as chloride, bromide, iodide, ammonia, imidazole and sulphur ligands. The stereochemistry of their complexes depends critically on the number of 'd' electrons. Thus d^8 ions of Pd(II), Pt(II) and Au(III) are predominantly square planar. This includes $PtCl_4^{2-}$, $Pt(NH_3)_4^{2+}$, $Pt(CN)_4^{2-}$ and $AuCl_4^-$. Occasionally these ions accept one further ligand to give a square pyramid or two ligands to give octahedral coordination but the fifth and sixth ligands are much more weakly bound. On the other hand Pt(IV) has a d^6 electron configuration and forms stable octahedral complexes such as $PtCl_6^{2-}$ with six covalently bound and equivalent ligands. However, Pt(IV) will tend to be reduced to Pt(II) in solutions containing proteins and other biological macromolecules.

In order to understand their complex protein chemistry we must consider the factors affecting the thermodynamic and kinetic stability of the complexes (Section 8.5). In particular we must consider not only the potential protein ligands but also the effect of salting-out agent, buffer and pH on the reaction. Sigler and Blow (1965) have drawn attention to the fact that NH_3 from $(NH_4)_2SO_4$ may displace chloride from the square planar $PtCl_4^{2-}$, and may alter the reaction with proteins. They suggested transferring the crystals into phosphate; with α-chymotrypsin this led to a faster and more reproduceable reaction with $PtCl_4^{2-}$. Wyckoff et al. (1967b) found further compounds in protein crystals

Time of soak	Site number	Z	Binding site	Authors' reference
40 days	1		His 231	Colman et al. 1972
	2		His 88	
30 days	1		His 166, Ser 158	Lipscomb et al. 1970
	2		His 120	
	3		His 29, Lys 84	
	4		His 303	
	1		His B-5 His GH-1	Bluhm et al. 1958
	1	80	Cys 104α (SH)	Cullis et al. 1961
	2	80	Cys 93β (SH)	
4 days	1		His 57, Asp 102	Chambers et al. 1974

TABLE 8.IX
The binding sites of palladium, platinum

Protein	Reagent	Conc. of reagent	Buffer	pH
Concanavalin A	K_2PtCl_4	0.5 mM	2.1 M phosphate	6.0
Chironomus haemoglobin	K_2PtCl_4		3.75 M phosphate	7.0
Chironomus haemoglobin	$Pt(NO_2)_2(NH_3)_2$		3.75 M phosphate	7.0
Ribonuclease S	Platinum ethylene diamine chloride	2 mM	3.2 M AS	8
Ribonuclease S	Platinum ethylene diamine chloride	2 mM	3.2 M AS	5.5
Lactate dehydrogenase	Platinum ethylene diamine chloride	2.5 mM		
Lysozyme	K_2PdCl_4 K_2PdBr_4		0.85 M NaCl	4.7
Lysozyme	K_2PdI_4		0.85 M NaCl	4.7
Concanavalin	K_2PtCl_4	1 mM		
Ferricytochrome c (horse)	K_2PtCl_4		4.6 M phosphate	6.2
Tuna ferrocytochrome c	K_2PtCl_4	0.1 mM	95% AS	6
Cytochrome c_{550}	K_2PtCl_4	1.3 mM		
α-Chymotrypsin	K_2PtBr_4 or K_2PtCl_4 or K_2PtI_4		3.5 M phosphate 2–4% dioxane	4.2
Subtilisin BPN'	K_2PtCl_4	0.65 mM	2.1 M AS 0.05 M acetate	5.9
Subtilisin novo	K_2PtCl_4			
Thermolysin	K_2PtCl_4	6 mM	5% OMS 0.01 $CaAc_2$ 0.01 M tris/acetate	5.8

8.6.2. COVALENT REACTION OF CLASS (B) IONS

and gold complexes in protein crystals

Time of soak	Site number	Z	Binding site	Authors' reference
3	1	61	Met 129, His 127	Hardman and Ainsworth,
	2	23	Met 129	1972; Hardman, 1974
	1	80	Met H17	Huber et al. 1971
	2	55	His G2, C-terminus	
	3	7	His G19	
	1	45	His G2	Huber et al. 1971
	2	74	His G19	
	1		His 119	Wyckoff et al. 1967b
30–50 hours. Fresh soln. every 10 hrs.	1	64	Met 29	Wyckoff et al. 1967b
	1	32	Cys (SH)	
	2	81	Cys (SH)	
	1		Arg 14, His 15, Asn 93, Lys 96, Arg 128	Blake, 1968
	1		Arg 13, Arg 13' (same as $IrCl_6^{3-}$, HgI_4^{2-}, $AuCl_4'$)	Blake, 1968
2 days	1	75	His 127, Met 129	Reeke, 1974
	2	11	Met 42	
	1	35	Met 65 \ Close together	Dickerson et al. 1971
	2	40	Met 65 /	
	3	8	His 33	
2 days	1	24	Met 65	Tanaka et al. 1975
7 days	1	69		Timkovich and Dickerson, 1973
	1, 2	95	N-terminus and S–S of Cys 1–127	Sigler et al. 1968
	3, 4	55	Met 192	
10–40 days	1	78	Met 50	
	2	14	His 64	
	1		Met 50	
	2		Trp 241 His 238 Trp 106	Hol (1971)
	3		Ala 1 (N-terminus)	
10 days	1		His 250	Colman et al. (1972b)
	2		His 216	

TABLE 8.IX

Protein	Reagent	Conc. of reagent	Buffer	pH
Carboxypeptidase A	K_2PtCl_4		0.2M LiCl 0.02M tris	7.5
Ribonuclease A	K_2PtCl_6			
α-Chymotrypsin	K_2PtCl_6		3.5 M phosphate	4.2
Papain	K_2PtCl_6	5 mM	Methanol/water	9.3
Concanavalin A	K_2PtCl_6	3 mM	2.1M phosphate	6.0
Chymotrypsinogen	K_2PtCl_6		10% ethanol	6.3
Lysozyme	K_2PtCl_6		0.85M Na	4.7
Thermolysin	K_2PtI_6	Sat. soln.	5% DMS 0.01 M Ca Ac$_2$ 0.01 M tris /acetate	5.6
High Potential Iron Protein: HiPIP	K_2PtBr_6	0.5 mM	3.2 M AS	6.5
High Potential Iron Protein: HiPIP	$K_2Pt(NO_2)_4$	10 mM	3.2 M AS	6.5
Adenyl kinase	$K_2Pt(NO_2)_4$	2 mM		
Carbonic anhydrase	$KAuCl_4$			
Ferricytochrome c_2 (Rhodospirillum rubrum)	$KAuCl_4$	10–100 × protein conc.	3.2 M AS	5.8
Lysozyme chloride	$KAuCl_4$		0.85 M NaCl	4.7
Lactate dehydrogenase	$NaAuCl_4$	1 mM		

8.6.2. COVALENT REACTION OF CLASS (B) IONS

(*continued*)

Time of soak	Site number	Z	Binding site	Authors' reference
42 days	1	74	Cys 161 (–S–S–)	Lipscomb et al. 1970
	2	45	Met 103	
	3	68	N-terminus: Ala 1	
	4	27	His 303	
	1	28		Carlisle et al. 1974
	2	84		
	3	48		
	1,2	55	Terminal amino group and S–S of Cys 1–127 (same as $PtCl_4^{2-}$)	Sigler et al. 1968
	1	32	N-terminus	Drenth et al. 1962, 1967, 1968
	2	11	His 159 (same as HgI_4^{2-}, $IrCl_6^{3-}$)	
13 days	1	78	Met 129 Same as	Hardman and
	2	23	Met 129 $PtCl_4^{2-}$	Ainsworth, 1972
	3	25	Met 42	
7–30 days				Kraut et al. 1962
	1		Arg 14, His 15, Asn 93, Lys 96, Arg 128 (same as $HgCl_2$ and $PdCl_4$)	Blake, 1968
15 days	1 major		His 250	Colman et al. 1972
	5 minor		Iodination of tyrosine	
12 hours	1		Met 49	Carter et al. 1974a,b
	2		(major site)	
	3			
7 days	1		Met. 49	Carter et al. 1974a,b
	2		(major site)	
68	1	40	His 36	Schulz et al. 1974
	2	17	(major site)	
	1		Zn, Thr 197, X139	Liljas, 1971
	2		H_2O on Zn, His 128	
	3		Arg 25, carbonyl of His 35	
7–14 days	1	38	His 42	Salemme et al. 1973
	2	2	Asp 3	
	1		Arg 12, Arg 13′ (same as HgI_4^{2-})	Blake, 1968
	1		Cys(SH)	Adams et al. 1970a
	2		Cys(SH)	
			Cys(SH)	Hill et al. 1972

8. PREPARATION OF HEAVY ATOM DERIVATIVES

TABLE 8.IX

Protein	Reagent	Conc. of reagent	Buffer	pH
Glycera haemoglobin	$HAuCl_4$	1.5 mM	2.6 MAS 0.06 M phosphate	6.8
Myoglobin	$KAuCl_4$	equimolar with protein	3 MAS	6.5
α-Chymotrypsin	$KAuI_4$			
Adenyl kinase	$HAuCl_4$			
Carbonic anhydrase	MMTGA + $K_2Pt(CN)_4$		2.3 MAS	8.5
Ribonuclease S	$K_2Pt(CN)_4$	5 mM	3 MAS 0.1M Acetate	5.5
Cytochrome c_{550}	$(NH_4)_2Pt(CN)_4$	11.2 mM		
Cytochrome c_{550}	$(NH_4)_2Pt(CN)_4$	11.2 mM		
Ferrocytochrome c (tuna)	$K_2Pt(CN)_4$	6 mM	95% AS	6
Liver alcohol dehydrogenase	$K_2Pt(CN)_4$	1 mM	0.05 M Tris/HCl	8.4
Adenyl kinase	$K_2Pt(SCN)_4$	2 mM		
Carbonic anhydrase	$KAu(CN)_2$	20 mM	8.5	
Lamprey haemoglobin	$KAu(CN)_2$	1 mM	3.6 MAS 20 μM NaCN	6.8

8.6.2. COVALENT REACTION OF CLASS (B) IONS

(*continued*)

Time of soak	Site number	Z	Binding site	Authors' reference
7–10	1		Cys 30 (B39) His 72	Padlan and Love, 1968
6–9 months	1		His B5 Same as Ag+ His GH 1	Bluhm et al. 1958
	1		Met 192, Cys 191–220, Same as $PtCl_4^2$	Tulinsky, 1974
	1		His 36	Schulz et al. 1974
	1		Zn, Thr 197, X139	Liljas, 1971 Liljas et al. 1972
	1 2 3 4 5	24 28 16 12 8		Wyckoff et al. 1967b
7 days	1 2 3 4 5	84 54 43 12 8		Timkovich and Dickerson, 1973
7 days	1 2 3	84 54 43		Timkovich and Dickerson, 1973
1 day	1	30	No near neighbour Same as HgI_4^{2-} Lys 53, Ala 4, Lys 7,	Takano et al. 1974
	2	15	Ser 100, Val 3 Glu 44, Gln 70, Lys 72, Lys 73	
	3	9	Lys 99, Lys 99', Ser 103,	
	4	5	Ser 103' Glu 21, Lys 7, Lys 25	
	5	6	Ile 269 (mainchain)	
Co Cryst.	1		Asp 223, Lys 228, Arg 47, Arg 369	Branden et al. 1973
8 days	1 2 3 4	90 38 27 19	Major site near His–36	Schulz et al. 1974
	1	70	H_2O on Zn, His 128	Liljas et al. 1972
6 days	1	32	Lys 106, Ser 107 Glu 92	Love, 1971
	2	8	Ser 107, Val 8	
	3	4	Cys 141 Not in presence of phosphate.	

8. PREPARATION OF HEAVY ATOM DERIVATIVES

TABLE 8.IX

Protein	Reagent	Conc. of reagent	Buffer	pH
Liver alcohol dehydrogenase	KAu(CN)$_2$	1 mM	Tris/HCl 0.05 M	8.4
Flavodoxin	KAu(CN)$_2$	10 mM	2.6 MAS	6.8

evidence for the displacement of chloride by ammonia in their study of ribonuclease S; only freshly prepared PtCl$_4^{2-}$ gave intensity changes in the X-ray pattern.

The platinum complexes found in the presence of ammonium sulphate and phosphate are illustrated in Fig. 8.6. The formation of the phosphate complex may be minimized if a large excess of chloride ions is present. Acetate may also form complexes in time if there is an excess of acetate and no chloride. These complexes vary in their ligands and in their charge; they will therefore react in very different ways with a protein (Petsko and Williams, 1975). The charged groups (PtCl$_4$)$^{2-}$, PtCl$_3$(PO$_4$)$^{4-}$, Pt(NH$_3$)Cl$_2$(PO$_4$)$^{3-}$ and Pt(NH$_3$)$_4^{2+}$ will not penetrate into a hydrophobic protein core. The anionic

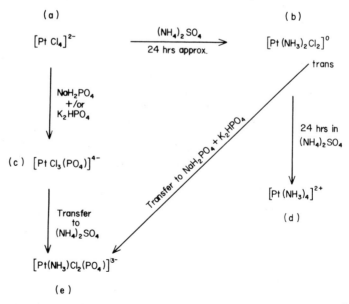

FIG. 8.6 Complexes of platinum which may be found when the PtCl$_4^{2-}$ ion is reacted with ammonium sulphate or phosphate. (Petsko, 1973)

8.6.2. COVALENT REACTION OF CLASS (B) IONS

(continued)

Time of soak	Site number	Z	Binding site	Authors' reference
Co cryst.	1		Phe 198, I6 224 Thr 274, Arg 271	Branden and Zeppezauer, 1974
	2		Lys 240	
	1	86	Cys 128	Burnett et al. 1974
	2	16	Cys 54	

groups will not react with anionic reagents such as RS⁻ but will be attacked more readily by neutral nucleophiles such as R–SH, imidazole or R–NH$_2$. The cation group Pt(NH$_3$)$_4^{2+}$ will be rather inert due to the weak trans effect of the ammonia ligands and it is most likely to form electrostatic complexes with anionic groups such as carboxylate. The neutral Pt(NH$_3$)$_2$Cl$_2$ molecule can penetrate into hydrophobic areas. It will require a stronger nucleophile but will be reactive to anionic nucleophiles such as R–S⁻.

These observations provide a rationale for the observed reactions of protein crystals when soaked in platinochloride (Petsko, 1973). At acid pH's platinochloride reacts with methionines, cystine disulphides, N-termini and histidine imidazole which all form stable complexes. These are good nucleophiles which can displace chloride from platinum complexes. Cysteine –SH groups are less nucleophilic. PtCl$_4^{2-}$ does not react with the cysteines of erythrocruorin or triosephosphate isomerase at about pH 7 in phosphate buffer or the cysteines of malate dehydrogenase at pH 5 in the presence of ammonium sulphate. However, prealbumin and triosephosphate isomerase at about pH 7 in ammonium sulphate react with PtCl$_4^{2-}$ through their cysteine groups, and these reactions occur in one or two days. In these cases the nucleophile is probably –S⁻ and the reaction occurs with PtCl$_4^2$(NH$_3$)$_2^0$ which will have formed within 24 hours.

The reaction of methionine and ionized cysteine appear to be faster than histidine; and so time of soaking may provide a further variable in the search for a specific reaction.

From the discussion above it can be seen that platinochloride is not really a very specific reagent but if the conditions can be varied, reactions with certain protein ligands might be enhanced relative to others. Petsko (1973) has pointed out that it may be possible to form more than one derivative using platinochloride under different cInditions with the same crystals. This has been achieved in the case of triosephosphate isomerase where different substitution occurs at pH 7 in ammonium sulphate and phosphate. Another example is ribonuclease S where (PtCl$_4$)$^{2-}$ binds to a methionine at pH 5.5 but a further site at a histidine is partially occupied at higher pH.

In most cases a square planar platinum complex is given. It is possible that a square pyramidal complex of platinum II is formed but unlikely that these

TABLE 8.X
The binding sites of osmium and

Protein	Reagent	Conc. of reagent	Buffer	pH
Ferricytochrome c_2 (Rhodospirillum rubrum)	$Os(NH_3)_6 I_3$	10–100 × Protein Concentration	3.2 MAS	5.8
Subtilisin novo	$Na_3 IrCl_6$			
Papain	$Na_3 IrCl_6$	5mM	Methanol water	9.3
Lysozyme	$K_3 IrCl_6$ $K_2 OsCl_6$		0.85M NaCl	4.7

complexes are oxidized to octahedral platinum IV complexes as suggested by Dickerson et al., (1969).

Wyckoff et al., (1967b) have suggested that cis-PtCl$_2$ (ethylenediamine) might be a useful reagent. This would prevent substitution by two protein ligands trans to each other. This kind of crosslinking by $PtCl_4^{2-}$ could lead to disorder of the protein crystals.

The rate of the reaction with the protein might be slowed down by using the other square planar anions $(Pt(NO_2)_4)^{2-}$ or $(AuCl_4)^-$. $(AuCl_4)^-$ provided a derivative of sperm whale myoglobin, but only after 6–9 months of soaking at pH 6.5 in ammonium sulphate. The reaction which occurred at two histidines may have occurred through an intermediate amine complex such as $AuCl_3(NH_3)$, $AuCl_2(NH_3)_2^+$ etc.

$(Pt(CN)_4)^{2-}$ does not allow nucleophilic substitution; in ribonuclease S the substitution is quite different from that of $PtCl_4^{2-}$. The binding of stable anions like $Pt(CN)_4^{2-}$ is described in Section 8.6.3.

8.6.2.5 Osmium and iridium reagents

Osmium and iridium are metals in the third row of the transition series which have many properties in common with platinum. They usually act as class (b) metals. They form covalent octahedrally coordinated compounds in oxidation states Ir(III) or Os(IV) giving stable anionic complexes such as $IrCl_6^{3-}$ and $OsCl_6^{2-}$ and stable cationic complexes such as $Ir(NH_3)_6^{3+}$. Osmium more than iridium

8.6.2. COVALENT REACTION OF CLASS (B) IONS

iridium reagents in protein crystals.

Time of soak	Site number	Z	Binding site		Authors' reference
7–14 days	1	39.6	Glu 37, Lys 112	I^- or	Salemme et al. 1973
	2	42	Lys 56, Met 55	anion	
	3	−21	Lys 109	binding	
	1		Lys–136		Hol, 1971
	2		Lys 27, Asn 118		
	3		Asn 25		
	4		Trp 241, His Trp 106		
	5		Gln.–103, Asn 240		
	1	32	N-terminus	Same as	Drenth et al. 1962,
	2	11	His 159	HgI_4^{2-} and $PtCl_6^{2-}$	1967, 1968, 1971a,b,c
	1		Arg 13, Arg 13' (same as HgI_4^{2-}, PdI_4^{2-} $AuCl_4{'}$)		Blake, 1968

also forms oxycompounds in the higher oxidation states, for example OsO_4 or $OsO_2(OH)_2$ but these are oxidizing agents (OsO_4 is a dangerous reagent killing epithelial cells; the eyes are especially vulnerable). OsO_4 will tend to be reduced to $OsO_2(OH)_2$ in most crystallizing media, and in the presence of ammonia or halide ions may become further reduced to give the cationic or anionic complexes described above.

In the high oxidation states, OsO_4 is used as a fixative and stain in electron microscopy, but this is probably due to its ability to form compounds with unsaturated fatty acid side chains of lipids, K_2OsO_4 or $OsO_2(OH)_2$ also bind cis-diols.

$$\begin{array}{c}\diagup\!\!\!-OH \\ \diagdown\!\!\!-OH\end{array} + K_2OsO_4 \longrightarrow \begin{array}{c}\diagup\!\!\!-O \\ \diagdown\!\!\!-O\end{array}\!\!Os\!\!\begin{array}{c}\diagup O \\ \diagdown O\end{array}$$

and this has been useful in the X-ray analysis of t-RNA (Suddath et al., 1974) where the binding is probably due to 3'-ribose; but this is clearly not the only mode of binding to t-RNA's (Schevitz et al., 1972).

In proteins, anions such as $IrCl_6^{3-}$ may bind through nucleophilic substitution by imidazole or amino groups as in papain where the binding sites are the same as those given by HgI_4^{2-} and $PtCl_6^{2-}$.

The anions may also bind basic groups as found in lysozyme and subtilisin novo. In ferrocytochrome c_2, $Ir(NH_3)_6I_3$ appears also to bind to basic groups, predominantly lysines, but this could possibly be due to I^- binding rather than the metal ions.

In conclusion it appears that osmium and iridium ions can bind in various ways to protein ligands which in many cases are similar to platinum, mercury and gold ions.

8.6.3 Electrostatic binding of heavy atom anions

Proteins contain a number of positively charged groups including the terminal α-amino and lysine ε-amino functions and the guanidinium group of arginine which may form ion pairs with heavy atom anions. The amino groups of

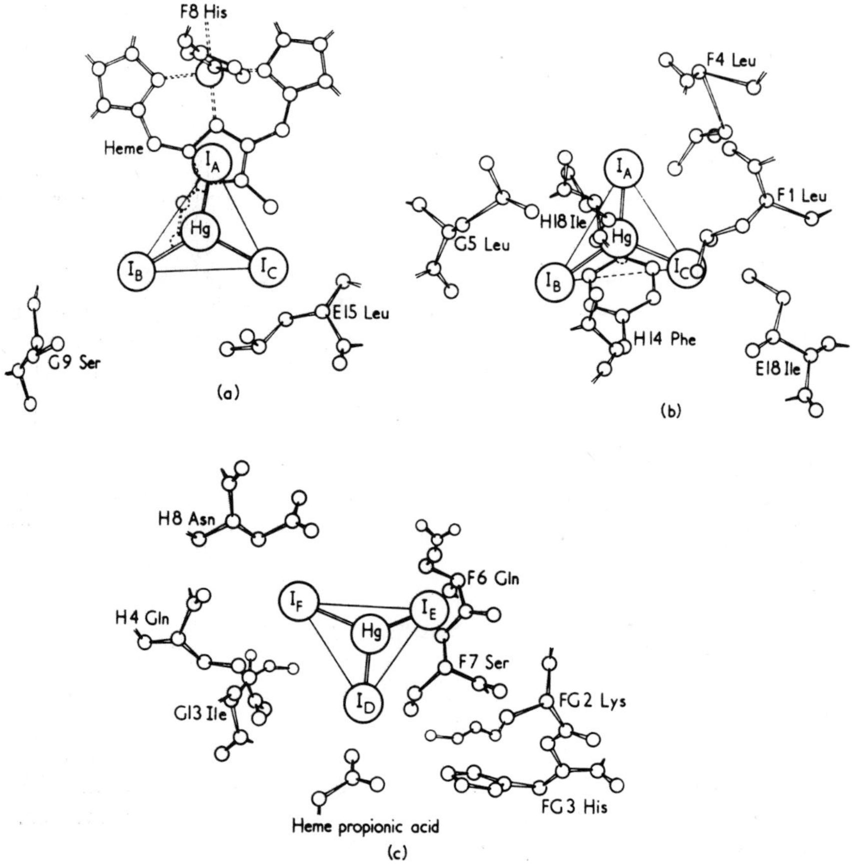

FIG. 8.7 The binding of HgI_3^- ions to myoglobin (Kretsinger et al., 1968). (a) and (b) show the residues on the two sides of the internal site; (c) shows the external site.

8.6.4. BINDING BY VAN DER WAALS INTERACTIONS

asparagine and glutamine may also bind anions and so may histidine especially at lower pH's where it is positively charged.

We have seen that ions like $PtCl_4^{2-}$ or HgI_4^{2-} tend to be bound covalently by soft ligands such as cysteine, methionine and histidine by displacement of the halide ligands. However, very often these ions are bound electrostatically. Thus in lysozyme and in the minor site of myoglobin, HgI_4^{2-} dissociates to HgI_3^- and binds ionically to the proteins. The myoglobin site is shown in Fig. 8.7c and involves a lysine, two glutamines and an asparagine (Table 8.VII). The lysozyme site involves two arginine guanidinium groups. The same site can be taken up by $PtCl_4^{2-}$, $AuCl_4^-$, $IrCl_6^{3-}$, $OsCl_6^{3-}$, and PdI_4^{2-}.

With the halide complexes discussed above the anionic binding gives rise to a certain lack of specificity. However, if the $Pt(CN)_4^{2-}$ ion is used the ligands are less likely to be displaced by protein ligands and so anionic binding becomes the most likely mode of interaction with the protein. Thus Table 8.IX shows that $Pt(CN)_4^{2-}$ binds at several sites involving lysines in ferricytochrome. In carbonic anhydrase and liver alcohol dehydrogenase (LADH), $Pt(CN)_4^{2-}$ binds at positively charged sites in the active site as we have seen in Section 8.2 and in Fig. 8.3. In ribonuclease, $Pt(CN)_4^{2-}$ binds at quite different sites from $PtCl_4^{2-}$. $Au(CN)_2^-$ will also tend to bind to anionic sites but this ion is two-coordinated and in the presence of soft ligands such as cysteine may give tetrahedral complexes. Similar anion binding sites may also be occupied by iodide and other halide ions as occurs in LADH (Norne et al., 1973). The fact that halide ions bind in similar ways implies that halide in the buffer or salting-out agent could interfere with binding of $Au(CN)_2^-$ or $Pt(CN)_4^{2-}$. Phosphate may also bind in a similar anion pocket. In chironomus haemoglobin (Huber et al., 1969a,b) $Au(CN)_2^-$ does not bind in the presence of phosphate, but can be bound in other buffers, such as acetate.

8.6.4 Binding by van der Waals interactions

Although the early experiments with HgI_4^{2-} on sperm whale myoglobin were designed to bind a mercury to a methionine (Bluhm et al., 1958), it was later found that one ion was bound as HgI_3^- in a hydrophobic pocket close to the haem group (Kretsinger et al., 1968) as shown in Fig. 8.7. This is not so surprising as the iodine ligands are very soft and would give rather good van der Waals interactions which would stabilize the binding. This description of the binding is consistent with the finding that AuI_4^- and I_3^- also bind at the same site but $HgBr_3^-$ does not. Even a single xenon atom can be bound (Shoenborn et al., 1965), when myoglobin is equilibrated with xenon at 2.5 atm., and this proves that the interaction can be through neither ionic nor covalent links but must be due to London interactions and induced dipole moments which make up van der Waals interactions.

The protein groups are slightly distorted by inclusion of these large groups or atoms and it is therefore not surprising that the binding depends critically on the nature of the globin. Seal myoglobin binds HgI_3^- in the hydrophobic pocket whereas haemoglobin does not.

8.6.5 The use of masked metal ions

In some proteins, there are many reactive groups which bind metal ions strongly but in doing so lead to a denaturaton of the protein. This is the case with liver alcohol dehydrogenase where the very many sulphydryl groups are extremely reactive and addition of class (b) metal ions usually gives rise to denaturation. In this situation the metal ion needs to be masked so that it cannot interact with the reactive groups, but rather binds at a site with different properties.

One technique for masking metal ions is to use them as stable, covalent anionic complexes such as $Pt(CN)_4^{2-}$ and we have already seen that this was a successful way of preparing a heavy atom derivative in the case of liver alcohol dehydrogenase. Zeppezauer (private communication) has suggested a different technique which involves binding the metal ion in a sandwich compound. Zeppezauer suggests the use of osmocene, the osmium analogue of ferrocene.

The osmium is quite unreactive to sulphydryls for instance. Various functional groups can be substituted onto the cylopentadienyl rings to change the solubility of the organometallic reagent, for instance

Further the osmocene can be modified in such a way as to allow a specific interaction with a protein functional group. The substitution of acylchloride or thiocaramyl functions on the ring could also lead to rather specific covalent reagents for amino groups. It has been suggested that such sandwich compounds might even bind through electronacceptor molecules such as tryptophan or tyrosine (Kornicker and Vallee, 1969) but this appears to be unlikely in liver alcohol dehydrogenase. The major problem, regardless of the nature of the binding, is that these sandwich compounds are rather large in size and may be difficult to diffuse into the crystals. Nevertheless hydroxy-methylferrocene appears to bind at a single site in liver alcohol dehydrogenase (Einarsson et al., 1972) and so this kind of binding may be useful in the preparation of derivatives.

8.6.6 Iodination

Many attempts to iodinate tyrosine sidegroups selectively have been reported; see for example papain (Drenth et al., 1968) and myoglobin (Bluhm et al., 1958; Kretsinger, 1968). The reaction of the protein with I_2 or KI_3 must be carried out above pH 5 to avoid oxidation, mild conditions are usually required to avoid cracking the crystals. Retrospective difference Fouriers have confirmed that accessible tyrosines have been mono- or di-iodinated, but in myoglobin (Kretsinger, 1968) and subtilisin (Wright et al., 1969) slight movements of sidechains are required to explain observed iodine positions. Similar sidechain movement may explain the crystal disorder and the change to a closely related space group on iodination of insulin and of cytochrome b_5. The tyrosine movement may be due to lowering of the pK of tyrosine on iodination which disrupts hydrogen bonds and also due to the large size of the iodine atom. This so-called "general method" appears to have involved more "trial and error" than some of the less systematic approaches, and certainly seems to have been of more value in chain tracing in electron density maps than in preparation of isomorphous derivatives. The iodination may not be as specific as at first thought as I^- or I_3^- may bind as an ion or even through hydrophobic interactions.

The iodine must be added very carefully. The reagent used is usually a mixture of KI and I_2 in equimolar amounts in 5% ethanol and in water, so that a ~0.4M KI_3 solution is obtained. The solution can be added to the buffer solution containing the crystals to give the desired molar ratio. Generally rather low levels of iodination give better results and it is best to add the KI_3 solution stepwise in small amounts. Follow the reaction by X-ray photography of the crystals and ensure that the solution is clear before further iodine is added.

8.7 THE INTRODUCTION OF NEW FUNCTIONAL GROUPS AS HEAVY ATOM BINDING SITES

Work in several laboratories has shown that it is possible to introduce a specific metal functional group into a protein, but these have not yet been used in a successful structure determination.

Benesch and Benesch (1965) have suggested the use of N-acetyl-DL-homocysteinethiolactone (AHTL) to give peptide linked sulphydryl groups by reaction with terminal amino groups:

$$CH_3CO-NH-CH-C=O \atop CH_2S \atop CH_2 \quad + R-NH_2 \longrightarrow \quad CH_3-CO-NH-CH-CO-NHR \atop CH_2 \atop CH_2 \atop SH$$

Shall and Barnard (1969) used this method to prepare two derivatives of ribonuclease. Each derivative involved reaction with one lysine and each gave a crystalline derivative on reaction with parachloromercuribenzoate. Avey and Shall (1969) have confirmed these results crystallographically at low resolution; 3.2% changes in one cell dimension restricted their use to low resolution work.

Benisek and Richards (1968) attempted to introduce a metal chelating group by reacting the amino groups with methyl picolinimidate, I. The reaction of lysozyme and guanidated lysozyme with methyl picolinimidate introduced 6.6 to 7 and 1.1 to 1.3 picolinimydyl groups, respectively, into the protein.

I

II

The picolinimydyl groups have a high affinity for metal ions such as the tetrachloroplatinum (II) ion to give a complex, II. Large crystals of the modified lysosymes were prepared. An alternative method of introducing a chelating agent is to convert tyrosines in a two-step process, described by Sokolovsky *et al.* (1967) to 3-aminotyrosine residues.

8.8 TECHNIQUES FOR CARRYING OUT MODIFICATION OF PROTEIN CRYSTALS

Heavy atom derivatives can be made by several different techniques. If the reagent forms covalent bonds with the protein, the reaction may be carried out in solution, the product purified to a monocomponent species, and then crystallized. Alternatively the heavy atom reagent can be added to the crystallizing medium so that cocrystallization of heavy atom and protein occurs. Although these processes might at first sight seem to be advantageous in maximizing the chance of a reaction and in the first case minimizing non-specific interactions, they have not been very useful. This is because the heavy atom

8.8. TECHNIQUES FOR MODIFICATION OF PROTEIN CRYSTALS

reagent often changes the interactions between proteins in solution which may prevent crystallization or give rise to crystals which are not isomorphous with those of the native protein.

Soaking crystals in a solution of the heavy atom reagent has been far more successful. The solvent channels in the protein crystals act as conduits along which the heavy atom reagents can diffuse and gain access to the protein in the crystal lattice. The intermolecular contacts often prevent reactions which would disturb the crystal packing, but at the same time provide binding sites which comprise ligands from adjacent protein molecules.

Special procedures may be needed to prepare derivatives from the volatile reagents. In the study of the Bence–Jones dimer (Schiffer et al., 1973) solid methylmercuric chloride was added past the point of saturation to 1.6M ammonium sulphate in a vessel which could be tightly sealed. Crystals of the protein in ammonium sulphate were placed in a glass cup on a pedestal over the reservoir of the mercurial solution.

The upper solution was kept saturated by vapour diffusion of the reagent without the presence of the solid to obscure observation of the crystals. Solid reagent as well as mother liquor were added to the capillaries to minimize diffusion from the crystal. Because of the toxicity of the reagent all steps were performed with care under a hood.

One of the major problems in preparing a heavy atom derivative is to find out how much heavy atom has bound and thus when to stop the reaction. There are a number of physical and spectroscopic techniques which might be useful. Because protein crystals have solvent in the lattice they often have the same density as the supernatant and therefore tend to "wander" in the solution. Substitution by a heavy atom increases the density and the crystals sink (Sigler and Blow, 1965). Alternatively neutron activation or X-ray fluorescence are quite useful probes for heavy atoms in small crystals, although a most useful and more simple technique is a change of colour of the crystals which is observed with many transition and actinide derivatives. Unfortunately these techniques give a measure of total metal atom taken up; they do not tell us how much is specifically bound and how much is disordered in the solvent channels of the crystals. However, the latter can often be removed by back soaking against mother liquor.

The extent of the reaction is best followed using X-ray photography. Changes in intensity of low angle reflections indicate non-specific and disordered binding sites. Changes at higher angles can be due to lack of isomorphism (see Section 6.6) but anomalous scattering is a good method of identifying metal ion contributions. The changes in X-ray diffraction pattern are conveniently followed using the flow cell of Wyckoff et al. (1967a). The crystal is mounted in a flow cell which enables it to be bathed continuously in crystallizing solvent, with or without heavy reagents or inhibitor molecules added. A diagram of this flow cell is shown in Fig. 8.8. The flow cell enables a wide range of different

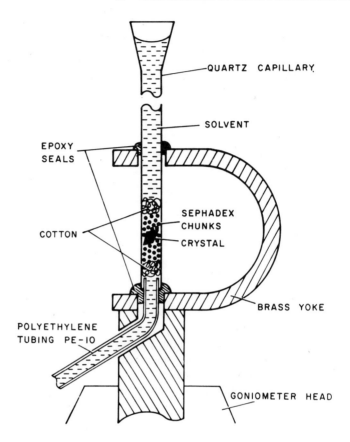

FIG. 8.8   The flow cell designed to study changes in the diffraction pattern of protein crystals due to soaking in buffer solutions containing heavy atoms (after Wyckoff *et al.*, 1967a).

heavy atom reagents to be diffused into the crystal lattice and the X-ray pattern examined for intensity changes. If no reaction occurs the heavy atom reagent can be washed out with solvent and a further reagent tested. The flow cell is very convenient for following the course of a reaction in a protein crystal, and choosing the optimal soaking time.

8.9 WARNING

The ability of the heavy atom reagents favoured by the protein crystallographer to bind to proteins in crystals is reflected in their strong affinity for many biological macromolecules in living organisms — including protein crystal-

8.9. WARNING

lographers! In other walks of life there have been many cases of death due to these reagents (especially due to some mercurials, lead and thallium) and there are occasional court cases including murder by poison. The resistance in a Dutch factory used thallium to murder the Nazi management; thallium gives rise to a slow death with signs of nervous disorder and hair falling out! So keep all these reagents away from food and drink. Wash your hands after preparing solutions and clean the bench and balance. Take care even if there is no warning on the reagent bottle.

9

DATA COLLECTION

9.1 INTRODUCTION

In his introduction to an International symposium on intensity measurements, Sir Lawrence Bragg wrote:

> "It is a curious feature of X-ray analysis that there are two points in the course of the investigation when it comes, as it were, to a focus. The first is the finale of the experimental measurements which are summed up in a list of values of F(hkl). All experimental effort and cunning has gone into making these values as accurate and as extensive as possible and they represent the raw material on which the whole analysis is based. The second is the list of the atomic co-ordinates in which a set of numbers constitutes a full account of all that the investigator has succeeded in establishing". (Bragg, 1969).

In this chapter we shall be concerned with the first of these objectives, namely the collection of intensity data, and with the special problems which confront the protein crystallographer in his efforts to make these data as precise as possible.

Protein molecules crystallize with large unit cells, in comparison with small molecules, and the number of reflections to be measured at any given resolution is correspondingly large. Moreover the intensities of these reflections are weak. An efficient method of data collection is therefore essential and in order to make the measurements as precise as possible, it would appear necessary to expose the crystals to X-rays for a long time while reflections are recorded. But this is seldom possible because protein crystals suffer from radiation damage. The protein crystallographer therefore needs to develop a strategy of data collection which will enable him to make intensity measurements to within the required precision and at the same time minimize the exposure of the crystal to the X-ray beam. Crystal size, absorption, the width of the diffracted beam and the separation of adjacent reflections must also be taken into account.

The earliest estimates of X-ray intensities from protein crystals were made by

9.1. INTRODUCTION

visual estimation of the optical densities of reflections recorded on rotation photographs. However the indexing of such photographs is by no means trivial and visual estimation of intensities is tedious. In 1944 Buerger introduced the precession camera which enables an undistorted representation of the reciprocal lattice to be recorded on a photographic film. The indexing of such photographs is therefore trivial and this factor, together with the development of semi-automatic microdensitometers capable of scanning along straight lines, meant that the precession camera became the instrument of choice for protein crystallography in the 1950's. The crystal structure of myoglobin, the first protein structure to be solved, was based on intensity data recorded in this way. Today the precession camera still enjoys great popularity both as a tool for assessing potential heavy atom derivatives and as a method of data collection. The development of proportional and scintillation counters as reliable detectors capable of counting individual X-ray quanta and of giving a more direct estimate of diffracted intensity than photographic film led to an increasing interest in this method of intensity measurement and the design of fully-automatic diffractometers. The high speed and accuracy of diffractometers and the fact that each reflection is recorded and measured at the same time and that the data output is in a form ready to be processed by a computer relieved much of the tedium involved in photographic work. During the 1960's, diffractometers, either of four circle or linear design, became increasingly popular in protein crystallography and many very precise structures have been solved with data collected with these instruments. However, diffractometers suffer from the disadvantage that only one reflection at a time is measured. The modification of some instruments to measure up to five reflections quasi-simultaneously has improved this situation slightly. But for crystals whose unit cell edges exceed 80 Å, even recording five reflections at a time is relatively inefficient compared with a method in which all lattice points that pass through the sphere of reflection for any one orientation of the crystal are measured at the same time. Both rotation photographs and screenless precession photographs enable such data to be recorded. The return in the 1970's to these earliest methods of data collection has been made possible through the development of fully automatic microdensitometers capable of digitizing the optical density of a whole film. In summary for the analysis of small to medium sized proteins ($M.W. < 60,000$), protein crystallographers now have a variety of methods of data collection open to them. For larger proteins it appears that rotation or screenless precession photographs are likely to represent the best way in which data can be collected. In the future we may see a return to counter methods as improved position sensitive detectors are developed (Arndt *et al.*, 1972; Cork *et al.*, 1974).

In this chapter we first discuss the production of X-rays and then consider the factors which govern the collection of X-ray intensity data from protein crystals. We then describe the geometry and operation of the basic instruments available for measuring such data.

9.2 X-RAYS

9.2.1 The production of X-rays

The production of X-rays has been treated comprehensively in a number of text books on physics and will not be described in detail here.

X-rays are produced when a beam of high energy electrons, which have been accelerated through a voltage, V, in a vacuum, strike a target. In protein crystallography the target is usually copper, but chromium, iron, cobalt, molybdenum and silver have been used in small structure work. Only a fraction of the energy of the electrons is converted into X-rays; the remainder is dissipated as heat. An X-ray tube run at a voltage, V, will emit a continuous X-ray spectrum with minimum wave length given by

$$\lambda \min = \frac{hc}{eV} = \frac{12398}{V}$$

where λ is in Å and V in volts. The critical voltage, V_0, which is required to excite the characteristic line of a particular target element can be calculated from the corresponding wave length for the appropriate absorption edge. For example for copper the K absorption edge $\lambda ae = 1.380$ Å. Hence

$$V_0 = \frac{12398}{\lambda ae} = 8.98 \text{ kV}$$

Provided $V > V_0$, the characteristic line spectra will be produced (Fig. 9.1). The wavelengths of the characteristic radiation of some of the more common targets are given in Table 9.1.

The intensity I of the characteristic radiation of a tube run at a current i varies according to $i(V-V_0)^n$ where n lies between 1.5 and 2 up to a voltage $V = 4 V_0$ and thereafter decreases to about 1. In order to obtain the highest intensity of the characteristic radiation relative to the continuous radiation the X-ray tube should be run at $V \simeq 4 V_0$. For Cu Kα radiation this corresponds to a voltage of 36 kV: in practice 40kV is frequently used.

Most X-ray tubes have line foci on a flat target area which is normal to the tube axis. The line is viewed at a small take-off angle (3–6°) to the target face so that the foreshortened spot appears approximately square. The power of the tube and hence the intensity of the radiation is limited by the rate at which heat can be dispersed from this area (Muller, 1927).

The great majority of protein structures have been solved using conventional fine focus sealed off X-ray tubes. These have the advantage of reliability and require no maintenance beyond occasional cleaning of the water cooling system. A Philips tube with a focus of 8 x 0.4 mm and run at a total loading of 1.2 kW (which is satisfactory for most projects) has a life time of about 1000 hours. However, greater resolution and greater intensity are required for small crystals

9.2.1. THE PRODUCTION OF X-RAYS

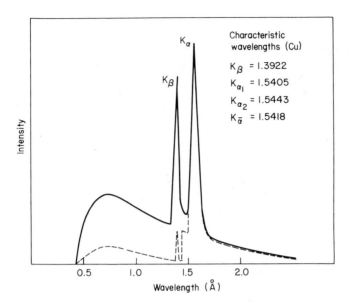

FIG. 9.1 Typical X-ray spectrum emitted by a tube with a Copper target operating at about 35 kV before (—) and after (– –) passage through a nickel filter.

with large unit cells. Although micro-focus tubes (Ehrenberg and Spear, 1951) provide a greater specific loading by means of a focussing effect of a simple electrostatic lens, they are not widely used in single crystal protein crystallography since the total amount of collimated radiation reaching the crystal remains small. The method of rotating the anode to diminish target heating by distributing the heat over a large area (Muller, 1929; Muller and Clay, 1939; Taylor, 1949) has found wider application. The commercially available Elliot rotating anode tube GX 6 has a focal spot 2 x 0.2 mm and may be run at a total power of 2 kW. The brilliance as estimated from the measured photon flux of a

TABLE 9.I
Wavelengths of characteristic radiation of various elements

	$\lambda K\alpha_1$	$\lambda K\alpha_2$	λK_β
Chromium	2.290	2.294	2.085
Iron	1.936	1.940	1.757
Cobalt	1.789	1.793	1.621
Copper	1.54051	1.54433	1.39217
Molybdenum	0.709	0.714	0.632

collimated X-ray beam (approximately 3×10^9 photons s^{-1}) is about four times that obtained from a fine focus sealed off tube and focussing of the beam on to a smaller target area provides greater resolution. The newer GX 13 model incorporates a larger anode (18" in diameter) and hence can be run at a higher power, up to a maximum of 4 kW for a 200 μ focus. Rotating anode tubes involve moving parts within a high vacuum system and maintenance can present problems.

In an attempt to realize an even more intense source, Rosenbaum *et al.* (1971) have explored the use of X-radiation emitted by an electron synchrotron. When an electron is accelerated it emits radiation. If the electron is travelling very near the speed of light on a curved path, the emitted radiation is confined to a narrow cone (1' of arc) around the instantaneous direction of flight of the electron. Thus an electron synchrotron radiates tangentially. [For a review of synchrotron radiation see Goodwin (1969)].

The useful photon flux $N(\lambda)$ in an electron synchrotron or storage ring is expressed as the number of photons emitted per second integrated over all vertical angles per 0.1% band width ($\delta\lambda/\lambda = 10^{-3}$) per milliradian of horizontal angle per ampere of circulating current and is given by

$$N(\lambda) = 2.46 \times 10^{13} E G(\lambda_c/\lambda)(\lambda/\lambda_c)^2$$

where $\overline{E}$ is the electron energy in GeV and $G(\lambda_c/\lambda)$ is a universal numerical function for all synchrotron radiation. The important parameter of the emitted radiation is the critical wavelength λ_c where $\lambda_c = 5.59R/E^3 = 18.6/BE^2$ Å (R is the radius of curvature of the electron orbit in metres and B is the magnetic field in tesla which controls the orbit of the electrons). Thus for a 2 GeV electron synchrotron with a storage ring of radius 5.5 m, $\lambda_c = 3.85$ Å. The radiation has a broad maximum in the region of interest to protein crystallographers (Fig. 9.2). Thus synchrotron radiation provides an intense source of X-rays whose spectral distribution is controlled by λ_c. The idea of using this radiation as an X-ray source in crystallography was introduced as long ago as 1959 (Parratt, 1959) but it is only comparatively recently that X-ray crystallographers and nuclear physicists have come together to exploit this phenomenon.

The actual advantage to be gained in a diffraction experiment depends on the optical system used to monochromate and focus the synchrotron radiation. In a discussion of their preliminary results with the DESY synchrotron in Hamburg (E = 7.5 GeV, $R \simeq 32$ m, beam current = 10 mA) Rosenbaum *et al.* (1971) have shown that the total flux density reaching the film after monochromatization and focussing of the synchrotron X-radiation by a Berreman monochromator (Berreman, 1955) is two orders of magnitude greater than the collimated filtered CuKα radiation from a rotating anode X-ray tube. This enormous gain in intensity may make feasible the investigation of microcrystals of protein molecules, the design of dynamic experiments with substrates in protein crystals and a whole host of other experiments related to fibre and virus diffraction

9.2.2. MONOCHROMATORS

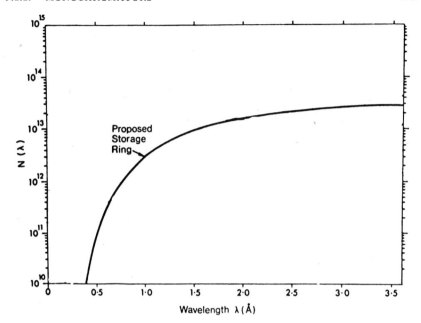

FIG. 9.2 Synchrotron Radiation. Plot of $N(\lambda)$ versus wavelength in the short wavelength region for the proposed storage ring at Daresbury. N is number of photons/S/mrad/A in 0.1% bandwidth for $E = 2.0$ GeV, $B = 1.2$T, $I = 1.0$ A and $R = 5.5$ m.

work. Investigations are underway with the DESY machine in Hamburg, Germany, with the synchrotron at Daresbury, England and with the SPEAR machine at Stanford, USA. Some preliminary results have been published (Phillips et al., 1976).

9.2.2 Monochromators

Monochromatization of the X-ray beam for crystals of small to medium sized proteins is usually accomplished by means of a nickel absorption filter (Fig. 9.1). The characteristics of nickel filtered CuKα radiation are shown in Table 9.II. A filter of thickness 0.018 mm is usually sufficient. Balanced filters have been used by some workers to give better monochromatization (Arnone et al., 1971) but such filters involve considerable care in their adjustment.

A higher degree of monochromatization is achieved by reflecting the direct X-ray beam from a single crystal set at the appropriate Bragg angle for the characteristic radiation. Such monochromators are advantageous in protein crystallography since removal of the unwanted radiation provides an improvement in peak to background ratios and a reduction in radiation damage for sensitive specimens. The general requirements and possibilities of focussing

TABLE 9.II
Characteristics of Nickel filtered Copper Kα radiation

Thickness of Ni		CuK$_{\alpha_1}$		CuK$_\beta$	
mm	mils	Relative Peak Intensity	Percentage Reduction	Relative Peak Intensity $\frac{IK_\beta}{IK_{\alpha_1}}$	Percentage Reduction
0		100		21.4	
0.013	0.5	66.0	34.0	1.9	98.1
0.018	0.7	55.0	45.0	0.6	99.4
0.025	1.0	41.4	59.6	0.1	99.9

1 mil = 0.001 inch

monochromators have been reviewed by Witz (1969), who also shows two photographs illustrating the startling improvement in the diffraction pattern of tobacco mosaic virus obtained by using a point focussing monochromator involving two quartz crystals. Although single crystal monochromators have been widely used in fibre diffraction and low angle X-ray work, their routine use in single crystal studies has been avoided partly because of loss of intensity and dangers in non-uniformity of the beam, and partly because of the extra adjustment required in setting up the monochromator.

However, single crystal monochromators using highly orientated graphite (002 reflection) have recently been developed (Gould et al., 1968; Ramussen and Henriksen, 1970) which provide a monochromatized X-ray beam of comparable intensity to that achieved with a filtered beam. Their use in protein crystallography has been assessed by Hagman et al. (1969) who report an intensity ratio I (graphite):I (nickel) of 0.8, an improvement in peak to background ratios, and a reduction in radiation damage for their cytochrome c peroxidase crystals by a factor of two. It is anticipated that graphite monochromators will be increasingly used in protein crystallography.

9.3 FACTORS WHICH AFFECT THE MEASUREMENT OF INTEGRATED INTENSITIES

9.3.1 Darwin's formula

Few crystals are perfect in that all unit cells are perfectly aligned with respect to one another over the whole volume of the crystal. Most crystals possess slight

9.3.2. UNIT CELL SIZE

irregularities and may be considered as if composed of small perfect blocks which are in slightly different orientations. Such a crystal is said to have a mosaic structure. Consequently the Bragg reflecting position will occur over a small angular range of crystal settings and the crystal must be rotated slightly to bring each element of volume into the diffracting position.

The total energy, $E(\mathbf{h})$, in a diffracted beam for a crystal rotating with uniform angular velocity, ω, through the reflecting position in a beam of X-rays of incident intensity, I_0, is given by Darwin's formula (Darwin, 1914; see also Woolfson, 1970).

$$E(\mathbf{h}) = \frac{I_0}{\omega} \lambda^3 \frac{e^4}{m^2 c^4} \frac{1 + \cos^2 2\theta}{2} \cdot \frac{L.A.V_x}{V^2} \cdot |F(\mathbf{h})|^2 \ldots \qquad (9.1)$$

where λ is the wave length of the X-rays; e the electronic charge; m the mass of the electron; and c the velocity of light; $1 + \cos^2 2\theta/2$ the polarization factor for a Bragg angle of θ. This factor takes into account the partial polarization of the diffracted beam if the incident beam is unpolarized. L is the Lorentz factor; this is a geometrical factor which takes into account the relative time each reflection spends in the reflecting position. $L = 1/\sin 2\theta$ when the rotation axis is normal to the scattering plane but is different for other geometries. A is the absorption factor; V_x the crystal volume; V the volume of the unit cell; and $|F(\mathbf{h})|$ the structure factor amplitude for the reflection $\mathbf{h}$. The equation was experimentally tested and confirmed for rock salt by Bragg et al. (1921).

In this formula $E(\mathbf{h})$ is the experimentally determined quantity and $|F(\mathbf{h})|$ is the required parameter. The precision of $|F(\mathbf{h})|$ depends on the precision with which $E(\mathbf{h})$ is measured. If we imagine that the diffracted energy is measured by a proportional or scintillation counter in which N counts are accumulated in a time t, then statistics tell us that the standard deviation of this measurement is given by $\sqrt{N}$. Therefore in order to achieve a fractional standard deviation $\sqrt{N}/N$ of 1% say, we need to accumulate 10,000 counts. In order to improve the fractional standard deviation to 0.1% we need to accumulate 10^6 counts. In general it is wise to aim for a set of intensity data for which the great majority of reflections have a background corrected intensity $N_0 = N - n_1 - n_2$ (where N is the number of counts in the peak and n_1 and n_2 the number of counts in the background) greater than three standard deviations where the standard deviation is given by $\sigma_{N_0} = (N_0 + n_1 + n_2)^{1/2}$.

We now examine Darwin's formula in order to see how the factors which contribute to the measured diffraction intensity restrict and guide the recording conditions.

9.3.2 Unit cell size

Perhaps the most serious consequence of Darwin's equation for the protein crystallographer is the inverse relationship between the energy of the diffracted beam and the unit cell volume. Many of the most interesting biological molecules crystallize in large unit cells and the diffracted intensities are correspondingly weak. The dependence is not quite as bad as $1/V^2$, however. If we remember that the average value of $\langle \overline{F(\mathbf{h})^2} \rangle$ is equal to $n\hat{f}^2$ where n is the number of atoms per unit cell and $\hat{f}$ their average atomic scattering factor. The number of atoms per unit cell, n, is directly proportional to V the unit cell volume. Hence the energy of the diffracted beam $E(\mathbf{h})$ is proportional to $1/V$. For example both lysozyme (M.W. 14,600) and phosphorylase (M.W. 100,000) form tetragonal crystal space group $P4_3 2_1 2$ containing eight molecules per unit cell. The volume of the lysozyme unit cell is 2.37×10^5 Å^3 and that of the phosphorylase unit cell is 1.94×10^6 Å^3. Hence the average intensity from a phosphorylase crystal is approximately one eighth of the average intensity from a lysozyme crystal. Clearly as larger and larger molecules are examined, the weakness of the intensity data and the number of reflections to be measured become more serious.

9.3.3 The number of reflections

The number of reflections to be measured will depend upon d_m, the minimum interplanar spacing for which F's are included in the Fourier summation of the electron density. Images of high resolution require the use of many more reflections than images of low resolution. A resolution of d_m corresponds to a sphere in reciprocal space of volume $(4/3)\pi(\lambda/d_m)^3$. To obtain an estimate of the number of reflections to be measured, this volume is divided by the volume of the reciprocal unit cell, i.e. λ^3/V. Hence the number of reflections to be measured is given by

$$N = \frac{4}{3} \frac{\pi V}{d_m^3} \div n$$

$n = 1$ primitive lattice; 2 body centred or single face centred; 4 face centred. For a set of data in which each reflection is measured only once, N must be divided by m where m is the multiplicity factor of the Laue group (International Tables for Crystallography Vol. I). This is not strictly correct as the multiplicity refers to the general reflections of the type hkl. The multiplicity of certain zones of reflections may be different and less than m (e.g. the multiplicity of reflections of the type h00 0k0, and 00l is 2). Hence the number of unique reflections estimated by this method will form an under-estimate, especially for high symmetry space groups. Precise values take into account the different multiplicities of the reflections. In Table 9.III values of N for different resolutions are given for three proteins of very different molecular weights and unit cell size.

9.3.4. ABSORPTION

TABLE 9.III
Number of Reflections to be measured

Enzyme		Lysozyme	Phosphoglycerate kinase	Phosphorylase
Molecular weight		14,600	38,000–40,000	100,000
Space Group		$P4_32_12$	$P2_1$	$P4_32_12$
Unit cell Volume V ($Å^3$)		2.37×10^5	1.97×10^5	1.94×10^6
Multiplicity m		16	4	16
Total number of reflections $N = \dfrac{4\pi V}{3 d_m^3}$	$d_m = 6Å$	4.60×10^3	3.83×10^3	3.74×10^4
	$d_m = 3Å$	3.70×10^4	3.06×10^4	3.01×10^5
	$d_m = 2Å$	1.24×10^5	1.04×10^5	1.02×10^6
Approx. number of unique reflections	$d_m = 6Å$	390	1,000	2,700
	$d_m = 3Å$	2,700	7,700	20,000
	$d_m = 2Å$	8,500	27,000	67,000

The very large number of intensities which need to be measured reflects the complexity of the protein structure. Nevertheless the structure is far from being overdetermined. For example the lysozyme molecule contains approximately 1000 non-hydrogen atoms. If each atom is characterized by three positional co-ordinates and a temperature factor then the ratio of parameters to be determined to the number of observations is 4000:8500 at 2 Å resolution. In the structure analysis of small molecules the structure is much more precisely determined and a typical ratio of unknown parameters to the number of observables may be of the order of 1:15. This point is referred to again in Chapter 15, Refinement.

9.3.4 Absorption

Equation 9.1 shows that the intensity of diffracted radiation increases as the crystal volume, V_x, increases. However there is an upper limit for the appropriate size of a crystal due to the *absorption factor A*. When a narrow beam of monochromatic radiation passes through a thickness t of the crystal, the emergent intensity is related to the incident intensity by

$$I = I_0 e^{-\mu t}$$

where μ (Units cm^{-1}) is the linear absorption coefficient. Values of μ for different elements are tabulated in Table 7.1 p. 169 and International Tables for X-ray Crystallography Volume III. They increase severely with the atomic number of the element. Jeffrey and Rose (1964) have calculated that a crystal of a molecular organic compound containing no atoms heavier than oxygen has a typical value of $\mu = 10$ cm^{-1} for copper Kα X-radiation. As the crystal thickness increases, the absorption of X-rays will diminish the amount of radiation transmitted through the crystal. At the same time the intensity of radiation in a diffraction spot increases as the volume of the crystal increases. From a simple calculation involving these two criteria, it may be shown (e.g. Buerger, 1942) that the optimum crystal thickness for an X-ray experiment in which the entire crystal is bathed in the X-ray beam is of the order of between $1/\mu$ to $2/\mu$. This suggests that the optimum size for a protein crystal is in the range of 1–2 mm.

It is not only the absolute magnitude of the absorption correction which is important, but also its relative variation from reflection to reflection. Plate-like crystals will exhibit extreme variation in absorption according to whether the paths of the incident and diffracted beams lie in the plane of the plate or normal to it. Although in principle it is possible to calculate these path lengths according to the geometrical shape of the crystal, such calculations are lengthy and not very precise for protein crystals where the crystal is sealed within a glass capillary and adheres to the wall by surface tension. For this reason semi-empirical methods are used in protein crystallography and these are described in Section 10.2.4 and 10.3.3. The empirical methods have been shown to be relatively accurate up to an absorption of about 40%. This suggests that, ideally, absorption should be kept to within this limit by the choice of suitable equi-dimensional crystals. If the crystal structure analysis is to be based on small differences in Friedel pairs of reflections due to anomalous scattering, then it is especially important to measure these reflections, wherever possible, under conditions in which they have the same absorption correction.

Finally examination of equation 9.1 might suggest that diffracted intensities could be increased by using X-radiation of longer wavelength since $E(\mathbf{h}) \propto \lambda^3$ However this is not as favourable as it might appear since absorption also increases as λ^3. The absorption will not only diminish the intensities of the X-rays transmitted by the crystal but air absorption will also diminish the intensities of the incident and reflected beams in their path from source to crystal and crystal to detector. Assuming an absorption coefficient of $\mu = 11 \times 10^{-3}$ cm^{-1} for air, absorption of copper Kα radiation is estimated to be 1% per cm. A path length of 10 cm of air will result in a 10% reduction of intensity. This problem can be partly alleviated by the use of a helium ($\mu \simeq 0.069 \times 10^{-3}$ cm^{-1}) filled tube. Helium absorption for CuKα is less than 0.1% in 10 cm.

Although CuKα ($\lambda = 1.5418$ Å) radiation is almost universally used in protein crystallography there may be advantages in using slightly longer wavelengths, such as CrKα radiation $\lambda = 2.291$ Å, when only very small crystals are available.

9.3.5 Radiation damage

Almost all protein crystals suffer from radiation damage during exposure in the X-ray beam, although their sensitivity varies considerably from protein to protein. Derivative crystals containing heavy atoms are usually more sensitive than the native protein. Radiation damage effects are characterized by a decrease in intensities of reflections during the time of the experiment. The decrease may or may not be isotropic with respect to the position of the reflection in reciprocal space but invariably radiation damage affects high angle reflections more than low angle reflections. Occasionally an initial increase in the intensities of very low angle reflections has been noted. Very little is known of the precise physical–chemical mechanism of radiation damage in protein crystals. There appear to be at least two processes involved: (i) the formation of free radicals, (ii) heating effects.

9.3.5.1 Formation of free radicals

The detection by electron spin resonance (ESR) spectroscopy of radiation-produced radicals in biologically important molecules and in living material has lent strong support to the hypothesis that radical intermediates form important primary steps in radiation damage. Müller (1962) has found that for amino acids, proteins and nucleic acids irradiated *in vacuo* in the *dry state* approximately one radical is produced per 100 eV of absorbed energy. (100 eV is roughly the mean energy transferred to irradiated matter per primary ionization.) Aromatic amino acids however appeared to be much less sensitive and this is in agreement with the general observation that aromatic hydrocarbon liquids exhibit a lower sensitivity to radiation damage than corresponding aliphatic hydrocarbons. Indeed the addition of a small amount of an aromatic substance to a liquid aliphatic hydrocarbon results in a reduction of the radiation sensitivity of the mixture which is much greater than that expected on the basis of a simple additive effect (Merlin and Lipsky, 1962).

Temperature appears to have a small but significant effect on the production of radicals in the dry state. For example gelatine (which is an especially radiation sensitive polypeptide chain containing approximately one third glycine residues) when cooled from $27°C$ to $-96°C$ exhibits a three-fold reduction in the radical yield (Müller, 1962). The effect of oxygen on the radiation sensitivity of biological materials appears to be less clear. Early hypotheses suggested that ionization radiation should be more effective in promoting damage when applied in the presence of oxygen than under anaerobic conditions but there are important exceptions. For example gelatine, glucose oxidase, catalase, and ribonuclease show a reversal of this effect; that is inactivation is greater in anaerobic than in aerobic irradiation. However there is no doubt that the presence of oxygen enhances the damage caused by ionizing radiation to cells in an aqueous environment. Some protection has been observed by the addition of hydrogen donors such as sodium dithionite, $Na_2S_2O_4$, and simple sulphydryl compounds. Equilibrium of buffer solutions with nitrogen and the use of inert gases such as argon, helium or nitrogen so as to prevent oxygen from coming into contact with the specimen during irradiation also reduces radiosensitivity (see discussion by Joy, 1973).

Protein crystals commonly contain about 50% solvent and the effect of ionizing radiations on water is probably the most significant primary step in the radiation damage process for protein crystals. Irradiation of water with Co^{60} γ-rays results in approximately 4.5 water molecules damaged per 100 eV of energy absorbed. The chemically active radiation products are principally the solvated electron e^-_{aq}, the hydrogen atom H, and the hydroxyl radical OH.

The discovery of the solvated electron revolutionized the field of radiation chemistry. The chemical activity of e^-_{aq} towards a wide range of substances is known principally from use of its intense absorption band ($\lambda = 7200$ Å; 1.72 eV) in pulse radiation studies, although often the nature of the products remain obscure. (For a discussion of some of the results in this field see The Solvated Electron, E. J. Hart, Chairman. Advances in Chemistry Series 50, 1965.) The solvated electron is undoubtedly a very strong reducing agent and appears especially reactive towards asymmetrical bonds causing heterolysis and the formation of free radicals

$$RX + e^-_{aq} \longrightarrow \dot{R} + \bar{X}$$

It may also convert to a hydrogen atom by reaction with H^+

$$H^+ + e^-_{aq} \longrightarrow H$$

There is some evidence that the ion ($H_2PO_4^-$) facilitates this reaction (Ebert and Swallow, 1965). Histidine, cysteine, tryptophan, phenylalanine and tyrosine appear to be very reactive with the solvated electron and this is in contrast to their behaviour following radiation in the dry state. Not surprisingly the interaction between solvated electrons and biological macromolecules depends on the conformation of the molecule. Long thin molecules like DNA appear to be more sensitive than globular protein molecules like ribonuclease.

The OH radical, which is the other reactive species formed on irradiation of water is strongly oxidizing and will attack both single and double bonds.

$$\text{e.g.} \quad OH + R-H \longrightarrow H_2O + \dot{R}$$

A possible method for protection against radiation damage involves the introduction of a scavenger molecule which will compete successfully with the protein for the active fragments of water. In this respect ethyl alcohol may prove effective since the reaction $CH_3CH_2OH + OH \longrightarrow CH_3\dot{C}HOH + H_2O$ is extremely rapid. Thiourea also reacts rapidly with solvated electrons and OH radicals and the reduced and oxidized free radicals so formed then react with each other to generate the original material. When a polymer is present, the OH radicals

9.3.5. RADIATION DAMAGE

can attack the polymer and the electrons can attack the thiourea. Reaction between the two types of radical so produced then appears to lead to a degree of protection. Recently the introduction of styrene in protein crystals has been reported to result in a considerable enhancement of crystal lifetime (Zaloga and Sarma, 1974).

How does this information relate to actual X-ray observations on protein crystals? In the most detailed study to date Blake and Phillips (1962) have observed that myoglobin crystals tended to deteriorate in the X-ray beam in a manner which indicates that an increasing proportion of the crystal volume became seriously disordered as irradiation proceeded. After a dose of some 20 m rad corresponding to about 104 hours exposure to CuKα radiation from an X-ray tube run at 40 kV 20 ma, about 10% of the crystal volume had been rendered amorphous, some 23% severely disordered and some 67% unchanged. Although there are difficulties in estimating quantitatively the relationship between dose and damage, the authors deduced that if the main effects are due to the major component of the CuKα radiation (8000 eV), then each quantum absorbed disrupts 70 molecules and somewhat disorders a further 90. This figure is in reasonable agreement with other radiobiological data.

The chain reaction initiated by free radical formation probably accounts for the common observation that radiation damage effects in protein crystals continue, even after the X-ray shutter has been closed.

9.3.5.2 Heating effects

For each quantum of radiation absorbed, some energy will be dissipated in the form of heat. Unless the heat is conducted away fairly efficiently, a temperature rise of a few degrees will result, which may increase the disorder in the crystal. The lifetimes of many protein crystals in the X-ray beam are improved on cooling. The reasons for this may be partly explicable in terms of the elimination of the heating effects described above and in terms of the slowing down of the formation of reactive radiation products, but it is likely that other effects, such as the stability of the crystal lattice and the nature of the stabilizing forces of the protein structure, also play a role. Elastase crystals are stable in the X-ray beam for up to at least 200 hours exposure at room temperature, whereas no progress could be made with crystals of carboxypeptidase, alcohol dehydrogenase or staphylococcus nuclease unless the X-ray measurements were made on crystals cooled to 4°C.

A simple and reliable low temperature device for cooling protein crystals has been described by Marsh and Petsko (1973). With this device, increases in crystal lifetimes of the order of 4 to 12 times were observed on cooling several different protein crystals from 20°C to 0°C. There are indications that cooling the crystal further results in even greater enhancement of lifetimes, although temperatures below 0°C will require the presence of some cryo-protective agent such as glycerol to prevent the formation of ice crystals. Crystals of lactate dehydrogenase have been cooled to $-75°C$ and an enhancement of crystal lifetime of 75 reported (Haas and Rossman, 1970).

A further reduction in radiation damage may be effected through the use of single crystal monochromators for the primary X-ray beam. The interaction of X-rays with matter increases markedly with wave length. The elimination of the unwanted radiation from the primary beam therefore results in an improved crystal lifetime. A reduction in radiation damage by a factor of 2 has been reported from CuKα radiation monochromatized by a single graphite crystal compared with radiation monochromatized using a nickel filter (Hagman et al., 1969; Tsernoglou et al., 1972).

In general, it is advisable to limit the exposure of a protein crystal to dosages for which a set of reference reflections decrease in intensity by no more than 15% during the time of the experiment. The reference reflections should be a representative sample of the volume of the reciprocal space which is being measured. These may then form a rough guide to the type of correction factor to be employed. Usually the correction factor is linear with time and a better estimate of the magnitude can be obtained by repeating the measurement of the first 50 reflections (say) every two or three days and at the end of the run. Comparison of the overall intensities of these reflections can then provide a fairly precise guide to the correction factor. It is worth noting that in the early crystal structure analysis of rubredoxin (M.W. 6000), where only small amounts of the protein were available, crystals were used up to the extent of 33% deterioration (Herriott et al., 1970).

9.3.6 The width of the reflection

In the discussion on the integrated intensity, we have assumed that every point of the crystal can see every point on the X-ray source. We now need to calculate the angle through which the crystal must be rotated in order for all the mosaic blocks in all parts of the crystal to have the opportunity to diffract X-rays from all parts of the source. This angle will depend on the finite dimensions of the crystal and the finite dimensions of the source (both of which are under the experimenter's control to some extent), and the mosaic spread of the crystal and the dispersion spread in wavelength of the X-ray beam, which are not usually under the experimenter's control.

The angular range over which a crystal diffracts as it rotates through the Bragg position is given by (Burbank, 1964; Arndt and Willis, 1966).

$$\Delta = \delta_c + \delta_F + \eta + \delta\theta$$

where δ_c = angle subtended by the crystal at the X-ray tube focus = a/s where a is a linear dimension of the crystal and s is the distance between the focus and crystal; δ_F = angle subtended by the X-ray tube focus at the crystal = f/s where f

9.3.6. THE WIDTH OF THE REFLECTION

is the linear dimension of the focus; η = mosaic spread of the crystal. This latter parameter is generally taken to be between 0.01 and 0.001 radians but obviously varies from crystal to crystal. Banner (1972) has estimated η to be 0.005 radians (0.29°) for triosephosphate isomerase crystals. The mosaic spread of a crystal is often anisotropic. $\delta\theta$ = dispersion spread in wavelength due to the $K\alpha_1$ $K\alpha_2$ doublet of $CuK\alpha$ radiation. $\delta\theta = \tan\theta \delta\lambda/\lambda = 0.008$ radians for a Bragg angle θ corresponding to an interplanar spacing of 2.5 Å.

Thus for a crystal of dimensions a = 1 mm on a diffractometer where typical values might be f = 0.4 mm, s = 200 mm, η = 0.005 and $\delta\theta$ = 0.008 the minimum rocking range is Δ = 0.0128 radians = 0.74°. Since the contribution of the mosaic spread is significant and is the least certain parameter, the value of the rocking angle should always be determined experimentally.

The size of the diffracted beam at the detector may be calculated in a similar way (Burbank, 1964; Alexander and Smith, 1962, 1964). The intensity distribution at the detector is a convolution of the distributions given by the projection of the crystal volume, the brightness variation across the focus and the scatter of the mosaic blocks. In the case where the crystal size and Bragg angle are small, the size of the diffracted beam is determined by the following contributions (Banner, 1972):

A_c = the contribution from the crystal size = $\dfrac{a(s+D)}{s}$ where D is the distance from the crystal to the detector,

A_F = the contribution from the focus size = $(f/s)D$,

A_M = the mosaic spread term = $2\eta D \sin\theta$ parallel to the rocking axis,

A_D = the dispersion term = $2\delta\theta D$ in the plane of diffraction.

In equatorial geometry (Section 9.61) the dispersion and mosaic spread terms are normal to each other and so the largest of these may be taken to determine the maximum size of the diffracted beam, i.e. $A = A_c + A_F + A_M$ if $\eta > 0.0025$ at $\theta = 18°$ (2.5 Å limit). With the same parameters as before, i.e. a = 1 mm f = 0.4 mm, s = 200 mm, η = 0.005 and D = 200 mm; A = (2.0 + 0.4 + 0.6) mm = 3 mm.

The resolution of neighbouring reflections may be assessed by differentiating Bragg's law

i.e. $d^* = \dfrac{\lambda}{d} = 2\sin\theta$

$\therefore \Delta d^* = 2\cos\theta \Delta\theta$

The maximum value of Δd^* is therefore $2\Delta\theta$. Reflections separated by a reciprocal lattice spacing p^* will be resolved at all values of θ if

$p^* > 2\Delta\theta = 2\Delta$

where Δ is the minimum rocking range defined previously. This condition is seldom met for crystals with large unit cells, and for diffractometers it is necessary to resolve reflections at the detector by means of an aperture. Let the diameter of the aperture be B (where $B > A$, the size of the reflection). Then we require the angle subtended at the crystal (B/D) to be less than $2\Delta\theta$

$$\frac{B}{D} < 2\Delta\theta = \frac{p^*}{\cos\theta}$$

i.e. $p^* > \dfrac{B\cos\theta}{D}$

9.3.7 Optimum collimation conditions

This discussion refers primarily to collimation of the X-ray beam for the recording of X-ray intensities on film. Collimation for diffractometers is discussed by Arndt and Willis (1966).

(i) Size of focus: Let us imagine that the crystal is illuminated by a point source. Then the width of the image of the film is $(D + s)a/s$ (Fig. 9.3), where D is the crystal to film distance (or crystal to detector distance) and s the source to crystal distance and a the width of the crystal. As the size of the focus is increased a series of overlapping images is produced. When the focal size exceeds a certain limit then images formed from points at opposite extremities no longer overlap. The spot size increases but the peak intensity per unit area within the image remains constant. The optimum focus size is expressed by (Huxley, 1953)

$$f \leqslant \frac{(D+s)}{(D)}a$$

In practice, this condition is almost always fulfilled since a is usually greater than f.

(ii) Crystal to film distance. The width of the spot on the film or detector is

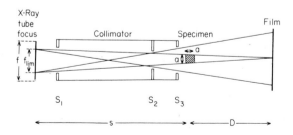

FIG. 9.3   Collimation geometry (after Huxley, 1953).

given by (Section 9.3.6)

$$A = \frac{a(s+D)}{s} + \frac{fD}{s} + \delta D$$

where $\delta = 2\eta \sin\theta + 2\tan\theta(\delta\lambda/\lambda)$ and is the sum of the mosaic spread and dispersion spread. The separation of the spots is given approximately by p^*D where p^* is the reciprocal lattice spacing. In order to resolve reflections, the separation of spots should be greater than the spot size by some distance α. The smallest value of α is chosen to suit the experimental conditions and this must be large enough to permit reliable background corrected intensity measurements of the spot (e.g. $\alpha = 0.2$ mm). Hence $Dp^* = A + \alpha$.

i.e. $$D = \frac{s(a+\alpha)}{s(p^* - \delta) - (a+f)}$$

This expression gives the optimum value for D in terms of the other parameters.

(iii) Total Intensity. If I_0 is the specific loading on the X-ray focus then it may be shown (Huxley, 1953) that the intensity at the image is given by

$$I \propto \frac{I_0 f^2 a}{(D+s)^2}$$

i.e. substituting for D from the expression given above

$$I \propto I_0 f^2 a \frac{\{s(p^* - \delta) - (a+f)\}^2}{\{s^2(p^* - \delta) + s(\alpha - f)\}^2}$$

Thus in order to obtain the highest intensities it is necessary to maximize I given the restrictions imposed by the experiment. In many situations the size of the crystal and focus are relatively invariant and it remains to choose the optimum values of D and s.

The source collimator should be designed so that all points on the crystal may see all points of the X-ray source, whilst avoiding as far as possible any radiation scattered within the collimator from reaching the film. Fig. 9.3 shows the usual form of collimator used in film and diffractometer work. S_1 limits the amount of radiation originating outside the focus itself which is seen by the crystal; S_2 forms the limiting aperture which restricts the beam to the crystal and S_3 is a guard aperture which prevents radiation scattered from S_2 from reaching the crystal. Fig. 9.4 illustrates a design for a precession camera collimator which incorporates some of these principles and which works well in practice.

9.4 CYLINDRICAL POLAR CO-ORDINATES

In the description of the diffraction geometries employed by X-ray cameras and diffractometers it is often more convenient to define the reciprocal lattice vector

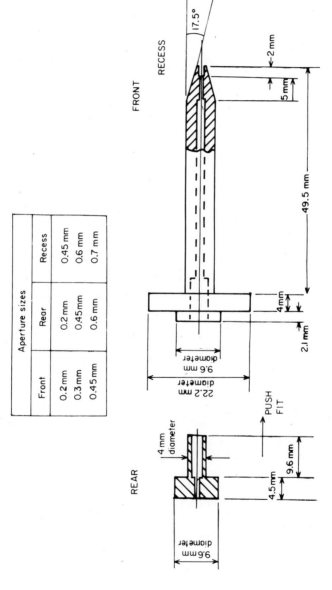

FIG. 9.4 Design of precession camera collimator. (M. F. Perutz, private communication.)

9.4. CYLINDRICAL POLAR CO-ORDINATES

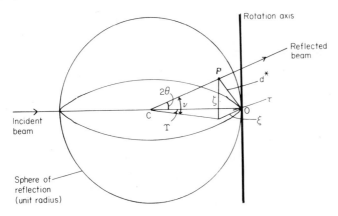

FIG. 9.5   Definition of cylindrical polar co-ordinates.

$d^* = ha^* + kb^* + lc^*$ in terms of cylindrical polar co-ordinates ξ, ζ, τ (xi, zeta and tau). The reciprocal lattice vector is resolved into two components one of which (ζ) is parallel to the rotation axis and the other (ξ) is perpendicular to the rotation axis. Zero τ is defined by the projection of the incident beam on to the plane normal to the rotation axis (Fig. 9.5).

For example for a crystal with orthogonal axes oriented with its **c** axis parallel to the rotation axis and with **a*** parallel to the trace of the incident beam at $\tau = 0$, then the following relationships may be easily found.

$$d^* = \xi + \zeta$$
$$d^{*2} = \xi^2 + \zeta^2$$
$$\xi = (h^2 a^{*2} + k^2 b^{*2})^{\frac{1}{2}}$$
$$\zeta = lc^*$$
$$\tan \tau = \frac{kb^*}{ha^*}$$

It is customary to express the direction of the reflected ray, which is of course defined by Bragg's Law, in terms of an azimuthal angle Υ (upsilon), normal to the plane containing the reflected beam and the rotation axis, and an inclination angle ν (nu) in the plane of the rotation axis.

For normal beam geometry, where the incident X-ray beam is normal to the rotation axis as in Fig. 9.5

$$\cos 2\theta = \frac{1}{2}(2 - \zeta^2 - \xi^2)$$
$$\cos \Upsilon = \frac{\cos 2\theta}{\sqrt{1 - \zeta^2}}$$
$$\sin \nu = \zeta$$

9.5 PHOTOGRAPHIC METHODS

9.5.1 The precession camera

The precession camera is the basic preliminary tool of the protein crystallographer. Precession photographs are used to determine the symmetry and space group of the crystal, to scan for potential heavy atom derivatives and in many cases to collect three-dimensional diffraction data. The camera was introduced by Buerger (1942, 1964) and permits the recording of an undistorted plane of the reciprocal lattice on a photographic film.

(i) The recording of zero level photographs. The crystal is oriented with the desired reciprocal lattice plane, *POQ*, normal to the incident beam, CO (Fig. 9.6a). This plane is then inclined through the chosen precession angle $\bar{\mu}$ so that the normal to the plane, ON, makes an angle $\bar{\mu}$ with the incident beam (Fig. 9.6b). The X-ray shutter is opened and the camera mechanism rotates the crystal so that the normal to the plane rotates about the incident beam. During this rotation the extreme positions of the normal will be first in direction CN_1, then in direction CN_2 (out of the plane of the page), then in CN_3 and then in CN_4 (into the plane of the page). These four extreme positions are shown in Fig. 9.6c. At each of these positions the reciprocal lattice level will cut the sphere of reflection in a circle diameter $2 \sin \bar{\mu}$. As the crystal is moved continuously from one of these positions to the next the area of the reciprocal lattice level which is moved through the sphere of reflection corresponds to a circle of radius $2 \sin \bar{\mu}$. The film cassette is mounted on a universal joint and its motion is linked to the movement of the crystal so that the plane of the film cassette is always parallel to the reciprocal lattice plane. Hence an undistorted image of the reciprocal lattice level is recorded on the film with a magnification F, where F is the crystal to film distance.

In order to prevent reflections from upper levels reaching the film it is necessary to insert an annular screen which selects the cone corresponding to the appropriate reciprocal lattice level. The movement of the screen is linked to the movement of the crystal and film. The position of the screen, s_0, is given by (Fig. 9.6d).

$$s_0 = r_s \cot \bar{\mu}$$

where r_s is the radius of the annulus.

(ii) Recording of upper levels. In the recording of upper levels the film has to be moved closer to the crystal so that the centre of the film follows the precessing movement of the centre of the reciprocal lattice layer. The distance moved is Fnd^* where n corresponds to the nth upper level of a reciprocal spacing d^*. In order to isolate the cone of the nth upper level the screen is set to a position given by (Fig. 9.6c).

$$s_n = r_s \cot \bar{\nu}_n$$

9.5.1. THE PRECESSION CAMERA

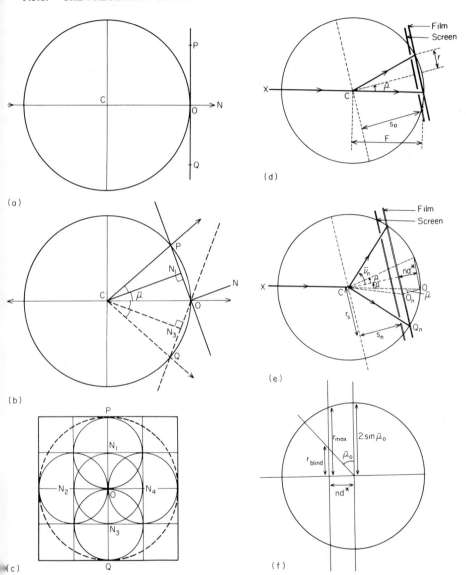

FIG. 9.6 Precession camera geometry. (a) Reciprocal lattice plane POQ is oriented normal to X-ray beam. (b) The plane is tilted through an angle $\bar{\mu}$ and the crystal rotated so that the normal to the plane (ON) rotates about the incident beam. In the position given by ON_1 the plane PO cuts the sphere of reflection in a circle. (c) The four extreme positions swept out as the crystal is moved. (d) Screen position for a zero layer precession photograph. (e) Film and screen position for a non-zero layer precession photograph. (f) Relationship between the maximum radius and the blind radius for upper level precession photographs for fixed resolution.

where

$$\cos \bar{\nu}_n = \cos \bar{\mu} - nd*$$

When upper levels are recorded on a precession camera there is a blind radius, corresponding to the distance $O_n Q_n$ in Fig. 9.6e, within which reciprocal lattice points will not cut the sphere of reflection. The radius is given by

$$r_{\text{blind}} = F(\sin \bar{\nu}_n - \sin \bar{\mu})$$

where $\cos \bar{\nu}_n = \cos \bar{\mu} - nd*$.

The radius of the outer limiting circle for upper levels is given by

$$r_{\text{max}} = F(\sin \bar{\nu} + \sin \bar{\mu})$$

A simple relationship between the blind radius and the maximum radius is shown in Fig. 9.6f. Suppose we require data to a resolution of d_m. Then the zero level precession angle required is $\bar{\mu}_0$ where $2 \sin \bar{\mu}_0 = \lambda/d_m$ (e.g. for 3Å resolution data $\bar{\mu}_0 = 15°$). For upper levels we require a smaller precession angle $\bar{\mu}_n$ since r_{max} is given by $\sin \bar{\nu}_n + \sin \bar{\mu}_n$ (neglecting the film magnification factor F). The maximum radius required for the nth upper level of a reciprocal axis spacing $d*$ is $(4 \sin^2 \bar{\mu}_0 - n^2 d*^2)^{1/2}$. The blind radius may readily be obtained from the point of interception on the nth upper level plane by the line drawn at an angle $\bar{\mu}_0$ with the projection of the zero level plane (Fig. 9.6f). Hence $r_{\text{blind}} = nd* \cot \bar{\mu}_0$. It is often preferable to use a larger value of $\bar{\mu}_n$ than is strictly required by the resolution criteria in order to decrease the area of the blind region.

9.5.1.1 Width of screen annulus

The width of the screen annulus is important for high quality precession photographs. An annulus that is too small will block part of the diffracted beam while an annulus that is too large will result in high background levels on the film and possibly overlap of reflections from adjacent levels.

The minimum width of the annulus may be estimated from the size of the diffracted beam at the screen (Section 9.3.6). This will depend upon the crystal and source parameters. A crystal of dimensions 0.5 mm and mosaic spread 0.005 situated 120 mm from a rotating anode X-ray source of focal dimensions 0.2 mm will give rise to reflections of width approximately 1 mm at the screen placed 50 to 70 mm from the crystal. Thus the width of the screen annulus must be greater than 1 mm. In order to ensure that reflections from adjacent levels are separated, the width of the screen annulus (w) must fulfill the condition

$$w \leqslant (r'_{n+1} - r_n) - \Delta$$

where $r_n = s_n \tan \bar{\nu}_n$, $r'_{n+1} = s_n \tan \bar{\nu}_{n+1}$ and Δ is the width of the reflection.

For example, in order to separate the zero and first levels on a 15° precession photograph with screen radius 15 mm placed at the required distance of

9.5.1. THE PRECESSION CAMERA

55.9 mm from the crystal described above we require

$w \leq 6.2$ mm $- \Delta$ for a 50 Å axis

$w \leq 3.3$ mm $- \Delta$ for a 100 Å axis

$w \leq 2.2$ mm $- \Delta$ for a 150 Å axis

In practice a width of 2 mm is suitable for all but the largest unit cells.

With protein crystals in which one unit cell dimension is very much larger than the other two, it is possible to record two levels at the same time by arranging this axis to be parallel to the X-ray beam. The screen and film positions are adjusted to the average value of the two individual settings for the adjacent levels and the appropriate width of the screen annulus calculated as described above.

9.5.1.2 The setting of a protein crystal on the precession camera

The setting of protein crystals on a precession camera is relatively straightforward because the large unit cell size ensures that many reflections are in the reflecting position for any orientation of the crystal. The procedure involves a series of still photographs to obtain an approximate orientation (to within 30' of arc say) followed by a series of small angle precession photographs which allow the crystal to be set to within 5' of arc.

Still photographs. The crystal is orientated on its goniometer head as accurately as possible under the polarizing microscope. It is most convenient to arrange the plane of one of the goniometer arcs parallel to the normal (i.e. ON Fig. 9.6a) of the desired reciprocal lattice plane. We will call this arc the "parallel arc". The plane of the other arc is then perpendicular to the normal of the reciprocal lattice plane. We will call this the "perpendicular arc". The crystal and goniometer are placed on the precession camera and the crystal centred in the X-ray beam. The dial setting, which corresponds to the rotation ϕ about the goniometer axis, is chosen so that the X-ray beam is directed approximately normal to the desired reciprocal lattice plane. A still photograph is taken. Typical exposure times for protein crystals are from 3 to 10 minutes, but depend very much on the size of the crystal and the X-ray source. For example with phosphorylase crystals dimensions 0.6 x 0.6 x 1.0 mm and unit cell edges 129.1 x 129.1 x 116.3 Å, still exposure times are of the order of 5 mins with a rotating anode X-ray source. These crystals are easy to orientate. For crystals which are difficult to set or are unknown it is advisable to use longer exposure times so as to obtain as much information as possible.

Provided that the crystal has been oriented to within about 20° of the desired setting, the still photograph will show a series of concentric circles (Fig. 9.7a). These circles correspond to the points where the reciprocal lattice layers cut the sphere of reflection. The circles are only seen with crystals of large unit cells where the reciprocal lattice planes are densely populated. From Fig. 9.8 it is

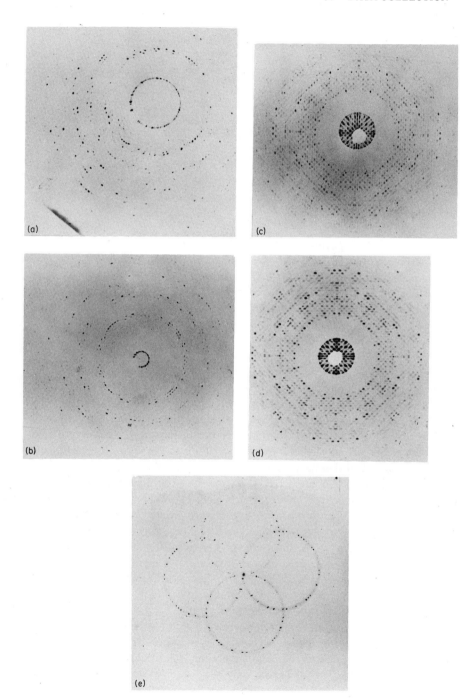

9.5.1. THE PRECESSION CAMERA

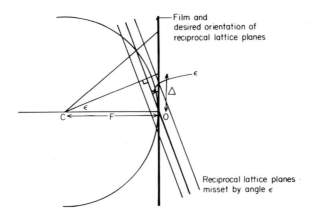

FIG. 9.8   Geometry of mis-setting of a still precession photograph.

apparent that there is a simple geometrical relationship between the mis-setting angle (ϵ) of the crystal and the displacement (Δ) of the centre of the circles from the straight through X-ray position, i.e.

$$\tan \epsilon \approx \frac{\Delta}{F}$$

where F is the crystal to film distance.

The crystal misorientation is resolved into two components; a vertical component which corresponds to a dial mis-setting and a horizontal component which corresponds to a parallel arc mis-setting. These two correction angles may be read off from the chart (shown in Fig. 9.9) and the corrections applied to the crystal. If the crystal is not at the centre of the arc, it will also be necessary to re-adjust the goniometer slides to bring the crystal back to the centre of the beam.

A second still photograph should show the circles approximately centred on the straight through position. Note that the zero level is now tangential to the sphere of reflection and does not appear on the photograph. The crystal is now set probably to within 15'.

FIG. 9.7   Setting photographs of a lysozyme crystal on a precession camera. The crystals are tetragonal space group $P4_32_12$ and unit cell dimensions $a = b = 79.1$Å, $c = 37.9$Å. When the crystal is set c(c*) is parallel to the X-ray beam at $\bar{\mu} = 0°$. (a) Still photograph: crystal mis-set; (b) still photograph: crystal almost set; (c) $\bar{\mu} = 3$: crystal mis-set; (d) $\bar{\mu} = 3°$ crystal set (note: backstop is displaced). (e) Screen test corresponding to the four positions of Fig. 9.1c. One position is usually sufficient. Crystal to film distance = 7.5 cm. Reduction x2.

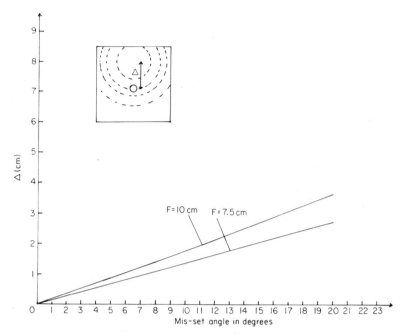

FIG. 9.9   Chart for mis-setting angle for still precession photograph.

9.5.1.3 Small angle precession

A small angle precession photograph ($\mu = 2° - 3°$) is then taken without inserting a screen and this is shown in Fig. 9.7c. The circular area of the zero level and the annular regions of the upper levels that have passed through the sphere of reflection are apparent. If the crystal were perfectly set, the centre of the zero level circle would correspond to the straight through position of the X-ray beam. The angle of mis-setting may be derived with reference to Fig. 9.10 (see also Henry *et al.*, 1961, p. 139). The thick line represents the correct position of the reciprocal lattice level and the thinner line represents the actual position of the reciprocal lattice level which is mis-set by a small angle ϵ. This means that, with a precession angle $\bar{\mu}$ and in one extreme position, the plane will be tilted an angle $\bar{\mu} + \epsilon$ and part of the precession pattern outside the limiting radius of $2 \sin \bar{\mu}$ will be developed. The distance from the straight through position to the edge of the circle is therefore *OJ* which gives a distance r_1 on the film of

$$r_1 = F\left(2 \sin \bar{\mu} + \frac{\sin 2\epsilon}{\cos(2\epsilon + \bar{\mu})}\right)$$

At the other extreme position the reciprocal lattice plane will be tilted $\bar{\mu} - \epsilon$ and part of the diffraction pattern normally developed within the limiting radius of

9.5.1. THE PRECESSION CAMERA

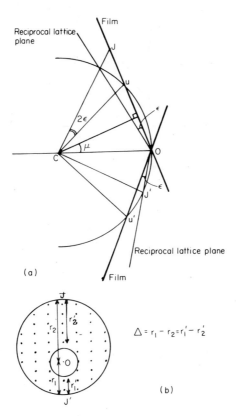

FIG. 9.10 (a) Geometry of mis-setting of a small angle precession photograph. (b) Schematic representation of the appearance of zero level for a small angle precession photograph of a mis-set crystal.

$2 \sin \bar{\mu}$ will not cut the sphere of reflection. The distance from the straight through position to the edge of the circle is therefore OJ' which gives a distance r_2 on the film of

$$r_2 = F\left(2 \sin \bar{\mu} - \frac{\sin 2\epsilon}{\cos(2\epsilon - \bar{\mu})}\right)$$

The difference $\Delta = r_1 - r_2$

$$= \frac{F 2 \sin 2\epsilon \cos 2\epsilon \cos \bar{\mu}}{\cos^2 2\epsilon \cos^2 \bar{\mu} - \sin^2 2\epsilon \sin^2 \bar{\mu}}$$

$\simeq 4\epsilon F$ if $\bar{\mu}$ and ϵ are small.

For $F = 100$mm, a rough guide is therefore

$$\epsilon = \Delta \times 8.5'$$

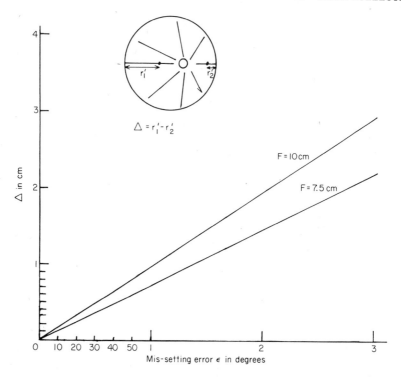

FIG. 9.11 Chart for mis-setting angle for small angle precession photograph.

The position of the straight through X-ray beam is always obscured by the backstop which may not be perfectly centred. Δ is best estimated by taking two equivalent reflections on either side of the X-ray beam as reference points and measuring the distances of these points to the edge of the limiting circle. A chart of the correction factor is shown in Fig. 9.11. As with still photographs, the correction factor is resolved into two components, one corresponding to the correction for the dial setting and the other for the correction to the arc setting. The precession angle should, of course, be reset to zero before any correction is made.

The appearance of a perfectly set crystal photograph is shown in Fig. 9.7d.

9.5.1.4 *Screen setting photograph*

Before the exposure for the large angle ($\bar{\mu} = 9 - 23°$) screened precession photograph is started, it is as well to check the position of the screen. The screen of appropriate radius is placed in the holder and its position set to the correct value for the precession angle $\bar{\mu}$. The precession angle is set but the motor is not switched on and a still photograph is taken. If both the crystal and screen are

9.5.2. SCREENLESS PHOTOGRAPHY

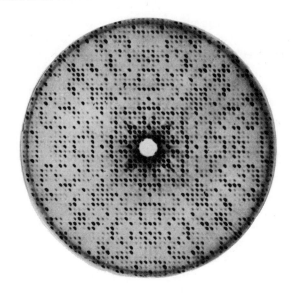

FIG. 9.12 A screened zero-level precession photograph of a lysozyme crystal $\bar{\mu} = 15°$.

well set, then the photograph will have the appearance of a circular annulus with a concentrically arranged ring of spots (Fig. 9.7e).

If the screen setting photograph is taken with the nickel filter in place, then all that remains to be done for the high angle precession photograph is to load the film, start the motor and open the shutter. A $\bar{\mu} = 15°$ photograph of a lysozyme crystal is shown in Fig. 9.12.

9.5.2 Screenless photography

As protein crystal unit cells become larger, it becomes increasingly important to develop methods capable of recording the intensity data most efficiently. The most efficient method may be defined as that which gives the maximum number of recorded reflections in a given time for the minimum amount of incident radiation on the crystal specimen. In both conventional screened precession photographs and Weissenberg photographs, a layer line screen is used to block out all reflections except those lying in a single plane of the reciprocal lattice. This gives rise to films which can be readily indexed by eye but is obviously a relatively inefficient method of recording data since a large proportion of reflections are not permitted to reach the film. In 1968 Xuong and his colleagues (Xuong *et al.*, 1968) and Arndt (1968) pointed out that, with the advent of the fully automatic densitometer directly linked either to a small computer or a magnetic tape disc, it was possible and most advantageous to dispense with layer line screens. The resulting screenless photographs are complicated to index but can be readily indexed by the computer.

In order that reflections will be resolved from one another it is necessary to restrict the precession or rotation angle to a small value (1.0° to 3°) and a complete three-dimensional data set is recorded by sweeping out the whole of the reciprocal space in a series of adjacent or slightly overlapping small angle photographs.

9.5.2.1 Screenless precession photography

Precession cameras are available in almost every protein crystallography laboratory and may be used for screenless work with no modification except to ensure that the collimator system produces a clean X-ray beam and no superfluous scattering. Collimators are discussed in Section 9.3.7.

The indexing of screenless precession photographs and measurement of relative intensities has been described by Xuong and Freer (1971). The treatment given below is an abbreviation of their more detailed discussion.

First it is necessary to derive the equations which relate the position of a spot on the film to the indices hkl. The position of a particular reflection is specified by the reciprocal lattice vector.

$$\mathbf{d}^* = h\mathbf{a}^* + k\mathbf{b}^* + l\mathbf{c}^*$$

where $\mathbf{a}^*$, $\mathbf{b}^*$ and $\mathbf{c}^*$ are in dimensionless reciprocal lattice units.

The position of a spot on the film is specified by the film co-ordinate system (X, Y) in which the origin is at $0'$, the point on the film which remains invariant during precession, and the X axis is parallel to the camera spindle axis (Fig. 9.13). It is most convenient to refer the reciprocal lattice point to a Cartesian system in which the origin 0 is the origin of the reciprocal lattice, the x axis is parallel to the film X axis and the y axis is parallel to the film Y axis. The z axis is perpendicular to the film and directed toward the X-ray source. Hence

$$x = h\mathbf{a}_x^* + k\mathbf{b}_x^* + l\mathbf{c}_x^*$$

$$y = h\mathbf{a}_y^* + k\mathbf{b}_y^* + l\mathbf{c}_y^*$$

$$z = h\mathbf{a}_z^* + k\mathbf{b}_z^* + l\mathbf{c}_z^*$$

where $\mathbf{a}_x^*$ is the projection of the a^* axis on to the x Cartesian axis etc. The x, y, z co-ordinate system remains fixed, while the reciprocal lattice co-ordinate system moves with the crystal.

Let P be a reciprocal lattice point on the sphere of reflection. The point P' at which the reflected beam strikes the film is specified by $\mathbf{O'P'} = \mathbf{O'S'} + \mathbf{S'P'}$ where SS' is the normal to the plane of the film which passes through the centre, S, of the sphere of reflection. Angle $SS'O'$ is therefore $\bar{\mu}$, the precession angle. The

9.5.2. SCREENLESS PHOTOGRAPHY

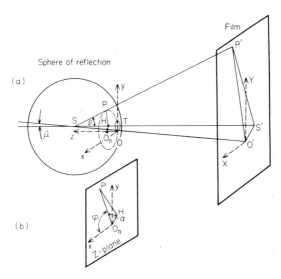

FIG. 9.13 The diffraction geometry of screenless precession photography. (a) Overall view. (b) Detailed geometry in rational plane of the diffracting reciprocal-lattice point P. (from Xuong and Freer, 1971)

film co-ordinates are

$$X = (O'S')_X + (S'P')_X$$
$$Y = (O'S')_Y + (S'P')_Y$$

where $(O'S')_X$ is the projection of $O'S'$ along the x axis etc. From simple trigonometry it can be shown that

$$X = F\left[\sin\bar{\mu}\cos\beta + \frac{(x - \sin\bar{\mu}\cos\beta)\cos\bar{\mu}}{\cos(\bar{\mu}) - z}\right]$$

$$Y = F\left[\sin\bar{\mu}\sin\beta + \frac{(y - \sin\bar{\mu}\sin\beta)\cos\bar{\mu}}{\cos(\bar{\mu}) - z}\right]$$

where F is the crystal to film distance, a^*, b^* and c^* are in reciprocal lattice units and the radius of the sphere of reflection = $SO = 1$.

β is the angle $x\, O_n H$ where O_n, H and P lie in the z plane, the rational reciprocal lattice plane which is parallel to the plane of the film. O_n and H are the intersections of Oz and SS' with the z plane (Fig. 9.13).

There are two possible values for the angle β

$$\beta_1 = \phi + \alpha \text{ and } \beta_2 = \phi - \alpha$$

where

$$\tan \phi = \frac{y}{x} \quad \text{and} \quad \cos \alpha = \frac{x^2 + y^2 + z^2 - (2 \cos \bar{\mu})z}{2(x^2 + y^2)^{1/2} \sin \bar{\mu}}$$

i.e.

$$\cos \alpha = \frac{\xi^2 + \sin^2 \bar{\mu} - \sin^2 \bar{\nu}}{2\xi \sin \bar{\mu}}$$

in cylindrical co-ordinates where $\cos \bar{\nu} = \cos \bar{\mu} - z$.

The two values of β correspond to the two positions in which each reflection is recorded and because these two values are not equal upper level reflections will be split. Xuong and Freer (1971) report that this splitting can be minimized by reducing the crystal to film distance by $\delta = 0.1F$ along the film normal. The equations then become

$$X_i = F \sin \bar{\mu} \cos \beta_i + \frac{(x - \sin \bar{\mu} \cos \beta_i)(F \cos \bar{\mu} - \delta)}{\cos \bar{\mu} - z}$$

$$Y_i = F \sin \bar{\mu} \sin \beta_i + \frac{(y - \sin \bar{\mu} \sin \beta_i)(F \cos \bar{\mu} - \delta)}{\cos \bar{\mu} - z}$$

where $i = 1,2$.

Thus it is possible to compute the position of any reflection in the film given the unit cell dimensions, the h,k,l indices of the reflection, the relative orientation of the crystal, the small angle precession $\bar{\mu}$, and the crystal to film distance.

These equations show us how to compute the position of a reflection in the film. In order for this reflection to be recorded the following conditions must be fulfilled:

(1) Angle $\bar{\nu}$ must be real, i.e. $|\cos \bar{\nu}| = |\cos \bar{\mu} - z| < 1$

(2) There will be a minimum and maximum value of the radius ξ within which spots will be recorded. These values correspond to the outer and inner limiting circles already given for conventional screened precession photography, i.e.

$$\xi_{max} = \sin \bar{\nu} + \sin \bar{\mu}$$
$$\xi_{min} = \sin \bar{\nu} - \sin \bar{\mu}$$

This gives rise to a blind region within which no reflections are recorded.

In addition to these geometrical factors two further aspects need to be considered to test whether a reflection has been fully recorded and is measurable. These are:

(1) A complete reflection cannot be recorded for reciprocal lattice points that lie too near the cut off limits ξ_{min} and ξ_{max}. A suitable criterion is that the

9.5.2. SCREENLESS PHOTOGRAPHY

distance c given by

$$c = F \, | \, (\xi_{max} - \xi) \, |$$

or

$$c = F \, | \, (\xi_{min} - \xi) \, |$$

should be greater than the spot size of the reflection on the film. The Lorentz factor varies very rapidly in this region and this forms an additional reason for rejecting reflections in the cut-off region.

(2) The optimum precession angle $\bar{\mu}$ should be chosen for a particular crystal orientation so that there is a minimum number of reflections which overlap in the film.

A useful film to film dial setting increment $\Delta \tau$ is given by

$$\Delta \tau = 2\bar{\mu} - 0.5°$$

This gives sufficient overlaps between films to ensure that reflections partially recorded on one film are fully recorded on the next.

In order for the computer to be able to predict the spot position precisely it is necessary to know the crystal orientation to within a few minutes of arc. This may be determined by taking a conventional small angle setting photograph (Section 9.5.1). The position of the film in the cassette must also be known and for this purpose it is most convenient to have three small holes drilled at the edge of the cassette in the shape of an L. Fiducial marks are recorded on the film by light shone through the holes and allow the automatic determination of the film orientation by the computer. Xuong and Freer (1971) describe a series of programmes which are used to measure the intensities. First the film is placed on the automatic densitometer. The film is scanned and a digitized image of the film recorded on magnetic tape or disc. The spot positions of all fully recorded reflections for the particular dial setting and precession angle are calculated from the known crystal orientation by means of the equations given above. The integrator programme then scans the densitometer tape for each reflection at the appropriate position using an empirical correction to find the precise centre of each spot. Intensities and background are measured and the data are put out in a form ready for the next stage in data processing.

For example in the three-dimensional data collection for the trypsin–trypsin inhibitor complex (Ruhlmann et al., 1973) crystals were oriented with their longest axis ($c = 122.9$Å) parallel to the spindle axis. The precession angle was 1.5° and photographs were taken at 2.5° intervals of the dial setting from 0° to 80°. In a typical film 2600 reflections were recorded, 360 were overlapped and 500 lay on the edge, so that 1740 measurable intensities remained. Within the 2.8Å resolution limit, approximately 600 reflections were missing largely due to the blind region which is unavoidable if the crystal is rotated about one axis only.

The general reproducibility of photographic data obtained by the screenless precession method (to date) is represented by reliability index values of between 7% to 9% where

$$R = \frac{\sum_h \sum_i |k_i I_i - \bar{I}|}{\sum_h \sum_i k_i I_i}$$

where I_i is the intensity of the ith individual measurement and $\bar{I}$ is the mean value of the intensity of this particular reflection. k_i is the scaling constant for a particular film. The summation is over all reflections. The value of R is generally rather worse than that for a corresponding set of data recorded on a diffractometer (where values 4% to 6% are typical). However further improvements are likely both in the recording and measuring of screenless precession data and with improved methods of data processing. There are now a number of structures (e.g. chymotrypsinogen, cytochrome c_2 trypsin—trypsin inhibitor complex) where data have been collected, for the greater part, by screenless precession techniques and where these data have led to interpretable electron density maps of high quality.

9.5.2.2 The rotation (oscillation) camera

The second method of screenless photography, the rotation or oscillation camera, has been slower in coming into use than the precession camera. This is because it was necessary to develop a new instrument or substantially modify existing equipment. The rotation camera enjoys certain advantages over screenless precession photography and with the advent of a commercial instrument on the market (Arndt et al., 1973) it is likely that the instrument will become more widely used in protein crystallography.

The principle of the rotation camera is very simple. The only movement is the rotation of the crystal about one axis. The rotation axis is normal to the X-ray beam and the reflections are recorded on a flat film (Fig. 9.14). Other geometries, such as inclination geometry (see Section 9.6.2.1) and possibly cylindrical films, might prove advantageous in the future, but the present design has the advantage of simplicity. As with screenless precession photography, the collimation of the X-ray beam is very important since any scattering arising from the collimator will increase the general background level of the film. A typical rotation photograph is shown in Fig. 9.15.

The geometry of reflection is shown in Figs 9.14 and 9.16. As the crystal is rotated about the goniometer axis ϕ, individual reciprocal lattice points will cut the sphere of reflection and these reflections will be recorded on the film. From Fig. 9.16 it can be seen that the normal beam geometry gives rise to a blind region of radius given by

$$\xi_{\min} = 1 - (1 - \zeta^2)^{1/2}$$

9.5.2. SCREENLESS PHOTOGRAPHY

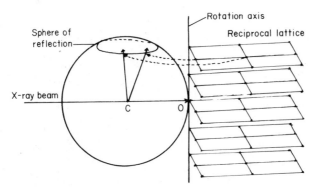

FIG. 9.14   The geometry of the rotation camera. The figure shows part of the reciprocal lattice and indicates the manner in which reciprocal lattice points can be brought into the reflecting position by a rotation about the rotation axis.

The actual proportion of reflections which are not recorded because they lie within the blind region is small (approximately 3% of the total number of reflections at 3Å).

As with screenless precession photographs, three-dimensional data are collected by taking a series of small angle rotation photographs ($\Delta\phi = 1°$ to $3°$) through an appropriate sweep of the dial axis. Spots which are partially recorded on one film, because they occur too close to the beginning or end of the rotation range, will have their remaining portion recorded on the adjacent film. Provided that the camera rotation or oscillation mechanism is sufficiently free from backlash, it should be possible to add these individual components of the spot and hence obtain an integrated intensity. This is not possible with screenless precession photographs since the volumes of reciprocal space recorded for two separate crystal orientations are not contiguous in this way. Moreover the Lorentz factor approaches infinity for such reflections in screenless precession photographs. The feasibility of addition of partially recorded reflections has been tested by Arndt *et al.*, (1973). They show that with their camera which was constructed in such a way that one photograph can start at an angle ϕ identical to within 0.005° with that at which the previous one finished, the sum of the partially recorded intensities of reflections recorded on two films did not differ significantly from the intensity of the corresponding fully recorded reflection. The ability to add partially recorded reflections is likely to prove one of the main advantages of the rotation camera, especially with protein crystals with very large unit cells where it is necessary to restrict the rotation angle to a very small value in order to prevent overlap of spots.

The Lorentz factor is given by

$$\frac{1}{L} = (\sin^2 2\theta - \zeta^2)^{1/2}$$

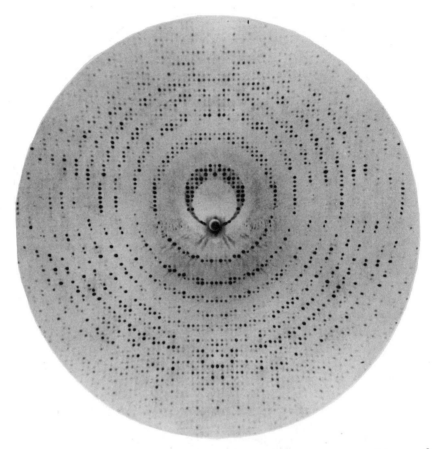

FIG. 9.15 A rotation photograph of a phosphorylase **b** crystal. $\Delta\phi = 1.25°$ The crystals are tetragonal space group $P4_32_12$ with unit cell dimensions $a = b = 129.1$ Å, $c = 116.2$ Å. The c (c*) axis is parallel to the rotation axis which is horizontal. Crystal to film distance = 10.0 cm.

which approaches zero as the blind region is approached but otherwise varies much less rapidly than the corresponding correction for screenless precession photographs. (see Chapter 10).

The total rotation angle through which it is necessary to move the crystal in order to record the whole of reciprocal space is given by $\phi = 180° + 2\theta_{max}$ (Fig. 9.17). If the crystal has symmetry then this angle is less than 180°. For example with a tetragonal crystal point group 422 mounted to rotate about the unique four fold axis, for data to a resolution of 3Å a rotation of $\phi = 45°$ will ensure that each reflection is recorded four times. (Reflections hkl and $hk\bar{l}$ will be recorded on the same film and others, e.g. reflections of the type $\bar{h}kl$ and $\bar{h}k\bar{l}$,

9.5.2. SCREENLESS PHOTOGRAPHY

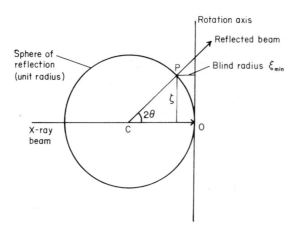

FIG. 9.16 The blind region radius ξ_{min} within which reflections cannot be recorded by normal beam geometry.

on a second film.) The area of reciprocal space swept out is shown in Fig. 9.18.

The optimum rotation angle $\Delta\phi$ for each film is chosen so as to allow the maximum number of recorded reflections with the minimum amount of overlap. $\Delta\phi$ should be at least as great as the minimum rocking range (Δ) (Section 9.3.6) which is determined by the mosaic spread of the crystal and the geometrical factors of source size and crystal size. Arndt et al., (1973) have shown that

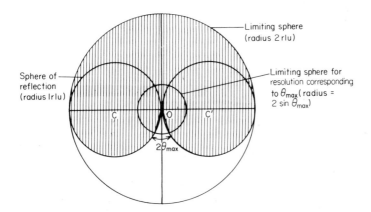

FIG. 9.17 Projection down rotation axis showing the area of reciprocal space swept out as the crystal is rotated through 180°. (Shaded area). In the diagram the reciprocal lattice has been kept fixed and the centre of the sphere of reflection rotated from C to C'. In order to explore the full sphere for data to a resolution corresponding to θ_{max} it is necessary to rotate the crystal through $180° + 2\theta_{max}$. The small circle corresponds to 3Å resolution.

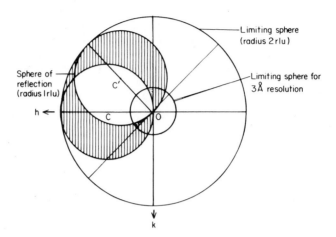

FIG. 9.18 Projection down rotation axis showing the area of reciprocal space swept out as the crystal is rotated through 45° (shaded area), i.e. as the centre of the sphere of reflection is rotated from C to C'. The small circle corresponds to 3Å resolution. For a tetragonal crystal point group 422 with c parallel to the rotation axis, a 45° rotation will permit measurement of 3Å data.

overlap of reflections is most likely to occur in the equatorial region at maximum Bragg angle where lines in the equatorial plane of reciprocal space are most highly populated. The maximum permissible angular range $\Delta\phi$ which avoids overlap is given by

$$\Delta\phi = \frac{p^*}{dm^*} - \Delta$$

where p^* is the relevant reciprocal lattice spacing and dm^* is the maximum resolution in reciprocal space. In some cases it may be preferable to use a slightly larger value of $\Delta\phi$ and accept some overlap of reflections in order to reduce the number of films in a data set. Exposure times are governed by the crystal size, unit cell dimensions and the X-ray source. For exposure times of longer than 1 hour, it is desirable to oscillate the crystal through the angle $\Delta\phi$ in order to average the effects which may lead to fluctuations in intensity (such as instability of the X-ray source or radiation damage).

The angular separation of adjacent spots is given by

$$\Delta_{2\theta} = \frac{p^*}{\cos\theta}$$

The crystal size and source to crystal distance need to be adjusted so that the angular width of the reflected beam $\delta < \Delta_{2\theta}$ (Section 9.3.6).

9.5.2. SCREENLESS PHOTOGRAPHY

(i) The positions of reflected spots on the film. First let us assume that the crystal has been oriented to rotate about a real cell axis (**c** say). The zero position of ϕ is defined at the position where the reciprocal lattice axis which cyclicly follows the rotation axis (**a***) is anti-parallel to the X-ray beam. It is most convenient to refer the reciprocal lattice axes to an orthogonal axial system (x_0, y_0, z_0), where z_0 is parallel to the rotation axis and x_0 is anti-parallel to the X-ray beam. A reciprocal lattice point P which is specified by the vector

$$\mathbf{d^*} = h\mathbf{a^*} + k\mathbf{b^*} + l\mathbf{c^*}$$

is then expressed as

$$x_0 = ha_x^* + kb_x^* + lc_x^*$$
$$y_0 = ha_y^* + kb_y^* + lc_y^*$$
$$z_0 = ha_z^* + kb_z^* + lc_z^*$$

where a_x^* is the projection of the **a*** reciprocal lattice axis on x_0 etc.

Consider the reciprocal lattice point P which cuts the sphere of reflection after a rotation of ϕ. This is illustrated in Fig. 9.19. Instead of showing the rotation of the crystal through an angle of $+\phi$ we show the rotation of the sphere of reflection through an angle of $-\phi$. The centre of the sphere of reflection moves from position $C(1,0,0)$ to a position C_1 ($\cos\phi$, $\sin\phi$, 0). Since P lies on the sphere of reflection, (x_0, y_0, z_0) must satisfy the equation of the sphere, i.e.

$$(x_0 - \cos\phi)^2 + (y_0 - \sin\phi)^2 + z_0^2 = 1$$

which gives the solution for ϕ in terms of

$$\tan\frac{\phi}{2} = \frac{2y_0 \pm \sqrt{4y_0^2 + 4x_0^2 - d^{*4}}}{2x_0 + d^{*2}}$$

where $d^{*2} = x_0^2 + y_0^2 + z_0^2$

The two real values for ϕ correspond to the two rotation angles at which the point P cuts the sphere of reflection.

To find the co-ordinates of the reflection on the film we define the orthogonal co-ordinate system (Y, Z) in which Z is parallel to the rotation axis and the origin is the point at which the main beam strikes the film. The relationship between the co-ordinate system (x, y, z) and (x_0, y_0, z_0) is given by (Fig. 9.20)

$$x = x_0 \cos\phi + y_0 \sin\phi$$
$$y = y_0 \cos\phi - x_0 \sin\phi$$
$$z = z_0$$

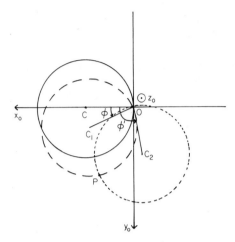

FIG. 9.19 Projection down rotation axis showing the reciprocal lattice point P (in zero level) in reflecting position for rotations ϕ and ϕ'. The reciprocal lattice has remained stationary, while the centre of the sphere of reflection has rotated to $C_1(\phi)$ and to $C_2(\phi')$.

Hence from Fig. 9.21

$$\frac{Y}{F} = \frac{y}{1-x}$$

where F is the crystal to the film distance and

$$\frac{Z}{z_0} = \left(\frac{F^2 + Y^2}{1 - z_0^2}\right)^{1/2}$$

i.e.

$$Y = F\left(\frac{y_0 \cos\phi - x_0 \sin\phi}{1 - x_0 \cos\phi - y_0 \sin\phi}\right)$$

$$Z = z_0 \left(\frac{F^2 + Y^2}{1 - z_0^2}\right)^{1/2}$$

Alternatively these co-ordinates may be given in terms of the cylindrical co-ordinates (ξ, ζ, ϕ) of the reciprocal lattice point as (Buerger, 1942)

$$Y = F \tan \cos^{-1} \frac{2 - (\zeta^2 - \xi^2)}{2(1 - \zeta^2)^{1/2}}$$

$$Z = F \frac{2\zeta}{2 - \zeta^2 - \xi^2}$$

9.5.2. SCREENLESS PHOTOGRAPHY

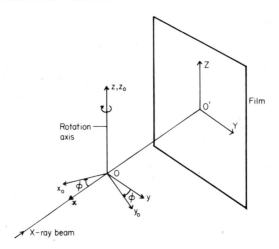

FIG. 9.20 The relationship between the crystal co-ordinate system (x_0, y_0, z_0) which rotates with the crystal, and the (x, y, z) co-ordinate system which is defined such that x is antiparallel to the incident beam and z is parallel to the rotation axis. The (Y, Z) co-ordinate system of the film is also shown.

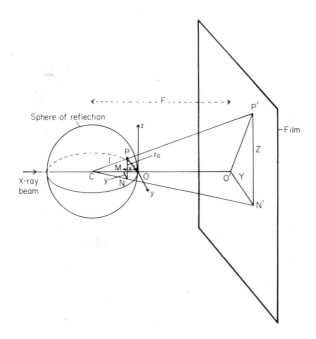

FIG. 9.21 The relationship between the film co-ordinate system (Y, Z) and the crystal co-ordinate system (x, y, z). Crystal to film distance = $\overline{CO'} = \overline{F}$.

Thus for any reciprocal lattice point and orientation of the crystal it is possible to calculate the position of the reflected spot on the film and the value of ϕ at which this reflection should occur.

(ii) Determination of crystal orientation. During the measurement of the intensities of the reflections on photographic film, it is necessary that the precise orientation of the crystal should be known to within $5'$ of arc so that all partially and fully recorded reflections and their positions may be calculated.

The crystal may be set to within about $5'$ of arc, by a series of still zone axis photographs similar to the still setting photographs on the precession camera. Consider a crystal with orthogonal axes mounted about the c axis with the a* axis anti-parallel to the incident X-ray beam. The still photographs comprise a series of circles corresponding to $h = 1, 2, 3$ etc. and centred on the position of the straight through beam. If the crystal is mis-set by a small angle ϵ, the centre of the $h = 1$ circle is displaced a distance Δ from the position of the direct beam, where the relationship between Δ and ϵ is given by (Mosley, 1973)

$$\Delta = \frac{F}{2} [\tan(2\theta + \epsilon) + \tan(2\theta - \epsilon)]$$

where the value of θ is given by

$$\cos 2\theta = 1 - \mathbf{a}^*$$

A plot of Δ against the correction angle ϵ may be constructed for a particular crystal system. Thus a series of still photographs may be used to adjust the goniometer arcs and ϕ rotation axis so that these angles are set to within $5'$.

In order to determine the setting more precisely, two small angle rotation photographs $90°$ apart on the rotation dial are taken. A number of reflections which are partially cut off at either end of the rotation range are indexed by means of a computer-generated list of reflections and their positions for this particular rotation range. If the crystal were perfectly set and the unit cell parameters known precisely, then these reflections should lie on the Ewald sphere for either the initial or final values of the rotation angle ϕ. The condition that a point P lies on the sphere of reflection may be expressed by the condition:

$$\Delta^2 = d^{*2} - 2x_0 \cos\phi - 2y_0 \sin\phi = 0$$

where $d^{*2} = x_0^2 + y_0^2 + z_0^2$ (Fig. 9.22).

Hence values of the three crystal mis-setting angles and precise estimates of the unit cell dimensions may be obtained from a least squares refinement of the observed distance Δ^2 (= zero) and the calculated distance, assuming a perfect crystal setting. These corrections may then be applied in the computation of the spot positions.

The precise film position and orientation may then be determined from the refinement of the experimentally measured co-ordinates of a few reflections.

9.5.3. FILM DEVELOPMENT

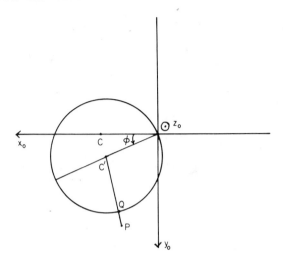

FIG. 9.22 The condition that the point P lies on the sphere of reflection after a rotation ϕ is given by $C'P = C'Q$. $C'P^2 = (x_0 - \cos\phi)^2 + (y_0 - \sin\phi)^2 + z_0^2$. $C'Q = 1$

This second refinement should allow spot positions to be predicted to within 0.1 mm, which is sufficiently accurate to permit automatic densitometry. Data processing is discussed further in Chapter 10.

Data obtained with the rotation camera compare favourably with those obtained with diffractometer measurements (Arndt et al., 1973). It is anticipated that this method will become more widely used in the future.

The rotation camera has the additional advantage that the flat film cassette can be easily replaced by an electronic area detector, when these new detectors become available. Arndt et al. (1972) have described a television recording system in which the X-ray diffraction pattern falls on to a ZnS phosphor screen deposited on a 40 mm diameter fibre optics plate which is optically coupled to an image isocon camera tube. A phosphor–image-intensifier–film combination has been explored by Minor et al. (1974) but although the overall exposure times were reduced, the agreement among the measurements was disappointing. An alternative approach is favoured by Cork et al. (1974) who have shown that a multi-wire proportional counter (of the type widely used in high energy physics) with an electronic read-out into a large core memory system may be specially suitable for protein crystallography. With both methods it appears that exposure times may be reduced 60 fold compared with photographic film.

9.5.3 Film development

In order to produce intensity data of high quality and reproducibility it is essential that a correct and standard routine development procedure should be adopted. The effective speed and sensitivity of the film depends on the developing conditions and the manufacturers recommendations should be followed. The steps involved are:

9.5.3.1 Development

The developer is a reducing agent, usually an alkaline solution of aromatic amines or phenols, which decomposes the exposed silver halide grains to metallic silver. Most developers contain a preservative to prevent oxidation but the developer tank should be kept covered when not in use and solutions changed regularly. The optimum developing time will be that for which exposed grains are developed and unexposed grains are not. This optimum time is temperature dependent and the developer tank should be kept at a constant temperature by means of a water bath. Uneven agitation during development can produce undesirable variations in density and contrast. A good method of controlled and gentle agitation is to bubble nitrogen gas through the solution.

9.5.3.2 Stop bath

The stop bath usually consists of a dilute solution of acetic acid which stops development by neutralizing the alkali in the developer-soaked emulsion.

9.5.3.3 Fixer

The fixer, which contains sodium thiosulphate, dissolves the unexposed silver halide grains which would otherwise blacken on exposure to light. A good guide is to fix for twice the time the film takes to clear and to keep the safe lights on for at least half this period.

9.5.3.4 Washing

The purpose of washing is to remove the residual chemicals which would otherwise stain the film.

The following procedure has been found satisfactory for Ilford Industrial G film:

Developer: Ilford Phenisol; 5 min at $20°C$ with continuous agitation by a gentle stream of nitrogen gas.
Stop Bath: 2% Acetic acid; 1 min.
Fixer: May and Baker Super Ampfix; 10 min.
Wash: Bath of running tap water; 30 min.
Dry: Oven at $37°C$; 45 min.

9.6 DIFFRACTOMETERS

The ease and precision with which X-ray intensity measurements may be made by means of fully automated diffractometers has greatly contributed to the

9.6.1. THE FOUR CIRCLE DIFFRACTOMETER

increased growth rate of successful protein crystallographic analyses. In a diffractometer we require a mechanical system which will enable us to orient the crystal so as to bring each reflection in turn into the reflecting position and to move the detector so as to receive this reflection. The crystal is then rocked through the reflecting position by rotating about a suitable axis and the counts accumulated by the detector. Instruments in use in protein crystallography are of two types which are based on different geometries.

(i) The four-circle diffractometer operates with "normal-beam equatorial" geometry. The crystal orientation is determined by three Eulerian angles which need to be calculated for each reflection. The detector is constrained to move in the equatorial horizontal plane which contains the incident beam and is set to the 2θ value appropriate for each reflection. The crystal and detector setting angles are calculated by a small on-line computer which also controls the digital diffractometer shaft settings. The great advantage of these instruments is that, for certain geometries, all reflections are accessible and may be measured fully automatically.

(ii) The linear diffractometer operates with inclination geometry in which both the incident beam and the diffracted beam are inclined at variable angles, $90°-\mu$ and $90°-\nu$ respectively, to the crystal goniometer head axis, ϕ. The crystal is rotated about the single axis ϕ to bring each reflecting plane to the reflecting position and the detector is tilted by an angle $90°-\nu$ with respect to the vertical axis and rotated through Υ to receive the diffracted beam. In the linear diffractometer the crystal and detector rotations ϕ and Υ are linked through a series of shafts. The settings are determined by a mechanical analogue computer. The angle $90°-\nu$ has to be set by hand for each level.

Diffractometers are discussed in detail by Arndt and Willis (1966). In this section we give a brief account of the basic geometries of the two types of diffractometers and discuss their special adaption to protein crystallography.

9.6.1 The four circle diffractometer

9.6.1.1 The geometry of the four-circle diffractometer

In the four circle diffractometer the crystal orientation is defined by three angular settings (ϕ, χ, ω). The detector position is defined by a fourth angle, 2θ, and the detector is constrained to move in the horizontal equatorial plane containing the incident X-ray beam. The goniometer head which carries the crystal is mounted on the ϕ circle, and ϕ, as usual, indicates a rotation about the goniometer head axis. The ϕ circle is carried on the χ circle, which lies in the vertical plane. The angle χ is the angle made by the ϕ axis with the vertical diameter. The χ-circle is carried on the ω-circle. The axes of these two circles are perpendicular to one another. Therefore ω indicates a rotation about a vertical axis (Fig. 9.23). The axis of the ω circle represents the axis about which the

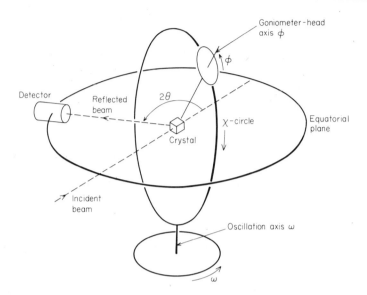

FIG. 9.23   Normal-beam equatorial geometry. The crystal is mounted on a goniometer-head attached to the ϕ-axis; the ϕ-circle moves round the vertical χ-circle, and the $\phi\text{-}\chi$ assembly rotates as a whole about the vertical ω-axis. The detector moves in the horizontal, equatorial plane and the incident beam is normal to the crystal oscillation axis ω (From Arndt and Willis, 1966). Zero positions are defined as follows: At $\omega = 0$ the plane of the χ circle is perpendicular to the X-ray beam. At $\chi = 0$ the axis of the ϕ circle is parallel to the vertical diameter of the χ circle. The zero position of ϕ, the angle of rotation about the goniometer head axis, is arbitrary but may be defined with respect to the crystal orientation (e.g. with crystal mounted to rotate about c, zero ϕ corresponds to a^* parallel to the X-ray beam).

crystal rotates during measurement. This axis is always normal to the X-ray beam and hence the four circle diffractometer operates with normal beam geometry.

The settings of the three crystal angles ϕ, χ and ω and the detector angle, 2θ, may be determined from the basic conditions for diffraction. The crystal will be in the reflecting position for a certain set of crystal planes when the glancing angle made by the incident beam with respect to these planes is equal to θ_{hkl}, the Bragg angle for the reflection hkl (Fig. 9.24). According to Bragg's Law the incident beam s_o, the diffracted beam s and the normal to the reflecting planes, S, all lie in the same plane which in the case of the four circle diffractometer is the equatorial plane.

The ω-axis, about which the crystal rotates is normal to the equatorial plane and therefore lies in the reflecting plane. Under these conditions the velocity with which a reciprocal lattice point passes through the Ewald sphere is a

9.6.1. THE FOUR CIRCLE DIFFRACTOMETER

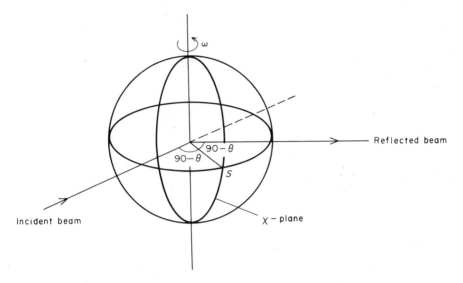

FIG. 9.24 Symmetrical A setting for the four-circle diffractometer $\omega = \theta$.

function of θ only and the view of the source is the same for all reflections (Lang, 1954).

Within certain limits, the normal to the set of *hkl* planes can make any angle with the χ plane and the conditions for diffraction, defined above, will remain satisfied. A further criterion is needed in order to fully define the geometry of the four circle settings for intensity measurements. The most widely used geometry is the *Symmetrical A setting* (Furnas and Harker, 1955). In this setting the χ-circle is arranged to bisect the angle between the incident and diffracted beams. The χ plane therefore contains the diffraction vectors **S** and the incident and diffracted beams are symmetrically arranged with respect to the χ-circle (Fig. 9.24). Under these conditions the setting angles may be shown to be:

$$\omega = \theta$$

$$\sin \chi = \frac{-\zeta}{(\xi^2 + \zeta^2)^{1/2}}$$

$$\phi = 90° - \tau$$

where (ξ, ζ, τ) are the cylindrical polar co-ordinates of the reflection *hkl* (Section 9.4). The detector arm is moved through an angle 2θ from the straight through position. The Lorentz factor is given by

$$L^{-1} = \sin 2\theta.$$

Other settings of the four circle diffractometer may be more appropriate to

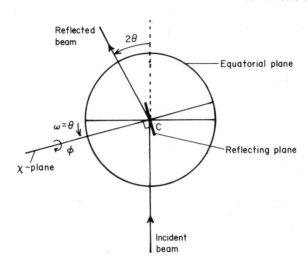

FIG. 9.25 View normal to equatorial plane showing crystal at $\chi = 90$ in symmetrical A setting. The crystal is in the reflecting position for this particular axial reflection for all values of ϕ.

certain experimental conditions such as the need for bulky apparatus in low temperature studies or the use of multiple counters.

From the relationships given above, the unit cell parameters and the orientation matrix of the crystal with respect to the instrument parameters may be calculated from the measured values of (ϕ, χ, ω) for at least three reflections. With these parameters the setting angles for any reflection may be determined.

A simple procedure for setting a crystal on the diffractometer is as follows: Align the crystal by means of the optical microscope, or by a precession camera if necessary, so that a reciprocal lattice axis (c^* say) is approximately parallel to the goniometer axis ϕ. Mount the crystal on the diffractometer and centre in the beam by means of the slides on the goniometer. Select a strong low order 00l reflection and drive the detector to the appropriate 2θ value. Set $\chi = +90$ or -90, $\omega = \theta$. In this position the reflection is ϕ independent (Fig. 9.25). Adjust the goniometer arcs so that the 00l reflection comes up at precisely $\omega = \theta$ for all values of ϕ. c^* is now aligned parallel to the goniometer axis. If this axis is normal to the other two reciprocal lattice axes, then $h00$ and $0k0$ reflections can be rapidly found in the $\chi = 0$ plane by setting $\omega = \theta$ and the detector $= 2\theta$ for the appropriate reflection and varying ϕ. Thus (ω, χ, ϕ) settings for 00l, $0k0$, $h00$ reflections may be found and these values used to calculate the cell dimensions and orientation matrix. Note that this method of setting applies only to orthogonal crystal systems in which a reciprocal lattice axis is parallel to the goniometer axis. In principle the crystal axes may be in any orientation and any three representative reflections used to define the orientation matrix.

9.6.1. THE FOUR CIRCLE DIFFRACTOMETER

The determination of unit cell parameters requires very precise values of θ. It is therefore preferable to determine θ both for hkl and $\bar{h}\bar{k}\bar{l}$ for reasonably strong high angle reflections.

9.6.1.2 Methods of scanning a reflection

There are three methods available for scanning a reflection on a diffractometer.
(i) Stationary crystal–stationary detector in which the peak intensity and the background in the vicinity of the reflections are measured with the crystal and detector both stationary. In this method it is assumed that all the mosaic blocks diffract radiation from different parts of a sufficiently broad uniform source to give an automatically integrated intensity. The technique is especially sensitive to errors arising from non-uniformity of the source and crystal mis-setting and hence it is seldom used in crystallography. An interesting modification of this method due to Wyckoff, which has found some application in protein crystallography, is described below.
(ii) Moving crystal–stationary detector ($\omega-$ scan). This is the most widely used method in protein crystallography. The crystal is moved slowly through the reflecting position and the profile of intensity of the reflection and the background are recorded by the stationary detector. The integrated intensity is the area of the profile of the reflection less the level of the background.
(iii) Moving crystal–moving detector ($\omega/2\theta$ scan). In this method both crystal and detector are moved (the detector at twice the speed of the crystal) and the counts accumulated as in the moving crystal–stationary detector technique.

A comparison of the ω scan and the $\omega-2\theta$ scan methods have been made by Burbank (1964), Alexander and Smith (1962) and Arndt and Willis (1966). In the ω scan the background is lower and more constant than in the $\omega/2\theta$ scan but the background corrected intensity may contain significant contributions from other wavelengths. In protein crystallography where the mosaic spread of the crystal may be large and the data are collected for relatively small Bragg angles, where the wavelength dispersion is small, the ω scan is generally preferred. The $\omega/2\theta$ scan, although more correct, ideally requires some knowledge of the reflection profile and a judicious choice of scan range. Rubredoxin and trypsin–trypsin inhibitor complex represent two protein structures for which data were collected by the $\omega/2\theta$ scan method.

A useful modification of the ω scan method for protein crystallography has been introduced by Watson (Watson *et al.*, 1970) under the title of *Ordinate Analysis* (Fig. 9.26). In this scheme the crystal is moved through the reflecting position in a series of 2N steps. The largest consecutive N intensities are taken as the peak and the remainder as the background. The method therefore allows for a slight mis-setting of the crystal without the need to widen the peak scan range during automatic data collection. The disadvantage of this procedure is that the weakest reflections will tend to be overestimated, since a peak will always be found.

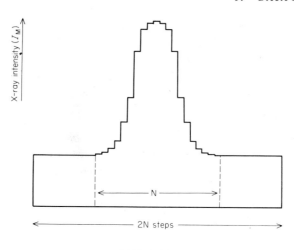

FIG. 9.26

An improved method for predicting peak positions from the observed profile has been proposed (I. J. Tickle, 1975). The peak centroid is computed as

$$\frac{\sum_{i=1}^{2N} (C_i - B) x_i}{\sum_{i=1}^{2N} (C_i - B)}$$

where C_i is the count and x_i is the angular position for the ith step. B is the background level which is taken as the mean of the C_i and the summation is for $C_i > B$ only. The peak is taken as the sum of the N steps centred on this position and the sum of the remaining N steps is taken as background. This method overcomes the tendency of the conventional ordinate analysis procedure to over-estimate weak reflections. It would not prove satisfactory for split reflections.

With multiple-counter diffractometers ordinate analysis has distinct advantages over a background–peak–background procedure. When the reflections are not precisely simultaneous, the peak of one reflection can be measured while the background of another is being measured. Further, the precise position of the centre of the peak can be defined from the multiple reflections, taking into account their quasi-simultaneous separation (Evans, 1973).

A further refinement of the ordinate analysis procedure has been described by Diamond (1969). In this *Profile Analysis Method* a curve fitting procedure is applied to the shape of the observed profile either by means of an on-line computer programme or by printing out each of the $2N$ intensities of an ordinate analysis sweep followed by subsequent computer analysis of the data. The shape of the profile of one reflection is learnt, applied to the next reflection

9.6.2. THE LINEAR DIFFRACTOMETER

and revised accordingly. In typical cases the standard deviation of the intensity is approximately half that given by alternative methods. The method has not yet found widespread use in protein crystallography, largely because of the need for fairly extensive on-line or data handling facilities, but may become more popular in the future as increasing accuracy in intensity measurements is sought.

The stationary crystal–stationary detector method of intensity measurement requires a uniform convergent incident beam. The problems associated with non-uniformity of the broad source and high background have been partially overcome by the method developed by Wyckoff (Wyckoff et al., 1967). In this method a limited number of intensity measurements is made in the vicinity of the peak of a reflection at small intervals of the most sensitive parameter (usually ω). A fraction of these, such as the lowest 3 out of 6, is ignored and the sum of the rest taken as the peak count. The small movement of the crystal helps to smooth out errors arising from non-uniformity of the source and increases the setting tolerance. Diffuse background scatter is reduced by using relatively large crystal to detector distances with a narrow receiving guard aperture placed a short distance from the crystal. Detector apertures are adjusted experimentally. For example, in the case of ribonuclease-S the source (1.2 mm high x 0.9 mm wide, 1200 W) and detector were placed 200 mm and 500 mm respectively from a crystal 0.5 mm in diameter. The counter slits were adjusted to 5 mm high x 1.2 mm wide. (The latter allowed for 0.08° mosaic spread.) The half intensity width of the reflections was about 0.2° on ω. During measurement the crystal was scanned through 6 steps of 0.03° on ω and the sum of the three highest contiguous counts taken to represent the integrated intensity. The background curve was measured separately as a function of 2θ by scanning radially in a region which did not include diffracted intensities at intermediate positions of χ and ϕ which were used in the data collection. The method has now been employed in the successful analyses of several protein crystal structures. It offers the advantage of rapid data collection and there is increasing interest in this technique.

9.6.2 The linear diffractometer

9.6.2.1 Inclination geometry

The linear diffractometer operates with inclination geometry which is the geometry used in many photographic methods such as the rotation and Weisenberg cameras. The reflections are measured by rotating the crystal about the goniometer head axis ϕ (in contrast to the normal-beam equatorial method for the four circle diffractometer where the rotation is about the ω-axis). In inclination geometry the incident beam is inclined at an angle $90°-\mu$ to the ϕ axis and the reflected beam at an angle $90°-\nu$ to the ϕ axis. The crystal is brought to the reflecting position by a rotation of ϕ and the detector to the accepting position by rotation of an angle Υ about the ϕ axis (Fig. 9.27). If the crystal is oriented so that a zone axis (e.g. c axis) is parallel to the rotation axis ϕ, then all the reflections in one reciprocal lattice level will lie on a cone of semi-angle $90-v$ where v is characteristic of each level.

For the general position, the three setting angles are given by:

$$\sin v = \zeta + \sin \mu$$

$$\cos \Upsilon = \frac{2\cos^2 \mu - 2\zeta \sin \mu - \xi^2 - \zeta^2}{2\cos \mu (\cos^2 \mu - 2\zeta \sin \mu - \zeta^2)^{1/2}}$$

$$\phi = 180° - \tau - \cos^{-1} \frac{\xi^2 + \zeta^2 + 2\zeta \sin \mu}{2\xi \cos \mu}$$

Three special settings are usually used.

(i) Normal beam setting ($\mu = 0$) (Fig. 9.28). The incident X-ray beam is normal to the rotation axis as in the rotation camera (Section 9.5.2.2).

The setting angles are given by:

$$\sin v = \zeta$$

$$\cos \Upsilon = \frac{2 - \xi^2 - \zeta^2}{2(1 - \zeta^2)^{1/2}}$$

$$\phi = 180° - \tau - \cos^{-1}\left(\frac{\xi^2 + \zeta^2}{2\xi}\right)$$

In this setting there is a blind region corresponding to a radius $\xi_{min} = 1 - (1-\zeta^2)^{1/2}$ within which no reciprocal lattice point will cut the sphere of reflection.

The Lorentz factor is given by:

$$L^{-1} = \cos v \sin \Upsilon$$

(ii) Equi-inclination setting ($\mu = -v$) (Fig. 9.28). In this setting the incident and the reflected beams are equally inclined to the positive direction of the rotation axis. The setting angles are given by

$$\sin v = \zeta/2$$

$$\Upsilon = 2 \sin^{-1}\left(\frac{\xi}{2 \cos \mu}\right)$$

$$\phi = 180° - \tau - \cos^{-1}\left(\frac{\xi}{2 \cos \mu}\right)$$

There is no blind region and all reflections are accessible. However the Lorentz factor is given by:

$$L^{-1} = \cos^2 v \sin \Upsilon$$

9.6.2. THE LINEAR DIFFRACTOMETER

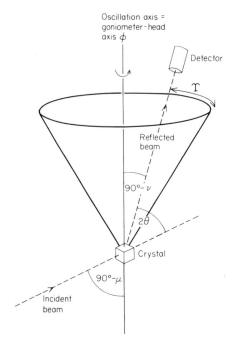

FIG. 9.27 Inclination geometry. The crystal is mounted on a goniometer-head attached to a single shaft ϕ. The detector moves around a cone with variable semi-angle, $90° - \nu$, and the incident beam likewise makes a variable angle, $90° - \mu$, with the crystal oscillation axis ϕ. (From Arndt and Willis, 1966).

which approaches as ξ approaches zero. This corresponds to the fact that reflections of the type 001 where c* is parallel to the rotation axis will be in the reflecting position for all values of ϕ. These reflections cannot therefore be measured by rotating the crystal about ϕ.

(iii) Flat cone setting $(\nu = 0)$ (Fig. 9.28). In this setting the diffracted beam is normal to the rotation axis and lies in a flat cone. The setting angles are given by

$$\sin \mu = -\zeta$$

$$\cos \Upsilon = \frac{(2 - \xi^2 - \zeta^2)}{2(1 - \zeta^2)^{\frac{1}{2}}}$$

$$\phi = 180° - \tau - \cos^{-1}\left(\frac{\xi^2 - \zeta^2}{2\xi \cos \mu}\right)$$

$$L^{-1} = \cos \mu \sin \Upsilon.$$

As in the normal beam setting there is a blind region of radius $1-(1-\zeta^2)^{1/2}$ at the centre of each level.

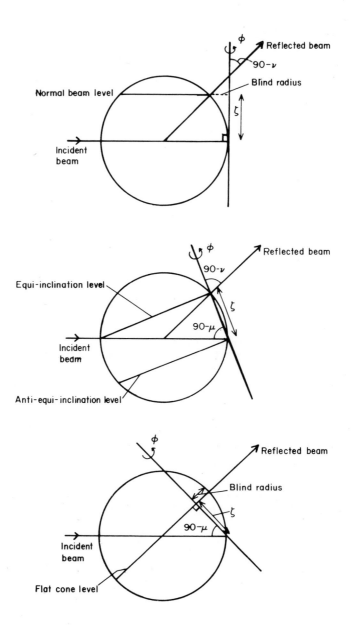

FIG. 9.28 Normal beam, equi-inclination and flat cone geometry for the same reciprocal lattice level ζ.

9.6.2. THE LINEAR DIFFRACTOMETER

When two levels are located symmetrically on either side of the flat cone level and separated from it by $\pm\zeta$, one will occur in the equi-inclination and the other in the anti-equi-inclination setting and this geometry forms the basis for multiple counter diffractometry for both the linear and the four circle diffractometers (Section 9.6.3).

9.6.2.2 The design and operation of the linear diffractometer

The design and principles of the linear diffractometer have been described by Arndt and Phillips (1961). The instrument utilizes inclination geometry and incorporates a mechanical analogue of the reciprocal lattice. The settings of the crystal and counter are generated automatically from the reciprocal lattice dimensions of the crystal by means of a mechanical analogue which is composed of three slides representing the reciprocal lattice axes. Once the crystal has been correctly oriented with respect to the slide system any reflection hkl can be found by driving to the co-ordinates ha^*, kb^*, lc^* of the corresponding reciprocal lattice point on the three slides. The crystal and counter take up their positions automatically. The advantage of this system is that computer calculation of angles is not necessary and only distances related to the different axial lengths need be set on the slides.

The principle of operation is shown in Fig. 9.29. YXO represents the direction of the incident beam with X the centre of the Ewald sphere and O the

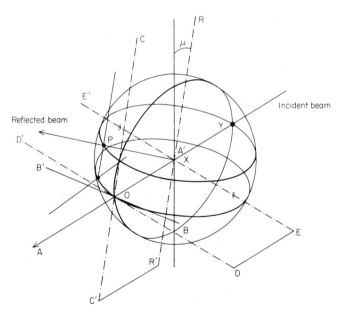

FIG. 9.29 The geometry of the linear diffractometer (from Arndt and Phillips, 1961).

origin of the reciprocal lattice. A'OA, B'OB and C'OC are the principal axes of the reciprocal lattice, here assumed to be orthogonal. XP is the direction of the diffracted beam corresponding to the reciprocal lattice point P. Any reciprocal lattice point may then be brought into the reflecting position by a rotation (ϕ) about C'OC and a rotation (μ) about the axis D'OD which is perpendicular to the incident beam. The linear diffractometer is a mechanical version of this diagram in which the three slides A, B and C are mounted parallel to A'OA, B'OB, C'OC respectively and rotate about C'OC. The link XP, which in effect can pivot freely about X and P, is of fixed length XP = XO. The scale of the instrument is fixed by the distance XO which is equivalent to 1 reciprocal lattice unit and is 5 inches. The crystal is mounted at X and may be rotated about R'XR, independently of the counter arm link XP. The axes R'XR and C'OC are held parallel by means of parallel linkages.

The slides A and B extend only to ±1 reciprocal lattice units and the slide C from 0 to ±1 r.l.u. The maximum value of $\Upsilon(2\theta$ for zero levels) $<60°$ and of $\mu < 30°$. These angular limits were chosen to be similar to the angular limits of the Buerger precession camera and they provide no serious limitation for most protein crystallographic measurements.

In the operation of the instrument the crystal is set so that the three reciprocal lattice axes are parallel to the three slides A, B, C. For non-orthogonal systems, the angle between the horizontal slides A and B may be adjusted. With monoclinic crystals mounted to rotate abour their c (or a) axis, the vertical distance between parallel levels is then $c^* \sin \beta$ and the origin of the level has to be offset by an amount ($c^* \cos \beta$).

In the automatic collection of data (Fig. 9.30), the diffractometer is driven to a reflection hkl at the limit of resolution of the data set to be collected. The diffractometer then moves in a series of equal steps along the scanning slide. At the end of each step the oscillation mechanism (described below) takes control for the measurement of the intensity and background. After each measurement the scan of equal steps is repeated until a limit switch is reached. The diffractometer completes the current translation, measures the last reflection, and then moves one step on the stepping slide to the next parallel row. This row is scanned, in the opposite direction, until the limit switch is again reached. In this way the whole of a particular reciprocal lattice level may be measured. In order to measure upper levels it is necessary to adjust the vertical slide (c) and the inclination angle μ by hand.

During measurement of the intensity of a reflection (Fig. 9.31), the connection between the crystal and the slide system is automatically interrupted. The crystal is offset by an angle $\phi - \Delta\phi$ from the reflecting position, ϕ, the shutter opened and background counts n_1 accumulated for time t with the crystal stationary. The oscillation mechanism drives the crystal through the reflecting position from $\phi - \Delta\phi$ to $\phi + \Delta\phi$ for a time $2t$ and the peak counts N are accumulated. Background counts n_2 at $\phi + \Delta\phi$ are accumulated for a further

9.6.2. THE LINEAR DIFFRACTOMETER

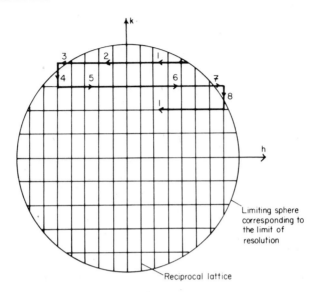

FIG. 9.30 The scanning and stepping sequences for automatic data collection on the linear diffractometer. The numbers 1 to 8 correspond to the numbers of the control programmes which control the direction of step or scan and the indices.

time t and the oscillation mechanism returns the crystal quickly to its original setting, $\phi - \Delta\phi$. The oscillation range may be varied from 1° to 5°, but the time period for one complete oscillation is fixed. In the measurement of upper levels the crystal is not oscillated about an axis perpendicular to the plane of reflection, and thus the effective angular range of reflection is greater for upper level reflections than for zero level reflections. It may be necessary to alter the oscillation range accordingly and to allow for this in the scaling of different levels.

The outer limit switch, which is used to control the limit of resolution of data, is set to correspond to the angle Υ_{max} for the particular level. An inner limit switch is also incorporated in order to prevent the instrument driving into the blind region in the flat cone of normal beam setting. This switch is set so that $\xi_{min} > 1 - (1 - \zeta^2)^{1/2}$.

An important disadvantage of the linear diffractometer is the relative inaccuracy of setting in the low angle region. This is illustrated in Fig. 9.32. Thus reflections with $\xi < 0.02$ (corresponding to a resolution of about 8 Å for the zero level) are best set by hand. This means that collection of low resolution data is especially tedious on the linear diffractometer and is probably better carried out on the four circle diffractometer.

A second disadvantage of the linear diffractometer is the relative inflexibility

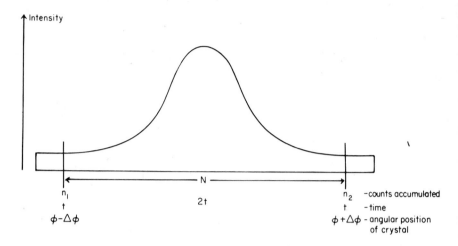

FIG. 9.31 The method of measuring the intensity of a reflection on the linear diffractometer.

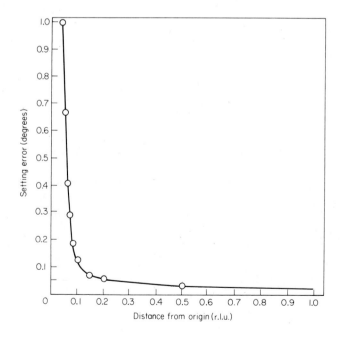

FIG. 9.32 Setting errors on the linear diffractometer (from Arndt and Willis, 1966).

9.6.3. MULTIPLE COUNTER DIFFRACTOMETRY

of the present oscillation mechanism for measuring the intergrated intensity. The usual method of counting for a fixed time on a stationary background requires great precision of setting especially in the low angle region. A small mis-setting for a large intensity can make the two backgrounds appear very unequal. This disadvantage is not inherent in the machine and could be overcome in a variety of ways. For example the linear diffractometer could be connected to an on-line computer and the profile analysis method used (Section 9.6.1.2). Any modification of this sort, however, reduces the great advantage of the linear which is its relative simplicity of design. The use of the analogue method for the determination of setting angles means that electrical faults that entirely prevent data collection are rare.

9.6.3 Multiple counter diffractometry

The major disadvantage of conventional diffractometers is their relative inefficiency. A single counter diffractometer measures only one reflection at a time, whereas for crystals with large unit cells there may be very many reflections which occur in the reflecting position simultaneously or quasi-simultaneously. The principles of systematic concurrence of reflections and the adaption of both the linear diffractometer and the four circle diffractometer to measure more than one reflection at a time have been described by Phillips (1964).

Let us first consider the general case with the special provision that the crystal is mounted to rotate about a crystal axis which is normal to the other two axes and thus coincident with a reciprocal lattice axis (c and c^* say). Then any pair of reciprocal lattice points hkl_1 and hkl_2 which are symmetrically related to the flat cone level at $\zeta = \zeta_f$ will lie simultaneously on the surface of the Ewald sphere at the same value of the crystal setting ϕ (Fig. 9.33). In the simplest case levels $\zeta = 0$ and $\zeta = 2\zeta_f$ will lie in the anti-equi-inclination and equi-inclination settings for any flat cone setting $\zeta = \zeta_f$.

During the measurement of a reflection the crystal is rocked through a range of ϕ of $1-2°$. When the restriction that simultaneous reflections must occur at identical values of ϕ is relaxed, it may be shown that a number of reflections will occur quasi-simultaneously during the movement of the crystal through its rocking range when the spacing between reciprocal lattice levels is small (as it always is for protein crystals).

Figure 9.34a shows an elevation of the sphere of reflection with a crystal set for the upper level at ζ to be recorded in the flat cone setting. Two general levels are shown at distances Z and $Z + z$ from the flat cone level. The corresponding circles of reflection are shown in Fig. 9.34b, together with the points at which they are intersected by reciprocal lattice points ξ from the axis of rotation.

Reflections in levels $\zeta + Z$ and $\zeta + Z + z$ occur at the same value but at slightly different ϕ and Υ values. The difference varies least with z when $z = -Z$. The difference in values of the crystal setting ϕ_0 for the flat cone level and the

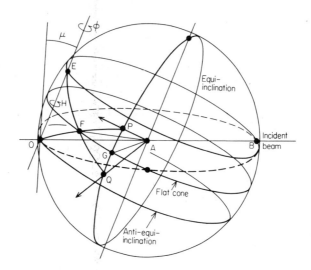

FIG. 9.33 Perspective drawing of the sphere of reflection showing inclination geometry with simultaneous reflections (reciprocal lattice points P and Q) from general levels symmetrically related by the flat-cone setting. (From Phillips, 1964)

setting $\phi_0 + \Delta\phi$ for the levels $\zeta \pm z$ is given by

$$\Delta\phi = \frac{-z^2}{2\xi(1-\zeta^2)^{1/2}\sin\phi_0}$$

where

$$\cos\phi_0 = \frac{\xi^2 - \zeta^2}{2\xi(1-\zeta^2)^{1/2}}$$

for the flat cone setting.

Suppose the criterion that $\Delta\phi < 0.2°$ is adopted for the condition of quasi-simultaneity levels ζ and $\zeta \pm z$, then Fig. 9.35 shows that there is a large region of reciprocal space within which the condition for quasi-simultaneity is satisfied. There is of course the blind region corresponding to $\xi_{min} = 1 - (1 - \zeta^2)^{1/2}$ within which no reflections can be measured.

In fact the criteria that $\Delta\phi < 0.2°$ can be relaxed slightly. The equation for $\Delta\phi$ shows that $\Delta\phi$ is always negative. Hence reflections in levels $\zeta \pm z$ will slightly precede those in the flat cone level ζ. Hence if the crystal is mis-set back from the flat cone level by 0.2° (for example), a value of $\Delta\phi$ of 0.4° may be tolerated, since reflections in the flat cone and neighbouring levels would only be displaced by 0.2° from the centre of the rocking range. The limitations on measurable reflections due to this criteria of quasi-simultaneity are shown in Fig. 9.36. The

9.6.3. MULTIPLE COUNTER DIFFRACTOMETRY

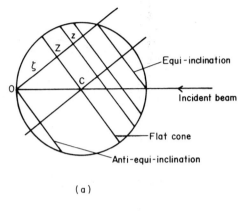

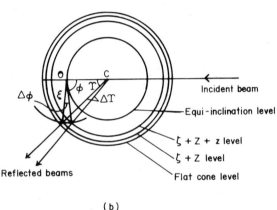

FIG. 9.34 (a) Elevation of the sphere of reflection showing levels at ζ and 2ζ in the flat cone and equi-inclination settings respectively and general levels at $\zeta + Z$ and $\zeta + Z + z$. (b) Corresponding circles of reflection with reflected beams corresponding to reciprocal lattice points at ξ in the general levels. (After Phillips, 1964)

region of reciprocal space which can be usefully measured becomes restricted when z, corresponding to the spacing along the mounting axis becomes greater than 0.04 (say) (corresponding to real spacings of less than 40 Å). The additional limitation in the minimum values of ξ for reflections which can be measured quasi-simultaneously corresponds effectively to an increase in the radius of the blind region.

The difference in the values of Υ for the different reflections is given by

$$\Delta\Upsilon = \frac{(\xi^2 + \zeta^2)z^2}{2\{4\xi^2 - (\xi^2 + \zeta^2)^2\}^{1/2}}$$

with $\Delta\Upsilon_{min} = \xi z^2/2(1 - \zeta^2)^{1/2}$ for constant z and ζ.

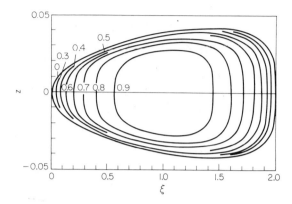

FIG. 9.35 Regions of reciprocal space within which $\Delta\phi < 0.2°$ near levels at $\zeta = 0$–0.9 r.l.u. in the flat-cone setting. (From Phillips, 1964).

This difference essentially imposes a restriction on the maximum resolution that can be obtained, without requiring different positions of the counters for different levels. The maximum value of $\Delta\Upsilon$ which may be tolerated is governed by the width of the detector apertures and the restriction is only important at high ξ. This is illustrated in Fig. 9.36 for $\Delta\Upsilon_{max} \leqslant \Delta\Upsilon_{min} + 0.1°$.

9.6.3.1 Modification of the linear diffractometer to measure reflections quasi-simultaneously

The adaption of the linear diffractometer to measure reflections simultaneously was described by Arndt et al. (1964). Since this instrument is already designed

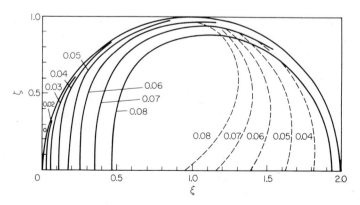

FIG. 9.36 Maximum and minimum values of ξ near the flat-cone setting when $z = 0.02$–0.08 r.l.u. Full lines show limitations due to $\Delta\phi \leqslant 0.4°$, broken lines those due to $\Delta\Upsilon°_{max} \leqslant \Delta\Upsilon°_{min} + 0.1°$. (From Phillips, 1964).

9.6.3. MULTIPLE COUNTER DIFFRACTOMETRY

to be used with inclination flat cone geometry, no modifications are required to the mechanical parts of the instrument in order to measure more than one reflection at a time. The modifications needed are to the counter assembly by which the single scintillation or proportional counter is replaced by an array of counters. With the instrument set in the flat cone setting for the centre counter, the length of the counter arm (crystal to detector distance = L) and the separation of the counters (x) are related by

$$L = \frac{x}{d^*}$$

where d^* represents the reciprocal lattice spacing along the vertical axis. The main technical problem was to mount the detectors close enough together for quasi-simultaneous measurements to be possible with a counter arm of tolerable length. Five side-window proportional counters have been mounted in echelon with their windows 7.5 mm apart and a draw tube fitted to the crystal-to-counter arm so that variable distance L may be achieved for crystals of different unit cells. A line of 5 circular apertures, (whose diameters can be changed to suit experimental conditions), is mounted directly in front of the counter assembly with the apertures fixed at 7.5 mm apart. Their centres lie on a line parallel to the crystal rotation axis. It has not been found necessary to make provision for the variation in Υ between reflections in adjacent levels.

The polarization factors are slightly different for the different levels. The values of P^{-1} for the central reflections is readily obtained from

$$4 \sin^2 \theta = \xi^2 + \zeta^2$$

Using the suffixes $-$, 0, and $+$ for the lower, central and upper reflections.

$$4 \sin^2 \theta_+ = \xi^2 + (\zeta + z)^2$$
$$4 \sin^2 \theta_- = \xi^2 + (\zeta - z)^2$$

The Lorentz factor is given by

$$L^{-1} = (\xi^2 + \zeta^2 z^2 - \tfrac{1}{4}(\xi^2 + \zeta^2 + z^2)^2)^{1/2}$$

The Lorentz factors for the upper and lower reflections are identical since these reflections lie symmetrically about the flat cone level. The difference between the Lorentz factors for the flat cone and other levels is very small except near the surface of the blind region. Since the diffractometer is not used in this region it is normally unnecessary to apply different Lorentz factors to the central and outer reflections. Figure 9.37 shows a scheme for the collection of a set of three-dimensional data to a resolution of 2.5 Å for a lysozyme crystal on a linear diffractometer modified to measure five reflections simultaneously. The crystals are tetragonal space group $P4_3 2_1 2$ with unit cell dimensions. $a = b = 79.1$ Å, $c = 37.9$ Å. The crystal is mounted about the **a** (**a***) axis (where

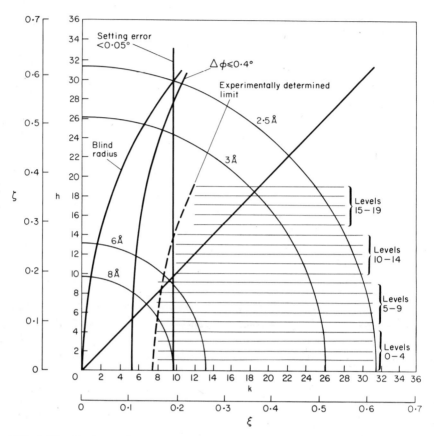

FIG. 9.37 Regions of reciprocal space measurable on the linear diffractometer for lysozyme. The limits shown are those imposed by the flat cone blind radius and the $\Delta\phi \leqslant 0.4°$ corresponding to a spacing of $z = 2a^* = 0.0392$. The limit imposed by the setting error of the diffractometer (Arndt and Willis, 1966) and the experimentally determined limit (Snape, 1974) are also shown.

$a^* = 0.0196$ rlu) so that with a counter separation of $x = 75$ mm the crystal to counter distance is $x/a^* = 384$ mm. The counter arm is fitted with a tube which contains helium at a pressure of 5 lbs per square inch in order to minimize the air absorption of the reflected beams. The use of the long counter arm tends to improve the background to peak ratio. For the measurement of five reflections simultaneously, $z = 2a^* = 0.0392$. The data are collected in four sets of five levels (Fig. 9.37). This leaves a small number of reflections for large h which are not measured. Because of the tetragonal symmetry, $I(hkl) = I(khl)$ so that most reflections are measured twice but in different levels. These data may be used to derive the relative scale factors for the individual levels in subsequent

9.6.3. MULTIPLE COUNTER DIFFRACTOMETRY

data processing programmes. For crystals of lower symmetry it would be necessary to collect data for crystals mounted about two different axes in order to collect a complete set of data and to provide data for scaling the different levels.

The limits imposed by the various factors discussed above are also shown in Fig. 9.37. It is seen that the major limiting factor is the setting error of the diffractometer. The low order reflections corresponding to spacings greater than 8 Å are not accessible by this method of automatic data collection and must be collected separately with a single counter mode or on a four circle diffractometer. The blind radius ($\xi_{min} = 1 - (1 - \zeta_f^2)^{1/2}$ where ζ_f corresponds to the flat cone setting for the particular set of levels) is not limiting and neither is the criterion for quasi-simultaneity ($\Delta\phi \leq 0.4°$ for $z = 0.0392$).

With this scheme of data collection a unique set of 8 Å – 2.5 Å data comprising 8000 reflections can be measured with less than 40 hours exposure of the crystal.

9.6.3.2 The five circle diffractometer – modification of the four circle diffractometer to measure reflections quasi-simultaneously

(i) *Flat cone geometry for the four circle diffractometer.* In the four circle diffractometer the movement of the counter is restricted to the equatorial plane of the instrument under the control of one circle and the normals to the reflection planes are brought into this plane by operation of the other three circles. The flat cone setting corresponds to the constraint $\omega = 2\theta$, in which the axis of the χ circle is parallel to the reflected beam (Fig. 9.38). (Compare this setting with the symmetrical A setting in which $\omega = \theta$ and the axis of the χ circle bisects the angle between the incident beams and the diffracted beams.)

As with the linear diffractometer, the measurement of multiple reflections is most easily achieved if the crystal is oriented so that the crystal axis, which is parallel to the line of reflections to be measured is also parallel to the goniometer and diffractometer axis ϕ. The line of detectors is then placed parallel to this spindle axis and is kept parallel to the ϕ axis by means of a fifth circle σ. The axis of the σ circle passes through the centre of geometry of the instrument and lies in the equatorial plane. The value of σ is necessarily equal to χ in the flat cone setting, so that only three angles need to be computed.

Either ϕ or ω scans may be used to measure the integrated intensities. When ϕ scans are used, equatorial flat cone geometry has all the features of inclination flat cone geometry and the Lorentz factors, the blind regions and the range of ζ values over which quasi-simultaneous reflections can be made are exactly the same as those given previously for the linear diffractometer. The peak width increases as the blind region is approached because in the flat cone setting the rotation axis is not, in general, perpendicular to the reciprocal lattice vector so that rotation through a small angle $\Delta\phi$ corresponds to a rotation by an angle $\Delta\omega$

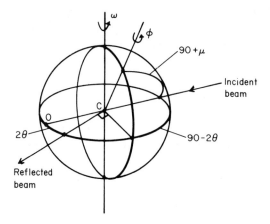

FIG. 9.38 Perspective view of the four-circle diffractometer in the flat cone setting $\omega = 2\theta$. The reflected beam is normal to the χ plane (which contains the rotation axis ϕ) so that $\nu = 0$.

where (Fig. 9.39)

$$d^* \Delta\omega = \xi\Delta\phi$$

i.e.

$$\Delta\phi = \Delta\omega/\sin\rho$$

Evans (1973) has confirmed experimentally that the peak width on a ϕ scan is proportional to $1/\sin\rho$. With a crystal of phospho-glycerate kinase the half-height peak width varied from $0.1°$ at $\rho = 90°$ to $0.3°$ at $\rho = 17°$. It is recommended that reflections for which $\rho < 20°$ should not be measured in the flat cone setting.

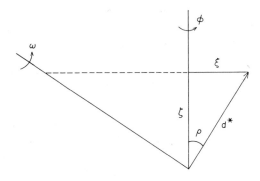

FIG. 9.39 Relationship between the flat-cone and perpendicular rotation axes. (From Evans, 1973).

9.6.3. MULTIPLE COUNTER DIFFRACTOMETRY

The contribution of the three factors ($\Delta\phi$, peak width and $\Delta\Upsilon$) are illustrated for two different mounts of a phosphoglycerate kinase crystal in Fig. 9.40. It is seen that with the crystal mounted about the short b^* axis ($b^* = 0.0144$ rlu) and measuring 5 reflections simultaneously ($z = 0.0288$ rlu) the main limitation is due to $\Delta\phi$ at low ξ, to peak width for most of the range and $\Delta\Upsilon$ only at high values of ζ. Most of the reciprocal space is accessible. On the other hand for the a^* mount ($a^* = 0.0307$ rlu), the $\Delta\phi$ limit is the most important factor but again the $\Delta\Upsilon$ limit becomes important at high ζ values. The peak width is not limiting.

(ii) Modification of the four circle diffractometer A Hilger and Watts Y 230 four circle diffractometer in the Laboratory of Molecular Biophysics, Oxford has been modified to measure five reflections simultaneously (Banner, 1972; Evans, 1973; Banner, Evans, Marsh and Phillips, to be published). The normal counter arm was replaced by a flat beam 1 m long on which the axis of the fifth circle, the σ circle, is positioned. The σ axis is a horizontally mounted ϕ axis assembly which carries the box containing the proportional counters. The pre-amplifiers and the signal, E.H.T. and power supply cables are carried on a slave arm which is pivoted above the centre of geometry of the instrument and is moved by a motor driving a small wheel on a circular track. The angular displacement of the slave arm follows that of the counter arm.

The proportional counters are stacked in echelon as in the linear diffractometer. A centre to centre spacing of 9.5 mm for the apertures was chosen so that all X-rays passing through the largest apertures (5.5 mm) pass through the 0.25" counter windows over the full range of 250 to 1000 mm crystal to counter distance (Banner, 1972). The 7.5 mm separation of the counters on the linear permits smaller crystal to counter distances but is only technically correct for a crystal to counter range of 250 to 500 mm for 5.5 mm apertures.

The Oxford five circle diffractometer is connected through an interface, to a Ferranti Argus 500 computer located in the next room. This computer has a 2 μsec cycle time and a core store of 20 K 24-bit words. The diffractometer programme requires rather less than 4 K of store and shares computer time with other programmes by means of a time-sharing Supervisor programme. The diffractometer control programme which incorporates various checks (e.g. to avoid driving into the blind region) has been written and implemented by Evans (1973). The control programme permits considerable flexibility in the sequence of data collection. Facilities are included for the measurement of reference reflections in order to monitor radiation damage, for checks on X-ray and shutter mechanism, and for drives to the instrument datum references.

The method of measuring each reflection is a modification of the ordinate analysis scheme devised by Watson (Watson *et al.*, 1970), in which $2N$ steps are taken across the peak and the largest consecutive N are taken as the peak, and the remainder as background. This method has considerable advantages with the five counter diffractometer, since the peak position may be more closely defined from the average peak positions for the five counters, taking into account their

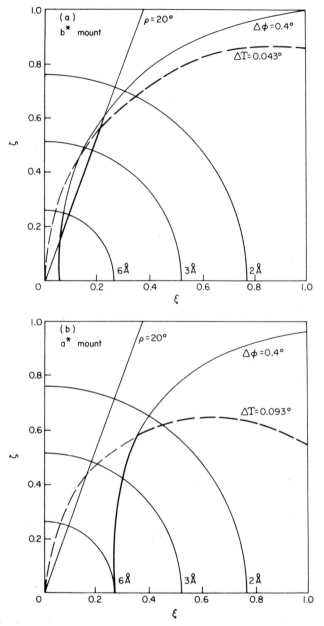

FIG. 9.40 Regions of reciprocal space measurable on the five circle diffractometer for phosphoglycerate kinase. The limiting criteria are $\Delta\phi \leqslant 0.4°$, $\rho \geqslant 20°$ and $\Delta\Upsilon°$ corresponding to a 0.5 mm displacement of the outer reflected beam from the straight line through the central reflection. For a 9.5 mm counter separation this corresponds to $\Delta\Upsilon° = 0.043°$ for a $b*$ mount ($z = 2b* = 0.0288$) (Figure a) and $\Delta\Upsilon° = 0.093°$ for an $a*$ mount ($z = 2a* = 0.0614$) (Figure b). (From Evans, 1973).

9.6.3. MULTIPLE COUNTER DIFFRACTOMETRY

calculated peak separation. Further the background in one counter may be measured while the peak in another counter is being measured so overcoming the disadvantages of a very wide peak scan in the conventional moving crystal method of measurement. An increase in scan width to account for the increase in peak width with $1/\sin \rho$ is also incorporated in the programme.

With this instrument Evans was able to collect three dimensional data to a resolution of 3 Å for phosphoglycerate kinase with only two crystals. Some 14,600 reflections were collected from one crystal in b^* mount and 3000 for a second crystal in an a^* mount in order to measure the blind region.

(iii) Comparison with the linear diffractometer. The modifications which are needed to adapt the four circle diffractometer to measure five reflections simultaneously are considerably more extensive than those required by the linear. They involve the incorporation of a fifth circle and ancillary equipment and, ideally, an on-line computer efficiently programmed. Once these modifications have been made, the five circle diffractometer exhibits several advantages.

(a) There is a greater accuracy in setting. The known displacement ($\Delta\phi$) of reflections from the setting position may be calculated and allowed for in the measurement scheme.

(b) The on-line control by the computer provides a greater flexibility in the method of measurement of integrated intensities and the order in which the reflections are measured. Thus for example Friedel pairs hkl and $\overline{hkl}$ can be measured close to each other in time and with the same absorption correction. Further it can be arranged that when χ is negative $\sigma = \chi$ for the hkl reflection and when χ is positive $\sigma = 180 - \chi$ for the $\overline{hkl}$ reflection. Hence Friedel equivalents are measured by the same counter.

NOTE ADDED IN PROOF

In their recent paper Banner, Evans, Marsh and Phillips (to be published) show that the disadvantages of variation in peak width associated with ϕ scans on the five circle diffractometer may be overcome by using ω scans. The ω axis is perpendicular to the reciprocal lattice vector for the central reflection and $\Delta\omega$ is always smaller than or equal to $\Delta\phi$. However with this setting reflections symmetrically disposed about the centre reflection do not pass through the sphere of reflection at the same value of ω. Simultaneity may be achieved by rotation of an angle ψ about the reciprocal lattice vector of the central reflection so that one outer reflection moves closer to the sphere and the other moves further away (ψ then defines the other setting angles (ϕ, χ, ω and σ)). This optimum setting still retains a blind region near the mounting axis but since the derivation of the multi-counter setting angles allows any mounting of the crystal, the missing reflections can be measured about another collection axis without remounting the crystal.

10

DATA PROCESSING

10.1 INTRODUCTION

The rapid growth of successful protein structure analyses in recent years owes much to the explosive growth of computer technology. In this chapter we shall be concerned with the computer processing of the intensity measurements. Fast computers with large store size and efficient programmes are required to handle the large number of reflections generated in any protein crystal structure analysis and to reduce trivial arithmetic errors. Since the amount of data prevents the personal checking of measurement, it is essential that the computer programmes incorporate checks on the data for systematic and random errors and that the users attention is drawn to those measurements that are suspect.

In the primary data processing, the set of raw intensity measurements are corrected for background, Lorentz, polarization, absorption and radiation damage factors to produce a set of F^2. Photographic data and diffractometer data are discussed separately. With a diffractometer, data are recorded and measured at the same time (except in the case of profile analysis). With photographic methods the data are recorded on film and the intensities measured later. We first discuss the use of photographic film as an intensity measuring device and the design of modern densitometers. We then outline the procedure for applying the corrections referred to above to the photographic data and to the diffractometer data. After this stage, data processing methods are similar for the two types of data. We describe the calculations which may be used for scaling different sets of data, for scaling heavy atom derivative data to the native set, and for the determination of the absolute scale.

10.2 PRIMARY DATA PROCESSING: PHOTOGRAPHIC DATA

10.2.1 Film characteristics

X-rays are recorded on a photographic film through the interaction of the X-ray quanta with the silver halide granules contained in the film emulsion. On developing and fixing the film these give rise to blackened areas of metallic silver. Experiments have shown that the absorption of one X-ray quanta

10.2. PRIMARY DATA PROCESSING: PHOTOGRAPHIC DATA

produces approximately one grain. The total X-ray quanta absorbed is therefore directly proportional to the number of blackened grains. Various methods have been used to determine this number. The almost universal method in X-ray crystallography is to measure the optical density of the blackened film where optical density D is defined by

$$D = \log \frac{I_0'}{I_t'}$$

where I_0' is the intensity of the incident light and I_t' is the intensity of light transmitted by the blackened spot on the X-ray film. In order to produce good intensity data with film techniques it is important to understand the basic properties of the X-ray film used and the relationship between D and the X-ray exposure.

These properties have been examined for a variety of films by Morimoto and Uyeda (1963) and are summarized in Table 10.I for the films most widely employed in protein crystallography. Ilford Industrial G and Eastman Kodak No Screen film are used in Great Britain and the United States, respectively, for recording X-ray intensity data. Kodirex is a fast film which is relatively cheap but has been found to be slightly too grainy for intensity measurements. It is useful for setting photographs. From Table 10.I it is seen that Kodirex is approximately 30% faster, and Kodak No Screen approximately 50% faster than Industrial G. In general there is an increase in granularity with speed but the increase for No Screen is not as great as for Kodirex. Aging and increase in fog density are also correlated with the speed of the film. In Morimoto and Uyeda's measurements Industrial G showed no change after 6 months but for the other two films there was a drop of approximately 10% in the speed of the film and a 15% rise in the fog density after storage for 6 months at a warm temperature. These values will depend on the way in which the film is stored. Storage in the refrigerator is recommended. The values for the film factor show that approximately 70% to 80% of the incident radiation is absorbed by the film. X-ray film is an efficient and sensitive detector. Even so it is not possible to record the full range of intensities from a crystal on a single film. It is customary therefore to use a film pack of 2 or 3 films so that the weaker reflections are recorded on the first film and the stronger reflections fall within measurable ranges on subsequent films.

Table 10.I indicates that the response of the film is not quite linear with exposure time. Matthews *et al.* (1972b) have shown that the non-linearity between intensity and optical density may be represented by a parabolic curve of the form

$$I = aD + bD^2$$

where I is the intensity of the scattered X-ray beam, D is the observed optical

TABLE 10.1

Film Characteristics (from Morimoto and Uyeda, 1963) for Cu Kα radiation

Film	Manufacturer	Speed (μ^2/photon)	Granularity G_s (μ)	Film Factor	Relative Exposure Deff =									
					0.2	0.4	0.6	0.8	1.0	1.2	1.4	1.6	1.8	2.0
Ilford Industrial G	Ilford (U.K.)	1.08	2.3	2.93	19	38	58	79	100	122	145	170	194	218
Kodak No Screen	Eastman-Kodak (U.S.A.)	1.59	2.7	3.89	19	38	59	79	100	122	144	166	190	214
Kodak Kodirex	Kodak (U.K.)	1.38	2.9	2.89	20	40	60	80	100	121	142	166	189	214

Density $D = \log \frac{I'_0}{I'}$; Deff = $D - D_F$ where D_F = Fog density. *Speed*: is defined as the reciprocal of the number of X-ray photons per μ^2 required to produce a $D_{eff} = 1.0$.; *Granularity* G is defined as the root mean square of a 100 deviations from the mean density over an illuminating area s, $G_s = G\sqrt{2s}$; *Film Factor* = I_0/I_T where I_0 is the intensity of incident x-radiation and I_T is the intensity of the transmitted X-radiation; *Relative Exposure* is defined as the exposure required to give a D_{eff} = 0.2, 0.4, 0.6 . . . 2.0 normalized to the exposure E = 100 at D_{eff} = 1.0.

10.2.2. DENSITOMETERS AND INTEGRATED INTENSITIES

density and a and b are constants. The application of the parabolic correction is discussed in the next section.

From Table 10.I it can be seen that the total number of quanta required to produce a certain blackening on the film depends on the size of the spot. For example in order to produce an optical density of 1, it requires approximately 10^6 quanta for a spot 1 mm x 1 mm but only 10^4 quanta for a spot 0.1 mm x 0.1 mm. Chemical fog sets the lower limit to the number of quanta per unit area that can be detected (rather than the absolute total number of quanta) and for this reason it is best to reduce the size of the reflection spots as much as possible by adjusting the collimator conditions (see Section 9.3.7). The lowest significant optical density that can be detected is probably of the order of 0.1.

10.2.2 Densitometers and integrated intensities

The revival of interest in photographic techniques in recent years has been due to the introduction of fully automatic microdensitometers. These have been based on two designs:

(i) The rotating drum densitometer in which the entire area of a selected portion of the film is scanned and the optical density determined at a very large number of grid points (Abrahamsson, 1966; Xuong, 1969). The film is placed over a hole in a cylindrical drum which can rotate. Light is transmitted through the film with a light beam of diameter 50, 100 or 200 μ focussed at the film and the drum rotates at 1, 2 or 4 rev s^{-1}. Transmitted light is measured at 50, 100 or 200 μ intervals in the direction of rotation and after each revolution the light source is advanced 50, 100 or 200 μ along the drum axis. The densitometer is adjusted so that the light path in air has $D = 0$, and the maximum integer reading 255 (= $2^8 - 1$) corresponds to a density $D = 2$ or 3. The sampled optical density is digitized and a digital image of the film stored either directly in the computer or on magnetic tape. The positions of the diffraction spots and their indices are subsequently calculated and the data processed. This type of instrument is now commercially available and the performance has been assessed by Nockolds and Kretsinger (1970) and Matthews et al. (1972b).

(ii) The flying spot densitometer (Arndt et al., 1968). This uses the spot of a high resolution cathode ray tube as the source of light. The spot position is under computer control and the film is scanned only at those parts where useful information is known to be. The film must be accurately oriented and the unit cell dimensions and the orientation of the crystal known. The cathode ray scanner has advantages of speed and flexibility but it is limited to a maximum optical density of about 1.2 because of stray light from the halo which surrounds the light spot on the film.

Integrated intensities are obtained by summing the individual densities within a certain box size centred on the centre of the reflection spot and subtracting a

52	50	50	55	49	51	54	59	53	52	53	56	51	51	52
55	52	53	47	51	54	52	59	54	55	56	52	53	55	53
55	49	49	56	51	53	60	69	64	60	59	53	53	53	53
53	52	54	56	54	71	93	102	99	82	66	57	57	57	54
52	56	50	53	67	97	132	149	152	120	86	61	57	52	55
55	55	52	55	79	126	157	175	169	154	98	70	52	55	50
52	52	53	56	90	138	174	185	171	165	123	80	57	55	57
54	51	54	61	85	134	171	180	178	161	122	81	57	52	55
52	51	55	52	73	113	156	178	163	156	94	70	59	60	57
49	49	50	53	58	81	118	131	125	113	78	64	58	51	58
52	49	53	51	55	56	78	85	85	79	61	61	53	55	52
54	50	52	51	55	55	60	55	60	57	54	51	52	50	55
50	49	52	53	55	55	51	54	52	53	54	52	53	54	53
50	54	53	50	52	53	52	52	52	51	49	51	54	53	50
53	49	45	50	51	57	50	48	48	52	50	52	48	47	55

FIG. 10.1 A digitized representation of a diffraction spot from a phosphorylase b rotation photograph showing the summation box for the peak and the surrounding area which is used to estimate background. Scale: Numbers 0–255 correspond to optical densities O– 2.

background density obtained by summing the same number of density points over the area which surrounds the reflection (Fig. 10.1). In order to obtain the most precise intensities a reasonably small light spot size must be used and the film scanned over a fine raster. Because of the graininess of the film it is probably not worth examining an area of less than 100 μ in diameter. If a larger spot is used the effects of film graininess will be averaged out but errors may be introduced when large changes in optical density occur within the sampling area (Wooster, 1964). This is because of the logarithmic dependence of optical density on intensity of light. (The logarithm of an average is not the same as the average of the logarithm.) Matthews *et al.* (1972b) have examined this effect and conclude that up to a spot size of 100 μ the "Wooster effect" is negligible.

In the recording of photographic data, a pack of two or three films is usually used so that measureable intensities for all reflections of the diffraction pattern may be recorded. These films must be scaled together and proper attention given to the non-linear relationship between optical density and exposure time as noted in Table 10.I. Two methods are available.

(i) An empirical curve for the non-linear response is constructed from measurements of a film on which the main X-ray beam (suitably attenuated) has been recorded for a series of fixed exposure times. The correction factor derived from this curve is applied to the optical density values and the films scaled together on the basis of common measured intensities. If the non-linear

10.2.2. DENSITOMETERS AND INTEGRATED INTENSITIES

response correction has been successful, the film scale factor should be independent of intensity. Its value should be approximately equal to the value shown in Table 10.I but some variation is expected arising from developing conditions, age of the film etc.

(ii) The non-linear response correction factor is taken into the film scale factor by means of a parabolic scaling procedure. Matthews *et al.* (1972b) have shown that the relationship between intensity (I) and optical density (D) may be represented by

$$I = aD + bD^2$$

where a and b are constants. The relation between X-ray diffraction intensity and integrated background–subtracted density can also be considered as parabolic if all reflections are of the same shape and if either the background density is low or the reflection density considerably exceeds that of the background.

Assuming these conditions to be met, then the intensity of the reflection h can be expressed as

$$I(\mathbf{h}) = AD(\mathbf{h}) + BD^2(\mathbf{h})$$

where $D(\mathbf{h})$ is the integrated background subtracted density and A and B are constants. The constant A may be set arbitrarily equal to one. The constant B may be obtained from the scaling factor k between successive films i and j in a film pack from the relationship

$$B = B'k/(k-1)$$

$$B' = \frac{\Sigma I_i^2 \, \Sigma I_i^2 I_j - \Sigma I_i^3 \, \Sigma I_i I_j}{\Sigma I_i^4 \, \Sigma I_i I_j - \Sigma I_i^3 \, \Sigma I_i^2 I_j}$$

$$k = \frac{\Sigma I_i^4 \, \Sigma I_i^2 - \Sigma I_i^3 \, \Sigma I_i^3}{\Sigma I_i^4 \, \Sigma I_i I_j - \Sigma I_i^3 \, \Sigma I_i^2 I_j}$$

where I_i and I_j are the intensities of the same reflection recorded on the ith and the jth films and the summation is taken over all common reflections.

For rotation photographs there are certain additional systematic corrections that need to be applied during the early stages of data processing.

(i) Oblique incidence absorption. The more widely scattered X-rays strike the film obliquely so that these rays encounter a longer path length and hence greater absorption in each of the films (Fig. 10.2). This results in an apparent increase in film factor with scattering angle, since the widely scattered X-rays suffer greater absorption in the first film and hence reach the second film with diminished intensity. It may readily be shown that the correction factor for the intensities recorded on the second film is $\exp(\mu t \sec 2\theta)$ where μ is the linear absorption coefficient of the film and t the thickness. The factor $\exp(\mu t)$ may be obtained from the value of the apparent film factor at $2\theta = 0°$.

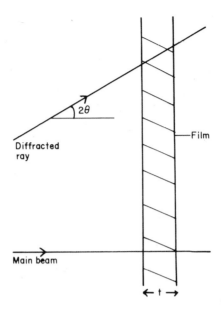

FIG. 10.2 Oblique incidence absorption in X-ray film. The increase in path length through the film for diffracted rays.

(ii) Oblique incidence "Cox and Shaw" factor. The more widely scattered X-rays strike the film obliquely so that a thicker layer of the sensitized material of the X-ray film is presented to the beam with a consequent increase in photographic action. The number of grains exposed is sec 2θ times as great for oblique incidence as for normal incidence. This gives rise to a correction factor (Cox and Shaw, 1930; Whittaker, 1953) of the type:

$$I_{corr} = \frac{I(2\theta)(1 - e^{-A})(1 + e^{-(A+B)})}{(1 - e^{-A\sec 2\theta})(1 + e^{-(A+B)\sec\theta})}$$

where $I(2\theta)$ is the observed intensity at 2θ and I_{corr} is the corrected value, and e^{-A} and e^{-B} are the absorption factors for a normal beam in a single emulsion and in the film base, respectively. For modern films $A \simeq 0.43$ and $B \simeq 0.20$. These constants can be readily determined from measurements of the total absorption of the film (emulsion plus celluloid base) and of the absorption of the celluloid base after the emulsion has been removed. A plot of the correction factor is shown in Fig. 10.3. It is seen that at $2\theta = 30°$ (approximately 3Å resolution) the correction is as great as 8%.

(iii) Variation in crystal to film distance. The crystal to film distance (F') for the scattered rays varies with scattering angle 2θ according to

$$F' = F \sec 2\theta$$

10.2.2. DENSITOMETERS AND INTEGRATED INTENSITIES

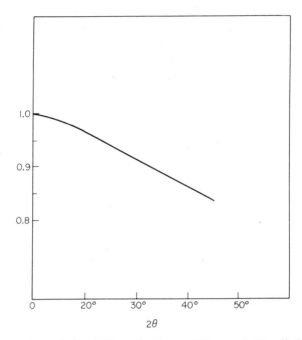

FIG. 10.3 A plot of the oblique incidence "Cox and Shaw" factor for a photographic film with absorption factors $A \simeq 0.43$ and $B \simeq 0.20$ for the emulsion and celluloid base respectively. (See text for further details).

where F is the normal crystal to film distance. The more widely scattered X-rays travel greater distances and their intensities will be decreased slightly by the greater air absorption. For example with a crystal to film distance of 10 cm the path travelled by rays scattered through an angle of 30° is 11.55 cm and assuming air absorption of 1% per cm, these rays will be diminished by 1.55% when compared with the straight through beam.

(iv) Variation in spot size. The increase in path length with scattering angle results in a concomitant increase in spot size with scattering angle which should be allowed for in the computer programmes.

Estimates of the standard deviations of photographic intensity data are not as straightforward as those for diffractometer measurements (Section 10.3.1) where the errors arise primarily from uncertainties in counting. For photographic data the effects of counting randomness are small in comparison with errors from other sources, since even the faintest reflections on a film result from many times the number of photons counted at ordinary scan rates by the diffractometer. Errors in film data will arise from graininess in the film, irregularities in development, instability of the X-ray source, limited precision of the densitometer as well as from crystal mis-setting and crystal shape parameters.

TABLE 10.II
Values of standard deviations of optical density obtained from repeated measurements of the same film. (From Matthews et al, 1972b)

Optical density range	Standard deviation
0 – 1.0	0.007
1.0 – 2.0	0.013
2.0 – 2.5	0.017

It is evidently desirable to obtain estimates of these errors so that proper weights may be attached to the measurements in subsequent calculations.

The precision of the densitometer may be tested empirically by repeated scans of the same film. Typically microdensitometers can measure a range of 256 grey levels with an accuracy of ±2 levels. Values of standard deviations with different optical density ranges are shown in Table 10.II.

The usual way to obtain an empirical estimate of a standard deviation of a measurement is to repeat the measurement several times and to determine the root mean square deviation from the mean value. This is not possible for photographic measurements on protein crystals, since it would involve the repeated photography of the same crystal. Apart from the labour involved, the crystal would suffer from radiation damage. However three procedures may be used to monitor different sources of error in the integrated intensities. Firstly, errors likely to arise from lack of precision in the densitometer may be derived from several scans of the same film. Secondly, comparison of the intensities of reflections recorded on different films in a film pack will reveal uncertainties in graininess, densitometer precision, as well as errors due to irregular film development and film to film scaling. Finally a comparison of symmetry related reflections on the same film will allow an estimate of errors due to absorption and crystal mis-setting.

The reliability index is defined as

$$R = \frac{\sum_h \sum_i |\bar{I}(\mathbf{h}) - I(\mathbf{h})_i|}{\sum_h \sum_i I(\mathbf{h})_i}$$

where $\bar{I}(\mathbf{h})$ is the mean of the measurements and $I(\mathbf{h})_i$ is the ith measurement of reflection $\mathbf{h}$. Three different reliability indices are obtained with respect to the analyses described above. (i) The level of agreement between several scans of the same film (R_m) (where $\bar{I}(\mathbf{h})$ is the mean intensity of reflection $\mathbf{h}$ obtained from the different scans). (ii) The level of agreement between reflections recorded on successive films of a film pack after scaling (R_{sca}) (where $\bar{I}(\mathbf{h})$ is the mean of the intensity of reflection **h** obtained from the scaled values recorded on the different films). (iii) The level of agreement between symmetry related reflections which have been recorded on the same film (R_{sym}) (where $\bar{I}(\mathbf{h})$ is the mean of the symmetry related reflections). Typical values are $R_m = 1\%$, $R_{sca} = 4\%$, $R_{sym} = 5\%$. These are the levels which should be aimed at in data collection. R_{sca} will be sensitive to any systematic error which is a function of intensity such as that which may be introduced in the development of the films. R_{sym} which measures the agreement between reflections of approximately equal intensity will be sensitive to crystal setting and crystal shape parameters.

In summary, for the initial stages of photographic data processing we require:

(a) A programme which will accurately predict the positions of the reflections on the film according to the methods described in the previous chapter.

(b) A programme which will calculate the integrated background corrected intensity of each reflection. This programme should incorporate checks for unequal backgrounds which may arise from radiation streaks, excessive background scattering by the collimator, incorrect prediction of reflection position or reflections too close together.

(c) A programme which will enable a series of films in a film pack to be scaled together and the corrections described above applied.

10.2.3 Lorentz–polarization corrections

The polarization correction is given by

$$p^{-1} = \frac{2}{1 + \cos^2 2\theta}$$

The Lorentz factors for different camera geometries are given in Table 10.III. The correction for the precession camera is complicated (Waser, 1951, Buerger, 1964) and is dependent on τ the angle which the projection of the reciprocal lattice vector **h** on to the zero layer makes with the spindle axis of the precession camera. This is because the nature of the suspension of the precession camera causes the reflecting sphere to sweep around the reciprocal lattice plane with an angular velocity that varies during the cycle.

For screenless precession photographs two factors should be taken into account: (i) If the spots appear at different positions according to whether $\beta_i = \tau \pm \alpha$ (i.e. the spots are split), then it will be necessary to use the appropriate value of $\sin \beta_i$ in the second part (β^{-1}) of the expression for the Lorentz factor and correct each of the spots separately. (ii) The Lorentz factor varies extremely rapidly for small angle precession photographs and it may be advisable to correct each of the density points separately before integration and background correction.

TABLE 10.III
Lorentz factors for different camera geometries

Rotation Camera $\quad L^{-1} = (\sin^2 2\theta - \zeta^2)^{1/2}$

Precession Camera $\quad L^{-1} = A^{-1} B^{-1}$

where $A^{-1} = \sin \Upsilon \sin \bar{\nu}_n \cos \bar{\mu} \sin \bar{\mu}$

$\qquad = 4 \cos \bar{\mu} [4\xi^2 \sin^2 \bar{\mu} - (\sin^2 \bar{\nu}_n - \sin^2 \bar{\mu} - \xi^2)^2]^{1/2}$

$$B^{-1} = \left[\frac{1}{1 + \tan^2 \bar{\mu} \sin^2(\tau + \alpha)} + \frac{1}{1 + \tan^2 \bar{\mu} \sin^2(\tau - \alpha)} \right]^{-1}$$

where $\bar{\mu}$ is the precession angle

$\pi/2 - \bar{\nu}_n$ is the semi-angle of the cone of diffraction

$\cos \bar{\nu}_n = \cos \bar{\mu} - \zeta_n$

Υ is the projection of 2θ on to the zero layer

$$\cos \Upsilon = \frac{2 - \zeta_n^2 - \xi^2}{2(1 - \zeta_n^2)^{1/2}}$$

τ is the angle between the horizontal axis of the universal joint and the projection of the reciprocal lattice vector **h** on to the zero layer.

α is an angle which describes the passage of the reciprocal lattice point through the reflecting sphere.

$$\cos \alpha = \frac{\xi^2 + \sin^2 \bar{\mu} - \sin^2 \bar{\nu}_n}{2\xi \sin \bar{\mu}}$$

10.2.4 Absorption

In the collection of three-dimensional data by photographic methods absorption is not usually considered a problem. It is assumed that the absorption remains constant over the precession or rotation angle of any individual photograph and that any variation in absorption due to different crystal orientations for different photographs is incorporated into the film to film scaling procedures. This assumption is probably valid for regularly shaped crystals. For irregularly-shaped crystals there may well be some variation in absorption for each individual film. If the crystal is smaller than the X-ray beam, an experimental determination of the absorption may be possible by diffractometer measurements (Section 10.3.3). Alternatively if several equivalent reflections are measured which have different absorption corrections, then the method of Katayama *et al.* (1972) may be appropriate (Section 10.3.3).

Schwager *et al.*, (1973) have described an experimental method of absorption

10.3.1. BACKGROUND CORRECTIONS

correction for the precession camera which is applicable when the crystal is larger than the X-ray beam. A transmission surface within a diffraction cone is constructed by recording the transmission of the unreflected primary beam for different orientations of the crystal. This is done by removing the beam stop and displacing the film holder so that the primary beam (suitably attenuated) produces a blackened ring on the film during the precession motion. After correction for Lorentz factor and oblique incidence, the variation in intensity in the ring is used as an empirical absorption curve for corrections to the intensities. The method has been applied to the three-dimensional data collected by screenless precession methods for the irregularly-shaped crystals of the trypsin–trypsin–inhibitor complex. Curves were recorded for precession angles of $\bar{\mu} = 10°$ and $20°$ corresponding to cones with $2\theta = 10°$ and $20°$. Application of the absorption correction substantially improved the agreement between equivalent reflections recorded on the same film. The method may also be applied to the rotation camera, provided that the camera is equipped with an additional circle to permit rotations about an axis perpendicular to the X-ray beam and rotation axis.

10.2.5 Radiation damage

Radiation damage corrections are seldom applied in photographic data collection since the intensities recorded on the photographic film will correspond to the average intensities from the crystal during the exposure time. In the collection of data by screenless methods, however, the crystal may well have suffered some significant damage by the time the last photograph is taken. Some measure of the damage may be determined by recording the first photograph again and an empirical correction derived from a comparison of the intensities on the two photographs. As in the case of the absorption correction, film to film scaling on the basis of equivalent reflections will also incorporate some compensation for radiation damage.

10.3 PRIMARY DATA PROCESSING: DIFFRACTOMETER DATA

10.3.1 Background corrections

If N are the total counts accumulated by the diffractometer counter for the peak and $n_1 + n_2$ the total background counts in the vicinity of the peak, then the background corrected intensity is given by

$$N_0 = N - n_1 - n_2$$

The emission of X-ray quanta from an X-ray tube is a random process in which the number of counts received in a standard time follows approximately a

Gaussian distribution so that if the mean value of a measurement is x, the standard deviation is $\sqrt{x}$. The standard deviation σ_{N_0} of the background corrected count is therefore estimated to be

$$\sigma_{N_0} = (N + n_1 + n_2)^{\frac{1}{2}}$$

This value is likely to be an underestimate of the probable error since it reflects only errors which arise from statistical fluctuations in the X-ray intensity measurements and ignores the effects of non-statistical variations in X-ray intensity, such as instability of the counting circuits, misalignment of the crystal and inaccuracy in absorption corrections. Nevertheless it provides a reasonable indication of the *relative* reliabilities of different reflections and may be used to weight ($\omega = 1/\sigma_{N_0}$) measurements in subsequent calculations (North, 1964) (Fig. 10.4).

The significance of a measurement may be assessed from the ratio σ_{N_0}/N_0 Reflections for which this ratio exceeds a certain value (usually taken as unity) have low statistical significance, although they may not be faulty. Such reflections should be monitored so that an idea of the percentage of significant

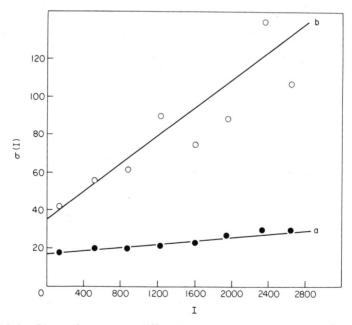

FIG. 10.4 Plots of errors in diffractometer measurements vs I for a phosphorylase **b** crystal: (a) standard deviations estimated from counting statistics alone (see text for further details); (b) differences observed between symmetry related reflections ($\sigma(I) = [\sum_{i=1}^{n} (\bar{I} - I_i)^2]^{1/2}$ for equivalent reflections. Values of $\bar{\sigma}(I)$ have been averaged in ranges of intensity.

10.3.2. LORENTZ–POLARIZATION CORRECTION

reflections in a data set may be determined. Occasionally the background corrected count may lead to negative values of N_0. For these weak reflections N_0 is generally set equal to zero.

Mis-setting of the crystal is frequently revealed by unequal backgrounds. For the background–peak–background method of measurement, reflections are rejected if the difference between the two backgrounds exceeds a certain number (usually 4) of standard deviations of the background i.e. if

$$(n_1 - n_2) > 4(n_1 + n_2)^{\frac{1}{2}}$$

For the ordinate analysis or Wyckoff method of measurement, a reflection is judged to be mis-set if, in a scan including $2n$ steps through the peak, the initial position of the peak includes steps $n = 0$ or $n = n$.

10.3.2 Lorentz–polarization correction

The polarization factor, p, is a function of θ only and the inverse is given by

$$p^{-1} = \frac{2}{1 + \cos^2 2\theta}$$

The Lorentz correction is dependent on the geometry which is used to record the X-ray intensity measurements. The different expressions are summarized in Table 10.IV. The Lorentz–polarization corrections are applied to the observed measurements as $(Lp)^{-1}$.

TABLE 10.IV
Lorentz Corrections for different diffractometer geometries

Geometry		L^{-1}
A.	Four Circle Symmetrical A setting	$\sin 2\theta$
B.	Inclination geometry	
	General	$\cos \mu \cos \nu \sin \Upsilon$
	Normal beam ($\mu = 0$, $\sin \nu = \zeta$)	$\cos \nu \sin \Upsilon = (\sin^2 2\theta - \zeta^2)^{1/2}$
	Equi-inclination	
	($\mu = -\nu$, $\sin \mu = -\zeta/2$)	$\cos^2 \nu \sin \Upsilon = \xi \cos \theta$
	Flat Cone ($\nu = 0$, $\sin \mu = -\zeta$)	$\cos \mu \sin \Upsilon = (\xi^2 - \frac{1}{4}(\xi^2 + \zeta^2)^2)^{1/2}$
	Flat Cone; multi-counter where z is the separation of the level from the level in the flat cone setting	$(\xi^2 + \zeta^2 z^2 - \frac{1}{4}(\xi^2 + \zeta^2 + z^2)^2)^{1/2}$

10.3.3 Absorption

The correction for absorption often represents the most serious correction factor for intensity data. This is especially true for diffractometer measurements where the crystal presents many different profiles to the X-ray beam during the collection of a set of three-dimensional data.

A discussion of the physical basis and the effects of absorption has been given in Section 9.3.4. In this Section we are concerned with the calculation of the correction factor. The computer methods (e.g. Busing and Levy, 1957), in which the path of the X-ray beam through the crystal is calculated from a knowledge of the crystal shape, have proved useful for small structure determinations but are not suitable for protein crystals where the shape of the crystal and the effects of the surrounding mother liquor and glass capillary tube are difficult to estimate and where the time involved in the calculation of path lengths for individual reflections for many thousands of reflections becomes prohibitive.

A simple semi-empirical method has been described by North et al. (1968) which is widely used in protein crystallography. In this method an empirical absorption curve is obtained from the variation in intensity of a strong Bragg reflection as the crystal is rocked about the normal to the corresponding planes. Suppose the crystal has orthogonal axes and is mounted for rotation about c. Then in the equi-inclination setting (which corresponds to $\sin \mu = lc^*/2$ for the linear diffractometer and the symmetrical A setting with $\chi = -90°$ for the four circle diffractometer), the reflection $00l$ is in the reflecting position for all values of ϕ, the rotation about the goniometer axis. In this setting the incident and reflected beams are equally inclined to the axis of crystal rotation so that the mean of these two directions lies in the plane perpendicular to the rotation axis (Fig. 10.5). The variation in intensity of the $00l$ reflection as the crystal is stepped through all values of ϕ with a stepping interval of 10 or 15 degrees provides a measure of the relative absorption suffered by the X-rays on passing through the crystal in a mean direction perpendicular to the rotation axis. Transmission coefficients for any general reflection hkl are then given approximately by $T(hkl) = T(\phi_i) + T(\phi_r)$ when ϕ_i and ϕ_r are the azimuthal angles of the incident and reflected beams, respectively. These angles may be calculated for each reflection and the value of the absorption at this angle determined from the empirical curve which is fed into the computer. A typical absorption curve for a phosphorylase crystal measured on the four-circle diffractometer is shown in Fig. 10.6.

The efficacy of the method rests on the assumption that the absorption is a function of ϕ only. The validity of this assumption depends, among other things, on the crystal shape and range of θ values explored. In order to minimize effects due to variation of absorption in θ, it is advisable to measure the absorption curve for a strong reflection which lies in or close to the reciprocal lattice levels

10.3.3. ABSORPTION

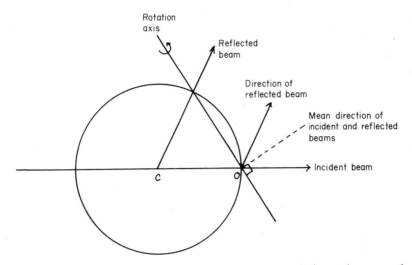

FIG. 10.5 Equi-inclination geometry for the recording of absorption curves by the method of North et al. (1968).

under investigation. There is an additional difficulty when crystals and their mountings are markedly asymmetric (Fig. 10.7). The semi-empirical method gives equal absorption corrections for reflections hkl and $\bar{h}\bar{k}l$ for a crystal rotating about c*. This will be appropriate for the orientation shown in Figure 10.7a but not for the orientation shown in Figure 10.7b. In some cases it may therefore be necessary to apply a secondary absorption correction based on an analysis of agreement of such equivalent reflections.

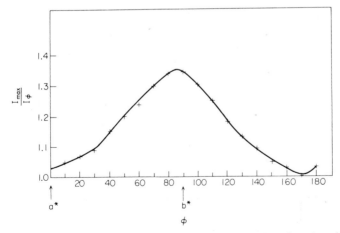

FIG. 10.6 A typical absorption curve for tetragonal phosphorylase b crystals.

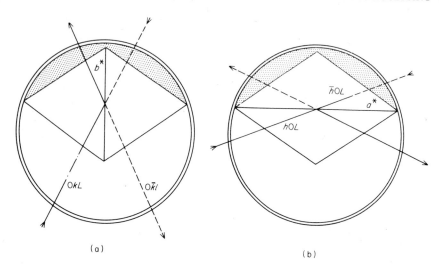

FIG. 10.7 Asymmetric mounting of protein crystal with mother liquor in capillary. Anomalous scattering differences would be seriously affected in (b) but not in (a). [From North et al. (1968)].

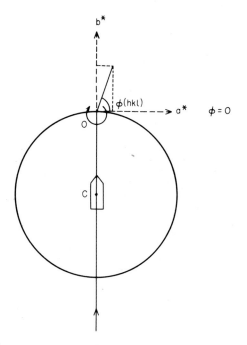

FIG. 10.8 Projection onto equatorial plane showing the $\phi = 0$ position of the crystal and the definition of $\phi(hkl)$.

10.3.3. ABSORPTION

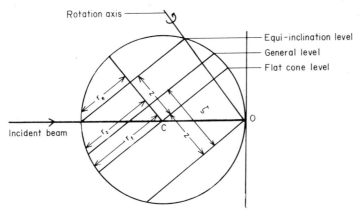

FIG. 10.9 Elevation showing the flat cone, equi-inclination and general settings.

Measurement of absorption curves on the linear diffractometer presents difficulties because this instrument only permits rotation of the crystal about the goniometer axis ϕ. Thus it is not possible to scan through an axial reflection in the equi-inclination setting and only values of peak heights can be obtained. Provided that the crystal is well set, this may not prove too much of a drawback. The high resolution data for lysozyme were corrected by absorption curves obtained in this way, but then the maximum absorption seldom exceeded 1.3 and the crystals showed little tendency for focusing–defocusing effects. Clearly for the majority of crystals the four circle diffractometer is better suited where the $\omega(\theta)$ or χ circles can be used for rocking the crystal, and a measure of the integrated intensity obtained.

Expressions for the angles ϕ_i and ϕ_r may be derived with reference to Figs. 10.8, 10.9, 10.10. When the crystal is mounted for rotation about the $c(c^*)$ axis, the origin of ϕ is defined as the angle at which a^* is perpendicular to the incident beam. For an orthogonal system of axes, $\phi(hkl)$ is given by

$$\tan \phi(hkl) = \frac{kb^*}{ha^*}$$

$\phi(hkl)$ is therefore the crystal setting in which the reflection planes hkl are parallel to the incident beam. In order to bring these planes into the reflecting position, the crystal is rotated through $\phi(hkl)$ and back through an angle α, where the value of α depends on the geometry used to record the reflection. When the crystal is in the reflecting position in a general setting of the diffractometer, the incident and reflected beams traverse the crystal in directions, projected on the equatorial plane, corresponding to the angles

$$\phi_i = \phi(hkl) - \alpha \quad \phi_r = \phi(hkl) + \gamma$$

From Fig. 10.10, it may be shown that

$$\sin \alpha = \frac{2\zeta z - \zeta^2 - \xi^2}{2\xi(1-z^2)^{1/2}}$$

and from Fig. 10.10

$$\sin \gamma = \frac{\zeta^2 - 2\zeta z - \xi^2}{2\xi(1-(\zeta-z)^2)^{1/2}}$$

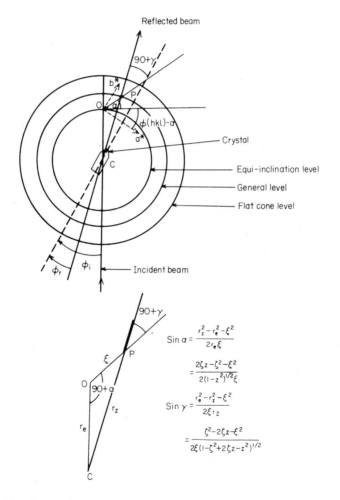

FIG. 10.10 Projection onto equitorial plane showing the reflection hkl in the reflecting position for general geometry and the calculation of $\phi_i = \phi(hkl) - \alpha$ and $\phi_r = \phi(hkl) + \gamma$.

10.3.3. ABSORPTION

For normal beam geometry $\zeta = z = 0$

$$\therefore \alpha = \gamma = -\theta(hkl)$$

For equi-inclination geometry and symmetrical A setting $z = \zeta/2$

$$\therefore \sin \alpha = \sin \gamma = \frac{-\xi}{(4-\zeta^2)^{1/2}}; \quad \text{i.e. } \alpha = \gamma = -\Upsilon/2$$

where Υ is the projection on the equatorial plane of 2θ.

For flat cone geometry $z = \zeta$

$$\sin \alpha = \frac{\zeta^2 - \xi^2}{2\xi(1-\zeta^2)^{1/2}}$$

$$\sin \gamma = \frac{-\zeta^2 - \xi^2}{2\xi}$$

Thus the expressions for ϕ_i and ϕ_r are not symmetrical with respect to the angles α and γ for the flat cone setting. The sign of these angles defined here is opposite to that defined by North et al. (1968). The change has been made to be consistent with the counter arm to the right hand side of the beam (when looking along the direction of the incident beam).

A more elaborate but inherently more precise method of empirical absorption correction has been proposed by Kopfmann and Huber (Kopfmann and Huber, 1968; Huber and Kopfmann, 1969). A transmission surface for the whole crystal is constructed from measurements of the intensities of several reflections as these are rotated about their reciprocal lattice vectors. Since these reflections are general reflections and not axial reflections parallel to the goniometer axis, the method requires a four circle diffractometer where the appropriate rotations can be accomplished by suitable adjustment of the ϕ, χ and ω circles. Model calculations suggest this method is reliable for crystals with μt as great as 3. But it requires considerably more time and measurements (approximately 300 measurements for the protein erythrocruorin) than the simpler method of North et al. (1968).

A further empirical method which overcomes the asymmetrical absorption sometimes observed for equivalent reflections has been proposed by Katayama et al. (1972). This method assumes that any observed differences in intensities among equivalent reflections are due to absorption. Differences due to anomalous scattering are treated as random errors. The function R is minimized by a least squares procedure where

$$R = \sum_{\mathbf{h}} \sum_{i=1}^{N} \omega_{\mathbf{h}i}(A(\mathbf{h}i) \cdot I^0_{\mathbf{h}i} - I^f_{\mathbf{h}})^2$$

where $\mathbf{h}$ and i represent an order of independent and equivalent reflections respectively; $A(\mathbf{h}i)$ is the absorption correction which is to be determined, I^0 and

I^t are the observed and true intensities, w is a weighting factor and N is the total number of equivalent reflections. Since the true intensity is not known it is replaced by the average value of the observed intensities so that R becomes

$$R = \sum_{h}\sum_{i} \omega_{hi} \left\{ A(hi) \cdot I^0_{hi} - \left[\sum_{j=1}^{N} \omega_{hj} A(hj) \cdot I_{hj}\right] \bigg/ \sum_{j=1}^{N} \omega_{hj} \right\}^2$$

It is assumed that $A(hi)$ is periodic so that it may be expanded in terms of the series

$$A(hi) = \sum_{n}\sum_{n'}\sum_{m}\sum_{m'} \{C_{nn'mm'} \cos(n\phi_p + n'\phi_s + m\mu_p + m'\mu_s)$$
$$+ S_{nn'mm'} \sin(n\phi_p + n'\phi_s + m\mu_p + m'\mu_s)\}$$

where C and S are to be determined and $\phi_p, \phi_s, \mu_p, \mu_s$ are angles which refer to the directions of the primary and secondary beams. Since the transmission varies smoothly with respect to these directions the expansion converges rapidly. The observational equations are linear in terms of the unknown parameters and hence may be solved by usual least squares methods.

In equi-inclination geometry

$$\mu_p = \mu_s = \mu, \quad \phi_p = \phi + \Delta\phi, \quad \phi_s = \phi - \Delta\phi$$

and

$$\sin\mu = \sin\chi \sin\theta$$

and

$$\tan\Delta\phi = \cos\theta/\sin\chi \sin\theta$$

where the symbols have their usual meaning in diffractometry. If cross terms of ϕ and $\Delta\phi$ are neglected, $A(hi)$ is given by

$$A(hi) \simeq \sum_{n}\sum_{m} C_{nm} \cos(n\phi + m\mu) + S_{nm} \sin(n\phi + m\mu) \cos n\Delta\phi$$

This method is likely to prove most effective when a large number of equivalent reflections are measured under different absorption conditions.

10.3.4 Radiation damage

Radiation damage correction curves may be constructed from the set of reference reflections measured throughout the run. It is generally assumed that the decrease in intensity arising from radiation damage is linear with exposure time, but it is relatively simple to apply some non-linear function or a totally empirical correction which also takes into account fluctuations in the performance of the X-ray tube and counters. Radiation damage undoubtedly affects the higher angle reflections more than the lower angle reflections. For high resolution data it is wise to incorporate measurement checks which allow this and any asymmetric variation to be detected. This may be done by measuring the intensities of the $h00$, $0k0$, or 001 rows of reflections (or similar rows if the axial rows contain too many systematic absences) at the beginning and end of data collection.

10.4 COMPARISON OF EQUIVALENT REFLECTIONS

Analysis of the agreement of symmetry related reflections as a function of h, k, l and $\sin^2\theta/\lambda^2$ at this stage allows any systematic errors due to imprecise crystal setting, incomplete absorption correction etc. to be detected and corrected. At the same time very bad disagreements between equivalent reflections (those which differ by more than six standard deviations, say) are monitored and the cause of these errors should be sought. If the equivalent reflections are Bijvoet pairs, then an estimate of the significance of the anomalous scattering for heavy atom derivatives is obtained from a comparison of the agreement between equivalent reflections for the centric and acentric reflections. Since anomalous scattering does not affect the centric terms, the agreement among these reflections is likely to be better than that for the acentric data.

The quality of the data is assessed by the reliability index R_{sym} where

$$R_{sym} = \frac{\sum_{\mathbf{h}} \sum_{i=1}^{N} |\bar{I}(\mathbf{h}) - I(\mathbf{h})_i|}{\sum_{\mathbf{h}} \sum_{i=1}^{N} I(\mathbf{h})_i}$$

where $I(\mathbf{h})_i$ is the ith measurement of reflection $\mathbf{h}$ and $\bar{I}(\mathbf{h})$ is the mean value of the N equivalent reflections. Values of R_{sym} of the order of 5% or less are acceptable.

10.5 SCALING DIFFERENT SETS OF DATA

Three-dimensional high resolution data for a protein crystal will almost certainly have been collected from several crystals, possibly with several counters, or, if collected photographically, from many films. It is necessary to scale these different sets of data together on the basis of the common reflections between the sets. A method is required which takes into account the fact that some sets may contain reflections which overlap with many other sets but that not every set will contain overlaps with all other sets. The method of Hamilton et al. (1965) [see also Rollett (1965)] has proved useful and reliable.

Let $I(\mathbf{h})_i$ be an observation of the intensity of reflection $\mathbf{h}$ for the crystal, film or level i. We require a set of scale factors, K_i, for each of the crystals etc. to be applied to the $I(\mathbf{h})_i$ so that we may determine the best least squares value $I(\mathbf{h})$ of the intensity. In the least squares equations it is more convenient to have the unknown parameters (K_i and $I(\mathbf{h})$ in this case) separate from the observations so that the minimization function is not altered if every K_i is multiplied by some scale factor. We therefore substitute $G_i = 1/K_i$ and solve the equations in terms of the reciprocals of the scale factors. The observational equations thus become

$$\phi(\mathbf{h})_i = \sqrt{w(\mathbf{h})_i}\,(I(\mathbf{h})_i - G_i I(\mathbf{h}))$$

where $\sqrt{w(\mathbf{h})_i}$ is the weight of the ith observation and is proportional to the reciprocal of the standard deviation of $I(\mathbf{h})_i$. Note that with the observational equations in this form with G_i rather than K_i, the weights are more easily calculated since they are based on uncertainties in $I(\mathbf{h})_i$ rather than $K_i I(\mathbf{h})_i$.

We wish to minimize

$$\psi = \sum_{\mathbf{h}} \sum_{i=1}^{N} \phi(\mathbf{h})_i^2$$

where N is the number of observations of the reflection $\mathbf{h}$. The best least squares value of $I(\mathbf{h})$ is found from the condition $\partial\psi/\partial I(\mathbf{h}) = 0$ i.e.

$$I(\mathbf{h}) = \frac{\sum_{i=1}^{N} w(\mathbf{h})_i G_i I(\mathbf{h})_i}{\sum_{i=1}^{N} w(\mathbf{h})_i G_i^2}$$

The solution of the observational equations for G_i is not straightforward since, with the substitution of the above value for $I(\mathbf{h})$, the equations become non-linear in G_i. The usual non-linear least squares approach (section 5.15) is adopted, in which approximate values of G_i ($= 1$ say) are used initially and corrections ΔG_i calculated in an iterative process. The observational equations are expanded by Taylor's theorem for values of ϕ_i at $G_i + \Delta G_i$ i.e. the observational equations become

$$\phi(\mathbf{h})_i = \sqrt{w(\mathbf{h})_i}\,(I(\mathbf{h})_i - G_i I(\mathbf{h})) + \sum_{k=1}^{N} \frac{\partial \phi(\mathbf{h})_i}{\partial G_k} \Delta G_k$$

Thus $I(\mathbf{h})$ and the partial derivatives are evaluated with the approximate values of G_i and the resulting equations solved for ΔG_i by the usual least squares procedure. Because ΔG_i are not independent, one of them is set arbitrarily equal to zero. The G_i are corrected by ΔG_i and the process repeated until the correction values are small. Convergence is usually achieved within 5 to 10 cycles.

10.6 ABSOLUTE SCALE AND THE SCALING OF DERIVATIVE TO NATIVE DATA

An approximate value of the absolute scale of the data may be derived from a Wilson plot. Wilson (1949) has shown that the probability of a reflection having an intensity between I and $I + dI$ is given by $P(I)dI$ where

$$P(I) = \frac{1}{\Sigma f_j^2} \exp - \frac{I}{\Sigma f_j^2}$$

and Σf_j^2 is the sum of the squares of the scattering factors of the atoms for all atoms in the unit cell. The mean value of the intensity is given by

$$\langle I \rangle = \Sigma f_j^2$$

These expressions were derived with the assumption that the structure consists of a large number of atoms which are randomly distributed. This assumption is not true for a low resolution protein structure and is only partially true for a high resolution structure where the presence of regular secondary features such as α-helices or β-pleated sheet produce a non-random structure. Nevertheless these expressions are useful for placing the data on an approximate absolute scale.

Suppose the atoms in the native protein structure suffer from thermal vibrations which may be represented by an average isotropic temperature factor B_P. Then we require a scale factor K_P such that

$$K_P \langle I_P \rangle = \Sigma f_j^2 \exp - 2B_P \frac{\sin^2 \theta}{\lambda^2}$$

where $\langle I_P \rangle$ and Σf_j^2 are the average values of the observed intensities and the sum of the squares of the scattering factors of the atoms in the protein, respectively, within the appropriate range of $\sin^2 \theta/\lambda^2$. (Note that we use $2B_P$ because we are dealing with intensities.) Hence

$$\ln \frac{\langle I_P \rangle}{\Sigma f_j^2} = - \ln K_P - 2B_P \frac{\sin^2 \theta}{\lambda^2}$$

Therefore a plot of $\ln \langle I_P \rangle/\Sigma f_j^2$ versus $\sin^2 \theta/\lambda^2$ should give a straight line slope $-2B_P$ and intercept $-\ln K_P$. A Wilson plot for insulin is shown in Fig. 10.11. It is seen that there is a significant deviation of the points from linearity especially in the low angle region, which indicates a breakdown in the assumptions underlying the Wilson equation, and that there is some flexibility in the assignment of the absolute scale and temperature factor. The high angle data (between 3 Å and 1.5 Å resolution, say) form a better guide to the absolute scale than the low angle data.

In the recording of intensities from the protein-heavy atom crystals, separate

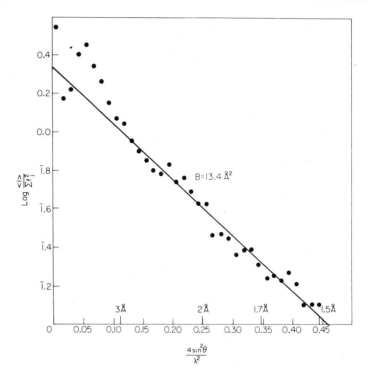

FIG. 10.11 A Wilson plot for 2Zn-insulin crystals. (Unpublished data provided by Dr. G. Dodson). The plot is based on 13,668 reflections to a resolution of 1.5 Å. The value of the temperature factor B has been determined from the best straight line for points between 3Å and 1.5 Å. The minimum and maximum in the low angle region occur at spacings corresponding to 8.2 Å and 4.5 Å respectively.

measurements of the Bijvoet pairs $F_{PH}(+)$ and $F_{PH}(-)$ are usually made. For some calculations we require the average value of F_{PH}. It has been shown (Petsko, 1973, 1975) that

$$F_{PH} = \frac{F_{PH}(+) + F_{PH}(-)}{2}$$

forms a more correct estimate of F_{PH} than the expression

$$F_{PH} = \left[\frac{F_{PH}(+)^2 + F_{PH}(-)^2}{2}\right]^{1/2}$$

The scaling of heavy atom derivative data to the native set requires careful analysis. Incorrect scaling often leads to "ghosts" (either holes or peaks) appearing at the heavy atom sites in the final electron density map. An

10.6. ABSOLUTE SCALE AND THE SCALING OF DERIVATIVE

approximate value of the derivative scale factor, K_{PH}, derived simply by equating $K_{PH}\Sigma F_{PH}^2/\Sigma F_P^2$ to unity may prove adequate in the initial stages of analysis of the heavy atom but is unlikely to prove inappropriate later.

The problem has been discussed by Green et al. (1954) and Blow (1958) who suggested that the scale factor K_{PH} for the F_{PH} terms may be adjusted to compensate for the additional scattering power of the heavy atoms by

$$\Sigma F_H^2 = K_{PH}^2 \Sigma F_{PH}^2 - \Sigma F_P^2$$

However unless the details of the heavy atom substitution are understood, the value of F_H is not known.

For centric data F_H may be replaced by $K_{PH}F_{PH} - F_P$ which leads to the expression $K_{PH}\Sigma F_{PH}F_P = \Sigma F_P^2$. An equivalent expression may be derived by noting that the origin peaks of the two different Patterson syntheses, one constructed from coefficients $(F_{PH} - F_P)^2$ and the other from coefficients $F_{PH}^2 - F_P^2$, should be equal and should consist solely of heavy atom self vectors if F_{PH} is properly scaled with respect to F_P. This leads to the condition

$$\Sigma F_H^2 = \Sigma(K_{PH}F_{PH} - F_P)^2 = \Sigma(K_{PH}F_{PH})^2 - \Sigma F_P^2$$

i.e.

$$K_{PH}\Sigma F_{PH}F_P = \Sigma F_P^2$$

where K_{PH} is the derivative scale factor and the summation is for all reflections of the centric zone.

For acentric data F_H may be replaced by F_{HLE} (Chapter 11) following the method described by Arnone et al. (1971) which is based on a concept suggested by Kraut et al. (1962). The equations become

$$\Sigma F_{HLE}^2 = \Sigma(K_{PH}F_{PH})^2 - \Sigma F_P^2$$

where, as shown in Section 11.2.2,

$$F_{HLE}^2 = F_M^2 + F_P^2 - 2\left[F_M^2 F_P^2 - \left(\frac{k}{4}\right)^2 (\Delta I)^2\right]^{1/2}$$

where

$$F_M^2 = \frac{1}{2}\left(F_{PH}(+)^2 + F_{PH}(-)^2\right), \quad \Delta I = (F_{PH}(+)^2 - F_{PH}(-)^2) \quad \text{and}$$

$$k = F_H'/F_H''H,$$

the ratio of the real to the imaginary scattering factors for the heavy atom derivative. An approximate scale for F_{PH} needs to be known before F_{HLE} is calculated. This may be obtained using the expression given for the centric terms. An improved value of the derivative scale factor is then obtained using the equation with F_H replaced by F_{HLE} and the process repeated until convergence is reached. The F_{HLE} scale factor may be subsequently refined during the least squares refinement of the heavy atom parameters (Evans, 1973) but experience suggests that the least squares minimum characterized by this parameter is not sharply defined. In all cases it is advisable to analyse the scale factor as a function of h, k, l and $\sin^2 \theta/\lambda^2$ in order to detect any systematic variation in the derivative data.

The ratio of the scaled absolute intensities for the derivative and native data may differ quite significantly from unity. From Wilson statistics

$$K_{PH} \langle I_{PH} \rangle = (\Sigma f_j^2 + \Sigma f_H^2) \exp - 2B_{PH} \frac{\sin^2 \theta}{\lambda^2}$$

where Σf_j^2 and Σf_H^2 are the sums of the squares of the scattering factors for the protein and for the heavy atoms, respectively. B_{PH} is the temperature factor for the protein plus heavy atom data. Therefore at $\sin^2 \theta/\lambda^2 = 0$,

$$\frac{K_{PH} \langle I_{PH} \rangle}{K_P \langle I_P \rangle} = 1 + \frac{\Sigma f_H^2}{\Sigma f_j^2}$$

e.g. a single mercury atom (80 electrons) in a protein molecular weight 28,000 (i.e. approximately 2000 atoms with an average of 7 electrons per atom) gives

$$\frac{K_{PH} \langle I_{PH} \rangle}{K_P \langle I_P \rangle} = 1 + 0.066$$

and a single mercury atom in a protein of molecular weight 12,000 (i.e. approximately 855 atoms with an average of 7 electrons per atom) gives

$$\frac{K_{PH} \langle I_{PH} \rangle}{K_P \langle I_P \rangle} = 1 + 0.153$$

Difference Wilson plots may also be used to determine the ratios of the absolute scales and temperature factors for the protein and derivative data. The intensity distribution of the two crystals are likely to be similar and therefore deviations from Wilson statistics should cause less problems. The greatest uncertainty lies in the degree of heavy atom substitution which is required for the calculation of Σf_H^2 and which is unlikely to be known in advance.

11

THE DETERMINATION OF HEAVY ATOM POSITIONS

11.1 INTRODUCTION

In order to use the method of isomorphous replacement we must first find the heavy atom contribution to the structure factor of the derivative crystals. The structure factors, F_P, F_{PH} and F_H of the protein, derivative and heavy atoms respectively are related together by

$$\mathbf{F}_H = \mathbf{F}_{PH} - \mathbf{F}_P \qquad (11.1)$$

The magnitudes F_{PH} and F_P can be measured, but this does not give rise to a direct estimate of F_H as 11.1 is a vector equation. The first step is to make the best estimate possible for F_H; this will be more straightforward in centric zones than for general reflections where the phase can have any value between 0 and 2π. The second step is to use the estimated value of F_H to determine the heavy atom positions. This can be achieved by using Patterson methods (Section 5.13) to estimate approximate positions and difference Fouriers (Chapter 14) and least squares refinement (Section 5.15) to improve these. The third step is relatively trivial; the heavy atom positions are used to calculate the heavy atom structure factor, $\mathbf{F}_H$. Alternatively the estimates of F_H can be used in direct methods (Section 5.14) to give the phases.

When an approximate estimate of the protein phases is available from other derivatives, difference Fourier techniques may be used to find the heavy atom positions. The protein phases can also be used in refining the heavy atom parameters. This technique has considerable advantages in interpreting complex heavy atom substitution patterns but may suffer from "feed back" errors if the phases used are not good estimates.

In this chapter we will describe the different techniques available in some detail and try to evaluate their relative advantages and disadvantages.

11.2 ESTIMATION OF F_H

11.2.1 Centric zones

In centric zones the vectors $\mathbf{F_P}$ and $\mathbf{F_{PH}}$ must be colinear so that

$$F_H = |F_{PH} \pm F_P| \qquad (11.2)$$

If the degree of heavy atom substitution is small, then F_H will usually be correctly estimated from the difference of F_{PH} and F_P, and this assumption is usually made. However if both F_{PH} and F_P are very small it may be that the correct estimate is $|F_{PH} + F_P|$. This is known as "crossing over". Situations where this might occur can be identified if there is some independent estimate of the degree of heavy atom substitution. Then the maximum possible value of F_H (F_H max) can be derived, and terms where $F_H\text{max} \geqslant |F_P + F_{PH}|$ can be eliminated from subsequent calculations. As soon as some information about the heavy atom positions is known from initial Patterson and refinement techniques, a better estimate of $F_H$max may be made.

11.2.2 Acentric zones

$|F_{PH} - F_P|$ will not be a good estimate of F_H for acentric zones where there is no correlation between the vectors $\mathbf{F_H}$ and $\mathbf{F_P}$. In fact, we have seen (Equation 6.5) that

$$\Delta F_{iso} = |F_{PH} - F_P| = \left| F_H \cos(\alpha_{PH} - \alpha_H) - 2F_P \sin^2\left\{\frac{\alpha_P - \alpha_{PH}}{2}\right\} \right| \qquad (11.3)$$

When the degree of heavy atom substitution is small, $(\alpha_P - \alpha_{PH})$ must be small and the second term approaches zero. Then

$$|F_{PH} - F_P| \simeq |F_H \cos(\alpha_{PH} - \alpha_H)| \qquad (11.4)$$

Again $|F_{PH} - F_P|$ will be a good estimate only when the vectors $\mathbf{F_{PH}}$ and $\mathbf{F_P}$ are almost colinear. It will be a very bad estimate when they are mutually perpendicular.

Anomalous scattering differences can also be used to estimate F_H but they have similar shortcomings. We have seen (Expression 7.15) that

$$\Delta F_{ano} = \frac{k|F_{PH}(+) - F_{PH}(-)|}{2} \simeq |F_H \sin(\alpha_{PH} - \alpha_H)| \qquad (11.5)$$

where $k = F_H/F_H''$.

ΔF_{ano} is only a good estimate of F_H when α_{PH} and α_H differ by about $\pi/2$. The method will have further disadvantages. First on average the differences $|F_{PH}(+) - F_{PH}(-)|$ will be much smaller than the isomorphous differences and so the estimate is likely to be rather inaccurate. Secondly the value of k may not

11.2.2. ACENTRIC ZONES

be easily derived for heavy atom agents which are complex ions or mercurials, or where the heavy atom displace solvent or ligands on binding. The situation is improved by using an empirical value, k_{emp}, which we discuss in Section 11.2.3.

Although estimates of F_H from both isomorphous and anomalous scattering differences are inaccurate, they are complementary in the information they give. Taken together they can be used to give a good estimate of F_H.

A simple way of deriving an expression for F_H was first suggested by North (see Phillips, 1966) and is apparent from the following relation:

$$\Delta F_{iso}^2 + \Delta F_{ano}^2 \simeq F_H^2 \cos^2(\alpha_{PH} - \alpha_H) + F_H^2 \sin^2(\alpha_{PH} - \alpha_H)$$

Thus

$$F_H \simeq (\Delta F_{iso}^2 + \Delta F_{ano}^2)^{1/2} \tag{11.6}$$

This expression (Kartha and Parthasarathy, 1965a) is not the best we can derive but it has the advantage of demonstrating the complementarity of the contributions from isomorphous and anomalous scattering differences. Less approximate expressions had already been derived by Harding (1962) and later by Matthews (1966a). An exact statement has been derived by Singh and Ramaseshan (1966) which will now be discussed. We will call this the "combined difference" as a consequence of its derivation from a combination of the isomorphous differences and the anomalous scattering differences.

From triangle OAB of Fig. 7.6 on p. 178

$$F_{PH} \sin(\alpha_{PH} - \alpha_H) = F_P \sin(\alpha_P - \alpha_H) \tag{11.7}$$

and

$$F_{PH}^2 = F_P^2 + F_H^2 + 2F_H F_P \cos(\alpha_P - \alpha_H) \tag{11.8}$$

From triangle OBC of Fig. 7.6

$$F_{PH}(+)^2 = F_{PH}^2 + F_H''^2 - 2F_{PH}F_H'' \cos\left(\alpha_{PH} - \alpha_H + \frac{\pi}{2}\right) \tag{11.9}$$

From triangle OBD

$$F_{PH}(-)^2 = F_{PH}^2 + F_H''^2 - 2F_{PH}F_H'' \cos\left(\alpha_{PH} - \alpha_H - \frac{\pi}{2}\right) \tag{11.10}$$

By combining Expressions 11.7, 11.9 and 11.10 and assuming $k = F_H/F_H''$ the Bijvoet difference, ΔI, is given by

$$F_{PH}(+)^2 - F_{PH}(-)^2 = \frac{4}{k} F_H F_P \sin(\alpha_P - \alpha_H) \tag{11.11}$$

Let

$$F_M^2 = \tfrac{1}{2}(F_{PH}(+)^2 + F_{PH}(-)^2)$$
$$= F_{PH}'^2 + F_H''^2 \ldots\ldots\ldots \text{using Expressions (11.9) and (11.10)}$$
$$= F_P^2 + F_H^2 + F_H''^2 + 2F_H F_P \cos(\alpha_P - \alpha_H) \ldots \text{using Expression (11.8)}$$
$$= F_P^2 + F_H^2 \left\{1 + \frac{1}{k^2}\right\} + 2F_H F_P \cos(\alpha_P - \alpha_H)$$
$$= F_P^2 + F_H^2 \left\{1 + \frac{1}{k^2}\right\} + F_H F_P \left\{4 - \frac{k^2 \Delta I^2}{4F_H^2 F_P^2}\right\}^{1/2} \ldots \text{using Expression} \tag{11.11}$$

This equation can be expressed as a quadratic in F_H^2 which has the following positive roots

$$F_H^2 = (1 + 1/k^2)^{-2} [(1 + 1/k^2)F_M^2 + (1 - 1/k^2)F_P^2]$$
$$\pm 2(1 + 1/k^2)^{-2} [F_P^2 \{(1 + 1/k^2)(F_M^2 - F_P^2) + F_P^2\}$$
$$- (k/4)^2 (1 + 1/k^2)^2 (\Delta I)^2]^{1/2} \tag{11.12}$$

This relation gives F_H^2 in terms of F_M^2 and F_P^2. The positive and negative signs before the second term in expression (11.12) correspond respectively to the negative and positive values of $\cos(\alpha_{PH} - \alpha_P)$.

In most cases $(\alpha_{PH} - \alpha_P)$ is acute and hence only the negative sign before the second term need be considered, and this corresponds to taking a negative sign with isomorphous differences for centric data and is related to the assumption that $(\alpha_{PH} - \alpha_P)$ is small when considering "isomorphous differences". This is the Heavy-atom Lower Estimate, F_{HLE}. The Heavy-atom Upper Estimate, F_{HUE}, is given by the positive sign (Harding, 1962). In practice, to a good approximation, $1/k^2$ may be neglected as compared with unity, and the Expression (11.12) reduces to

$$F_{HLE}^2 = F_M^2 + F_P^2 - 2\left[F_M^2 F_P^2 - \left(\frac{k}{4}\right)^2 (\Delta I)^2\right]^{1/2} \tag{11.13}$$

Expression (11.12) is exact, but expression (11.13) is a reasonable approximation usually introducing less than 3% error. This is closely related to the equation derived by Matthews (1966a) and Harding (1962). Expression (11.6), of Kartha and Parthasarathy (1965a), can be derived from the others if $(\alpha_{PH} - \alpha_P)$ is assumed to be small and $F_{PH}/F_P = 1$.

11.2.3 The treatment of errors

Our estimates of F_H depend on the differences of two large quantities, either F_{PH} and F_P for isomorphous differences or $F_{PH}(+)$ and $F_{PH}(-)$ for anomalous differences. Therefore it is not surprising that ΔF_{iso}, ΔF_{ano} and F_{HLE} have

11.2.3. THE TREATMENT OF ERRORS

rather large errors, and this is particularly true of ΔF_{ano} as the anomalous differences are very small. If we are to use these as coefficients in Patterson functions, least squares refinement and direct methods to find the positions of the heavy atoms, we need to have an estimate of the errors so that we can correct for bias in their values and introduce proper weighting schemes.

The most simple approach to errors is to consider the relative error in the isomorphous and anomalous differences. This can be estimated by deriving an empirical value k_{emp} for the ratio of the real and imaginary parts of the heavy atom structure factor, i.e. $k = F'_H/F''_H$. (Matthews, 1966a, Kartha and Parthasarathy, 1965a). If F_{PH} and F_P are large compared with F_H, then the equations approximate to

$$|F_{PH} - F_P| = |F_H \cos(\alpha_{PH} - \alpha_H)|$$

and

$$|F_{PH}(+) - F_{PH}(-)| \simeq \frac{2}{k} |F_H \sin(\alpha_{PH} - \alpha_H)|$$

By averaging over a large number of reflections, and taking the ratio

$$k_{emp} = \overline{|F_H \cos(\alpha_{PH} - \alpha_H)|} / \overline{|F''_H \sin(\alpha_{PH} - \alpha_H)|}$$

$$2\overline{|F_{PH} - F_P|} / \overline{|F_{PH}(+) - F_{PH}(-)|}$$

Alternatively k_{emp} can be calculated as

$$2\sqrt{\overline{(F_{PH} - F_P)^2} / \overline{(F_{PH}(+) - F_{PH}(-))^2}} \tag{11.15}$$

The reflections over which the average is taken must of course include only acentric reflections. Inclusion of the centric zones will give an unexpectedly high value of k_{emp}!

Figure 11.1 shows a plot of $2\ |F_{PH} - F_P|/|F_{PH}(+) - F_{PH}(-)|$ against $\sin^2\theta/\lambda^2$; this can be used to calculate k_{emp} as a function of resolution. Occasionally k_{emp} is larger than the theoretical value. This may be due to lack of isomorphism, incorrect scaling of F_{PH} relative to F_P or to ligand atoms bound to the metal atom which are also introduced into the crystal near to the heavy atom site. However, k_{emp} is usually less than the theoretical value owing to random errors which tend to increase the estimate of the anomalous difference. If the derivative is isomorphous the ratio becomes smaller at higher angles for two reasons. First the imaginary part of the heavy atom structure factor F''_H has less $\sin\theta$ dependence than F_H. Secondly, the intensity measurements become smaller at higher angles and consequently the error in $F^2_{PH}(+)$ and $F^2_{PH}(-)$ is larger. This leads to an increased overestimation of the anomalous difference.

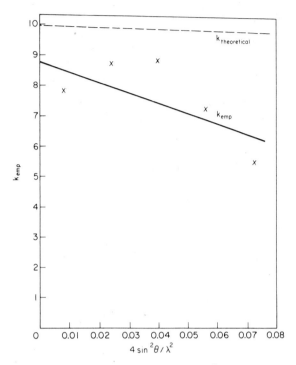

FIG. 11.1   Plot of k_{emp} against $\sin^2 \theta/\lambda^2$ for the AgNO$_3$ derivative of glucagon.

The use of k_{emp} in the expression 11.13 for F_{HLE} then gives a relative weighting for combination of the isomorphous and anomalous differences which accounts not only for counting statistics but also for the effect of ligands bound to the heavy atom or displaced by the heavy atom. It has been used very successfully in many analyses (Matthews, 1966a; Blundell *et al.*, 1971b; Arnone *et al.*, 1971).

We have seen that most calculations of k_{emp} indicate that the anomalous difference is overestimated. In order to make a better estimate of the errors we need to consider more carefully why this should be so. The distribution of errors in the structure factor amplitudes F_P, $F_{PH}(+)$ and $F_{PH}(-)$ used in calculating the ΔF_{iso} and ΔF_{ano} can be assumed to be distributed symmetrically about zero, and the errors in Δ are normally distributed about the mean. However, if ΔF_{iso} or ΔF_{ano} have magnitudes comparable with the standard deviation then the observed value will occasionally have the opposite sign to the true value. Tickle (1973, unpublished results) showed that the expected value of Δ^2 will be an overestimate given by

$$E(\Delta^2) = \{E(\Delta)\}^2 + \sigma^2(\Delta) \tag{11.16}$$

11.3.1. ISOMORPHOUS DIFFERENCES PATTERSONS

As the ratio of the isomorphous difference to its standard deviation is usually higher than the ratio of the anomalous difference to its standard deviation the overestimation is most significant for the anomalous difference. This is the reason why k_{emp} is usually less than the theoretical value. Statistical theory (see Appendix 1 of Dodson, Dodson et al., 1975) suggests that for $f(P_1, P_2, \ldots)$, where the P_i are normally distributed uncorrelated parameters with means $\bar{P}_i$ and variances:

$$\mathrm{Var}((P_i)) \approx \Sigma_i \left(\frac{\partial f}{\partial P_i}(\bar{P}_i)\right)^2 \sigma_i^2 \qquad (11.17)$$

$$\mathrm{Bias}((P)) \approx \sum_i \tfrac{1}{2} \frac{\partial^2 f}{\partial P_i^2}(\bar{P}_i)\sigma_i^2 \qquad (11.18)$$

For F_{HLE} the expression is difficult to derive and rather complex in its formulation. Dodson et al. (1975) have suggested that the sum differences estimate for F_H^2 (expression 11.6) leads to an adequate indication of the variance and bias which are approximately given by:

$$\sigma^2(F_H^2) \approx 4(F_{PH} - F_P)^2 \sigma_{iso}^2 + k^4 (F_{PH}(+) - F_{PH}(-))^2 \sigma_{ano}^2 \qquad (11.19)$$

$$\mathrm{Bias}\,(F_H^2) \approx \sigma_{iso}^2 + \left(\frac{k}{2}\right)^2 \sigma^2\mathrm{ano} \qquad (11.20)$$

where σ_{iso} and σ_{ano} are the standard deviations of the isomorphous and anomalous differences respectively. Expressions for $\sigma^2(F_H)$ and bias (F_H) can also be derived. Model calculations indicate that these formulae give good estimates of the predicted values, and confirm that errors will usually be large compared to their mean values and that F_H will tend to be seriously overestimated (Dodson et al., 1975). We will consider the use of these estimates of errors in Patterson syntheses, least squares refinement and direct methods in the following sections.

11.3 HEAVY ATOM VECTOR MAPS

We have seen in Section 5.13 that a Patterson synthesis gives peaks at the ends of vectors between atom locations, and therefore the positions of the atoms can be found by deconvolution of the vector map. We must now consider the Patterson functions obtained by using the different estimates of F_H discussed in the previous section.

11.3.1 Isomorphous differences Pattersons

In the centric zones the problem is quite straightforward. A Patterson function with coefficients $(F_{PH} - F_P)^2$ will give a good vector map (Perutz, 1956; Blow,

1958). We must, of course, remove values where $|F_{PH} + F_P|$ could be the correct estimate of F_H (see Section 11.2.1). For general reflections we have seen that $|F_{PH} - F_P|$ is a less good estimate of F_H; it will only be an accurate estimate when F_P, F_H and F_{PH} lie in the same direction. This will be the situation for most of the large values of $|F_{PH} - F_P|$. Thus most of the large terms in the summation will be accurate and, if sufficient terms are included, the Patterson can be very useful. (Blow, 1958; Rossmann, 1960). Examples of such Patterson functions are given in Figs. 11.2b and 11.3a. It is helpful to consider in more detail the nature of the approximations in this Patterson summation (Phillips, 1966). From expression 11.3 it can be seen that:

$$(F_{PH} - F_P)^2 = 4F_P^2 \sin^4 \left(\frac{\alpha_P - \alpha_{PH}}{2} \right) \quad \text{(i)}$$

$$+ F_H^2 \cos^2 (\alpha_{PH} - \alpha_H) \quad \text{(ii)}$$

$$- 4F_P F_H \sin^2 \left(\frac{\alpha_P - \alpha_{PH}}{2} \right) \cos(\alpha_{PH} - \alpha_H) \quad \text{(iii)}$$

The angles $(\alpha_P - \alpha_{PH})$ will tend to be small if values of F_H are small and consequently term (i) which gives protein–protein interactions will be of low weight. The transform of term (iii) is zero if sufficient terms are included. However, if $F_H \ll F_P$ (which may not always be true) $(\alpha_{PH} - \alpha_H)$ is effectively random and term (ii) will give heavy atom vectors with half the expected peak heights.

As F_H becomes larger relative to F_P, $(\alpha_P - \alpha_{PH})$ tends to increase. Term (i) will become larger and the noise due to protein–protein interactions will increase. However, as F_H becomes larger, $(\alpha_{PH} - \alpha_H)$ will tend to decrease and so the contribution of the heavy atom vector map will also increase. In model calculations using from one to six mercury atoms per protein of molecular weight 12000 (Dodson and Vijayan, 1971) a deterioration of peak to noise ratio was not apparent.

Problems with isomorphous differences Patterson occur mainly when there are several sites of low occupancy or even where there are several fully occupied sites on a large protein, i.e. M.W. $\sim$ 40,000. In these situations noise due to terms (i) and (iii) does often give rise to peaks larger than the heavy atom vector peaks and the Patterson becomes very difficult to deconvolute.

Strictly speaking the true difference Patterson would have coefficients $(F_{PH}^2 - F_P^2)$. The resulting vector map would include peaks corresponding not only to heavy-atom–heavy-atom vectors but also to heavy-atom–protein interactions. Both Pattersons will be sensitive to lack of isomorphism. However, the true difference Patterson will be more sensitive to errors of scaling and measurement as it depends on the difference of measurements rather than the difference of their square roots.

11.3.2 Anomalous differences Pattersons

An anomalous differences Patterson with coefficients ΔF^2_{ano}, i.e. $k^2/4[F_{PH}(+) - F_{PH}(-)]^2$, will also give a heavy atom vector map which is approximately equivalent to coefficients of $F_H^2 \sin^2(\alpha_{PH} - \alpha_H)$ (see Section 11.2.2) (Rossmann, 1961). In this Patterson there will also be noise and the peaks will be half weight due to the $\sin^2$ term multiplying F_H^2. Examples of anomalous differences Patterson are shown in Figs. 11.2.a and 11.3b. These are equivalent to the isomorphous differences Patterson in Figs. 11.2b and 11.3a. They are much more noisy than the isomorphous differences Pattersons although they contain the expected heavy atom vector peaks. Quite often the ratio of the heavy atom vector peaks to the origin peak will be three or four times less in the

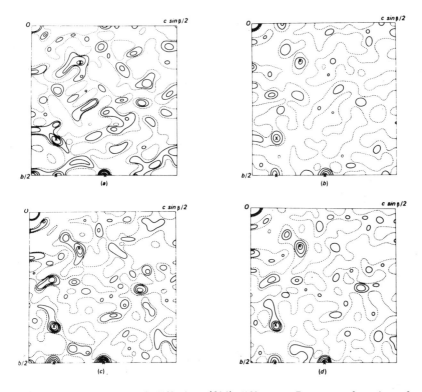

FIG. 11.2 Comparison of different (0kl) difference Patterson functions for the K_2PtCl_4 derivative of α-chymotrypsin: (a) anomalous differences synthesis; (b) isomorphous differences synthesis; (c) sum differences synthesis; (d) combined differences synthesis. The average peak height to backgrounds are given in Table 11.I. The position and size of the small crosses indicate the position and approximate height of the expected vector peaks. (From Matthews, 1966a).

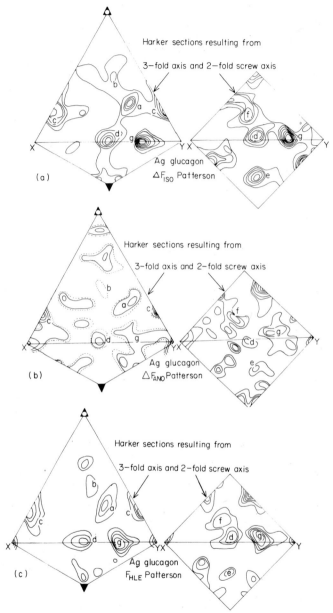

FIG. 11.3 Heavy atom Patterson functions for the $AgNO_3$ derivative of glucagon. (a) Isomorphous differences; (b) anomalous differences; (c) combined differences. The letters a–g. indicate the positions of the Harker peaks corresponding to the vectors between the silver atoms, one in each asymmetric unit. The spacegroup is $P2_1 3$. [Blundell, T. L., Sasaki, K., Dockerill, S. and Tickle, I. J. (1975) unpublished results.]

anomalous differences Patterson than in the isomorphous differences Patterson, and there may be several erroneous peaks as large as the correct vector peaks.

Both the isomorphous and anomalous differences Pattersons can be improved by proper weighting using the expressions for variance and bias given in section 11.2.3. Weighting by the inverse of the variance appears to have a rather drastic effect on the high angle terms in the anomalous differences Patterson, giving a map which is effectively rather low in resolution. Better results have been obtained by subtracting the bias from the estimates of the difference and using these as coefficients in the Patterson (Tickle and Blundell, unpublished results).

11.3.3 "Sum" and "combined" differences Pattersons

An improved heavy atom vector map may be obtained by combining the information from the isomorphous and anomalous scattering differences.

We have seen in section 11.2.2 that a better estimate for F_H^2 is given by $\Delta F_{iso}^2 + \Delta F_{ano}^2$. A Patterson function with these coefficients was originally proposed by North (see Phillips, 1966) and, independently, by Kartha and Parthasarathy (1965a); we will call it the "sum" difference Patterson function. A Patterson function using these coefficients is shown in Fig. 11.2c. It is clearly a better heavy atom vector map than the equivalent isomorphous differences and anomalous differences functions.

Alternatively we can use the "combined" difference expressions derived by Harding (1962), Matthews (1966a) or Singh and Ramaseshan (1966) (Equations 11.12 and 11.13). As discussed in Section 11.2.2, these estimates have an ambiguity in sign. The lower estimate, F_{HLE}, is usually used, and reflections where the upper estimate F_{HUE} is possible are removed from the summation. Figures 11.2d and 11.3c show the Patterson using coefficients F_{HLE}^2. They are very similar to the "sum" difference functions but are probably a slight improvement. Matthews (1966a) has suggested that the anomalous scattering contribution should be further decreased by multiplying k by 0.75. This somewhat arbitrary procedure gave better peak to noise ratios with the α-chymotrypsin derivatives but workers with other proteins have not always found this to be the case. Matthews (1966a) has found that with such a weighted combination difference Patterson the vector peaks were enhanced by an average of 27% relative to the vector peaks in the isomorphous differences Patterson and the higher background peaks are in fact reduced. With a good estimate of the standard deviation and bias of F_{HLE}^2 this should provide the basis of a more rational weighting scheme. The best Patterson is probably calculated with coefficients $F_{HLE}^2 - \text{bias}(F_{HLE}^2)$. Summations of the peak heights and backgrounds for a series of difference Patterson syntheses are given in Table 11.I.

Including information from the anomalous scattering differences can improve the heavy atom vector map, but it also can lead to difficulties if it is not used carefully. We have already seen that the anomalous scattering differences are relatively inaccurately measured; it is very easy to include contributions which are mainly due to errors rather than anomalous scattering.

For this reason it is important to calculate the isomorphous differences and anomalous differences Pattersons as well as combined differences Patterson. Clearly the anomalous differences Patterson must have some peaks in common with the isomorphous differences Patterson although there is likely to be a lot of noise. A "noisy" anomalous Patterson is quite usual; the data is probably still usefully included in F_{HLE} calculation and used for phasing. However, if there are no common peaks it probably means that there are serious errors in the anomalous differences. and the differences need to be remeasured.

TABLE 11.I

Comparison of difference Patterson syntheses for $PtCl_4^{2-}$ derivative of α-chymotrypsin. The Pattersons are shown in Fig. 11.2. (From Matthews, 1966a)

	Anomalous Differences Patterson	Isomorphous Differences Patterson	Sum Differences Patterson	Combined Differences Patterson
Average Peak Height	7.1	384	483	488
Average Background	1.52	42.9	44.7	40.2
Peak to Background Ratio	4.7	9.0	10.8	12.1
Relative improvement	0.52	1.0	1.21	1.35

If the k_{emp} is very large it probably means that the scaling of F_{PH} and F_P are wrong or alternatively there is lack of isomorphism (see Section 6.6). If k_{emp} is less than the theoretical value it means that the combined difference function will be more like the isomorphous differences Patterson than the anomalous difference Patterson. In this case it is useful to check whether peaks chosen as heavy atom vectors are present in the anomalous differences Patterson also, and this may provide a way of discriminating between different heavy atom arrangements chosen from the combined difference map.

Heavy atom vector maps can be dominated by low angle terms. There are often large differences between F_P and F_{PH} as a result of movement of solvent or addition of heavy atom salt in a rather disorganized way. It is therefore advisable to compute a Patterson without the terms of $d < 20\text{Å}$ and to compare this with the Patterson containing these terms. The heavy atom derivative is clearly not useful if the major peaks assumed to be heavy atom vector peaks are removed when the low angle terms are removed.

11.4 DIFFERENCE FOURIER TECHNIQUES FOR FINDING HEAVY ATOM POSITIONS

If estimates of the protein phases are available from other heavy atom derivatives, then difference Fourier techniques can be used to locate the heavy atom positions. The Fourier synthesis used has the coefficients (Stryer *et al.*, 1964)

$$m(F_{PH} - F_P) \exp(i\alpha_P) \tag{11.21}$$

These coefficients give the vector contribution of the heavy atom in the direction of the protein structure factor specified by α_P and are weighted by the figure of merit, m (see Section 12.2).

Difference Fourier syntheses have been used to advantage in most protein structure analyses. They are particularly useful if there are many sites of heavy atom substitution when the difference Patterson syntheses become difficult to interpret. However, they must be used with great caution. In particular, the phases, α_P, must not be calculated from any model for the heavy atom derivative being tested (Dickerson *et al.*, 1967). This is because Fourier syntheses tend to be dominated by the phases rather than the magnitudes.

Srinivasen (1961) showed that a Fourier calculated with the correct phases, but with amplitudes either equal to unity or of random values, gave peaks at the expected positions; however, a Fourier with the correct amplitudes but incorrect phases gave no correct peaks. This implies that if the phases are calculated from a particular heavy atom arrangement, they are likely to reproduce this arrangement in a difference Fourier regardless of whether they are right or wrong. Thus difference Fouriers used in this way will tend to confirm the initial assumptions, and are worthless. As $\alpha_P(\text{calc})$ is sometimes correlated with α_H, the difference Fourier coefficients must include an α_P calculated from other heavy atom derivatives. These are known as "cross phase" difference Fouriers. If the phases are computed from only one derivative and are not very accurate this may also give rise to difficulties. In particular "ghost" peaks tend to appear in the difference Fourier for the second derivative at the position of some or all of the heavy atoms in the first derivative. These are probably due to errors of the occupation number, overall scale or thermal parameter. These peaks often appear as minor sites in the second derivative. Therefore sites in common between two derivatives indicated by difference Fourier techniques should be treated with scepticism. In all cases the atomic positions indicated by the difference Fourier should be crosschecked against the Patterson synthesis. The positions can only be accepted if the expected heavy atom vector peaks occur.

Unlike the difference Patterson, the difference Fourier will become less useful when the degree of heavy atom substitution is large. This becomes clear if we consider the coefficients in the following way. From equation 11.8 we can see that

$$F_{PH} - F_P = \frac{F_H^2 + 2F_P F_H \cos(\alpha_H - \alpha_P)}{F_{PH} + F_P}$$

But $\cos(\alpha_H - \alpha_P) = \tfrac{1}{2}[\exp\{i(\alpha_H - \alpha_p)\} + \exp\{-i(\alpha_H - \alpha_p)\}]$
and therefore

$$(F_{PH} - F_P) \exp(i\alpha_P) = \frac{F_H^2}{F_{PH} + F_P} \exp(i\alpha_P) \quad \text{(i)}$$

$$+ \frac{F_P F_H}{F_{PH} + F_P} \exp(i\alpha_H) \quad (11.22) \text{ (ii)}$$

$$+ \frac{F_P F_H}{(F_{PH} + F_P)} \exp\{i(2\alpha_P - \alpha_H)\} \quad \text{(iii)}$$

In expression 11.22, term (iii) will give rise to noise as $(2\alpha_P - \alpha_H)$ will assume random values. Term (ii) will give rise to a weighted heavy atom map, but term (i) is dominated by the protein phases and will tend to give the protein structure in the difference map. When F_P is much greater than F_H the contribution of the first term to the difference Fourier will be small, but as F_H becomes larger so term (i) will become relatively more important. The difference Fourier tends to resemble a highly distorted protein map and the height of the heavy atom peaks decreases. Dodson and Vijayan (1971) have found from model calculations that the addition of two mercury atoms to a protein of molecular weight 12000 daltons gives rise to difficulties in two dimensional analysis due to peaks resembling the protein structure although the problem is less serious in three-dimensional analyses.

We have already observed that in the difference Fourier, the coefficients $m(F_{PH} - F_P) \exp(i\alpha_P)$ give a vector contribution in the direction of α_P. It includes no contribution perpendicular to it. The necessary information is not available from isomorphous differences, but may be supplied from anomalous scattering differences (Matthews, 1966a).

Let the coefficients of the best Fourier be defined as

$$m'F_H \exp(+i\alpha_H) = m'F_H \exp(+i\alpha_P) \exp(+i[\alpha_H - \alpha_P])$$
$$= m'F_H\{\cos(\alpha_H - \alpha_P) + i\sin(\alpha_H - \alpha_P)\} \exp(+i\alpha_P) \quad (11.23)$$

The weight m' will depend to a large extent on the error of the protein phase and the figure of merit, m, can be used in this expression. The terms $F_H \cos(\alpha_H - \alpha_P)$ and $F_H \sin(\alpha_H - \alpha_P)$ can be estimated from the isomorphous and anomalous scattering differences as shown in expressions 11.4 and 11.5.

Matthews (1966a) has shown that difference Fouriers calculated with these coefficients can give heavy atom peaks of twice the height while the background is on average only 19% higher relative to isomorphous differences Fourier.

11.5 THE USE OF DIRECT METHODS IN FINDING HEAVY ATOM POSITIONS

If the solution of the difference Patterson synthesis is not straightforward, direct methods (see Section 5.14) may be used to find the heavy atom positions.

This was first exploited by Steitz (1968) for the centrosymmetric projection of carboxypeptidase at 2.8Å resolution. He used the Sayre equation (Section 5.14)

$$s(\mathbf{E_h}) = s\left\{\sum_k (\mathbf{E_k} \mathbf{E_{h-k}})\right\}$$

where $s(\ )$ means "sign of" and $\mathbf{E_h}$, $\mathbf{E_k}$, $\mathbf{E_{h-k}}$ are the normalized structure factors for reflections $\mathbf{h}$, $\mathbf{k}$ and $\mathbf{h-k}$. For centrosymmetric projections the magnitudes of the structure factors, E_h, can be estimated from isomorphous differences ($\Delta F_{iso} = |F_{PH} - F_P|$). For a centrosymmetric projection the signs of two large values are used to fix the origin. In his study Steitz assumed four additional signs to initiate phasing, and using the Sayre formula generated sixteen possible sets of signs. He chose between these by calculating a consistency index (equation 5.13). The four ΔE maps with the largest consistency indices are shown in Fig. 11.4 for a four site mercury derivative of carboxypeptidase. The correct solution was found to have the fourth highest index and could be distinguished by comparing the heavy atom vector sets predicted by each atomic arrangement to the Patterson function. This has to be done very carefully as Patterson and direct methods are very closely related and the incorrect ΔE maps will contain atomic positions related by vectors occurring in the Patterson. This is evident in the three incorrect solutions shown in Fig. 11.4. It is not enough to find some of the expected cross vectors in the Patterson; all expected peaks should occur. It is also wise to check the positions with a "cross-sign" difference Fourier when signs become available.

Direct methods may also be useful with acentric data. In this case the normalized structure factors may be derived from $\Delta F_{iso} = (F_{PH} - F_P)$ values (Neidle, 1973) or from F_{HLE} values (Expression 11.13). The tangent formula (Section 5.14) may be used to generate phases from starting sets including the origin and enantiomorph defining phases, and some arbitrarily fixed phases of large normalized structure factors. The value of the Karle R value (equation 5.18) and the internal consistency can be used as a guide to the correct set.

Neidle (1973) has shown that the method could be used successfully for derivatives of elastase even with low resolution data ($d \simeq 8$Å).

Direct methods are useful in finding heavy atom positions primarily because the number of heavy atoms is small and therefore the probabilities of sign relationships are high (see equation 5.14). However, low resolution data may give rise to "negative electron density" due to series termination effects and also

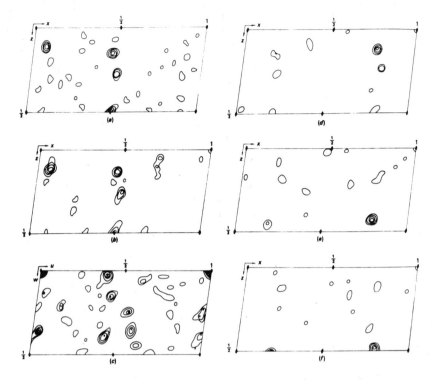

FIG. 11.4 The use of direct methods to determine the heavy atom positions in a four site mercury derivative, [010] projection, at 2.8 Å resolution. (a) Difference E map calculated using 85 ΔE's phased with Sayre's equation. (b) Difference Fourier map calculated applying protein phases deduced from a lead derivative. (c) Difference Patterson map. The expected interactions are marked by crosses. (d), (e) and (f) Incorrect ΔE maps. (Steitz, 1968).

the heavy atom sites may have different occupancies and so the structure is not strictly of equal atoms. In theory these factors invalidate the mathematical relationships, but in practice they are still workable.

11.6 REFINEMENT OF ATOMIC PARAMETERS FOR HEAVY ATOMS

The use of Patterson syntheses, cross phase Fouriers and direct methods described in the preceding sections give approximate positions for the heavy atoms but these may not be the best parameters which can be derived from the data available. The next task is to improve these positions and pull in any minor heavy atom sites which have been missed altogether.

In the analysis of small molecular structures this is achieved by Fourier

11.6.1. REFINEMENT SCHEME A: F_{HLE} REFINEMENT

analysis and by least squares techniques of refinement (see Section 5.15). Similar techniques may be used to improve the heavy atom positions by using either $|F_{PH} - F_P|$ in centric zones or F_{HLE} in acentric zones as the observed heavy atom structure amplitudes. However, F_{HLE} values are likely to be rather inaccurate estimates especially for large proteins or for heavy atoms with small anomalous scattering and so the refinement may not be very satisfactory. Alternatively if approximate protein phases are available from other derivatives, these may be used in "double" difference Fouriers and refinement by "lack of closure" procedures. We will now describe these procedures and discuss their relative advantages and disadvantages.

11.6.1 Refinement Scheme A: F_{HLE} refinement

Let us first consider refinement when there is a good estimate of the heavy atom structure amplitude, $F_H(\text{obs})$. The heavy atoms may be refined by minimization of $w[F_H(\text{obs}) - F_H(\text{calc})]^2$. $F_H(\text{calc})$ is computed in the normal way from the heavy atom positions and from the occupation and thermal parameters. Initial relative values of the occupation numbers can be estimated directly from the cross-phase Fouriers or from the ΔE maps or by consideration of the self vector peak heights in the Patterson. The overall thermal parameters of the heavy atoms can be estimated from a Wilson plot of $F_H(\text{obs})$.

We have seen in Section 11.3 that errors in the estimates of F_H from either isomorphous or anomalous differences (see Section 11.2.3) tend to cancel in the Patterson synthesis. In least squares this is less true and a better estimate is required. In a centric zone isomorphous differences $(F_{PH} - F_P)$ can be used successfully in conventional least squares analysis (Rossmann, 1960) but a trial and error least squares refinement due to Hart (1961) has been very popular. In the Hart technique the quantity minimized is

$$E = \Sigma[(F_H(\text{calc}) - F_H(\text{obs})]^2$$
$$= \sum_{\eta}[z_j \hat{f}_{oj} t_j \cos 2\pi(hx_j + lz_j) \pm F_P \pm KF_{PH}]^2 \qquad (11.24)$$

where η refers to one h0l reflection, z_j is an effective atomic number, $\hat{f}_{oj}$ is a unitary scattering factor for heavy atom j, t_j is the temperature factor, x_j and z_j are the heavy atom coordinates and K is the scale factor between protein and derivative data. The parameters are varied one at a time by $-2, -1, 1, 2$ units, the size of which is varied according to the stage of refinement. Error sums, E, are accumulated for each set of parameters plus the starting set. In each case the sign ambiguity in Equation 11.24 is resolved by taking the signs which give the smallest value of the term E. Full shifts are applied only to the parameter which is most influential in decreasing the error sum, and partial shifts are calculated for the others to compensate for the fact that only one parameter is varied at a time.

This trial and error procedure is particularly advantageous if there are large errors in the positional parameters. If the parameters are accurately defined then conventional least squares cycles refining simultaneously all parameters will be useful, but this can more easily lead to a false minimum if the initial parameters are inaccurate. Rossmann (1960) first suggested the use of $|F_{PH} - F_P|$ in least squares refinement with acentric data and this has proved successful in some structure analyses. Kartha (1965) used equation 11.6 to estimate F_H in the refinement of heavy atom positions of ribonuclease. Since then least squares refinements of the heavy atom positions in rubredoxin (Herriot et al., 1970), nuclease (Arnone et al., 1971), cytochrome b_5 (Matthews et al., 1971), calcium binding protein (Kretsinger and Nockolds, 1973) and insulin (Blundell et al., 1971b) have been achieved using F_{HLE} expressions of either Matthews (1966a), or Harding (1962) (Expression 11.3).

The success of least squares refinement using F_{HLE} depends critically on the proper estimation of errors and in the use of weighting schemes. We have seen in Section 11.2.3 that F_{HLE} values tend to have rather large errors (especially where they are mainly dependent on the anomalous differences) and in general they tend to be overestimated. Often many of the very large values are too large because of overestimation of the anomalous terms. The small terms are generally inaccurately estimated and may also suffer from crossover, i.e. F_{HUE} (the upper estimate rather than the lower estimate) should be taken. Thus a simple weighting scheme which weights down the smaller and the larger F_{HLE} values may be useful (Adams et al., 1969; Blundell et al., 1971b). A suitable weighting scheme is

$$\frac{1}{w} = \left[1 + \frac{(F_{HLE} - b)^2}{a}\right]^{1/2}$$

where a and b are chosen to make $w(F_{HLE} - F_H(\text{calc}))^2$ reasonably constant in different ranges of F_{HLE}. This weighting scheme proved useful in the refinement of insulin heavy atom positions.

Alternatively, the value of F_{HLE} may be estimated using k_{emp} (expression 11.14 or 11.15) or better by subtracting the bias (expression 11.18), i.e. (F_{HLE} − bias) is used. The weights can then be made equal to the inverse of the variance (expression 11.17) or alternatively refinement can be carried out with only those reflections for which the standard deviation is small.

Difference Fouriers with coefficients $|F_{HLE} - F_H(\text{calc})| \exp\{i\alpha_H(\text{calc})\}$ may also be calculated to pick up minor sites and to check the refinement of positions. In fact it is sensible to start by calculating a difference Fourier to check that there are no obvious heavy atom sites that have been omitted.

The major problem is assessing the correctness of the heavy atom parameters. A mean square error can be followed throughout the refinement

$$E(F_{HLE}) = \sum_h \frac{(|F_{HLE} - F_H(\text{calc})|)^2}{n}$$

11.6.1. REFINEMENT SCHEME A: F_{HLE} REFINEMENT

Conventional indices of agreement

$$R(F_{HLE}) = \Sigma \mid (F_{HLE} - F_H(\text{calc})) \mid / \Sigma F_{HLE}$$

may also be used. Table 11.II gives some final values of $\bar{R}(F_{HLE})$ for different structures. They range between 0.31 and 0.56, and clearly depend on accuracy of data, degree of isomorphism between protein and derivative, the nature and extent of the heavy atoms substitution, the size of the protein, and the resolution. A completely random structure should give $R = 0.58$ for acentric terms but most proteins will have centric zones (random structure $R = 0.80$) and the expected R value for the data will be somewhat higher. R values in the range

TABLE 11.II.
Examples of $R(F_{HLE})$ reported for useful derivatives in various successful protein structure analyses. $R(F_{HLE}) = \dfrac{\Sigma \mid F_{HLE} - F_H(\text{calc}) \mid}{\Sigma F_{HLE}}$

Protein and reference	Assymetric Unit MOL. WT.	Resolution Range	Heavy Atom Derivative	Number of Sites	R(FHLE)
Bacterial ferredoxin Adman et al. (1973)	6,000	2.8Å	UO_2^{2+} Sm^{3+} Pr^{3+}	11 4 8	0.44 0.43 0.44
Flavodoxin (Desulphovibrio Vulgaris) Watenpaugh et al. (1972)	6,000	2.5Å	Sm^{3+}	1	0.415
Rubredoxin Herriott et al. (1970)	6,000	2.8Å 3Å	UO_2^{2+} HgI_4^{2-}	6 1 + iodine included	0.34 0.39
Nuclease Arnone et al. (1971)		2Å 2Å	Ba^{2+} I	1	0.45 0.40
Insulin Adams et al. (1969)	12,000	2.8Å 2.8Å 2.8Å 4.5Å	$Pb^{2+}(I)$ $Pb^{2+}(II)$ UO_2^{2+} $UO_2F_5^{3-}$	5 4 2	0.40 0.44 0.43 0.28
PGK (phosphoglycerate kinase) Blake and Evans (1974)		3Å	EMP	4	0.44

0.31 to 0.45 are probably alright but it is easy to obtain incorrect heavy atom solutions (usually with some interatomic distances in common with the correct solution) with an R value of around 0.50.

It is very easy to obtain an incorrect solution by careless use of least squares. With small molecular structures the crystallographer rarely uses least squares until the R value is $\sim$0.20. If refinement is begun with only half the heavy atomic positions included the atomic positions found from the Patterson often shift to a false position in such a way that the R value is decreased and the difference map calculated using these incorrect positions is featureless. This may often happen if there are two or more partially occupied heavy atom sites clustered together. In summary R values are rather unreliable criteria for the choice of a correct solution. Careful consideration of the heavy atoms against the Patterson or with cross-Fouriers must provide the final check.

11.6.2 Refinement Scheme B: "Phase" refinement

If the protein phases, α_P, can be calculated from other heavy atom derivatives by the method of isomorphous replacement and the use of anomalous scattering (Chapter 12), these can be used in the refinement of the new derivative (Dickerson et al., 1961, 1968; Kraut et al., 1962; Lipscomb et al., 1966; Muirhead et al., 1967). Difference Fourier and or least squares techniques can be used.

Let us consider the vector triangle formed by $\mathbf{F_{PH}}$, $\mathbf{F_P}$ and $\mathbf{F_H}$. At this stage in the analysis F_{PH} and F_P have been measured, while F_H and the relative orientation of $\mathbf{F_P}$ and $\mathbf{F_H}$ given by $(\alpha_P - \alpha_H)$ have been estimated. If the estimates of the phases are correct and the magnitudes are measured accurately, the triangle will close. In cases where α_P has been estimated from several derivatives, there will usually be a lack of closure, ϵ_j. This lack of closure is defined as

$$\epsilon_j = F_{PH(obs)} - F_{PH(calc)}$$

where

$$F_{PH(calc)} = |\mathbf{F}_{P(calc)} + \mathbf{F}_{H(calc)}|$$

In each refinement cycle, the sum over all reflections

$$\Sigma w (F_{PH(obs)} - F_{PH(calc)})^2$$

is minimized. The desired shifts of the scale factor and positional and thermal parameters are computed using conventional least squares techniques. The initial value of the overall thermal parameter may be calculated from a Wilson plot of the isomorphous differences. The weight is taken as the inverse of the r.m.s. lack of closure error over all derivatives used in the calculation of α_P for the

11.6.2. REFINEMENT SCHEME B: "PHASE" REFINEMENT

reflection in question, i.e.

$w = 1/E$ where $E^2 = \langle \epsilon \rangle^2$

Many quantities can be calculated to follow the course of the refinement. For general reflections the R value due to Kraut is most usually quoted:

$$R_K = \frac{\Sigma |F_{PH(obs)} - F_{PH(calc)}|}{\Sigma F_{PH(obs)}}$$

The value of this reliability factor depends on the relative values of F_{PH} and F_H. When the degree of heavy atom substitution is large, R_K will be large. Thus R_K is not very useful in comparing different derivatives, but it is very helpful as a guide to refinement. At the end of the refinement, the estimate of F_H from the heavy atom positions may be used to calculate the quantity $R_K \Sigma F_{PH}/\Sigma F_H$ which should be independent of the degree of substitution. This is more comparable to $R(F_{HLE})$. For centric zones the R_c (Cullis et al., 1961) may be calculated from

$$R_c = \frac{\Sigma ||F_{PH} \pm F_P| - F_{H(calc)}|}{\Sigma |F_{PH} - F_P|}$$

The root mean square of the error should also be computed as a guide to the progress of refinement. Examples of R_c values found in successful structure analyses are given in Table 11.III. If the refinement is not weighted properly, the substitution parameter will be underestimated (Dodson and Vijayan, 1971). An alternative way of overcoming this problem is to use only reflections which have good estimates of α_P (i.e. $m > 0.8$, see Section 12.2) in the refinement.

The method was first used by Dickerson et al., (1961) to refine the y coordinates of myoglobin heavy atom positions. It has been used successfully for most high resolution X-ray analyses of proteins. In many cases it has been used in a theoretically incorrect way. If a contribution to the protein has been calculated from the heavy atom parameters which are currently being refined, then $F_{PH(calc)}$ is dependent on the parameters refined and it is incorrect to treat it as a constant. Including the parameters in the phase calculation will not cause any problems if the heavy atom positions are good estimates but if they are badly estimated, the refinement will be much slower in convergence (Blow and Matthews, 1973) and may even lead to an incorrect solution. Phases should be recalculated in the refinement of each derivative omitting the contributions of the heavy atom set under refinement. Used correctly, the method produces a fast convergence. Positions will refine in position correctly if their initial displacement is as much as half the nominal resolution (Dickerson et al., 1968).

Double difference maps may also be calculated using coefficients

$|F_{PH(obs)} - F_{PH(calc)}| \exp(i\alpha_P)$

These are very useful in pulling in minor sites.

TABLE 11.III

Values of R_c for heavy atom derivatives which have been useful in various successful protein structure analyses. $R_c = \dfrac{\Sigma \, || F_{PH} \pm F_P | - F_H(\text{calc}) |}{\Sigma \, | F_{PH} \pm F_P |}$

Protein and reference	Mol. Wt.	Resolution	Heavy Atom Derivative	Number of Sites	R_c
Thermolysin Colman et al. (1972b)		2.3Å	Dimercury Acetate (DMA)	1	0.50
			Hg^{2+}	2	0.55
			PtI_6^{2-}	6	0.54
Carboxy-peptidase Lipscomb et al. (1970)		2Å	Pb^{2+}	2	0.53
			Hg^{2+}	1	0.66
			Hg^{2+}	3	0.61
			$PtCl_4^{2-}$	4	0.73
Bacterial ferredoxin Adman et al. (1973)	6,000	2.8Å	UO_2^{2+}	11	0.48
			Sm^{3+}	4	0.48
			Pr^{3+}	8	0.47
Flavodoxin (Desulphovibrio vulgaris) Watenpaugh et al. (1972)	6,000	2.5	Sm^{3+}	1	0.61
Nuclease Arnone et al. (1971)		2Å	Ba^{2+}	1	0.65
			I	1	

11.6.3 Comparison of the refinement schemes

In the high resolution studies of rubredoxin (Herriott et al., 1970) staphylococcyl nuclease (Arnone et al., 1971) and insulin (Adams et al., 1969; Blundell et al., 1971b), both refinement schemes have been used and it is now possible to compare these different techniques. The most encouraging fact to emerge is that the heavy atom positions determined by the two schemes are in very good agreement. This in a sense justifies the use of either.

However, the schemes clearly have certain specific advantages and disadvantages relative to each other. Refinement scheme A in centric zones gives very good parameters as it is independent of anomalous scattering. Use of F_{HLE} values (scheme A) depends critically on the accurate estimation of anomalous

11.6.3. COMPARISON OF THE REFINEMENT SCHEMES

differences and on the choice of the correct scale and relative temperature factor for F_{PH}. It will also be affected by the anomalous differences which are intrinsic to the native protein, but these errors are minimized if native Friedel pairs are averaged, i.e. $F_P = [F_P(+) + F_P(-)]/2$. The inaccuracies in the measurement of the anomalous differences give rise to an overestimation of F_{HLE} which leads in turn to an overestimation of the occupation numbers for the heavy atoms refined. This is largely overcome by using a proper weighting scheme as described in Section 11.6.1. The scale and relative temperature factor for F_{PH} can also be refined using scheme A but they are better estimated using refinement scheme B. In most cases the errors in F_{HLE} seem to lead to errors in the heavy atom occupancies and thermal parameter rather than in the positional parameters.

In both refinement schemes the thermal parameters and occupancies are not very meaningful. This is because the heavy atom often displaces a solvent molecule or amino acid sidechain which has a different shape, or it sometimes introduces ordered ligand atoms into the crystals. These changes in the vicinity of the heavy atom cannot be described as either spherical or elliptical electron density. The Fourier transforms of electron densities have maxima and minima at certain values of $\sin \theta$. Thus it is best in both refinement schemes first to refine the thermal and occupation parameters to find approximate values, and then to fix the thermal parameter and refine the occupancy in different ranges of $\sin \theta$. These occupancies can then be used to calculate a new "group" scattering curve or alternatively they can be used directly in the phase calculation in the relevant $\sin \theta$ range.

The great strength of scheme A is its independence of other derivatives. For centric zones it can be used to refine the initial heavy atom parameters before any phases are calculated. The use of this scheme with F_{HLE} values as coefficients means that this can also be achieved even if there are no centric zones. This was the case in the insulin and rubredoxin studies where the spacegroup was R_3. The method gives good parameters with which initial single isomorphous derivative phases can be calculated (see Section 6.2). The method is also invaluable when the heavy atoms of several derivatives occupy closely related but not precisely the same sites, as was true for the lead, cadmium and uranyl derivatives of insulin and for the silver and platinum derivatives of glucagon. Refinement of the heavy atom parameters in the latter case using scheme B refinement always led to positions which were obviously slightly wrong as they were inconsistent with the Patterson (Sasaki *et al.*, 1975).

The independence of refinement of the different heavy atoms may not always be advantageous. Scheme B refines jointly all the data about the isomorphous differences including scale factors, relative occupancies and relative origins between derivatives. Dickerson *et al.* (1968) have shown by model calculations that inclusion of a derivative with incorrect heavy atom positions does not greatly spoil the refinement as long as there are good derivatives also included.

The substitution numbers of the incorrect positions tend to move towards zero. However, this may depend critically on the nature of the incorrect heavy atom derivative and its effect on the phase calculation. The importance of having at least one high quality derivative has been clearly demonstrated in the comparison of the uninhibited nuclease and the inhibitor-nuclease complex studies at low resolution (Arnone et al., 1971). As Phillips (1966) and North and Phillips (1968) have pointed out, this scheme in its assignment of all errors to F_{PH} relies heavily on accurate measurement of F_P. Arnone et al., (1971) have found that refinement of derivatives of nuclease by this method was considerably improved on making new and more extensive measurements of native protein data.

11.7 CORRELATION OF ORIGINS

When the arrangement of heavy atoms in the unit cell has been determined for a number of derivatives, the next task is to refer them all to the same origin. This is most difficult if the spacegroup has low symmetry for normally the position of the origin would be defined relative to the rotation axes. Thus in $P1$ the position of the origin is completely arbitrary, and in $P2_1$ the position of the origin in the direction of the screw axis is arbitrary. Here we are concerned with methods used to determine the relative origins. These methods include Patterson functions or double derivatives (i.e. containing two kinds of heavy atom) which can be used when there is no estimate of the protein phase, but difference Fourier methods are probably more useful when a phase can be calculated from one heavy atom set, using isomorphous and anomalous scattering differences.

The Patterson-like correlation functions have been discussed fully by Phillips (1966). The first function, suggested by Perutz (1956) and developed by Blow (1958), uses coefficients:

$$|F_{H_1} F_{H_2}| \cos(\alpha_{H_1} - \alpha_P) \cos(\alpha_{H_2} - \alpha_P)$$

where the angles are derived from the observed and calculated quantities

$$F_{PH}^2 = F_P^2 + F_H^2 + 2 F_P F_H \cos(\alpha_H - \alpha_P)$$

each heavy-atom being referred to its own arbitrary origin. This function gives a peak representing the difference of positions of the two arrays of heavy atoms, but is complicated if there are more than one heavy atom per asymmetric unit. To overcome this problem Perutz (1956) suggested a second synthesis with coefficients

$$(F_{PH_1}^2 - F_P^2)(F_{PH_2}^2 - F_P^2)/F_P^2$$

11.8. GENERAL APPROACH TO FINDING HEAVY ATOM POSITIONS

but this function is very dependent on the choice of the correct scale of $F_{PH_1}^2$, $F_{PH_2}^2, F_{PH_2}^2$ and F_P^2. It tends to give a high background.

Rossmann (1960) has suggested a function which has coefficients

$$(F_{PH_2} - F_{PH_1})^2$$

This gives positive peaks at the ends of self vectors for each of the derivatives, and negative peaks corresponding to the cross-vectors. Hol (1971) has shown that anomalous differences can be included in this function to give a better approximation

$$(F_{PH_2} - F_{PH_1})^2 + \{(F_{ano})_2^2 - (F_{ano})_1^2\}$$

Alternatively another useful correlation function due to Steinrauf (1963) has coefficients

$$(F_{PH_1} - F_P)(F_{PH_2} - F_P) \text{ or } (\Delta F_{iso})_1 (\Delta F_{iso})_2$$

This leads principally to a positive term giving the cross-vectors between the two heavy atom sets. Again this can be improved by the use of anomalous scattering measurements (Kartha and Parthasarathy, 1965) using coefficients

$$(\Delta F_{iso} + i\Delta F_{ano})_1 \times (\Delta F_{iso} + i\Delta F_{ano})_2$$

This functions has peaks only at the ends of vectors $(r_{H_2} - r_{H_1})$ and not at the inverse. It is probably the best function to use.

Preparation of a double derivative — one that contains two kinds of heavy atoms, each of which appears separately in another derivative — may also be useful. A difference Patterson of such a derivative contains vector peaks relating the two sets which allow the relative positions of the origins to be determined.

Calculated phases based either on a single isomorphous replacement in projection or combined with anomalous scattering for general reflections can be used in several ways to correlate origins. Difference maps with phases from one derivative but amplitudes from the other derivatives are often employed. The origins are often refined by scheme B as mentioned above. Alternatively a significant feature common to all electron density maps (based on the derivative separately) such as a metal atom or a local symmetry axis may be helpful.

11.8 GENERAL APPROACH TO FINDING HEAVY ATOM POSITIONS

The decisions which have to be made at the beginning of the study of a potentially useful derivative depend on the individual protein but usually include: (i) whether to work in projection or use three dimensional data; (ii) whether to measure anomalous scattering differences; (iii) at which resolution to make the initial study.

Unfortunately all of these decisions depend ultimately on the nature and extent of the heavy atom binding to the protein, and this will largely remain unknown until the heavy atom substitution pattern is known. However, the decisions also depend on the space group, the size of the protein, whether approximate phases are already calculated from another derivative, and whether the data is measured by a diffractometer or photographically.

For diffractometer data it is usual to work at low resolution, i.e. 5Å–6Å resolution, to survey the derivative before proceeding to high resolution. For photographic data the difficulty in going to high resolution once the photograph is recorded is less. For the preliminary survey on a diffractometer, 6Å resolution data are often recorded but this can lead to difficulties is some multi-site derivatives where the heavy atoms cluster round one amino acid sidechain and are unresolved at 6Å. It is useful to collect data to 5Å or even 4.5Å for preliminary study if at all possible. For the first derivative it is particular advantageous to measure three-dimensional data including Friedel pairs for a full analysis using F_{HLE} coefficients. The use of F_{HLE} terms in finding heavy atom positions is most effective for samarium, uranium, lead or mercury derivatives which have large anomalous scattering with CuK_α radiation. For the first derivative a clean covalently bound single site heavy atom would be very nice! This appears to be most likely found in a mercurial reaction with a sulphydryl. Perhaps uranyl is best avoided for the first derivative as it is most often multi-site. For the subsequent derivatives, the heavy atom positions may be found using difference Fourier techniques, and the need for anomalous scattering data is less important.

For all derivatives it is very important to carefully assess the quality of the derivative. The intensity changes should be significant and reproducible. Changes in intensity should be observable at low angle as well as high angle. Changes at low angle only indicate disordered heavy atom binding while changes at high angle probably only mean that the derivative crystals are not isomorphous. Clearly there is a need for care in all measurements but especially in the native set which is going to form the basis of the structure.

12

THE CALCULATION OF PHASES

12.1 INTRODUCTION

In Chapters 6 and 7 we outlined the general principles of the method of isomorphous replacement and the use of anomalous scattering in the solution of the phase problem. The methods involved five stages: 1. the preparation of heavy atom derivatives; 2. the measurement of X-ray data of native and derivative crystals; 3. the reduction and correction of the measured intensities; 4. the determination of the heavy atom positions; 5. the calculation of phases.

In Chapters 8, 9, 10, 11 we have considered the first four stages in detail. In this chapter we discuss the calculation of phases using the structure amplitudes of natives and derivatives, the anomalous scattering differences and the contribution of the heavy atom to the derivative structure factor. In particular we are concerned with the proper treatment of errors which result both from inaccuracies in the intensity measurements and also from lack of isomorphism of native and derivative crystals.

We have seen that the method of isomorphous replacement is most easily illustrated by the graphical construction of Harker, which is shown in Fig. 6.5 (see page 157). With perfect isomorphism and no experimental errors, $\mathbf{F}_P$, $\mathbf{F}_H$ and $\mathbf{F}_{PH}$ form a closed triangle in which the magnitudes of the three sides and the direction of one of them are known. Harker suggests that $-\mathbf{F}_H$ is plotted from the origin in an Argand diagram and two circles are drawn, one of radius F_P centred at the origin and one of radius F_{PH} centred at the head of the vector $-\mathbf{F}_H$. The two intersections of the circles (G and H of Fig. 6.5) will determine the possible orientations of F_P. With only a single isomorphous derivative there is an ambiguity in phase for F_P. In order to solve the ambiguity, at least two heavy atom derivatives are required as shown in Fig. 6.7. Ideally the point of intersection of the three circles should define the vector $\mathbf{F}_P$.

Separate information is provided by the anomalous scattering differences (see Section 7.6). The structure amplitudes of the Bijvoet pairs of reflections, $F_{PH}(+)$ and $F_{PH}(-)$, are considered separately, so that instead of the two intersecting circles there are in fact three, constructed as shown in Fig. 7.7. Again the point of intersection of the three circles defines $\mathbf{F}_P$.

The successful combined use of isomorphous and anomalous scattering differences in Patterson syntheses rest on their "anticomplementarity". We have seen that this is also true in phase determination. Expression (7.15), using anomalous scattering differences alone, gives

$$\sin(\alpha_{PH} - \alpha_H) \simeq \frac{F_{PH}(+) - F_{PH}(-)}{2F_H''} \tag{12.1}$$

However, from Fig. 6.4 using isomorphous differences we get

$$\cos(\alpha_{PH} - \alpha_H) = -\frac{F_P^2 + F_{PH}^2 + F_H^2}{2F_{PH}F_H} \tag{12.2}$$

Combination of these two Expressions allows the ambiguity of sign to be overcome, and hence ($\alpha_{PH} - \alpha_H$) is determined. The angle α_H and the structure amplitudes, F_P, F_{PH} and F_H are known. Consequently the angle α_P can be found from the vector triangle defined by F_P, F_{PH} and F_H.

12.2 THE TREATMENT OF ERRORS IN THE METHOD OF ISOMORPHOUS REPLACEMENT

In practice ideal conditions will not apply. Errors arise from inaccuracy in the measurements of intensities, lack of isomorphism and incorrect heavy atom positions. As a result the three circles of Figs. 6.7 or 7.7 will not intersect at one point, and the phase determination will again be ambiguous. The question of errors in isomorphous replacement has been considered by Blow and Crick (1959), who have described the proper treatment of errors and the best choice of weights in the calculation of the electron density map. Extension of the method to the use of anomalous differences has been discussed by Blow and Rossmann (1961), North (1965) and Matthews (1966b).

Blow and Crick consider errors to arise from two distinct causes. There is an error, e, in the calculated heavy atom contribution, $\mathbf{F}_{H(calc)}$, to the derivative structure factor. This is not only a consequence of errors in the positional, occupancy and thermal parameters attributed to the heavy atoms, but also results from lack of isomorphism both by translations and rotations of the protein molecules and by introduction or movement of light atoms in the vicinity of the heavy atoms. Experimental inaccuracies in the determination of the amplitudes, F_P and F_{PH} are a further source of error. Their r.m.s. value $\langle \delta \rangle$ may be estimated by comparing values of the structure amplitudes of the same derivative estimated from different crystals or by comparison of symmetry related reflections. The total error is given by

$$\langle E \rangle^2 = \langle \delta \rangle^2 + \langle e \rangle^2 \tag{12.3}$$

Blow and Crick have pointed out that the total error can be estimated from a

12.2. THE TREATMENT OF ERRORS OF ISOMORPHOUS REPLACEMENT

centrosymmetric projection by the expression:

$$\langle E^2 \rangle = \langle (|F_{PH} \pm F_P| - F_H)^2 \rangle \tag{12.4}$$

Thus, the knowledge of $\langle E \rangle$ and $\langle \delta \rangle$ permit $\langle e \rangle$ to be estimated for centric reflections. However, in the non-centrosymmetric case $\mathbf{F_P}$, $\mathbf{F_{PH}}$ and $\mathbf{F_H}$ have arbitrary phases and there is no direct way of comparing observed and calculated differences as in expression (12.4). Non-centrosymmetric spacegroups whose symmetry includes dyads or screw dyads possess centric zones, and Blow and Crick suggest use of these zones to estimate E for the whole structure.

Several proteins including rubredoxin and insulin crystallize in space-groups which have no centric zones. In this case, two further ways of estimating E may be considered. E may be estimated from the r.m.s. difference between the "combination difference" estimate F_{HLE}, and $F_{H(calc)}$. Alternatively E may be obtained by comparing the observed and calculated values of F_P. F_P is calculated from

$$F_P^2 = F_H^2 + F_{PH}^2 - 2F_{PH}F_H \cos(\alpha_{PH} - \alpha_H) \tag{12.5}$$

where $(\alpha_{PH} - \alpha_H)$ has the magnitude estimated from the anomalous difference alone as in expression (12.1). These methods have been found to give comparable estimates.

The probability distribution of the errors in the non-centrosymmetric case has a very complicated form. Blow and Crick make the assumption that F_P is accurately determined and show that the two probability functions representing $\mathbf{F_{PH}}$ and $\mathbf{F_H}$ can be combined into one. With reference to Fig. 12.1(b) $\mathbf{F_{PH}}$ and $\mathbf{F_H}$ are represented by probability functions and the point 0 is regarded as fixed. The two probability functions may be combined into one which gives the probability that the point marked P falls at that particular point on the diagram. This distribution is a convolution of the previous two distributions and is close to a Gaussian of elliptical contour with major axis $(\langle e \rangle^2 + \langle \delta \rangle^2)^{1/2}$ and minor axis $\langle e \rangle$ (Fig. 12.1c). Moving P along the minor axis of the ellipse has little effect on the phase of $\mathbf{F_P}$ relative to that of $\mathbf{F_H}$. To a good approximation, therefore, the probability function may be represented by its projection on its major axis of the ellipse, which is very nearly a Gaussian of breadth $(\langle e \rangle^2 + \langle \delta \rangle^2)^{1/2} = \langle E \rangle$ (Fig. 12.1d). Thus the whole error is regarded as residing in $\mathbf{F_{PH}}$. So long as $\mathbf{F_H}$ is small compared with $\mathbf{F_P}$ and $\mathbf{F_{PH}}$, the magnitude of the error in phase introduced by errors in $\mathbf{F_P}$ is of the same magnitude as an error in $\mathbf{F_{PH}}$ and the assumption that all the error lies in the determination of $\mathbf{F_{PH}}$ does not affect the magnitude of the phase error.

The "lack of closure" error defined by

$$\epsilon_j = (F_{PH}(\text{obs}) - F_{PH}(\text{calc})) \tag{12.6}$$

has been discussed in relation to the refinement of the heavy atoms by scheme B (Chapter 11) which depends on the same assumptions, namely that F_H is small

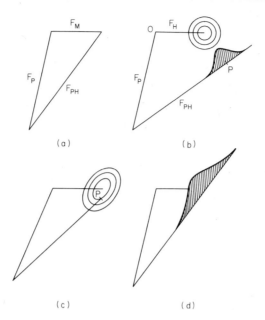

FIG. 12.1 Effect of errors on phase determination: non-centrosymmetric case. (See text for further details) (After Blow and Crick, 1959).

compared with F_P and F_{PH}. Assuming a Gaussian distribution of errors, the probability that a phase angle, α for the protein structure factor, is correct is related to the "lack of closure" of the phase triangle for this angle, and is given by

$$P_j(\alpha) = \exp[-\epsilon_j(\alpha)^2/2E_j^2] \qquad (12.7)$$

for the jth heavy atom derivative. (Note that in the following discussion α refers to the protein phase, α_P.) When several heavy atom derivatives are used simultaneously the total probability of the phase angle is given by the product of the individual probabilities,

$$P(\alpha) = \prod_j P_j(\alpha) = \exp[-\sum_j (\epsilon_j(\alpha)^2/2E_j^2)] \qquad (12.8)$$

Typical phase circle diagrams along with their probability function plots are shown in Fig. 12.2. These are smoothly varying functions, and on Fourier transformation could lead to a continuum of structures, each with an assigned probability.

The most obvious Fourier would be the "most probable" in which $\mathbf{F}_P$ for each reflection is chosen as that with the highest value of $P(\alpha)$. However, although this function would be useful for almost unimodal probability

12.2. THE TREATMENT OF ERRORS OF ISOMORPHOUS REPLACEMENT

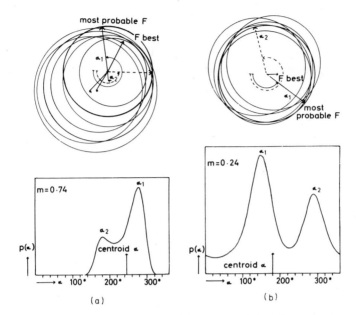

FIG. 12.2 Phase circles for the 112(a) and 317(b) reflections from horse oxy-haemoglobin, illustrating the difference between the most probable F and the "best" F. The curves show the corresponding phase probability distributions for the two reflections (from Cullis *et al.*, 1961).

functions as in Fig. 12.2a, it would give too much weight to uncertain phases with bimodal distribution as in Fig. 12.2b.

Blow and Crick have shown that the centroid of the distribution is a better choice since this results in a synthesis with the least mean square error in electron density over the unit cell. To show this let us consider the error in electron density arising from errors in one reflection. If the value of the coefficient used in the synthesis is $\mathbf{F}_s$ and if the true value is $\mathbf{F}_t$, then the mean square error over the unit cell from this reflection is

$$\langle \Delta \rho^2 \rangle = \frac{1}{V^2} (\mathbf{F}_s - \mathbf{F}_t)^2$$

$\mathbf{F}_t$ has a probability $P(\alpha)$ of having a phase angle α and may be represented by

$$\mathbf{F}_t = F \exp i\alpha$$

The mean square error then becomes

$$\langle \Delta \rho^2 \rangle = \frac{1}{V^2} \int_{\alpha=0}^{2\pi} (\mathbf{F}_s - F \exp i\alpha)^2 P(\alpha) d\alpha \Big/ \int_{\alpha=0}^{2\pi} P(\alpha) d\alpha \quad (12.9)$$

The numerator in this expression is equivalent to the moment of inertia of a ring, mass $P(\alpha)$ and radius $F \exp i\alpha$, about an axis which is perpendicular to its plane and which passes through the end of the vector $\mathbf{F}_s$. By the parallel axis theorem, this integral has a minimum value when $\mathbf{F}_s$ is at the centre of gravity of the ring, i.e.

$$\mathbf{F}_{s(best)} = F \int_{\alpha=0}^{2\pi} \exp i\alpha P(\alpha) d\alpha \bigg/ \int_{\alpha=0}^{2\pi} P(\alpha) d\alpha$$

$$= mF \exp(i\alpha_{best}) \tag{12.10}$$

Equation 12.10 is the expression for the centre of gravity of the probability distribution with polar co-ordinates (mF, α_{best}). Where m is defined by

$$m = \frac{\int_{\alpha=0}^{2\pi} P(\alpha) \exp(i\alpha) d\alpha}{\int_{\alpha=0}^{2\pi} P(\alpha) d\alpha} \tag{12.11}$$

$\mathbf{F}_s(best)$ is illustrated in Fig. 12.2. If the probability is sharp, **m** will have a value close to unity, but if the probability is fairly uniform around the circle, **m** will be close to zero. The magnitude of **m** is therefore equivalent to a weighting function and is known as the "figure of merit". The Fourier calculated using mF and α_{best} is expected to have the minimum mean square error from the true Fourier when averaged over the whole unit cell. It is known as the "best Fourier".

In a phase calculation programme the value of the probability function $P(\alpha_i)$ is computed in fixed intervals (5 degrees say) around the phase circle at angles α_i, for each reflection. The components (m, α_{best}) of the centre of gravity of the probability distribution around a circle of unit radius are calculated from

$$m \cos \alpha_{best} = \frac{\sum_i P(\alpha_i) \cos \alpha_i}{\sum_i P(\alpha_i)}$$

$$m \sin \alpha_{best} = \frac{\sum_i P(\alpha_i) \sin \alpha_i}{\sum_i P(\alpha_i)} \tag{12.12}$$

The error in phase angle at a given α_i is defined as $\Delta\alpha_i = \alpha_{best} - \alpha_i$. If the origin is shifted so that the origin axis is taken along the direction of the centre of gravity, then

$$\alpha_{best} = 0 \quad \text{and} \quad |\alpha_i| = |\Delta\alpha_i|$$

$$m = \frac{\sum_i P(\alpha_i) \cos \Delta\alpha_i}{\sum_i P(\alpha_i)} = \langle \cos \Delta\alpha_i \rangle \tag{12.13}$$

12.2. THE TREATMENT OF ERRORS OF ISOMORPHIC REPLACEMENT

Hence m is the mean value of the cosine of the error in phase angle for the reflection. In theory, a value of $m = 1$ corresponds to a zero error in phase angle while a value $m = 0.74$ corresponds approximately to a 42 degree error, and a value $m = 0.24$ corresponds to an error of 76 degrees.

Substitution of $\mathbf{F}_{s(best)} = m\mathbf{F}$ in equation (12.9) for the mean square error in the electron density map from a single reflection gives

$$\langle \Delta \rho^2 \rangle = \frac{1}{V^2} \int_{\alpha=0}^{2\pi} (mF - F r(\alpha))^2 P(\alpha) d\alpha \bigg/ \int_{\alpha=0}^{2\pi} P(\alpha) d\alpha$$

$$= \frac{F^2}{V^2} \int_{\alpha=0}^{2\pi} (m - r(\alpha))^2 P(\alpha) d\alpha \bigg/ \int_{\alpha=0}^{2\pi} P(\alpha) d\alpha$$

$$= \frac{F^2}{V^2} \int_{\alpha=0}^{2\pi} (m^2 - 2m \cos \Delta\alpha + 1) P(\alpha) d\alpha \bigg/ \int_{\alpha=0}^{2\pi} P(\alpha) d\alpha$$

$$= \frac{F^2}{V^2} \left\{ m^2 - 2m \int_{\alpha=0}^{2\pi} \cos \Delta\alpha P(\alpha) d\alpha \bigg/ \int_{\alpha=0}^{2\pi} P(\alpha) d\alpha + 1 \right\}$$

$$= \frac{F^2}{V^2} (1 - m^2)$$

where $\mathbf{r}$ is a vector of unit length in the direction α and $\Delta\alpha$ is the angle between $\mathbf{m}$ and $\mathbf{r}$. The total error in the "best Fourier" is therefore

$$\langle \Delta \rho^2 \rangle = \frac{1}{V^2} \sum_{\mathbf{h}} F^2(\mathbf{h})(1 - m^2) \tag{12.14}$$

The order of magnitude of this error may be illustrated by reference to a specific example. In the structure determination of lysozyme the mean "figure of merit" for the data to 2Å resolution was 0.6 and the corresponding root mean square error in the Fourier was 0.35 $e/\text{Å}^3$. The electron density map was contoured at intervals of 0.25 $e/\text{Å}^3$ and the heights of most of the peaks were between 1 and 1.5 $e/\text{Å}^3$ (excluding the $F(000)$ term).

In addition to the errors in the Fourier synthesis, it is also useful to consider the relative heights of peaks in a Fourier synthesis whose coefficients are weighted by their figures of merit. If the phases were all correct ($m = 1$) then these heights would be close to their true values, while if the phases were poorly determined then features in the map would be barely above noise level. The importance of the phases in a Fourier synthesis has been demonstrated by Ramachandran and Srinivasan (1961), who showed that a synthesis based on correct phases and incorrect amplitudes gave a recognizable structure but that a synthesis based on incorrect phases and correct amplitudes gave nonsense. Suppose α_B, the best phase, differs from the correct but unknown phase by $\Delta\alpha$, then the contribution of the observed coefficient $mF \exp i\alpha_B$ to the unknown correct structure is $mF \cos \Delta\alpha$ (Fig. 12.3). The mean contribution of

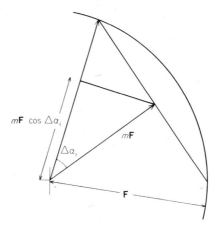

FIG. 12.3 Argand diagram showing the contribution of the vector mF to the structure factor whose phase is separated from it by an angle $\Delta\alpha_i$ (from Henderson and Moffat, 1971).

$mF \exp i\alpha_B$ to the true structure is given by

$$\frac{\sum_i P(\alpha_i) \, mF \cos \Delta\alpha_i}{\sum_i P(\alpha_i)} = mF \frac{\sum_i P(\alpha_i) \cos \Delta\alpha_i}{\sum_i P(\alpha_i)} = m^2 F \qquad (12.15)$$

Thus if all the coefficients have an equal value of the figure of merit, m, then the resulting Fourier synthesis will have peak heights of m^2 relative to the true structure (Henderson and Moffat, 1971). In practice of course m varies from reflection to reflection and is also dependent on $\sin^2 \theta/\lambda^2$ (Fig. 12.4). However, as Henderson and Moffat comment, their analysis does give some feel for the effect of m on peak heights in a Fourier synthesis. As the phases become worse, so the effect on peak heights is magnified by the m^2 dependence.

Finally it should be stressed that although in theory m represents the mean value of the cosine of the error in phase angle, in practice its value is highly dependent on the estimate of E_j. If E_j is underestimated, then m tends to be larger than its correct value and *vice versa*. It is more appropriate to consider m as a measure of the sharpness of the joint probability distribution which results in resolving the phase ambiguity. In order to check that m does indeed have a realistic value with respect to the measurements, a further index has been proposed by Dickerson *et al.* (1968). They define a mean residual error which is the mean value of $\sum_{j=1}^{n} (\epsilon_j(\alpha_B)^2 / 2E_j^2)$ where the summation is for all the isomorphous derivatives. If values of E_j have been estimated correctly then the mean residual error $= n \times 0.5$ where n represents the total number of derivatives, since $\langle \epsilon_j^2 \rangle = E_j^2$.

12.3. INCLUSION OF ANOMALOUS SCATTERING

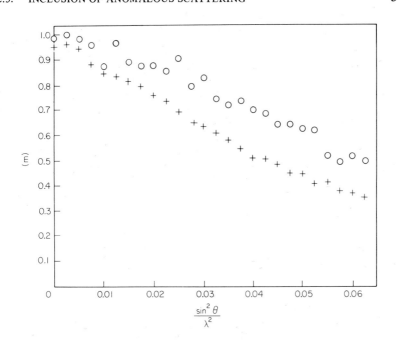

FIG. 12.4 Variation of the mean figure-of-merit (m in equation 12.12) with $\sin^2\theta/\lambda^2$ in the study of hen egg-white lysozyme, +: non-centric reflections, ○: centric reflections (from North and Phillips, 1968).

12.3 INCLUSION OF ANOMALOUS SCATTERING

Extension of the method to include the effects of anomalous scattering was first considered by Blow and Rossmann (1961). If ϵ_+ and ϵ_- represent the lack of closure for the direct and inverse reflections, the overall phase probability distribution for one isomorphous derivative including the anomalous scattering data is given by

$$P(\alpha) = \exp\{-(\epsilon_+^2 + \epsilon_-^2)/2E^2\}! \tag{12.16}$$

where E is the total r.m.s. error in determining ϵ_+ and ϵ_-. North (1965) and Matthews (1966b) have pointed out that this procedure does not take into account the inherently greater accuracy of the anomalous differences. Matthews (1966b) has shown that the lack of closure terms may be rewritten in a different form

$$\epsilon_+^2 + \epsilon_-^2 = \tfrac{1}{2}\{(\epsilon_+ + \epsilon_-)^2 + (\epsilon_+ - \epsilon_-)^2\} \tag{12.17}$$

In this expression the "total isomorphous lack of closure", ϵ, is given by

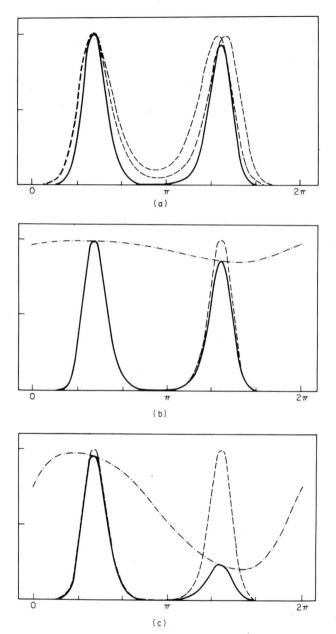

FIG. 12.5 (a) Phase probability curves for a Bijvoet pair of reflections (broken lines) with the joint probability curve (full line) derived by the method of Blow and Rossmann, (1961). (b) Isomorphous replacement phase probability curve derived from the mean of F^+_{PH} and F^-_{PH} (broken line); anomalous scattering probability curve (chain line); and joint probability curve (full line) derived by the method of North, (1965) using E'(anomalous) = $\bar{E}$ (isomorphous). (c) As (b) but with $E' = \frac{1}{3} E$. (from North, 1965.)

12.4. ABSOLUTE CONFIGURATION OF THE MOLECULE

$(\epsilon_+ + \epsilon_-) = 2(F_{\text{PH(calc)}} - F_{\text{PH}})$, (where $F_{\text{PH}} = (F_{\text{PH}}(+) + F_{\text{PH}}(-))/2$). By analogy the "anomalous scattering lack of closure", ϵ', is considered to be

$$\epsilon' = \epsilon_+ - \epsilon_- = -F_{\text{PH}}(+) + F_{\text{PH}}(-) - \frac{2F_P F_H''}{F_{\text{PH}}} \sin(\alpha_{\text{PH}} - \alpha_H) \qquad (12.18)$$

The probability distribution for the phase of F_P derived from anomalous scattering measurements, from crystals containing heavy atoms of only one kind is given by

$$P_{\text{an}}(\alpha) = \exp[-(\epsilon_+ - \epsilon_-)^2 / 2E'^2] \qquad (12.19)$$

where E' is the corresponding root mean square error in the anomalous scattering measurements. This function is symmetrical about the $\mathbf{F}_H''$ vector and will tend to reduce the symmetry of the isomorphous probability function which is symmetrical about the $\mathbf{F}_H$ vector.

Estimation of E' often presents problems. North (1965) has suggested that E' might be determined from agreement between reflections in centric zones or between equivalent reflections. Alternatively E' can be determined from different measurements of the same anomalous difference or by calculating the anomalous difference using Expressions (12.1) and (12.2) in which the magnitude of the angle $(\alpha_{\text{PH}} - \alpha_H)$ is calculated from the isomorphous difference alone and the sign ambiguity is resolved by the sign of the anomalous difference (Adams, 1968). This method is analogous to the method of determining E, and is useful when there are no centric projections. Generally, it has been found that E' is about one-third of E, confirming the inherent greater accuracy of the anomalous difference. The phase probability function for a reflection where anomalous scattering is useful is given in Fig. 12.5.

12.4 THE DETERMINATION OF THE ABSOLUTE CONFIGURATION OF THE MOLECULE

In principle, the combination of isomorphous and anomalous scattering information allows the determination of a protein structure from a single heavy atom derivative. Further, the use of anomalous scattering permits the identification of the absolute configuration of the molecule for small crystal structure determination (see Section 7.7, page 180). However there is a difficulty in the case of a single heavy atom derivative of a protein.

In the initial allocation of co-ordinates to the heavy atom derivative, an arbitrary choice has to be made about their absolute configuration while the reciprocal lattice is conventionally indexed on a right-handed system. These two choices may or may not be consistent. First let us consider the effect on the isomorphous contribution alone. It is evident that either choice of configuration for the heavy atom contribution (corresponding to vectors $F_H \exp i\alpha_H$ or $F_H \exp -i\alpha_H$) form solutions to the phase triangle (Fig. 6.5). As well as the ambiguity in phase there is also an ambiguity in enantiomorphs. This may be expressed analytically with the following equation

$$\cos(\alpha_{PH} - \alpha_H) = \frac{F_P^2 - F_{PH}^2 - F_H^2}{2F_{PH}F_P} = \cos\phi \text{ (say)}$$

Then

$$\alpha_{PH} = \alpha_H + \phi \text{ or } \alpha_H - \phi \quad \text{corresponding to the phase ambiguity}$$

or

$$-\alpha_H + \phi \text{ or } -\alpha_H - \phi \quad \text{corresponding to the enantiomorph ambiguity.}$$

The anomalous scattering information (Fig. 7.7) is incorporated into the equation

$$\sin(\alpha_{PH} - \alpha_H) = \frac{F_{PH}(F_{PH}(+) - F_{PH}(-))}{2F_p F_H''} = \sin\phi' \quad \text{(say)}$$

where the absolute sign of $F_{PH}(+) - F_{PH}(-)$ is fixed by the choice of the right-handed reciprocal lattice. Then

$$\alpha_{PH} = \alpha_H + \phi' \quad \text{or} \quad \alpha_H + 180 - \phi' \quad \text{or}$$
$$-\alpha_H + \phi' \quad \text{or} \quad -\alpha_H + 180 - \phi'$$

Suppose that $\alpha_{PH} = \alpha_H + \phi = \alpha_H + \phi'$ is the correct solution (i.e. $\phi = \phi'$), then these equations will also give $-\alpha_H + \phi$ as a possible solution. This second solution is totally wrong but the phase equations allow no way of distinguishing the true solution from the false solution. If the data have been collected to high resolution it is possible to resolve this ambiguity by calculating two electron density maps, one based on the set of phases with the co-ordinates of the heavy atom in one configuration and the other based on the enantiomorphic solution for the heavy atom. It may then be possible to recognize the correct map by inspection. The correct map will be interpretable and should show L-amino acids and right-handed α-helices. This method was used with flavodoxin (Watenpaugh et al., 1972).

If two heavy atom derivatives have been obtained, the correct enantiomorph may be found by calculating three different Fouriers based on the difference in structure factor amplitudes for the second derivative and the phases determined from the first derivative in the following ways: (i) the isomorphous contribution alone with coefficients weighted by the figure of merit; (ii) the isomorphous and anomalous scattering contributions together; (iii) as in (ii) but with the hand of the heavy atom derivative reversed. It is anticipated that the isomorphous map

12.5. A REPRESENTATION OF PHASE PROBABILITY DISTRIBUTIONS

TABLE 12.I

The determination of the absolute hand of the structure. Observed peak heights for the two EMTS sites in difference Fourier syntheses based on different sets of phases in space group $P4_1 2_1 2$.

Difference Fourier	Phases derived from:	Observed peak heights at EMTS sites (arbitrary units)	
		Hg1	Hg2
(i)	Single heavy atom isomorphous contributions for the platinum derivative alone.	391	826
(ii)	Single heavy atom isomorphous contributions and anomalous scattering data for the platinum derivative.	342	616
(iii)	Single heavy atom isomorphous contributions and anomalous scattering data for the platinum derivative, but with the sign of the observed anomalous contributions reversed.	501	1036

(i) will show peaks at the sites of the second derivative and that these heights will be reinforced when the anomalous scattering information is included with the correct hand and diminished when the hand is wrong. This method was applied to determine the absolute configuration of phosphorylase crystals at low resolution (Table 12.I). It is apparent that the wrong hand has been assigned to the platinum derivative and that the correct space-group is $P4_3 2_1 2$.

12.5 A SIMPLIFIED REPRESENTATION OF THE PHASE PROBABILITY DISTRIBUTIONS. THE HENDRICKSON–LATTMAN METHOD

The Blow and Crick (1959) formulation of the phase probability function is given by

$$P(\alpha) \exp - \sum_j \left[\epsilon_j(\alpha)^2 / 2E_j^2 \right]$$

where the summation j is over all the isomorphous derivatives and their anomalous scattering contributions. A major disadvantage of this formulation is that each time some new information is obtained (e.g. a new derivative, direct methods information, calculated phases from a partial structure), the summation $\Sigma_j[\epsilon_j(\alpha)^2/2E_j^2]$ has to be re-evaluated and the whole probability distribution re-calculated. Storage of all the data for each of the heavy atom derivatives is required for any new calculation. Hendrickson and Lattman (1970) have shown that the probability function can be cast in the form

$$P(\alpha) \propto \exp(A \cos \alpha + B \sin \alpha + C \cos 2\alpha + D \sin 2\alpha) \tag{12.20}$$

where the coefficients A, B, C and D constitute a complete record of the phase information for a reflection. Inclusion of newly obtained information only requires additions to these coefficients. In effect, the new representation separates the phase information parameters from the functions of phase angle and it is this which facilitates combinations of different or new phase information.

The major difference between the Blow and Crick and the Hendrickson and Lattman derivations occurs in their respective definitions of the lack of closure error for the isomorphous contributions. Blow and Crick treat the lack of closure error as residing in the derivative amplitude, F_{PH}. Thus

$$|\mathbf{F}_P + \mathbf{F}_H| = F_{PH} + \epsilon \tag{12.21}$$

where ϵ is the lack of closure error.

Hendrickson and Lattman propose a definition, which is no less legitimate, namely the total error lies in the derivative intensity, F_{PH}^2. Thus

$$|\mathbf{F}_P + \mathbf{F}_H|^2 = F_{PH}^2 + \delta \tag{12.22}$$

i.e.

$$F_P^2 + F_H^2 + 2F_P F_H \cos(\alpha - \alpha_H) = F_{PH}^2 + \delta$$

where δ is the new lack of closure error and $\alpha = \alpha_P$.

Assuming a Gaussian distribution of errors, the probability function for the isomorphous data is given for the jth derivative by

$$P_{\text{iso}j}(\alpha) = \exp - \left\{ \frac{\delta_j^2(\alpha)}{2D_j^2} \right\} \tag{12.23}$$

where D_j is the standard deviation of the errors associated with the distribution δ_j. In an *ab initio* structure, determination of the D_j may be estimated by calculations similar to those already described for estimating E_j. However, for many proteins the phase probability functions have already been calculated by the Blow and Crick method and it may be desirable to re-cast these functions in terms of the Hendrickson and Lattman representation in order that new information may be incorporated without *de novo* computations. Hendrickson (1971) has proposed a least squares method in which the best set of parameters

12.5. A REPRESENTATION OF PHASE PROBABILITY DISTRIBUTIONS

A, B, C and D are derived from minimization of the discrepancy between the observed probability function and that of the calculated probability function based on the parameters. Alternatively it is possible to derive the values of D_j from the values of E_j. Since $\delta_j = \epsilon_j(\epsilon_j + 2F_{PHj})$, D_j is a function of structure factor amplitude while E_j is only slightly dependent on intensity.

By definition

$$E_j^2 = \langle \epsilon_j^2 \rangle$$

$$\delta_j^2 = \epsilon_j^4 + 4F_{PHj}\epsilon_j^3 + 4F_{PHj}^2\epsilon_j^2$$

$$\therefore \langle \delta_j^2 \rangle = \langle \epsilon_j^4 \rangle + 4F_{PHj}\langle \epsilon_j^3 \rangle + 4F_{PHj}^2\langle \epsilon_j^2 \rangle$$

Assuming a Gaussian distribution for ϵ_j, then $\langle \epsilon_j^3 \rangle = 0$ and $\langle \epsilon_j^4 \rangle = 3\langle \epsilon_j^2 \rangle^2$ (Brookes and Dick, 1951, p 129). Hence

$$\langle \delta_j^2 \rangle = 3\langle \epsilon_j^2 \rangle^2 + 4F_{PHj}^2\langle \epsilon_j^2 \rangle$$

i.e.

$$D_j^2 = 3E_j^4 + 4F_{PH}^2 E_j^2 \tag{12.24}$$

which suggests that values for D_j can be obtained from values of E_j. An experimental investigation of the variation of D_j^2 with F_{PHj}^2 for lysozyme data (Stubbs, 1972) indicates that the relationship may not be simple, however, although it could be reasonably adequately represented by a step function for small F_{PH} and a linear regression curve for larger F_{PH}.

We now derive the expressions for the coefficients A, B, C and D for the isomorphous and the anomalous scattering phase information.

For the isomorphous data the δ_j^2 are given by

$$\delta_j^2(\alpha) = (F_P^2 + F_{Hj}^2 - F_{PHj}^2)^2 + 4(F_P^2 + F_{Hj}^2 - F_{PHj}^2)F_P F_{Hj}\cos(\alpha - \alpha_H)$$
$$+ 4F_P^2 F_{Hj}^2 \cos^2(\alpha - \alpha_H)$$

Then by substituting

$$\cos^2(\alpha - \alpha_H) = \tfrac{1}{2}(1 + \cos 2(\alpha - \alpha_H))$$

$$F_{Hj}\cos(\alpha - \alpha_H) = a_j \cos\alpha + b_j \sin\alpha$$

$$F_{Hj}\cos 2(\alpha - \alpha_H) = (a_j^2 - b_j^2)\cos 2\alpha + 2a_j b_j \sin 2\alpha$$

the exponent of the probability distribution (expression 12.23) can be written

$$\frac{-\delta_j^2(\alpha)}{2D_j^2} = K_{isoj} + A_{isoj}\cos\alpha + B_{isoj}\sin\alpha + C_{isoj}\cos 2\alpha + D_{isoj}\sin 2\alpha$$

where

$$K_{isoj} = -\frac{(F_P^2 + F_{Hj}^2 - F_{PHj}^2)^2 + 2F_P^2 F_{Hj}^2}{2D_j^2}$$

$$A_{isoj} = -\frac{2(F_P^2 + F_{Hj}^2 - F_{PHj}^2)F_P a_j}{D_j^2}$$

$$B_{isoj} = -\frac{2(F_P^2 + F_{Hj}^2 - F_{PHj}^2)F_P b_j}{D_j^2}$$

$$C_{isoj} = -\frac{F_P^2(a_j^2 - b_j^2)}{D_j^2}$$

$$D_{isoj} = -\frac{2F_P^2(a_j b_j)}{D_j^2} \tag{12.25}$$

The expression for the lack of closure error for the anomalous scattering is

$$\epsilon'(\alpha) = \epsilon_+ - \epsilon_- = -\frac{2F_P F_H''}{F_{PH}} \sin(\alpha_H - \alpha) - \Delta$$

where $\Delta = F_{PH}(+) - F_{PH}(-)$

which is the same expression as that defined previously. The probability distribution is given by

$$P_{anok}(\alpha) = \exp - \frac{\epsilon_k'^{\,2}(\alpha)}{2E_k'^{\,2}} \text{ for the } k\text{th anomalous scatterer.}$$

The standard deviation E_k' of the distribution of errors can be estimated as described previously. Using the relationships

$$\sin^2(\alpha_H - \alpha) = \tfrac{1}{2}[1 - \cos 2(\alpha_H - \alpha)]$$

$$\sin(\alpha_H - \alpha) = \frac{1}{F_{Hk}}[b_k \cos \alpha - a_k \sin \alpha]$$

$$\cos 2(\alpha_H - \alpha) = \frac{1}{F_{Hk}^2}[(a_k^2 - b_k^2)\cos 2\alpha + 2a_k b_k \sin 2\alpha]$$

it may readily be shown that

$$-\frac{\epsilon_k'^{\,2}(\alpha)}{2E_k'^{\,2}} = K_{anok} + A_{anok} \cos \alpha + B_{anok} \sin \alpha + C_{anok} \cos 2\alpha + D_{anok} \sin 2\alpha$$

12.5. A REPRESENTATION OF PHASE PROBABILITY DISTRIBUTIONS

where

$$K_{anok} = -\frac{\Delta_k^2 + (2F_P^2 F_{Hk}''/F_{PHk}^2)}{2E_k'^2}$$

$$A_{anok} = -\frac{2F_P F_{Hk}'' \Delta_k b_k}{F_{Hk} F_{PHk} E_k'^2}$$

$$B_{anok} = \frac{2F_P F_{Hk}'' \Delta_k a_k}{F_{Hk} F_{PHk} E_k'^2}$$

$$C_{anok} = \frac{F_P^2 F_H''^2 (a_k^2 - b_k^2)}{F_{Hk}^2 F_{PHk}^2 E_k'^2}$$

$$D_{anok} = \frac{2F_P^2 F_H''^2 a_k b_k}{F_{Hk}^2 F_{PHk}^2 E_k'^2} \tag{12.26}$$

It is worth noting that ϵ_k' represents the difference between the observed and calculated anomalous scattering contribution, while δ_j represents the difference between the squares of the observed and calculated protein plus heavy atom contributions.

The overall probability is the product of the individual phase probabilities. Hence

$$P(\alpha) = [\prod_j P_{isoj}(\alpha)] [\prod_k P_{anok}(\alpha)]$$

and since each probability function has been put in the form

$$P(\alpha) = \exp(K_s + A_s \cos\alpha + B_s \sin\alpha + C_s \cos 2\alpha + D_s \sin 2\alpha)$$

the combination of phase information is achieved by simply adding the coefficients in the exponent.

$$P(\alpha) = \exp(\sum_s K_s + \sum_s A_s \cos\alpha + \sum_s B_s \sin\alpha + \sum_s C_s \cos 2\alpha + \sum_s D_s \sin 2\alpha) \tag{12.27}$$

The phase independent term $\exp(\sum_s K_s)$ may be included in the normalization factor which cancels out when the weighting functions are calculated from the expression

$$m = \frac{\int_{\alpha=0}^{2\pi} \exp(i\alpha) P(\alpha) d\alpha}{\int_{\alpha=0}^{2\pi} P(\alpha) d\alpha}$$

Comparison of the phases calculated by the Hendrickson and Lattman method with those calculated by the method of Blow and Crick show small differences (Hendrickson and Lattman, 1970). However, both sets of phases appear to result in the same sort of magnitudes of the root mean square error in the electron density map when these maps are compared with that computed on the basis of calculated phases from the known structure (Stubbs, 1972). In their paper Hendrickson and Lattman show how phase probability functions from information from direct methods and partially known structures can also be cast in the four coefficient form. The method therefore allows a unified approach which permits the incorporation of all phase information from different sources. The record of the phase information is contained in the four parameters and newly obtained information can be combined with the previous information by simple addition of the coefficients. It is anticipated that this method will become more widely used in the future, especially when the experimenter wishes to include information from a variety of sources.

13

THE INTERPRETATION OF THE ELECTRON DENSITY MAP

13.1 Introduction

When the best phases and figure of merit, m have been calculated using isomorphous and anomalous scattering measurements, the computation of an electron density map is relatively straightforward. It is given by the expression

$$\rho(\mathbf{r}) = \Sigma\, mF_P \exp(i\alpha_{best})\, \exp\{-2\pi i\, \mathbf{h}\cdot\mathbf{r}\}$$

In Chapter 12 we considered the assessment of reliability of the phases and the mean square error in electron density of the "best" Fourier synthesis. In this chapter we are concerned with the interpretation of the electron density maps and the relation of the information obtained to the minimum interplanar spacing used in the calculation of the electron density.

13.2 Resolution

We have seen in section 5.4 that the smallest separation, r, which will be resolved in the X-ray image is given by

$$r \approx 0.71\, dm = 0.71\, \lambda/2 \sin\theta(\max)$$

where $\sin\theta(\max)$ is the maximum $\sin\theta$ value of the data included in the Fourier summation and dm is the corresponding minimum interplanar spacing (James, 1957).

However, the resolution attainable for protein crystals is limited by the inherent disorder of the molecules due to thermal vibrations and static disorder. The actual resolution will depend on the quality of the data as $\sin\theta(\max)$ is approached and in general on the quality of the calculated phases. For this reason the resolution actually attained in an electron density map will rarely be as high as $0.71\, dm$ and protein crystallographers usually quote the resolution of their electron density maps as the minimum interplanar spacing, dm.

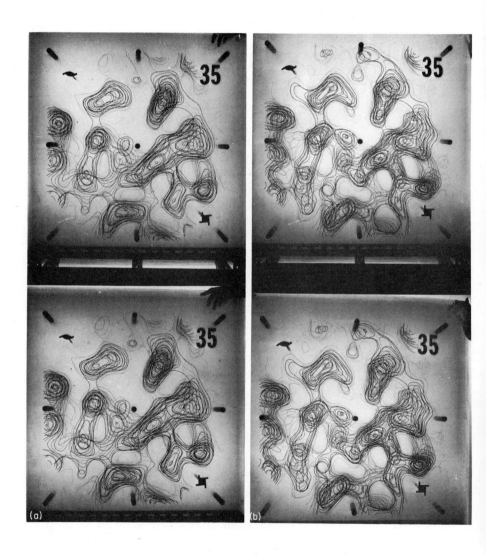

FIG. 13.1 Stereo views of equivalent sections of the hen egg white lysozyme electron density map at different resolutions: (a) 5.5Å, (b) 5Å, (c) 4.5Å, (d) 4Å, (e) 3.5Å, (f) 3Å, (g) 2.5Å, (h) 2Å. (North and Phillips, 1968).

13.2. RESOLUTION

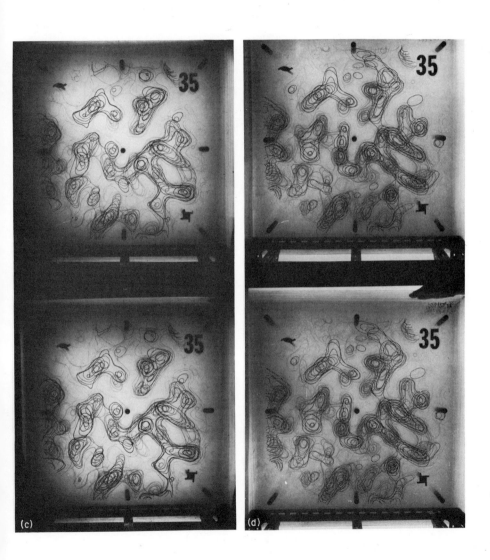

FIG. 13.1 *continued*

384 13. THE INTERPRETATION OF THE ELECTRON DENSITY MAP

FIG. 13.1 *continued*

13.2. RESOLUTION

FIG. 13.1 *continued*

386   13. THE INTERPRETATION OF THE ELECTRON DENSITY MAP

The first questions we must ask concern the quality of the electron density map at different resolutions. To illustrate this we show a number of different electron density maps showing the same features at different resolutions. Figure 13.1 compares the electron density maps of lysozyme calculated at 5.5, 5, 4.5, 4, 3, 2.5 and 2Å (North and Phillips, 1968).

Figure 13.2 shows parts of the insulin electron density maps at 6, 4.5, 2.8 and 1.9 Å resolution (Blundell *et al.*, 1972). The figure illustrates the coordination sphere of a zinc atom including a histidine imidazole and one turn of α-helix; the atomic arrangement found from the dm = 1.9Å study is drawn superposed on the electron density of lower resolutions.

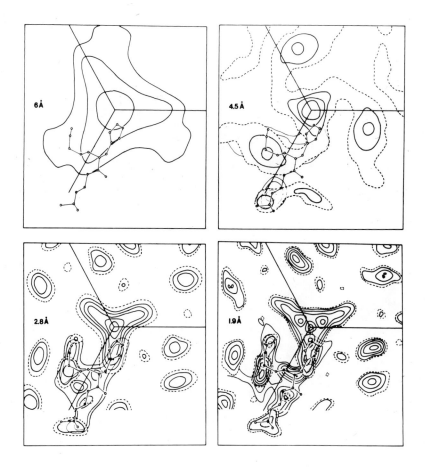

FIG. 13.2 Equivalent views of electron density maps of B9 Ser, B10 His, B11 Leu of insulin at different resolutions. (Blundell *et al.*, 1972)

13.3 LOW RESOLUTION STUDIES, $dm > 4.5$Å

The first protein electron density map ever calculated was that of myoglobin at $dm = 6$Å (Kendrew et al., 1958). It was realized that an α-helix would appear at this resolution as a rod with electron density of about 1.0 electron/Å^3 along its axis while regions containing liquid of crystallization and sidechains would have mean electron densities near 0.3 e/ Å^3. Fortunately for the history of protein crystallography myoglobin contains an unusually high percentage (about 70%) of helix. The 6Å electron density map was interpretable in terms of somewhat irregular packing of stretches of α-helix and provided sufficient encouragement for the studies at a higher resolution. The next low resolution study that of chymotrypsinogen (Kraut et al., 1962) and most subsequent studies have been less rewarding mainly because the proteins contain less α-helix.

The general features of low resolution electron density maps are apparent from Figs. 13.1 and 13.2. Helix is indeed usually apparent at $dm = 6$ Å as rods of electron density. Other parts of the polypeptide chain can often be followed but there are very often breaks in the electron density especially where the chain is near the surface of the globular protein. Also where chains are contiguous such as in β-pleated sheet the density often runs between the chains and makes the task of unambiguously following the chain impossible. Metal atoms like the zinc of insulin or the haem iron of myoglobin which are bound tightly have a high electron density and are extended over a large volume. Sidechains are often not represented by electron density, as can be seen in Figs. 13.1 and 13.2, although two sidechains close together may occasionally give rise to higher electron density than the polypeptide backbone. The most electron dense sidechains, the cystine disulphide bridges, are not obvious in most protein electron density maps at this resolution. Clearly disulphides cannot be used as a marker in the interpretation of low resolution electron density maps.

The main difficulty in interpreting the myoglobin electron density may lay in deciding the boundary of individual molecules which were in close contact with one another at a number of points in the crystal structure. However, in many proteins the problem of finding the molecular boundary has been quite straightforward particularly when the solvent of the crystals is high and the liquid of crystallization is salt free and consequently of lower electron density than the mean in the protein molecules (North and Phillips, 1968). The solvent region close to the two-fold screw axis of the sections shown in Fig. 13.1 of the 5.5Å resolution electron density map of lysozyme is quite empty. Within insulin the boundaries between the monomers in the dimer were difficult to distinguish although dimers and the hexamer itself could be quite clearly recognized. The solvent regions were again empty. Empty regions in the low resolution electron density map representing solvent of crystallization are a good indication of the correctness of the phases used.

In summary, low resolution electron density maps do not usually allow the polypeptide chain to be traced unambiguously although they may often give a good impression of the general shape of the protein and sometimes the symmetry relations of subunits in an oligomeric structure. This is best demonstrated by making an electron density model of the type shown in Fig. 13.3. Sheets of balsa wood or polystyrene of the appropriate thickness are fashioned into representations of the electron density greater than say 0.5 $e\text{Å}^{-3}$. These are stacked together and often the model is sand-papered to give a smooth surface.

FIG. 13.3 A low resolution (6Å) balsa wood model of lysozyme with an inhibitor (black) in the active site cleft.

Although low resolution electron density maps are of little use in the interpretation of the chain folding of most proteins, it is still current practice to compute low resolution maps; but such studies are increasingly considered as proving experiments on the way to investigations at higher resolution.

13.4 MEDIUM RESOLUTION ELECTRON DENSITY MAPS
3.5 Å $> dm >$ 2.5 Å

Electron density maps calculated with a minimum interplanar spacing between 3.5Å and 2.5Å usually allow the polypeptide chain to be followed, but their interpretation depends on a prior knowledge of the stereochemistry of the amino acid residues.

Although the polypeptide chain can be followed at 3.5Å it may be difficult to orient the peptide groups as the carbonyls tend to become clearly visible between 3.5Å–3Å in the more ordered parts of the polypeptide; in the most flexible parts of the chain they may not be visible even at 2.5Å.

The positions of the carbonyl peaks enable the direction of the polypeptide chain to be established. Figure 13.4 shows the electron density characteristic of an α-helical polypeptide while Fig. 13.5 shows a section of β-pleated sheet. These secondary structures are easily recognized in the electron density map.

At 3Å the electron density of the sidechains is clearly visible except in the case of amino acids which are on the surface of the protein and are consequently rather disordered. Surface groups, in particular lysines, may be so disordered that only electron density to the β-carbon is visible. Aromatic sidechains of phenylalanine, tyrosine, histidine and tryptophan are clearly defined at this resolution, but there is no depression at the centre of the aromatic ring. Further, the electron density tends to be towards the centre of the triangle formed by the α-, β- and γ-carbon atoms so that the β-carbon has to be placed on the edge of the density (see Fig. 13.6).

The electron densities of the aliphatic valine, leucine and isoleucine and the polar aspartate, asparagine, glutamine, glutamate and threonine sidechains tend to bifurcate at 2.8 Å resolution (for example see the leucine in Fig. 13.2). The disulphides are represented by the highest electron density (above 2.8 Å) but the sulphur atoms are not resolved. The conformation of the disulphides may be difficult to define as the β-carbon electron density is not always clear.

13.5 HIGH RESOLUTION ELECTRON DENSITY MAPS $dm <$ 2 Å

Although the lysozyme electron density maps shown in Fig. 13.1 do not show any appreciable increase in detail at $dm = 2$ Å compared to the map calculated at $dm = 2.5$ Å, this has not been the experience of all protein crystallographic studies. In the small proteins insulin and rubredoxin the electron density maps are much improved at $dm = 2$ Å.

In particular the electron density at $dm = 1.9$ Å of the disulphide bond in insulin can just resolve the two sulphur atoms and allows the positioning of the β-carbon atoms rather precisely. The aromatic residues also become better defined and depressions are evident in the centre of the rings (Fig. 13.7). The aliphatic sidegroups are generally more clearly bifurcated. The conformation of the polypeptide backbone is defined by better electron density for the carbonyl oxygens. It is difficult to describe in writing all the features of such an electron density map; it is much better to look at the maps themselves and for this purpose we reproduce in Fig. 13.8 the electron density of all the sidechains of rubredoxin as calculated by Herriott *et al.* (1973).

13.6 THE DISPLAY OF ELECTRON DENSITY MAPS AND MODEL BUILDING FOR MEDIUM AND HIGH RESOLUTION STUDIES

For medium and high resolution studies an efficient display of electron density and models can considerably speed up the interpretation of the electron density

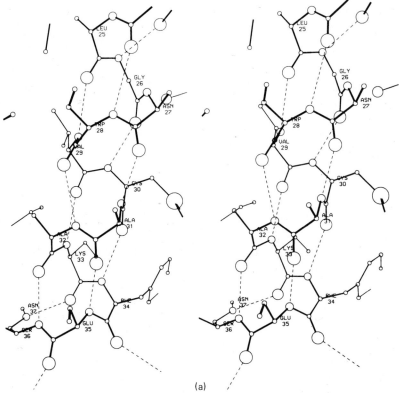

FIG. 13.4 A stereo-view of the electron density (b) at 2.8 Å of an α-helix in lysozyme, and the molecular structure (a) corresponding to this density (Snape, 1974).

13.6. THE DISPLAY OF ELECTRON DENSITY MAPS AND MODEL BUILDING

map. Further by properly relating the model to the electron density, the model building system can minimize the chance of mistakes in the interpretation of the electron density.

In medium and high resolution studies the electron density map is often first drawn out on a scale of 1 cm = 3 Å. The purpose of this exercise is to check the general character of the map. This mini-map will immediately reveal an obvious mistake such as use of the wrong hand of the heavy atoms. On this scale it is easy to "take in" the electron density of the protein as a whole, to define the boundaries of the protein and perhaps follow the rough course of the polypeptide chain. Once the quality of the map has been established and the molecular boundaries roughly delineated, the map is best redrawn to a scale of 2 cm = 3 Å. This scale is very advantageous as it allows direct comparison with Watson–Kendrew model parts (wire models). Even when the whole protein is built the interior is still accessible and clearly visible: this is most important in

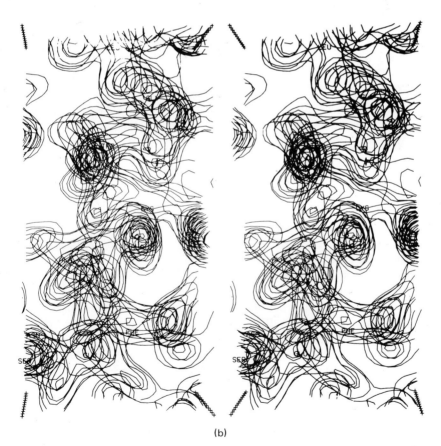

(b)

FIG. 13.4 (*continued*)

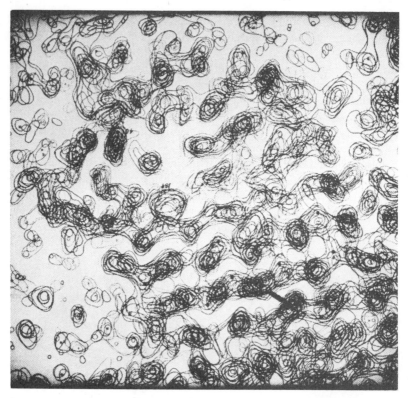

FIG. 13.5 Electron density at 2.8Å corresponding to the β-pleated sheet in concanavalin A (Hardman and Ainsworth, 1972).

model building as the protein structure needs constant revision as the fit between the model and electron density is improved. The provision of an adequate supporting framework presents a serious problem. Rather than using rigid supporting rods, a system of wires held under tension has the advantage of being less obstructive.

The model may be compared to the electron density using the optical device shown in Fig. 13.9 due to Richards and known variously as an "optical comparator", "Richards' Box" or "Fred's Folly", (Richards, 1968). It consists of a half-silvered mirror held at 45° to the electron density sections and also at 45° to the equivalent sections of a Watson—Kendrew model. The image of the model and the electron density then appear superposed when viewed in the half silvered mirror. This technique has facilitated interpretation of most protein structures which have recently been studied at medium or high resolution. The superposition of the model in the electron density using the Richards' box is shown in Fig. 13.9a.

13.6. THE DISPLAY OF ELECTRON DENSITY MAPS AND MODEL BUILDING 393

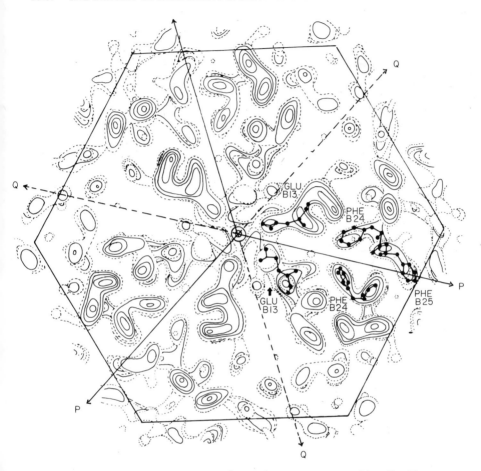

FIG. 13.6 Some sections of the 2.8Å electron density map of insulin. The map is viewed along the three-fold axis and the sections are in the plane of the approximate two-fold axis. The atomic arrangement of some residues in these sections is superposed (Blundell *et al.*, 1971b).

Colman *et al.* (1972a) have suggested an alternative version of this optical comparison technique. In their device the half-silvered mirror is placed parallel to the map sections as shown in Fig. 13.9b. It is desirable to have the rear illumination wider than the map. This construction has the advantage of simplicity of alignment and easy access to both model and map. The vertical position of the mirror means that it is under less strain than in the original device due to Richards. The model and the map sections are also closer to each other and so a more precise comparison is possible. The disadvantages of this arrangement are that the model cannot be easily photographed in the electron

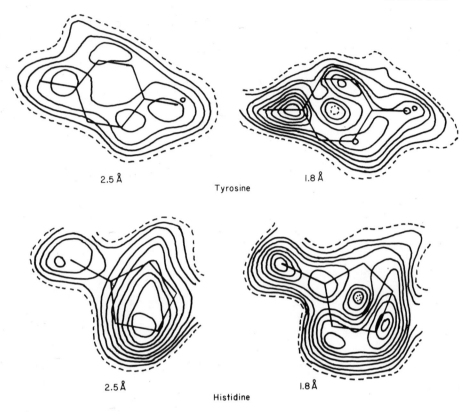

FIG. 13.7 The electron density of a tyrosine and a histidine at 2.5Å and 1.8Å (Insulin Research Group, Peking 1975).

density as the operator has to look through the model to the mirror, and for some directions of view the operator is confronted with his own image as well as that of the model. Perhaps the best device is one in which the mirror can be placed in either a vertical or 45° angle position so that both arrangements may be used.

The major problems in the use of this system are (1) movement of the model parts during the building of the protein; (2) the rigidity of the skeletal parts which are built to the "best" bondlengths and angles; (3) the assessment of the spacefilling properties of the amino acid residues; (4) the reading of the atomic coordinates.

Many devices have been developed for reading off the atomic coordinates. A point light source can be positioned behind the electron density and superposed on the atomic positions of the model using the mirror. This can be automated so that the coordinates are recorded directly (Salemme and Fehr, 1972), but the

13.6. THE DISPLAY OF ELECTRON DENSITY MAPS AND MODEL BUILDING

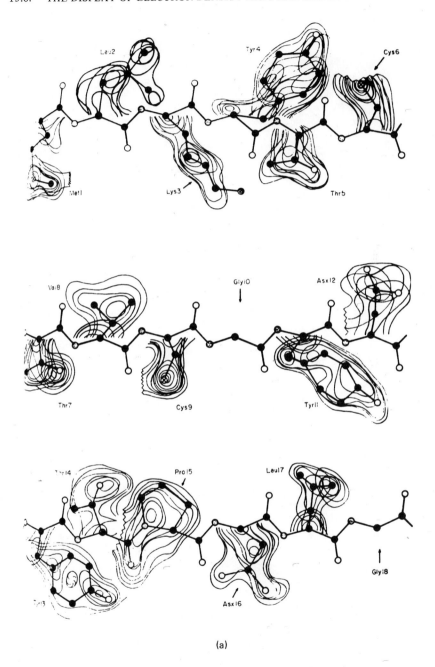

(a)

FIG. 13.8 The electron density of the side chains of rubredoxin at 2Å resolution (From Herriott *et al.*, 1973).

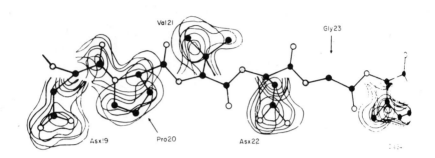

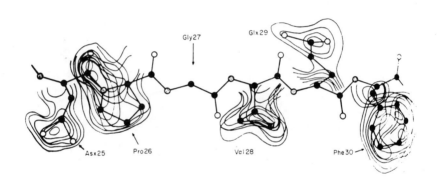

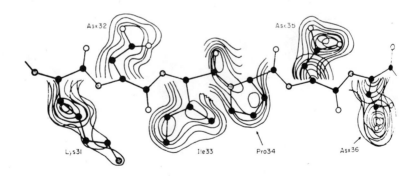

(b)

FIG. 13.8 (*continued*)

13.6. THE DISPLAY OF ELECTRON DENSITY MAPS AND MODEL BUILDING 397

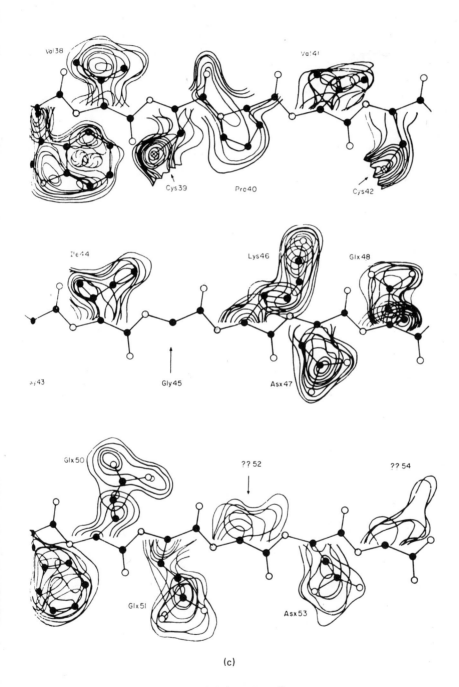

(c)

FIG. 13.8 (*continued*)

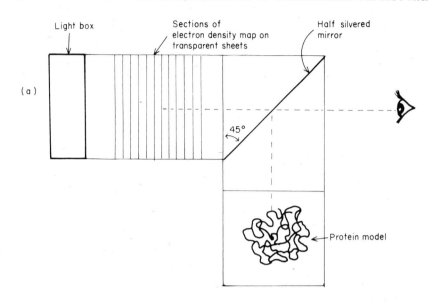

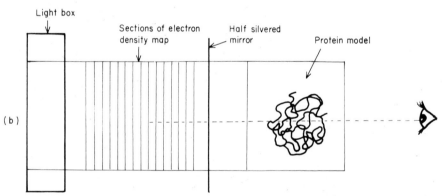

FIG. 13.9 The "Fred's Folly" or optical comparator to enable comparison of the molecular model with the electron density maps. (a) Mirror at 45° to the electron density maps and (b) mirror parallel to the electron density maps.

coordinates perpendicular to the electron density maps are still difficult to measure precisely. Perhaps the best way to overcome this is to compute the electron density map in sections in a direction perpendicular to the original map. By using two sets of electron density maps and two vertical mirrors at right angles to each other, a model may be built most precisely.

The spacefilling aspects of the model can be assessed by attaching Courtould spacefilling models to the Kendrew–Watson framework. However, with practice undesirable interatomic contacts can be quickly identified and modified using the Kendrew–Watson parts and space filling models are not usually necessary.

Computer graphics offer the versatile solution to the problem of fitting the protein model to the electron density. Techniques of computer graphics have been described by several groups of workers (Levinthal, 1966; Levinthal *et al.*, 1968; Barry *et al.*, 1969; Barry and North, 1971; Jacobi *et al.*, 1972). Barry and North (1971) and Jacobi *et al* (1972) have represented an electron density map conveniently by computing single contours in three orthogonal directions and displaying the appropriate projections. Molecular conformations can also be represented by the use of computer controlled graphical displays. The pioneer work of Levinthal (1966) has shown that large computers are not necessary for the effective manipulation of such displays. Programmes have been developed (Barry and North, 1971) to build up proteins from their constituent amino acid residues and to manipulate them by rotations and translations of the whole molecule or by partial rotation around single bonds. The illusion of three dimensions may be obtained by viewing a stereoscopic pair of projections or by displaying orthogonal projections simultaneously.

The programmes have been used to fit a model to an electron density map (bond angles and lengths may be varied at will). Barry and North (1971) have calculated interatomic distances and indicated close contacts by means of broken lines. This endows the system with some of the properties of space filling models. This method of model building allows storage of atomic coordinates for future use, and can be used to calculate a "fit" between the model and the electron density. The method adequately overcomes the shortcomings of the optical techniques. It has already been used with success in interpreting the electron density difference maps from lysozyme crystals containing inhibitor molecules (Ford *et al.*, 1974).

13.7 INTERPRETATION OF ELECTRON DENSITY MAPS AT MEDIUM AND HIGH RESOLUTION

High resolution electron density maps are initially interpreted on the assumption that the amino acid residues have the standard dimensions found in the X-ray analyses of peptides. The peptide bond is assumed planar with a "trans" configuration, except in the case of peptides where the nitrogen of the imide is part of a proline residue; there "cis" and "trans" configurations have similar energies, and either may occur (see Section 2.4). The main chain is treated as rigid except for rotations around $C\alpha$–C and $C\alpha$–N. The problem is where to begin fitting the model to the electron density. In most cases, the interpretation of electron density map has not been started at the chain termini, which are usually on the outside of the molecule, and are often subject to large vibrations or disorder. For instance in the case of nuclease a few residues at the chain termini cannot be seen in the electron density and are therefore presumably completely disordered (Arnone et al., 1971). Generally, the chain is interpreted from a marker and in many cases this has been a cysteine residue which has been labelled by reactions of a mercurial or other sulphydryl reagent (see Section 8.6.2). With enzymes such as the serine proteases the identification of the active serine by the use of labelled inhibitor molecules has been helpful. In other cases, parts of the sequence must be tentatively identified from the X-ray data and then compared to the chemical sequence. Large residues like tryptophan or cystine bridges provide very useful guides. In the case where a haem or a metal ion exists in the native protein, this may be readily identified and the residues binding these groups will help to restrict the possibilities of the sequence in these regions.

13.8 DETERMINATION OF THE AMINO ACID SEQUENCE

In several structure analyses the electron density map has been calculated before the sequence is known and the workers have attempted to determine the amino acid sequence from the electron density. In fact, for myoglobin (Kendrew et al., 1960) carboxypeptidase (Reeke et al., 1967) lamprey haemoglobin (Love et al., 1971) and carp calcium binding protein (Kretsinger et al., 1971) only the sequences of some tryptic peptides were known. For rubredoxin (Herriott et al., 1970), lactate dehydrogenase (Adams et al., 1970a), flavodoxin (Ludwig et al., 1971), thermolysin (Matthews et al., 1972a) and for several others the sequence was completely unknown. It is constructive now to consider the success of these studies.

In many cases it is possible to get a good estimate of the number of amino acids in the polypeptide chain but occasionally mistakes have been made at the chain termini and also in loops where the polypeptide extends into the solvent. At these points the thermal and static disorder will be high and the electron density correspondingly low. Occasionally a carbonyl with a hydrogen bonded water will look large and misleadingly like a serine sidechain. Great success has been achieved in fitting tryptic peptides. In the case of papain (Drenth et al., 1958) the incorrect placing of the peptide fragments by the chemists, who were unable to obtain overlapping sequences, was clearly recognized by the protein crystallographers. Drenth and his colleagues were able to rearrange these properly and tentatively identify thirteen missing residues.

The ability to correctly identify amino acid sidechains can be assessed from the studies of carboxypeptidase, rubredoxin and thermolysin where the primary structures have since been determined by chemists. Consider first carboxypeptidase. A complete sequence was published by Lipscomb et al. (1968) based on the X-ray data at 2 Å resolution and the sequences of nine fragments. Bradshaw et al. (1969) published the full sequence which confirmed the number

13.8. DETERMINATION OF THE AMINO ACID SEQUENCE

TABLE 13.I
Amino acid sequence for *C. pasteurianum* rubredoxin

No. of R group	Ident. from 2Å map	Confidence	Ident. from refinement	Ident. from Chem.
1	Met	3		fMet
2	Leu	1	Lys	Lys
3	Lys	3		Lys
4	Tyr	3		Tyr
5	Thr	2		Thr
6	Cys	3		Cys
7	Thr	2		Thr
8	Val	2		Val
9	Cys	2		Cys
10	Gly	3		Gly
11	Tyr	3		Tyr
12	Asx	3	?	Ile
13	Tyr	3		Tyr
14	Thr	2	Asx	Asp
15	Pro	3		Pro
16	Asx	2	?	Glu
17	Leu	1	Prob. Asx	Asp
18	Gly	3		Gly
19	Asx	3		Asp
20	Pro	3		Pro
21	Val	2	?	Asp
22	Asx	3		Asp
23	Gly	3		Gly
24	Ile	2	Val	Val
25	Asx	2		Asn
26	Pro	3		Pro
27	Gly	3		Gly
28	Val	3	Thr	Thr
29	Glx	2	Prob. Asx	Asp
30	Phe	3		Phe
31	Lys	2		Lys
32	Asx	3		Asp
33	Ile	2		Ile
34	Pro	3		Pro
35	Asx	3		Asp
36	Asx	2		Asp
37	Trp	3		Trp
38	Val	2		Val
39	Cys	3		Cys
40	Pro	3		Pro
41	Val	2	Leu	Leu
42	Cys	3		Cys
43	Gly	3		Gly
44	Ile	2	Val	Val
45	Gly	3		Gly

TABLE 13.I (*continued*)
Amino acid sequence for *C. pasteurianum* rubredoxin

No. of R group	Ident. from 2Å map	Confidence	Ident. from refinement	Ident. from Chem.
46	Lys	3		Lys
47	Asx	2		Asp
48	Glx	3		Glu
49	Phe	3		Phe
50	Glx	2		Glu
51	Glx	3		Glu
52	?	1	Val	Val
53	Asx	3	Prob. Glx	Glu
54	?	1	?	Glu

of residues, established as 307 in the X-ray studies. The sequenced fragments had been placed in the electron density correctly. By comparison with the original sequence it is now clear that 56 of the remaining 93 amino acid residues (60%) were identified correctly (Lipscomb *et al.*, 1969). Many of the mistakes involved confusion of valines and threonines, and leucines and asparagines. These are isosteric pairs and are easily confused. Several other leucines and some isoleucines and serines were incorrectly identified. The fact that histidines and tyrosines were also misidentified indicates that aromatic rings may not be so obvious as is often considered especially if they are on the surface of the protein. In contrast all the phenylalanines were identified correctly but identification here may have been aided by considering their non polar environments.

In the study of thermolysin at 2.3 Å resolution, Matthews and his coworkers (Matthews *et al.*, 1972 ; Colman *et al.*, 1972a) also found that the polar surface residues were more difficult to identify. In their study which did not take into account the environment of the sidechain as an aid to its identification, 53% of the residues were identified correctly. Particularly difficult to identify were: leucines which were confused with valine, isoleucine, threonine, aspartate and glutamate; threonines which were confused with alanine, serine, valine, glycine, aspartate; glutamines and glutamates which were confused with lysine, arginine, threonine, glycine, leucine, aspartate and serine; and histidines which were confused with glutamine, methionine and glycine. Of particular interest were the lysines which were four times identified as glycines presumably because the sidechains were disordered. Conversely three serines and two asparagines were identified as lysines presumably as a result of solvent molecular hydrogen bonded to them which made their appearance larger.

The most successful sequence determination has been that of rubredoxin but this protein is very much smaller than the others we have discussed. 40 of the 54 amino acids (i.e. 74%) were correctly identified, and the remaining fourteen sidechains differed from the correct ones only by an "atom or two"

13.8. DETERMINATION OF THE AMINO ACID SEQUENCE

(Watenpaugh et al., 1973). The electron density for the sidechains is shown for rebredoxin in Fig. 13.9 and the errors of identification are summarized in Table 13.1 so that the reader may make an assessment of the errors for himself.

In summary it appears that the use of the X-ray method, except in the electron density maps of small proteins, might be of limited use in identification of amino acids. However, the errors in the X-ray method are likely to be different from those of the normal chemical techniques where the problem is usually to obtain overlapping sequences to align tryptic peptides. The X-ray method has shown itself to be rather useful in this aspect of sequence determination, and so it seems that a combination of chemical and X-ray techniques may often be most advantageous.

14

DIFFERENCE FOURIERS

14.1 INTRODUCTION

Once a protein structure has been solved, the study of the association of small molecules with the protein may be accomplished relatively easily by means of difference Fourier syntheses. The difference Fourier technique was first used in protein crystallography to study the binding of azide ion to myoglobin (Stryer et al., 1964). Since then the method has been widely applied in the study of binding of inhibitors and pseudo-substrates to a large number of proteins and provides the means by which the active sites of enzymes may be located. The method has been extended to include the determination of the structures of small molecules when bound to the protein surface (Ford et al., 1974). Difference Fouriers also provide a sensitive monitor of conformational changes in the native protein structure and shifts of atoms of less than 1Å can be detected. In this chapter we shall be especially concerned with the use of difference Fouriers to locate small molecules. Their use in the study of heavy atom binding sites has already been discussed (Chapter 11).

The diffusion of a small molecule, such as a competitive inhibitor or co-factor of an enzyme, into a protein crystal will result in the binding of the molecule probably at a unique site on the protein surface (unless, in unfortunate cases, the packing of the protein molecules in the crystal lattice makes this site inaccessible). Often rather higher concentrations of inhibitors are required than is indicated by their equilibrium binding constants. For example the equilibrium binding constant of the inhibitor tri-N-acetylchitotriose ($GlcNAc_3$) for lysozyme is 10^{-5} M. Soaking crystals at a concentration of 6 mM produced no detectable binding after 24 hrs, whereas with a concentration of 50 mM diffusion and binding were extremely rapid and accomplished within ten minutes. Co-crystallization of lysozyme and $GlcNAc_3$ at 1:1 molar ratios (corresponding to a concentration of $GlcNAc_3$ of approximately 3 mM) produced the same result as the high concentration diffusion experiment. The ease with which small metabolites bind to proteins in a crystal is dependent upon the accessability of the binding site and this will differ for different protein crystals.

Because the structure of the active sites of enzymes are extremely responsive to the binding of substrates and substrate analogues, conformational changes

14.1. INTRODUCTION

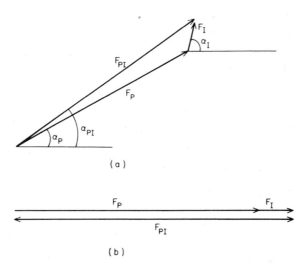

FIG. 14.1 Vector triangles showing the contributions to the structure factor following the addition of an inhibitor molecule to a protein crystal for (a) the acentric case and (b) the centric case.

may take place which give rise to cracking or disorder of the crystal. Cracking can sometimes be alleviated by adjusting the salt concentration of the mother liquor.

The addition of a small molecule, containing only light atoms, to a crystal will give rise to a change in structure factor amplitude as indicated in Fig. 14.1, when F_P and F_{PI} are the structure factor amplitudes of the native protein and the protein plus inhibitor, respectively, and F_I is the structure factor amplitude of the inhibitor molecule. This diagram is exactly the same as the phase triangle used to illustrate the binding of a heavy atom to a protein, the only difference being that F_I is likely to be smaller than F_H. The Fourier synthesis for the native protein is given by

$$\rho_P = \frac{1}{V} \sum_h F_P \, \text{expi} \, \alpha_P \, \text{exp} - 2\pi i \, \mathbf{h} \cdot \mathbf{x}$$

and that for the protein plus inhibitor by

$$\rho_{PI} = \frac{1}{V} \sum_h F_{PI} \, \text{expi} \, \alpha_{PI} \, \text{exp} - 2\pi i \, \mathbf{h} \cdot \mathbf{x}$$

Since F_I is small, α_{PI} will not be very different from α_P so that ρ_{PI} may be approximated by

$$\rho_{PI} \simeq \frac{1}{V} \sum_h F_{PI} \, \text{expi} \, \alpha_P \, \text{exp} - 2\pi i \, \mathbf{h} \cdot \mathbf{x}$$

Hence, the difference in electron density between the protein plus inhibitor and the native protein, which corresponds to the electron density of the inhibitor, is given by

$$\rho_{PI} - \rho_P \triangleq \frac{1}{V} \sum_h (F_{PI} - F_P) \exp i\alpha_P \exp -2\pi i \mathbf{h} \cdot \mathbf{x}$$

This is illustrated in Fig. 14.2.

The reliability of the protein phases which have been determined by the heavy atom isomorphous replacement method is represented by their figure of merit, m. Hence in the difference Fourier it is conventional to weight the difference terms also by m. Thus the difference Fourier $\Delta\rho$ is defined by:

$$\Delta\rho = \frac{1}{V} \sum_h m(F_{PI} - F_P) \exp i\alpha_P \exp -2\pi i \mathbf{h} \cdot \mathbf{x}$$

Difference Fouriers have the advantage that series termination errors cancel out (Lipson and Cochran, 1968).

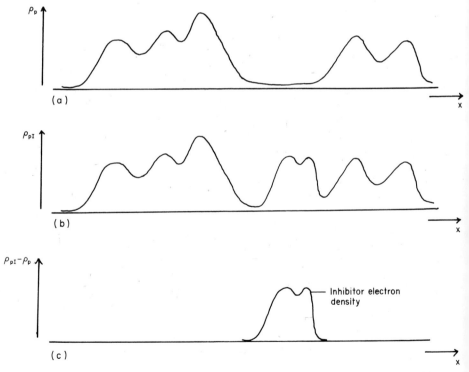

FIG. 14.2 Schematic diagrams of the one dimensional projection of (a) the protein, (b) the protein plus inhibitor and (c) the difference electron densities. There are no conformational changes in the protein on addition of the inhibitor.

14.2 PEAK HEIGHTS IN DIFFERENCE FOURIERS

In order to assess the reliability of a difference Fourier, the errors involved in the assumption that $\alpha_{PI} \simeq \alpha_P$ need to be known. Define

$$\Delta\rho_{obs} = \frac{1}{V} \sum_{h} (F_{PI} - F_P) \exp i\alpha_P \exp - 2\pi i \, \mathbf{h} \cdot \mathbf{x}$$

We wish to know how this difference Fourier, which we can calculate, differs from the true structure represented by

$$\Delta\rho_{true} = \frac{1}{V} \sum_{h} F_I \exp i\alpha_I \exp - 2\pi i \, \mathbf{h} \cdot \mathbf{x}$$

In the case of a centrosymmetric structure or projection, where the phases are either 0 or π, it can readily be seen (Fig. 14.1) that $(F_{PI} - F_P)\exp i\alpha_P$ is equivalent to $F_I \exp i\alpha_I$, except in the very rare cases of "cross-over". The difference Fourier will therefore contain an accurate representation of the inhibitor molecule with the atoms at full height.

For the acentric case the situation is rather more complicated. With reference to Fig. 14.1, and using the cosine rule,

$$F_{PI} - F_P = \frac{F_I^2}{F_{PI} + F_P} + \frac{F_P F_I \cos(\alpha_I - \alpha_P)}{F_{PI} + F_P}$$

Hence by using the same manipulation described in section 11.4, equation 11.22

$$(F_{PI} - F_P)\exp i\alpha_P = \frac{F_I^2 \exp i\alpha_P}{F_{PI} + F_P} \quad \text{(i)}$$

$$+ \frac{F_P F_I \exp i\alpha_I}{F_{PI} + F_P} \quad \text{(ii)(14.2)}$$

$$+ \frac{F_P F_I \exp i(-\alpha_I + 2\alpha_P)}{F_{PI} + F_P} \quad \text{(iii)}$$

Term (i) represents an image of the protein structure but since $F_P \simeq F_{PI}$ and both F_P and $F_{PI} \gg F_I$, its contribution will be very small. Term (ii) gives rise to the transform of $\mathbf{F}_I/2$ in the electron density difference map. Term (iii) gives rise to noise but since there is unlikely to be any correlation between α_I and $2\alpha_P$ this contribution to the transform is small.

Hence it is seen that a difference Fourier synthesis based on coefficients $(F_{PI} - F_P)\exp i\alpha_P$ contains features which accurately represent the additional atoms not included in the phasing, but that these atoms will appear at half their true heights and that there will be some background noise.

The problem of peak heights in difference Fouriers for different ratios of known to unknown atoms has been discussed by Luzzati (1953). His results,

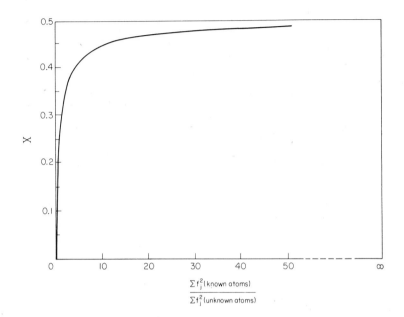

FIG. 14.3 Plot of χ versus Σf_j^2 (known atoms)/Σf_j^2 (unknown atoms), after Luzzati (1953), where χ is the fractional peak height expected for unknown atoms not included in the phase determination.

which are based on statistical methods, are shown in Fig. 14.3 for centric and acentric distributions. The expected peak heights of the unknown atoms are relatively insensitive to the ratio of known to unknown atoms within the ranges usually encountered in protein crystallography. Thus for example the addition of $GlcNAc_3$ to lysozyme represents an addition of 43 light atoms to a rather small protein of approximately 1000 atoms. The peak height for $GlcNAc_3$ is therefore expected to be 0.469 of its true height in a difference Fourier based on acentric terms. These theoretical results are only applicable to cases where the difference in phase angle between α_P and α_{PI} represents the main cause of the difference between the observed and true electron densities.

In order to bring the scale of the electron densities closer to their true values, it is conventional to use coefficients $2\Delta F = 2(F_{PI} - F_P)$ in the difference Fourier, so that the basic definition becomes

$$\Delta\rho = \frac{2}{V} \sum_{\mathbf{h}} m\Delta F \; \exp i\alpha_P \; \exp -2\pi i \, \mathbf{h} \cdot \mathbf{x}$$

In protein difference Fouriers the peak heights of the additional atoms will be further weighted down due to the uncertainties in the native phases. We would expect that the poorer these phases the lower the heights of the atoms in the

14.3. ERRORS IN DIFFERENCE FOURIERS

resulting difference Fourier. If these uncertainties are reflected by the figure of merit m, then the mean contribution of the structure factor $m(F_{PI} - F_P)\exp i\alpha_P$ to the unknown correct structure has been shown to be $m^2 (F_{PI} - F_P)$ (Henderson and Moffat, 1971) (Section 12.2). If it is assumed that m is the same for all reflections then the Fourier synthesis will have peaks whose heights are m^2 those of the true structure. In practice m is a function of $\sin^2\theta/\lambda^2$ and varies from one reflection to the next depending upon the contributions of the heavy atoms to the phasing so that the above analysis only gives a very approximate feel for the effects of the average value of m on the peak heights.

14.3 ERRORS IN DIFFERENCE FOURIERS

The quality and usefulness of a difference Fourier depends on three factors.

(i) The quality of the protein phases. There is obviously little to be gained from the computation of difference Fouriers in the search for binding sites of small light atom molecules until the experimenter is satisfied that the protein phases are the best that can be obtained. The use of a poor set of phases can give rise to spurious features in difference maps. This is especially true when the structural studies are still at the low resolution stage and the ultimate test of the phases (i.e. the provision of an interpretable high resolution map) has not been applied. Incomplete refinement of heavy atom parameters and imprecise determination of scale factors are likely to give rise to ghosts in difference syntheses, which confuse interpretation.

We have already seen that features in a difference map are weighted by m^2 so that as the protein phases become worse so the signal to noise error in the map deteriorates.

(ii) The errors in the intensity measurements. The mean square error produced in a Fourier synthesis by a mean square standard deviation σ_F^2 in the coefficients is given by

$$\delta_\rho^2 = \frac{1}{V^2} \sum_h \sigma_F^2$$

Hence in a difference Fourier synthesis based on coefficients $2(F_{PI} - F_P)$

$$\sigma_F^2 = 4\sigma_{F_{PI}}^2 + 4\sigma_{F_P}^2$$

So that the mean square error introduced into the difference Fourier is given by

$$\delta_{\Delta\rho}^2 = \frac{4}{V^2} \sum_h (\sigma_{F_{PI}}^2 + \sigma_{F_P}^2)$$

It is especially important therefore that the measurements are as precise as possible so that the small changes in structure factor amplitude are greater than the standard deviation of their measurement, i.e.

$$|F_{PI} - F_P| > (\sigma_{FPI}^2 + \sigma_{FP}^2)^{1/2}$$

Further the crystals of protein plus inhibitor must be closely isomorphous with the native protein crystals, at least with respect to cell dimensions, so that the molecular transform is sampled at the same reciprocal lattice points.

(iii) Errors introduced by the use of the difference Fourier coefficients $2m(F_{PI} - F_P) \exp i\alpha_P$ instead of the true but unobservable structure factors $F_I \exp i\alpha_I$.

From equation 14.2 and assuming F_P, $F_{PI} \gg F_I$ so that Term (i) may be ignored

$$2(F_{PI} - F_P)\exp i\alpha \simeq F_I \exp i\alpha_I + F_I \exp i(2\alpha_P - \alpha_I)$$

So that the error introduced by use of coefficients $2(F_{PI} - F_P)\exp i\alpha_P$ is

$$\Delta\Delta\rho = \frac{1}{V}\sum_h F_I \exp i(2\alpha_P - \alpha_I) \exp - 2\pi i \mathbf{h} \cdot \mathbf{x}$$

The mean square error is found by averaging the square of this expression over the entire unit cell, i.e.

$$\langle \Delta\Delta\rho^2 \rangle = \frac{1}{V} \int_v (\Delta\Delta\rho)^2 dv$$

$$= \frac{1}{V^3} \int_v \left\{ \sum_h F_I \exp i(2\alpha_P - \alpha_I) \exp - 2\pi i \mathbf{h} \cdot \mathbf{x} \times \sum_{h'} F_I \exp i(2\alpha_P - \alpha_I) \exp - 2\pi i \mathbf{h}' \cdot \mathbf{x} \right\} dv$$

Since

$$\int_v \exp - 2\pi i(\mathbf{h} + \mathbf{h}') \cdot \mathbf{x} \, dv = 0 \quad \text{if} \quad \mathbf{h} \neq -\mathbf{h}'$$
$$= V \quad \text{if} \quad \mathbf{h} = -\mathbf{h}'$$

$$\langle \Delta\Delta\rho^2 \rangle = \frac{1}{V^2}\sum_h |F_I|^2$$

Now $(F_{PI} - F_P) \simeq F_I \cos(\alpha_I - \alpha_P)$ and since the average value of $\cos^2(\alpha_I - \alpha_P) = \frac{1}{2}$

$$\langle (F_{PI} - F_P)^2 \rangle \simeq \frac{1}{2}\langle F_I^2 \rangle$$

Hence

$$\langle \Delta\Delta\rho^2 \rangle \simeq \frac{2}{V^2}\sum_h (F_{PI} - F_P)^2$$

14.4. CONFORMATIONAL CHANGES

If it is assumed that the errors in measurements are not correlated with the errors introduced by use of phases α_P then the total mean square error in the difference map based on coefficients $2(F_{PI} - F_P) \exp i\alpha_P$ is given by

$$\langle \Delta\Delta\rho^2 \rangle_{total} = \langle \Delta\Delta\rho^2 \rangle + \delta^2_{\Delta\rho}$$

In this derivation, the contributions of errors from the native phases have been ignored.

The problem of errors in difference Fouriers has been discussed by Henderson and Moffat (1971), but their final equation, equation 14, is not quite correct as it was obtained by an arbitrary division by 2. In a recent and more rigorous derivation Henderson (1973, private communication) suggests a similar formula to that given above. The derivation is based on the probability distribution of the unknown structure and also takes into account errors in phasing.

Both derivations suggest the error is proportional to $\langle \Delta F^2 \rangle^{1/2}$ and since ΔF is normally a small fraction of the average protein structure factor, F_P, difference Fouriers have a much lower error level than the corresponding map for the native protein.

14.4 CONFORMATIONAL CHANGES

The difference Fourier provides a sensitive monitor of the conformational changes between the protein–inhibitor complex and the native protein. The movement of a group of atoms in the native protein to a new position in the protein–inhibitor complex gives rise to a characteristic pattern of troughs and peaks in the difference Fourier (Figs. 14.4 and 14.5), which indicate the direction in which the atoms have moved. Difference Fouriers have been used from the very early days in the refinement of small molecule structures (where the coefficients of the synthesis are $(F_{obs} - F_{calc}) \exp i\alpha_{calc}$). More recently, the first stages in the refinement of the small protein, rubredoxin, were carried out by difference Fouriers (Watenpaugh et al., 1971) (see Section 15.3.4).

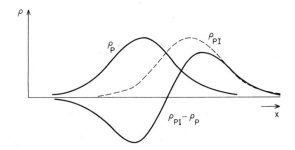

FIG. 14.4 Schematic diagram of a simple conformational change which involves one atom.

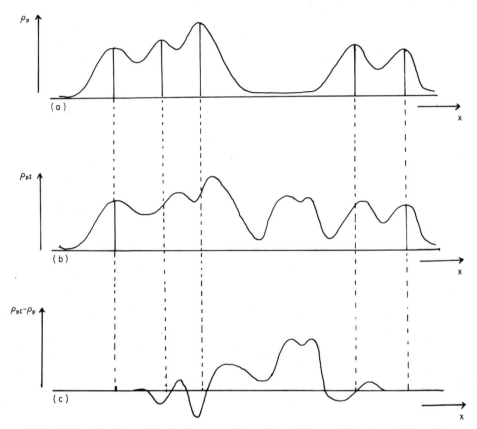

FIG. 14.5 Conformational changes observed in difference Fouriers. Schematic diagrams of the one dimensional projection of (a) the protein, (b) the protein plus inhibitor and (c) the difference electron densities. Two groups of protein atoms on the left have moved towards the inhibitor and a group on the right have moved away. This gives rise to a pattern of peaks and troughs in the difference Fouriers.

From conventional Fourier analysis it can be shown (e.g. Lipson and Cochran, 1968) that the shifts, Δx, are given by

$$\Delta x = - \frac{\partial \Delta \rho}{\partial x} \bigg/ \frac{\partial^2 \rho_P}{\partial x^2}$$

where $\partial \Delta \rho / \partial x$ is the gradient of the difference Fourier in the direction x and $\partial^2 \rho_P / \partial x^2$ is the curvature of the electron density in the native protein at this position. This method of calculation has been applied to the determination of the shift of a tryptophan residue (Trp 62) whose movement is observed on the binding of $GlcNAc_3$ to lysozyme. In the difference Fourier (calculated at a

14.4. CONFORMATIONAL CHANGES

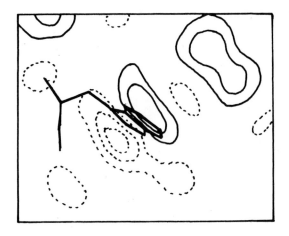

FIG. 14.6 The movement of residue Tryptophan 62 in the lysozyme-tri-N-acetylglucosamine complex. The difference map has been calculated at a resolution of 2.5 Å and contoured at intervals of 0.25 e/Å^3. The zero contour is omitted; negative contours are shown dotted. The section shown corresponds to a portion of section $Z = 33.5$ Å of the difference map. Tryptophan 62 extends from approximately $Z = 31$ Å to $Z = 35.0$ Å. The peak with two contours in the top right corner is part of the sugar bound in site B. The movement corresponds to shift of 0.87 Å.

resolution of 2 Å) there is a clear trough and peak indicating a movement of the indole side chain towards the inhibitor (Fig. 14.6). The gradient of the difference Fourier in this direction was $54/1.86$ e/Å^4. The electron density in the region of the peak in the native protein was approximated by a parabola $\rho_P = ax^2 + bx + c$ in which the three constants were determined from observations of ρ_P at three positions. The curvature of the native density (which is given by $\partial^2 \rho_P / \partial x^2 = 2a$) was found to be $-116/(1.86)^2$ e/Å^5. The shift in the tryptophan side chain, calculated from the gradient of the difference Fourier divided by the curvature of the native Fourier, was therefore 0.87 Å.

This simple calculation illustrates the fact that very small shifts of atoms are readily identified in difference Fouriers. However this method of calculation is not universally applicable. It is only satisfactory when the movements of the protein correspond to shifts of a discrete number of atoms. In cases where there are complicated changes over extensive regions of a polypeptide chain, the calculation of meaningful gradients is no longer possible (see Fig. 14.5). Secondly the method is reliable only when the group of atoms in the native Fourier is relatively well-defined. If these atoms have large thermal vibrations, then calculations of the curvature of the electron density can prove misleading and give rise to anomalously large shifts.

In cases where the conformational changes involve well-defined groups of atoms and are not complicated by shifts in adjacent atoms, the integrated electron count within the peak should equal the corresponding integrated electron count within the trough. Under these circumstances, Henderson (1970) has shown that the shifts may be calculated from a formula involving vector moments, i.e.:

$$\Delta x = \frac{\int_v x \Delta \rho(\mathbf{x}) dv}{\int_v \rho_P(\mathbf{x}) dv}$$

where the integration is taken over the volume containing the peak and the trough. Since there is the possibility of compensating shifts which would have no effect on a difference electron density map, the vector moment represents the minimum electron shift which could account for the observed pairs of positive and negative features. This procedure is clearly only appropriate where pairs of positive and negative features containing the same integrated density can be identified. In the denominator in the above equation the integral is evaluated over the same volume as the numerator for the native protein electron density and this integral should be equivalent to the electron count for the group of atoms. For example on reaction of α-chymotrypsin with toluenesulphonylfluoride, histidine 57 shows up with a vector moment of 17 eÅ in the difference map, which since the imidazole contains 35 electrons, gives a movement of 0.45 Å (Henderson, 1970). This procedure calculates a purely translational movement. If the real movements involve rotations, then this procedure would calculate the average translational movement. This method overcomes some of the difficulties associated with estimation of the curvature in the native electron density maps but it requires a fairly accurate estimate of the electron count.

If the shifts of a discrete group of atoms are sufficiently large (of the same order as the nominal resolution of the map, say) then the integrand electron counts in the peak and the trough should be approximately equal to the total number of electrons expected for the group of atoms (or at least equal to the integrated count observed in the native Fourier). Under these circumstances, the shifts are simply calculated from the separation of the trough and the peak. For example the oxidation of lysozyme by the stoichiometric amounts of iodine results in the formation of an oxindole—ester link between tryptophan 108 and glutamic acid 35 (Beddell and Blake, 1970; Beddell, 1970). In the difference Fourier (2.5 Å resolution, mean figure of merit = 0.78) between the modified and native protein there is a hole at the position formerly occupied by the carboxyl group of Glu 35 and a peak near the position of Trp 108 (Fig. 14.7). The number of electrons found in the peak was 18.6 with the limit of the peak defined by the contour level 0.2 e/Å^3. The reaction was found, from activity

14.4. CONFORMATIONAL CHANGES

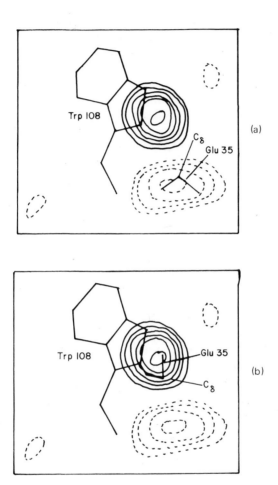

FIG. 14.7 The most significant section of the difference Fourier of iodine oxidized lysozyme against native lysozyme showing a relatively large conformational change of a discrete group of atoms. The map is contoured at intervals of 0.25 e/Å^3, positive contours are shown by full line, negative contours are dotted. The positions of the side chains of trp 108 and the carboxyl group of glu 35 in the native enzyme are shown superimposed in (a). Trp 108 is about 1 Å above the section and the carboxyl group of glu 35 about 1 Å below the section. The new location of the carboxyl group is shown in (b) where the Cδ atom of glu 35 has moved a distance of 2.3 Å. The new position of the carboxyl group places it extremely close to the δ_1 carbon position of trp 108. A covalent bond has been formed between one of the carboxyl oxygens and the δ_1 carbon of trp 108 resulting in an oxindole-ester (from Beddell and Blake, 1970).

measurements, to have been 87% complete. The number of electrons expected for the carboxyl group in the ester is therefore $22 \times 0.87 = 19.2$. The number of electrons in the trough was -11.7. The number expected for a displaced protonated carboxyl group is $-23 \times 0.87 = -20.0$. The number of electrons corresponding to the carboxyl group in the native map is 23.5 when the limit to the peak is taken as 0.35 e/Å^3. If the carboxyl group is replaced by an ordered water molecule (10×0.87 electrons), then the number of electrons lost is expected to be 11.3. Hence these values are in good agreement with the carboxyl group shifting to a new position and having been replaced by a water molecule. The number of electrons calculated for the peak is consistent with the number expected for a carboxyl group, and the number of electrons in the trough is consistent with the number expected for a protonated carboxyl group replaced by a water molecule. The conformational shift for the Cδ atom of Glu 35 was calculated to be 2.3 Å from coordinates taken from a model built to the difference Fourier. This relatively large movement is accomplished with little cost in energy by a rotation of 120° about the α–β bond of the glutamic acid side chain.

It is clear that shifts in well-defined groups of atoms are easily identified and measured in difference Fouriers. More complicated movements, involving whole stretches of backbone and side chains, can also be detected from the pattern of peaks and troughs but are much more difficult to quantitate. For many proteins there are a number of such small changes observed and it is desirable to be able to quantitate these, especially in view of controversial theories of enzyme action. So far there has been little progress in the automatic calculation of the overall changes observed for a protein inhibitor complex, although it would appear that the Fourier refinement methods of Diamond (1974a) (Section 15.3) are likely to be applicable. The refinement of the trypsin–trypsin inhibitor complex (Huber et al., 1974) forms an outstanding example of precise protein crystallography, although in this case the analysis involved a complete *de novo* structure determination of the complex.

14.5 OTHER TYPES OF DIFFERENCE FOURIERS

14.5.1 Compensation of displaced water molecules

One of the major difficulties in the precise interpretation of difference Fouriers involves the treatment of water molecules which are displaced on binding the inhibitor. In the native structure these water molecules may be firmly bound and occur as discrete peaks in the native map. If they are displaced, then there will be a corresponding hole at this site in the difference Fourier. If the displaced water molecule is replaced by atoms of the inhibitor molecule the negative density will be partially compensated by the positive density of the inhibitor.

14.5.3. SIM WEIGHTING

The compensation will not be complete because the phases used result in the atoms of the inhibitor molecule appearing at half heights while the displaced water molecules appear at full height. This complicates the location of the inhibitor atoms. It is therefore desirable to add back the water molecules in a manner which takes into account their relative heights. It has been proposed (D. C. Phillips, unpublished) that a synthesis based on coefficients

$$[2(F_{PI} - F_P) + F_P]\exp i\alpha_P = (2F_{PI} - F_P)\exp i\alpha_P$$

should overcome this problem. The resulting electron density will, of course, also contain the density of the native protein which may complicate the location of the inhibitor. But when used in combination with the conventional difference Fourier, this modified synthesis should permit a proper compensation for displaced water molecules and location of inhibitor atoms.

14.5.2 Calculated phases

Once a protein structure has been solved and the atomic coordinates refined by the methods discussed in Chapter 15, then the phases of the structure factors for the native protein may be calculated. These phases may then be used in the calculation of the difference Fourier in place of those determined by the heavy atom isomorphous replacement technique and should help to overcome some of the errors inherent in the experimental phases. This approach was tried with carboxypeptidase A in the study of the binding of glycyl-1-tyrosine (Lipscomb et al., 1970) in a synthesis which also took into account the fractional occupancy of the additional molecules. Although the modified difference synthesis did not lead to any new interpretation, certain features, notably those in the region of displaced solvent molecules and those involving conformational changes of the protein, were more distinct.

14.5.3 Sim weighting

The success of the difference Fourier technique rests on the assumption that the phase, α_{PI}, of the protein-plus-inhibitor are not too different from the phases, α_P, of the native protein. This assumption is valid if the number of atoms included in the native protein is large in comparison to the number of atoms in the inhibitor, as is usually the case in protein crystallography. However the approximation may well be more precise for some reflections than for others. If we consider a fixed contribution $\mathbf{F}_I$ from the inhibitor (Fig. 14.8), then α_{PI} is more likely to be close to α_P if F_P is large than if F_P is small. Clearly a weighting scheme which takes into account the probability that α_P is correct for that particular reflection could prove advantageous.

This problem has been discussed by Sim (1959). The probability that the difference in phase, $\alpha_{PI} - \alpha_P$, lies between ξ and $\xi + d\xi$ for structure factors

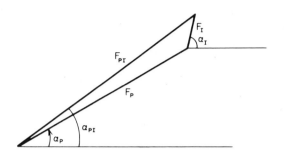

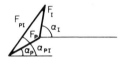

FIG. 14.8 Vector triangles showing the contribution to structure factor from the addition of an inhibitor molecule to a protein crystal. If $F_P \gg F_I$ then α_P forms a good approximation to α_{PI}. If $F_P \simeq F_I$ then the approximation is poor.

with fixed values of F_P, F_{PI} and α_P is given by

$$P(\xi)d\xi = \frac{\exp(X \cos \xi)d\xi}{2\pi I_0(X)}$$

where I_0 is a modified zero-order Bessel function (Watson, 1922, p 77) and

$$X = \frac{2F_P F_{PI}}{\Sigma}$$

In Sim's discussion Σ is taken to be Σf_j^2 where the summation is for those atoms not included in the phase determination. The term represents the mean square error in the structure factor calculated on the basis of the partial structure (in our case, the native protein). The summation provides an adequate representation of this error for the analysis of the structure of small molecules, where the intensity distribution obeys Wilson statistics, but is not appropriate for a protein–inhibitor complex. In the complex there may be small conformational changes of the protein, displaced water molecules, fractional occupancy of the inhibitor and minor binding sites. These changes will not be reflected in the simple expression Σf_j^2 and an empirical estimate of the mean square error such as $\langle |F_P - F_{PI}|^2 \rangle$ probably forms a better value for Σ.

Following Blow and Crick (1959), Sim (1960) suggests that the best set of weights are those which result in the least mean square error in the electron density map. The weights are therefore defined by

14.5.3. SIM WEIGHTING

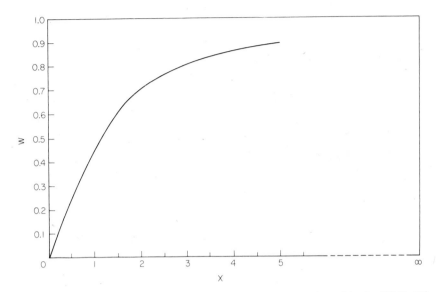

FIG. 14.9 Plot of Sim weighting factor W against X where $W = I_1(X)/I_0(X)$ and $X = 2F_P F_{PI}/\Sigma$. After Sim (1960).

$$W = \int_0^{2\pi} \cos \xi P(\xi) d\xi$$

$$= \frac{I_1(X)}{I_0(X)}$$

where I_0 and I_1 are zero order and first order modified Bessel functions of the first kind. Values of W as a function of X are shown in Fig. 14.9.

A similar weighting scheme has been used by Stubbs (1972). The probability that a phase angle α is correct was determined by the simpler expression

$$P(\alpha) = N \exp(X \cos(\alpha - \alpha_P))$$

where N is a normalizing factor. Stubbs then took the centroid phase as the "best phase" and coefficients $F_{PI} \exp i\alpha_P$ were replaced by the coefficients

$$F_{PI} \frac{\int_0^{2\pi} \exp i\alpha \, P(\alpha) \, d\alpha}{\int_0^{2\pi} P(\alpha) \, d\alpha}$$

This method was applied to a lysozyme-inhibitor complex. The Sim weighted Fourier showed greater continuity and greater contrast in the region of inhibitor binding than the corresponding unweighted F_{PI} synthesis. Evidently this method of weighting deserves further investigation for the analysis of protein inhibitor complexes.

15

REFINEMENT AND IMPROVEMENT OF RESOLUTION

15.1 INTRODUCTION

In Chapter 13 we discussed the quality of high resolution electron density maps calculated using phases from the method of multiple isomorphous replacement. These electron density maps are usually interpreted by comparison with polypeptide models of standard bondlengths and angles and variable dihedral angles. This comparison is achieved either by computer graphics or by using wire models and a half-silvered mirror to accomplish the superposition of the images of the model and map as described in Section 13.6. Protein models obtained in this way have often represented the last stage in the X-ray analysis. They indicate the general secondary and tertiary structure of the protein and have produced important changes in our thinking about enzymes, protein hormones, carrier proteins and immunoglobulins.

Nevertheless it is becoming increasingly clear that such models are far from satisfactory. In order to provide answers to many of the important biochemical questions we require a detailed knowledge of hydrogen bonds and other stabilizing interactions, and also a precise description of distortions from the geometry of "small molecules" giving rise to strain energy in the protein. For instance, the redox potential of the small iron protein rubredoxin which is important in electron transport may well depend critically on strain in the iron sulphur complex and an understanding of this depends on precise estimates of the iron sulphur bondlengths and angles. The inhibitory action of the protein inhibitors of trypsin rest on a precise fit of the two proteins and the interaction probably leads to small distortions of bond angles. A description of these small structural distortions requires an accuracy of the atomic positions better than 0.1 Å. However, the atomic coordinates derived from the electron density maps will be on average at least 0.3 Å – and probably 0.5 Å – in error. Clearly the protein crystallographer must improve his model.

There are several deficiencies in the crystallographic analysis: (1) The phases determined from multiple isomorphous replacement and anomalous scattering contain both random errors due to inaccuracies in the measurements and

15.2. TECHNIQUES AVAILABLE FOR THE REFINEMENT

non-random errors due to lack of isomorphism of the native and derivative protein crystals. (2) The phases may not have been determined by multiple isomorphous replacement for $d < 2$ Å although the observable data of the native protein may extend to a much higher resolution. (3) The model may not represent the best fit to the calculated electron density.

Thus there are opportunities for the protein crystallographer to refine the protein model: to refine the phases, to extend them to a higher resolution and to ensure that the model is the best fit to the calculated electron density. The studies on lysozyme (Diamond, 1974), on rubredoxin (Watenpaugh et al., 1973; Sayre, 1974), on bovine pancreatic trypsin inhibitor (Deisenhofer and Steigemann, 1974) on the porcine trypsin inhibitor–trypsin complex (Huber et al., 1974) and on high potential iron protein (Freer et al., 1974) have demonstrated the effectiveness of various techniques of refinement. Important biochemical information has emerged as a result of the studies. In this chapter we will discuss the crystallographic techniques which have been successful in the refinement of these proteins.

15.2 TECHNIQUES AVAILABLE FOR THE REFINEMENT OF PROTEIN MODELS

There are in principle many ways in which the protein model can be refined. Most of these are techniques which have proved successful in the structure analyses of much smaller molecules. It has long been evident that some of these methods are difficult to apply to proteins due to the magnitude of the computations. However, many of them also involve mathematical relationships which are not strictly correct for protein X-ray data at spacings of $d > 2$ Å. The problem of refinement of the protein model must then be approached with caution, and the limitations of the method carefully defined by practice.

The most cautious approach, and probably the best in the initial cycles of the refinement, is to treat both the protein phases, $\alpha_{P(calc)}$, calculated from isomorphous replacement and the structure factor amplitudes, F_P, as observables. The problem is then to adjust the model to the observed electron density. However, the protein phases may be later calculated from the atomic positions and used in place of those from isomorphous replacement. It is now accepted that this leads to the best protein model, although at first it was not clear that phase calculated from atomic positions could ever be better estimates than those from multiple isomorphous replacement. The calculation of phases in this way allows not only improvement of the phases already estimated but also extension of the phasing to higher resolution.

These methods involve some knowledge of the chemistry of the protein, and depend critically, especially in the early stages of analysis, on the correctness of the interpretation of the electron density. They are therefore "indirect methods". We have seen in Chapter 5 that as every part of the molecular structure contributes to each part of its transform, mathematical relationships can be expressed between different structure factors. These relationships may be used to refine values of the estimated phases and to extend the phasing to other structure amplitudes which have been measured. These are "direct methods" as they do not depend on a chemical interpretation. For this reason they may be advantageous in the refinement of the electron density.

Direct methods ultimately depend on certain assumptions about the electron density of the molecule (see Section 5.14). The electron density should be non-negative and comprise randomly distributed but resolved peaks of similar size. These assumptions are most commonly used to give mathematical relations in reciprocal space, but they have their real space equivalents (involving modification of the electron density followed by a Fourier transformation) and these turn out to have many computational advantages. For larger proteins at high resolution reciprocal space calculations become impossible even on the largest of modern computers as they increase in size with the *square* of the number of reflections – they show a dependence on n^2 – whereas real space calculations increase linearly with the number of reflections.

Similarly the refinement of atomic positions may be made in reciprocal space against the structure factor amplitudes or in real space against the electron density. Again reciprocal space operations become very consuming in computer time for larger proteins at high resolution, but they do have some advantages which may make them preferable in some cases.

In conclusion it can be seen that we may distinguish methods of refinement in terms of three criteria: (i) does the method treat α_p as an observable or is it recalculated during refinement? (ii) is the method "direct" involving only mathematical relationships between structure factors or does it depend on an interpretation of the electron density? (iii) is the calculation in real space or in reciprocal space?

We must now consider the advantages and limitations of the methods in more detail.

15.3 REFINEMENT OF ATOMIC POSITIONS

Methods for refinement of the atomic positions are summarized in Table 15.I, due to Diamond (1974b). They are generally categorized as being "real" or "reciprocal" space methods. All may be carried out keeping the phase as an observable and with restraints on the geometry of the model although in the real space refinement this is achieved more easily. We first describe the different methods and then consider how they have been used, often in combination, to refine the structures of some proteins.

15.3.1 Real space least squares refinement

A method for real space refinement of the polypeptide chain has been developed by Diamond (1971, 1974). The object of the method is to find the best fit of a polypeptide to the electron density. Thus the electron density is considered to be an objective expression of experimentally measured amplitudes and phases. The quantity minimized, R, is $(\rho_0 - \rho_m)^2$ where ρ_0 is the observed electron density and ρ_m is the electron density calculated from the model. The model is derived from approximate atomic positions estimated by optical comparison techniques [Richards (1968) as described in Section 13.6] by constructing a polypeptide chain of standard parameters using a mathematical procedure described by Diamond (1966). The method uses the parameters for the 20 different amino acid sidechains and of the peptide link, which have been taken from the results of X-ray analysis of the amino acids and some small peptides. The residues are assembled into a chain which is rigid for all bond lengths and all bond angles except the angle at the α-carbon τ (NCαC) and sometimes the bond angle at Cβ. Free rotation is allowed around all the single bonds. Thus the polypeptide chain is fitted closely to the guide points (approximate atomic positions), by allowing the dihedral angles ψ and ϕ in the main chain, the χ angles defining the sidechains and some bond angles τ to vary. The model building programme proceeds by successively adding one sidechain and one main chain link to the structure already built. These links are first adjusted alone to fit guide points, but they are then adjusted in conjunction with the two preceding amino acids in the polypeptide chain until a good fit for all three amino acids is achieved. The model building procedure does not require estimates for all the atomic positions, and works very well with three or four atomic positions for each residue. Alternative methods for geometrizing the co-ordinate set have recently been described by Dodson et al. (1976) and Ten Eyck et al. (1976).

The electron density $\rho_m(\mathbf{r})$ at a position defined by the vector $\mathbf{r}$ is calculated from this model by the expression,

$$\rho_m(\mathbf{r}) = K \sum_i Z_i a_i^{-3} \exp[-\pi |\mathbf{r} - \mathbf{r}_i|^2 / a_i^2] + \mathbf{d}$$

in which $\mathbf{r}$ is a position vector, $\mathbf{r}_i$ is the position vector of the i, the atom, Z is its atomic number and a_i its apparent radius, assuming Gaussian atoms. K is an overall scale factor and $\mathbf{d}$ is a background level a which adjusts for the omission of $F(000)/V$. The refinement then minimizes the volume integral $\int(\rho_0 - \rho_m)^2$ by varying any or all of K, Z_i, a_i, $\mathbf{r}_i$ and $\mathbf{d}$. The position vectors are adjusted so as to maintain the geometry of the polypeptide and vary only the dihedral angles of single bonds and certain bond angles. The main difficulty in this procedure is that a variation in one dihedral angle implies a rotation of the whole polypeptide chain and not just a local adjustment. This is overcome by working with a "molten zone" of about five amino acid residues at a time. The position vectors

TABLE 15.I
Characteristics of Refinement Methods (Diamond, 1974b)

Attribute	Linked atom real space refinement 1	Free atom gradient/curvature METHOD Whole map 2	Free atom gradient/curvature Difference map 3	Free atom reciprocal space refinement 4
Quantity minimized	$\int(\rho_0 - \rho_m)^2 dv$ $= \frac{1}{v}\Sigma\|\Delta F\|^2$	$\Sigma(\delta r)^2$ $= \frac{1}{v}\Sigma \hat{f}^{-1}\|\Delta F\|^2$	$\Sigma(\delta r)^2$ $= \frac{1}{v}\Sigma f^{-1}\|\Delta F\|^2$	$\Sigma\sigma^{-2}\Delta\|F\|^2$ or $\Sigma\sigma^{-2}\|\Delta F\|^2$
Weights on reflections	1	$\hat{f}^{-1}$	f^{-1}	σ^{-2}
Fits to	volume	point	point	primary data
Phases	fixed†	fixed†	fixed†	free or weighted
Minimum R factor	$1 - \bar{m}$	$1 - \bar{m}$	$1 - \bar{m}$	$\sigma/\bar{F}$
Derivatives : atoms	linear	linear	linear	quadratic*
Allowance for overlap	yes	difficult	more difficult	easy
Curvature and weight	refinable	apparent	obscure	refinable

					no[‡]
Proper stereochemistry		yes	no	no	
Parameters/atom	α	.75, 1.75 or 2.75	3 or 4	3 or 4	3 or 4
Observations/parameter (based on $r = 2\text{Å}$, $v = 30\text{Å}^3$)	$\beta = 1$	10.5, 4.5 or 2.8	2.6 or 1.9	2.6 or 1.9	2.6 or 1.9
	$\beta = 2$	21.0, 9.0 or 5.6	5.2 or 3.8	5.2 or 3.8	5.2 or 3.8
Convergence radius		$2r$	r	r	$r/4$?
Final error signal		vanishes	vanishes	$\neq 0$	vanishes
Estimation of coordinate errors		difficult	easy but unreliable	unreliable	easy and reliable
Mistake detection		fair	fair	excellent	poor
Space group generality		general	general	general	difficult to generalize
Speed		medium	fast	fast	slow

*New techniques involving fast Fourier transforms may make reciprocal space refinement have a less than quadratic dependence. Thus the speeds of the methods may become more comparable than indicated in the bottom line of the table.
†If the calculated electron density is transformed, a new set of phases may be calculated for methods 1, 2 and 3 (see text, page 426).
‡Hendrickson has recently refined haemerythrin using geometric restraints, which are much less expensive than rigid body refinement in reciprocal space.

TABLE 15.II
Table 15.II. Real space refinement of lysozyme—angles refined and stiffness parameters used

Angles	ϕ, ψ	$\chi_{1,2,3,4}$	τ†	ω	χ_5	$\theta_{1,2,3}$‡
Constants w	3.7	3.2	0.33	0.50	0.25	0.10

† $\tau(N-C^a-C)$ all residues
$\tau(C^a-C^\beta-C^\gamma)$ in Phe, His, Tyr, Trp
$\left. \begin{array}{l} \tau(N-C^a-C^\beta) \\ \tau(C-C^a-C^\beta) \end{array} \right\}$ in cystine

‡ $\left. \begin{array}{l} \theta_1 = \theta(C_{i-1}-C_i^a-N_i-C_i^\beta) \\ \theta_2 = \theta(C^a-N-C^\beta-C^\delta) \\ \theta_3 = \theta(N-C^\beta-C^\delta-C^\gamma) \end{array} \right\}$ in proline

for these atoms are varied in such a way as to vary the angles in this part of the chain but leave the positions of the extremes of the "zone" unchanged. The "molten zone" is then moved along the polypeptide chain and further adjustments are made. The programme meets difficulties when the electron density is continuous between adjacent polypeptide chains which are widely separated in the primary structure. For cystine disulphide bridges this is overcome by refining the bridge twice, once for each half-cystine. The difference between the two sets of refined values gives some indication of the errors in the refined parameters.

Problems tend to arise where the electron density is weak for a sidechain. The occupancies and radii must then be carefully refined to avoid the sidechains invading the electron density of other sidechains.

The radii of adjacent atoms will in general increase towards the end of the sidechain. Deisenhofer and Steigemann (1974) suggest making the radii increase linearly with the distance X_i from Cα according to $a_i = a_0 + kX_i$ where k is determined for each side chain by a least squares procedure. Diamond uses a bond stiffness parameter in the refinement of the parameters. Table 15.II indicates the angles refined and the parameters used. The larger the value of w the less stiff is the variation of that angle. The ϕ, ψ and χ angles are thus most readily varied. There is also a facility to filter out large alterations in the structure which result in little gain in the fit to the electron density, but in later cycles of refinement lowering the filter level may serve to identify some major changes required in the model.

Although Diamond's procedure considers both the amplitudes and phases to be observables, the electron density can be one calculated using phases either from multiple isomorphous replacement or from Fourier transformation of the protein model. The reader should refer to Diamond (1974) for a description of the refinement of lysozyme using the phases from multiple isomorphous

15.3.2. REAL SPACE GRADIENT – CURVATURE REFINEMENT

replacement and to Deisenhofer and Steigemann (1974) and Huber et al., (1974) for refinement of the trypsin inhibitor and the trypsin–trypsin inhibitor against phases calculated from the protein model. In the latter case refinement is carried out against the electron density map calculated from the Fourier-summation with coefficients

$$\{nF_P(\text{obs}) - (n-1)F_P(\text{calc})\} \exp(i\alpha(\text{calc}))$$

If $n = 1$, the map calculated will have all those atoms not included in the calculation of $\alpha(\text{calc})$ with a maximum of half the correct electron density. It is therefore preferable to use an electron density map where $n = 2$, in which the atoms not included in the Fourier summation should be of more equivalent weight to the others. (see Section 14.5.)

15.3.2 Real space gradient – curvature refinement

An alternative technique for real space refinement is to use the ratio of the gradient to the curvature of the electron density at a particular point as an estimate of the shift required.

The shift vector x_i is given by $x_i = (d^2\rho/dx_i dx_j)^{-1} (d\rho/dx_i)$. Methods based on this assumption have been widely used in the refinement of structures of small molecules (Lipson and Cochran, 1966). Unfortunately they tend to give extra weight to the higher angle reflections which are usually much less well phased for a protein (Diamond, 1974).

In order to calculate the gradient $d\rho/dx_1$ it is necessary to interpolate the electron density and so the electron density must be calculated on as fine a grid as possible. The curvature $d^2\rho/dx_i dx_j$ for any atom depends on the resolution of the electron density map. It is quite difficult to calculate even at 1.5Å because of the lack of resolution of adjacent atoms. It is usually sufficient to estimate the curvature of a representative atom and use this to estimate curvatures for other atoms. As in Diamond's real space refinement, a Gaussian distribution is usually assumed according to the equation

$$\rho = Ae^{-pr^2}$$

where A is the peak density, r is the distance from the atomic centre and p is a constant to be determined. In the work on rubredoxin the constant p was determined from a carbonyl oxygen. The curvatures for sulphur, oxygen, nitrogen and carbon atoms were taken in the ratio 12:6:5:4, and were assumed equal for all atoms of a given type. Alternatively the curvature can be estimated by differentiation of the model density.

The gradient/curvature technique can be used either with Fourier or difference Fourier maps. On the whole it has been used with difference maps as the real space least squares refinement of Diamond offers considerable advantages with ordinary Fourier maps. In the first place the Diamond method requires no interpolation and does not depend critically on the high angle reflections. The gradient/curvature method is not useful if the atom is rather inaccurately positioned for this method involves a point to point fitting. It will only be useful if the position is within an atomic radius of the correct peak; in any case the estimated shift will be rather inaccurate if the initial position is on the fringe of the electron density peak where the curvature has the opposite sign to that at the centre (Diamond, 1971). On the other hand the Diamond method involved a volume fitting approach and so convergence is to be expected provided some part of the calculated density overlaps with the observed density. The convergence radius is thus twice the value for gradient/curvature techniques.

In all real space methods the calculation of the standard deviations of the atomic parameters is difficult. One procedure is to use the expression given by Cruickshank (1949) for a noncentrosymmetric structure:

$$\sigma(x) = 4\pi(\Sigma h^2 (F_P(\text{obs}) - F_P(\text{calc})))^{1/2} / aV |C|$$

where a is the length of the unit cell in the x-direction, V is the volume of the unit cell, and C is the central curvature of the electron density at the peak position (Deisenhofer and Steigemann, 1974). There are of course equivalent expressions valid for $\sigma(y)$ and $\sigma(z)$. The curvature is calculated in the way described for calculating the shifts. This estimation of the standard deviation can be useful for Diamond's method also. However, in this case it is not strictly valid because this is volume fitting rather than point to point fitting and also the positional errors of adjacent atoms will be highly correlated. An alternative estimate of the error is given by a plot of the R-value against $\sin\phi/\lambda$ (Fermi, 1975) as suggested by Luzzati (1952) and illustrated in Fig. 5.21, but this overestimates the errors. Thus Luzzati and Cruickshank estimates of error may be regarded as upper and lower limits respectively.

15.3.3 Difference Fouriers in refinement

Although the use of difference Fouriers in refinement has the disadvantages discussed above, difference Fouriers are very useful in identifying sidechains which are quite incorrectly placed (i.e. positional errors greater than two atomic radii) and in finding positions of the solvent atoms.

The calculation of structure factors from a protein model for use in difference Fouriers must be approached carefully. The use of the atomic positions calculated from the optical comparison techniques and an overall thermal parameter from a Wilson plot usually gives rise to a value of the residual

$$R = \frac{\Sigma |F_P(\text{obs}) - F_P(\text{calc})|}{\Sigma F_P(\text{obs})}$$

in the region of 0.4. This was the case for initial structure factor calculations with myoglobin (North and Phillips 1968), lysozyme (Blake *et al.*, 1967), carboxypeptidase (Lipscomb *et al.*, 1970) and rubredoxin (Watenpaugh *et al.*,

15.3.4. RECIPROCAL-SPACE LEAST-SQUARES REFINEMENT

1971). Lipscomb *et al.*, (1970) found that the R factor had a value of 0.866 for carboxypeptidase for the data in the 10 to 6Å resolution range. However in the range 5 to 2Å the R factor was remarkably independent of the resolution and has a value of approximately 0.36. In a similar way the R factor for rubredoxin was found to rise steeply for $d > 10$Å. The poor agreement between calculated and observed structure factor amplitudes at low resolution is mainly due to the lack of inclusion of the solvent. This is rather disordered and will contribute mainly to the very low angle data. It is best to omit these data ($d > 10$Å) from the difference Fourier calculations. A further error in the calculated structure factors often arises from incorrect thermal parameters. A plot of the ratio $\Sigma F_P(\text{obs})/\Sigma F_P(\text{calc})$ against $\sin \theta$ should give an indication of a gross error in the overall thermal parameter. Measurement of the electron density at the different residue positions also can give an indication of the relative thermal parameters. It is generally found that the thermal parameters increase towards the surface of the protein. Main chain thermal parameters of atoms near the chain termini and in non-hydrogen bonded structures tend to be higher than those of atoms in α-helices or β-sheet.

Watenpaugh *et al.*, (1973) found that difference Fouriers could be used in conjunction with the gradient/curvature technique to give good predictions of shifts to about 1Å for 1.5Å resolution data of rubredoxin. For large errors no shift was indicated but usually a negative peak occurred at the erroneous site equal in electron density to the peak which appeared at the correct atomic position. Gross errors are easy to identify; of all the refinement methods, difference Fouriers offer the most straightforward identification of mistakes. All successful refinements of protein structures have used difference Fouriers at some stage in the analysis.

15.3.4 Reciprocal-space least-squares refinement of positional parameters

Most small molecular structures are refined using a least squares technique which fits the primary data by minimizing

$$\sum_{h} w / F(\text{obs}) - F(\text{calc})/^2$$

where w is a weight which is often taken as the inverse of the variance. The expression is minimized with respect to the overall scale, and the atomic positional and thermal parameters. The refinement is usually initiated when a satisfactory agreement between the calculated and observed structure factors has been obtained i.e. $R(= \Sigma_h |F(\text{obs}) - F(\text{calc})| / \Sigma_h F(\text{obs}))$ usually $\approx 0.2 - 0.3$. The number of observed structure factors usually exceeds the number of parameters which are to be determined by a factor of 10; the structure is highly overdetermined. As the structure factor amplitudes are rather inaccurately estimated in many cases, it is advisable that their number should greatly exceed the number of positional and thermal parameters needed to define the structure.

For protein structures, there are many genuine worries about this method of refinement. The R value based on the initial coordinates is usually ≈ 0.4 and the structure is underdetermined. For instance for rubredoxin (M.W. ≈ 6100) there are over 400 atoms in the protein and a further ≈ 100 atoms of ordered or semi-ordered solvent atoms in the asymmetric unit. At $d < 1.5$Å there are 7345 observed structure factor amplitudes, only 5033 of which have $I > 2\sigma$. For each atom, there are three positional parameters and one isotropic thermal parameter so that there are a total of more than 2000 parameters to be determined, but only ≈ 5000 observables. At 2Å resolution, the number of structure factors will normally be close to the number of parameters. For these reasons the refinement technique was considered of doubtful use for most proteins certainly for $d \geqslant 2$Å although the refinement of rubredoxin at 1.5Å has shown that many of the worries were not fully justified.

The method also presents considerable computational difficulties. Because of the magnitude of the problem it is impossible to take advantage of full matrix least squares (see Section 5.15). With 1.5Å resolution data, which is not the atomic resolution for most atoms in the protein, the positions of covalently bound atoms are highly correlated and block diagonal matrix least squares is clearly dangerous. The best compromise would seem to be that adopted by Watenpaugh et al., (1971, 1973) for rubredoxin. They partitioned the full matrix into blocks of 200 parameters; one cycle of least squares needed about 10 passes through the computer for rubredoxin and in each pass the parameters in one block were varied by full matrix least squares. The parameters of covalently bound atoms were blocked together, and the blocks were overlapped to introduce some correlation between atoms in different blocks which are covalently bonded.

Even with partitioning of the matrix the computations are extremely time consuming. One cycle of refinement for all the atomic parameters may take almost ten times as long as the calculation of a difference electron density map. On a CDC 6400 computer one cycle took Watenpaugh et al., (1973) as long as 15–18 hours. The number of derivatives that must be calculated depends on the product of the number of atoms, N, and the number of reflections and the number rises as N^2 for a given resolution. However, D. Sayre has recently suggested that fast Fourier transform methods can make the dependence less than N^2.

Refinement of rigid groups appears to be computationally expensive in reciprocal space (Diamond, 1974b) although Hendrickson (1976, unpublished results) has shown that geometric restraints (as against constraints) may be incorporated into the refinement. A possibility for minimizing the number of parameters to be varied is to link the thermal parameters of adjacent atoms. However this technique has the serious disadvantage of making identification of incorrectly placed atoms very difficult. The thermal parameter of incorrectly placed atoms usually increases sharply. This "spreading out" of the atom density often allows the calculated atomic position to converge on the correct position in a way that is not possible if the thermal parameter is tied to that of the other atoms.

Reciprocal space least squares refinement has three very definite advantages over its real space equivalent. First, the reciprocal space technique can be arranged to refine either against $F_P(\text{obs}) \exp(i\alpha_P)$ where α_P can be a phase calculated from isomorphous replacement or against $F_P(\text{obs})$. In the real space refinement the phases are always treated as observable quantities. Secondly, in reciprocal space the observations, $F(\text{obs})$, may be weighted according to their reliability whereas in real space the electron density is assumed everywhere equally reliable. Thirdly, reciprocal space refinement allows a direct and reliable calculation of the standard deviations of all the parameters from the inverse matrix. Watenpaugh et al., (1973) have found the estimates of the standard deviations given by the inverse matrix agree very well with the standard deviation calculated from the distribution of calculated lengths of equivalent bondlengths.

15.3.5 Energy refinement

The constrained refinement method of Diamond produces an atomic model with standard values of bondlengths and most bond angles but variable dihedral angles. The variation in the dihedral angles may lead to interatomic contacts between non-covalently bound atoms which are physically impossible because of the large repulsion energy which would result. In order to minimize these unlikely conformations energy refinement can be carried out.

The most useful approach to this problem is that of Levitt and Lifson (1969). They use a simple empirical function that is supposed to have a minimum value when the coordinates are refined. The function used is:

$$E = \sum_{\text{bonds}} \tfrac{1}{2} K_b (b_i - b^0)^2 + \sum_{\substack{\text{bond} \\ \text{angles}}} \tfrac{1}{2} K_\tau (\tau_i - \tau^0)^2$$

$$+ \sum_{\substack{\text{torsion} \\ \text{angles}}} K_0 \{1 + \cos(m\theta_i + \delta)\} + \sum_{\substack{\text{non-bonded} \\ \text{distances} < R \text{ max}}} \{A/r_i^{12} + B/r_i^6\}$$

where K_b and K_τ are force constants and b^0 and τ^0 are equilibrium bond lengths and bond angles respectively. K_0 is the torsion barrier, m the periodicity of the barrier and δ the phase. A and B are repulsive and long range non-bonded parameters respectively. The energy function is minimized by shifting Cartesian coordinates in the direction of steepest descent. The bond stretching and bond-angle bending force constants are derived from an analysis of the vibrational spectra of small molecules. The torsion constants are derived from theoretical calculations.

The use of this energy refinement technique on lysozyme showed that it improved the initial fit of the model to the electron density but gave less good agreement than the Diamond real space refinement (Levitt, 1974).

A similar technique applied to rubredoxin led to an increase of the R value (Rasse *et al.*, 1974). The method is probably most useful when used in conjunction with the real space refinement as a constant check against unfavourable interatomic interactions (Levitt, 1974). It may also be useful in distributing the distortion due to strain throughout the structure rather than letting the distortions from ideal geometry accumulate in $\tau(N-C\alpha-C_1)$ and the peptide dihedral angle, ω.

15.4 EXAMPLES OF REFINEMENT OF PROTEIN STRUCTURES

Given that the various techniques in refinement of protein structures all have certain relative advantages as well as rather distinct limitations we must now consider combinations of these which constitute a sensible procedure for improving the model. There are three procedures which have been used successfully and the flow diagrams for these are illustrated in Figs. 15.1, 15.2 and 15.3. The first of these has been devised by Jensen and his coworkers (Watenpaugh *et al.*, 1971, 1973) for the refinement of rubredoxin. It is a free atom refinement using gradient/curvature techniques with difference maps followed by reciprocal space least squares refinement. The second procedure has been used by Kraut and his coworkers to refine high potential iron protein (Freer *et al.*, 1974). It is a constrained atom refinement using the Diamond model building procedure and gradient/curvature techniques on difference maps. The third procedure is due to Huber, Deisenhofer and Steigemann and has been

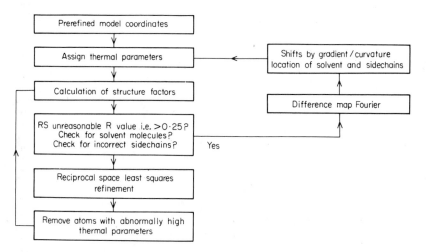

FIG. 15.1 Flow diagram for refinement of the atomic positions of a protein structure using unlinked atom reciprocal space refinement.

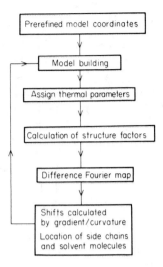

FIG. 15.2 Flow diagram for refinement of the atomic positions of a protein structure using the gradient/curvature technique and model building to constrain the geometry of the model.

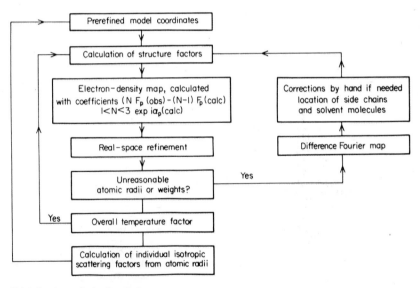

F_p(obs) = observed structure factor
F_p(calc) = calculated structure factor
α_p(calc) = calculated phase angle

FIG. 15.3 Flow diagram for refinement of the atomic positions of a protein structure using real space refinement and calculation of electron density. (After Deisenhofer and Steigemann, 1974).

used to refine the trypsin inhibitor structure (Deisenhofer and Steigemann, 1974) and the trypsin–trypsin inhibitor complex (Huber *et al.*, 1974). The method uses the Diamond real space refinement in conjunction with electron density maps calculated with the phases of the approximate protein model and difference maps to identify errors. We will consider here in more detail the refinements of rubredoxin and the trypsin inhibitor as examples of free atom and constrained atom refinements.

15.4.1 The refinement of rubredoxin

The native protein data for rubredoxin (MW 6100) were collected on a crystal of 0.7 x 0.7 x 0.35 mm. Watenpaugh *et al.*, (1971, 1973) decided to collect all the data (one equivalent only) to 1.5Å on one crystal during a period of a week in order to minimize the systematic errors due to radiation damage and scaling together different crystal data. The refinement was carried out using 5033 reflections with $I > 2\sigma$. The initial structure factors at $d = 1.5$Å were calculated using the positions of 401 protein atoms and 23 water atoms derived from the 2Å electron density map calculated from phases determined by isomorphous replacement. This indicated that the overall temperature factor should be adjusted to $B = 12$Å^2 and the terms with $d > 17$Å removed. This led to an R value of 0.372. The initial difference Fourier indicated shifts which were calculated by hand using the gradient/curvature technique as described in Section 15.3.2, and these reduced the R factor to 0.321. Subsequent difference Fouriers revealed the positions of 127 water molecules; oxygen atoms representing water were included only if peaks were present in the 2Å isomorphous phased Fourier as well as in the difference Fourier. Occupancy factors were made proportional to the peak height. Shifts from four successive difference Fourier calculations led to an R value of 0.224.

Three cycles of least squares refinement with the matrix partitioned as described in Section 15.3.4 were carried out. Despite the fact that the number of positional and isotropic thermal parameters was exceeded by the number of structure factors by a factor of about two, there was little tendency for the parameters to oscillate or diverge. The B values for atoms deep inside the molecule were in the range 4–8Å^2, not much greater than that found in small molecules. Toward and on the outside of the molecules, B values of 15–25Å^2 were common with a few ranging up to 40Å^2 or more. Abnormally high thermal parameters were associated with incorrectly placed atoms or with surface residues which were disordered. No constraints were placed on bond lengths and the standard deviations of the C, N and O atoms were in the range of 0.1 to 0.2Å when the R factor had fallen to 0.132. The atomic positions were based on a sequence derived from the electron density map. A few of the amino acid sidechains (see Section 13.8) were incorrectly identified and these gave rise to difficulties in the refinement. Finally the hydrogen atoms were introduced in the

15.4.2. REFINEMENT OF THE TRYPSIN INHIBITOR

structure factor calculation. A further refinement of the other atomic parameters led to a final R value of 0.126. The R value showed a strong dependence on the resolution, being lowest at 4Å and highest at low angle. The solvent continuum did not appear to contribute to reflections with $d < 10$Å.

The refinement of rubredoxin established that the least squares technique in reciprocal space was valid for proteins at $d = 1.5$Å and that the phases calculated from atomic positions can be better estimates than those from multiple isomorphous replacement. Information was obtained which was not available from the 2Å electron density map calculated with phases from multiple isomorphous replacement. Further refinement has since been carried out at $d = 1.2$ Å.

15.4.2 Refinement of the trypsin inhibitor

The collection of the native data for the crystals of pancreatic trypsin inhibitor was carried out using a rather different philosophy from that of rubredoxin (Deisenhofer and Steigemann, 1974). The data to $d = 1.5$Å were measured from nine crystals. The measurements were made for times inversely proportional to the peak height with a maximum scan time of ten minutes. About 15,000 measurements were merged to give 8079 unique reflections, of which 6988 were of significant intensity to $d = 1.5$Å. Thus precision in measurement was emphasized. The concomitant introduction of systematic errors due to radiation damage and intercrystal scaling were minimized by monitoring the fall-off after every fifteen reflections and by measuring sets of data at all resolutions on each crystal in order to establish not only relative scale factors but also relative thermal parameters.

The starting point for the refinement was the atomic coordinates derived from an electron density map at 1.9Å phased by the method of isomorphous replacement. A structure of standard geometry was derived using the Diamond model building procedure and this was refined using the real space refinement of Diamond (see page 423). This reduced the R value from 0.52 to 0.45. The refined model allowed the calculation of an electron density map with coefficients

$$\{nF_P(\text{obs}) - (n-1) F_P(\text{calc})\} \exp(i\alpha_P(\text{calc}))$$

where n was usually 2. This electron density map was then used for further Diamond real space refinement. The flow diagram is shown in Fig. 15.3. Fifteen cycles were completed to bring the R value to 0.225 and to $R = 0.197$ when the solvent dependent reflections with $d > 7$Å were omitted.

During this process five difference Fourier maps were calculated and these were used to identify incorrect sidechain positions and solvent molecules. The final difference map showed very few peaks most of which appeared to relate to incorrect assignment of thermal parameters. In the final cycle there were 1041 parameters, namely: 358 conformational angles, three degrees of freedom each for translation and rotation of the whole molecule, 489 thermal parameters, 141 positional parameters and 47 occupancies of solvent molecules. The ratio of observations to parameters was never less than 6.7, a situation which is more theoretically justifiable than that of the reciprocal space method. In fact the errors due to the incorrectness of the model were allowed for by making angles $\tau(N-C\alpha-C_1)$ flexible.

The higher R factor for trypsin inhibitor compared to rubredoxin may derive from several factors. These include the systematic errors due to incorrect scaling of the nine different crystals. However, the major contribution must be the constraints placed on the geometry of the molecule. (R values tend to decrease with an increase in the number of variables in the refinement).

The refined model allowed a better description of the trypsin inhibitor molecular structure. It established the existence of a number of hydrogen bonds and non-planar peptides. In particular the susceptible Lys 15–Ala 16 bond which is cleaved in the complex appears non-planar, and the carbonyl carbon is shifted towards the position that it would occupy were it tetrahedrally coordinated as is observed in the complex with trypsin. The refinement has clearly led to important new information which was not available from the electron density map calculated using phases from isomorphous replacement. The method has the very great advantage that it is useful for much larger proteins and has already been effectively used for the trypsin–trypsin inhibitor complex (MW ~ 30,000).

15.5 DIRECT METHODS OF REFINEMENT OF PROTEIN PHASES

We have shown in Section 5.14 that the normalized structure factor for the squared structure E'_h is given by

$$E'_h = K_1 \sum_k E_k E_{h-k}$$

When the structure contains only non-negative electron density representing equal and resolved atoms, $E_h = K_2 E'_h$, we may write

$$E_h = K_1 K_2 \sum_k E_k E_{h-k}$$

which is essentially the same as the equation derived by Sayre (1952).

The Sayre equation applies when the summation is over all reciprocal space to a resolution which is better than atomic resolution. It would in principle represent a useful system for extending a phase set or for phase refinement of protein data. Unfortunately protein crystals are usually comparatively disordered and the data do not extend to atomic resolution. It is then interesting to ask whether the Sayre equation is of any use in the refinement of protein phases.

Electron density maps at resolutions with $d > 2$Å have features which indicate that the Sayre equation may not be valid. First they may contain negative regions (due to series termination) and therefore the non-negativity criterion may not be a good approximation. Secondly the atoms are not resolved. In these circumstances the structure factor for the squared structure given by the Sayre equation will not be equal to that of the structure itself. Negative regions in the electron density map will give rise to positive peaks in the squared structure. Further the overlapping electron density of the unresolved atoms will give rise to peaks in the squared structure which are of varying heights emphasizing parts of the structure such as α-helices and aromatic sidechains. Of course the phase of the squared structure is usually combined with the observed structure factor amplitude and so the resulting Fourier is not exactly equivalent to squaring the structure. Nevertheless, it will have many of the features of the squared structure and at low resolutions (as with the small molecule heavy atom structures) the recycling of phases through the Sayre equation usually leads to a divergence of the phase set from the correct phases, corresponding to a building of electron dense parts of the structure.

15.5.1 Use of the tangent formula

Most of the first experiments with protein data used the tangent formula (see Section 5.14):

$$\tan \varphi_k = \frac{\sum_k |E_k E_{h-k}| \sin(\varphi_k + \varphi_{h-k})}{\sum_k |E_k E_{h-k}| \cos(\varphi_k + \varphi_{h-k})}$$

to estimate the phases of structure factors. (Coulter, 1971; Weinzierl, et al., 1969, Reeke and Lipscomb, 1969). The structure factors in the R.H.S. of the equation can be weighted with the figure of merit.

Weinzierl et al., (1969) have made some trial calculations at 4 Å and 5 Å on a model protein having structure factors calculated from the backbone chain of myoglobin out to and including the β-carbon atoms. The mean error for 50 largest E values was 6°. However, error for the calculated phases increased to about 20° for E values approaching unity at 4 Å resolution and the error is significantly greater for data cut off at 5 Å resolution. Phases included with an initial random mean error of 56° were improved in the first two cycles of refinement but then diverged to a self-consistent, incorrect set. Calculations with carboxypeptidase (Reeke and Lipscomb 1969) at resolutions between 6 Å and 2 Å also showed an initial improvement of phases in the first cycles of refinement followed by deterioration. However, in contrast, Reeke and Lipscomb found that improvement of the phases was greatest at low resolution.

Examination of insulin electron density maps at 3 Å based on consecutive cycles of tangent formula refinement (Blundell, unpublished results) showed that a clearing up and sharpening of the maps accompanied the improvement of phases in the first cycle but subsequent cycles caused electron density to pile up in places corresponding to large electron density peaks in the original map. This effect is similar to the increase of electron density at the haem and cysteine sulphurs observed by Wienzierl et al., (1969) in the refinement of cytochrome c.

These trial experiments clearly demonstrate that the tangent formula with data of $d > 2$ Å converges to a self-consistent set of structure factors which is not a helpful approximation to the correct set. At this resolution there is no unequivocal evidence that the tangent formula has introduced new structural information into the electron density map. Its most obvious contribution is to clean up the electron density map by removing much of the spurious *low* electron density. On the other hand the technique seems unable to remove *large* spurious peaks at this resolution; on the contrary, it tends to build up their density probably due to contributions from the squared structure.

One approach to minimizing the effect of the squared structure is to include a contribution from the cubed structure which can have negative regions (Hoppe and Gassmann, 1968, Gassmann and Zechmeister, 1972). This is the basis of their "phase correction" method which has been used with some success to extend the phases of myoglobin.

Despite the limitations of the tangent formula at spacing of $d > 2$ Å, the method does seem to be more useful with data of close to atomic resolution. Phase extension from 2 Å to 1.5 Å appears to have given an electron density map which is resolved properly into atomic peaks. Phases generated with the tangent formula could in this situation be used in combination with estimates from multiple isomorphous replacement (Hendrickson and Karle, 1973) for resolving the ambiguity in phase determination by single isomorphous replacement.

15.5.2 Real space direct methods

At very high resolutions the reciprocal space calculations become very expensive in computing time. Here real space techniques become advantageous. Even with conventional Fourier programmes in three dimensions, the Beevers–Lipson procedure has less than an N^2 dependence. However, with a Cooley–Tukey "fast Fourier" algorithm (Cooley and Tukey, 1965; Zwick, 1968), there are much greater savings (Barrett and Zwick, 1971). The simplest real space equivalent is transformation of the electron density with negative regions equated to zero. This is an expression of the non-negativity criterion and has been shown by Kartha (1969) to be useful for small molecules. Alternatively the electron density can be squared after minimum-level truncation, then the

15.5.3. SAYRE'S METHOD

structure factor is given by

$$E_h = C_h T^{-1}[\{T(E_k)\}^2]$$

The squaring of the structure introduces a degree of atomaticity in to the calculation (Barrett and Zwick, 1971). This method is closely related to but not analytically equivalent to the tangent formula. Both the transformations can be carried out using the Cooley–Turkey algorithm. It is important to sample the electron density map more finely than its transform in order to yield accurate structure factors. Lipson and Cochran (1968) suggest sampling the electron density at three times the highest index observed in that direction.

Apart from the advantages of speed in the computations the real space approach has the advantage of allowing more simple modification to the electron density. The electron density can be truncated at a maximum value to avoid the build up of electron density (Hoppe et al., 1970) or a contribution from the cubed structure subtracted from the squared structure (Collins et al., 1975).

15.5.3 Sayre's method

The techniques described above use only half the information given by the Sayre equation. They use the estimate of the phase, but ignore the estimate of the structure factor amplitude. Sayre (1972, 1974) has demonstrated that many of the problems inherent in recycling the phases with the tangent formula can be overcome by constraining $E_k E_{h-k}$ to approach E_h in both magnitude and phase. To achieve this Sayre (1972) suggests minimizing the quantity $\Sigma_h \mid a_h F_h - \Sigma_k F_k F_{h-k} \mid$ as a function of the phases, by least squares.

Sayre has carried out tests on model structures (Sayre, 1972) and on the small protein rubredoxin (Sayre, 1974). Initial values of the phases of spacings 2.5 Å $> d >$ 1.5 Å are estimated using the Sayre equation (p. 142). Only structure factors with intensities greater than 2σ need be used. An estimate of the term $F_{(000)}$ is added into the calculations and data processed to correspond to Gaussian atoms of shape $\exp(-4r^2)$ rather than point atoms as is the usual procedure with the tangent formula. The value of a_h in the Sayre formula is chosen to reduce the effect of omission of data with $d <$ 1.5 Å. a_h is given by an empirical factor of the form $p - qR - rR^2$ where p, q, and r depend on the amount and type of data incompleteness, but are independent of the structure.

For rubredoxin eleven cycles of least squares refinement were carried out and the refined phases were used to calculate an electron density map at 1.5 Å resolution. Figure 15.4 shows sections of the electron density map calculated with the direct method phases, with the phases obtained by difference maps and refinement by Watenpaugh et al. (1973) (see Section 15.4.1) and the 2.5 Å isomorphous replacement phases. The quality of the direct-method map is not

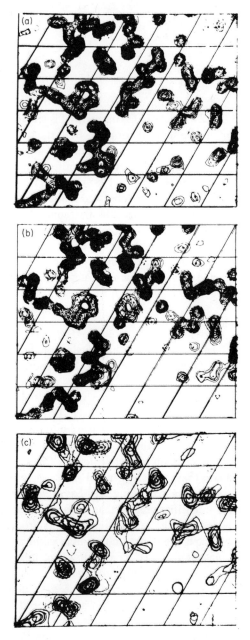

FIG. 15.4. Refinement of rubredoxin electron density using Sayre's direct method: (a) the electron density map at 1.5 Å calculated using phases calculated by Sayre's method; (b) the electron density map at 1.5 Å calculated using phases obtained from refinement by Watenpaugh et al. (1973); and (c) the electron density map at 2.5 Å resolution calculated using isomorphous replacement and anomalous scattering (From Sayre, 1974).

15.5.3. SAYRE'S METHOD

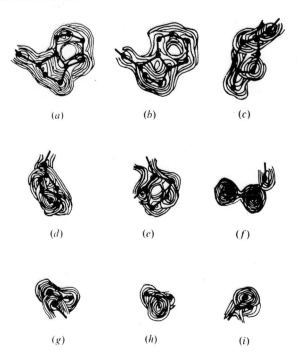

FIG. 15.5 Various sidechain electron densities calculated for rubredoxin using phases refined by Sayre's direct method (From Sayre, 1974). All are taken from the direct map except (b) which is from the map using the phases calculated from positions refined by reciprocal space least squares by Watenpaugh et al., 1973 (discussed in section 4.1.). (a) Trp. 37 (b) Trp. 37 (c) Tyr 11 – this has an atom missing and compares unfavourably with the Watenpaugh et al. map (d) Phe 49 (e) Pro 20 (f) Cys 9 showing also one of the Fe-S bonds. The peak heights of the Fe and S atoms are lower than they would be in nature, reflecting the attempt on the part of the phase refinement to produce an equi-atom structure. (g) Asp. 35 (h) Val 8 (i) Val 44.

quite as clean as the Watenpaugh et al. map, but is nevertheless sufficient to allow approximately 400 of the 424 atoms in the molecules to be correctly located. The method shows a very good agreement for those residues which were erroneously sequenced by Watenpaugh et al. (1973). It thus demonstrates how a direct method, not depending on the subjective interpretation of the electron density map, may offer some substantial advantages over techniques involving refinement of atomic positions. Some sidechains are shown in Fig. 15.5.

Sayre (1974) has found that his method is not useful when the phase extension is started from 3 Å data. In this case the electron density map contains resolved peaks but these do not correspond to the atomic positions. There was no advantage in starting with the 2 Å phases rather than 2.5 Å phases. The major problem with the method is the size of the computations. Even after optimizing the computing techniques the refinement of rubredoxin is estimated to cost U.S.$7500 and with present day computing facilities it has a practical limit for proteins in the 30,000 molecular weight range. Nevertheless the process reduces considerably the number of intensities (i.e. of derivatives) which need to be measured and has the great advantage of objectivity, and Sayre has recently suggested techniques which lead to increased speed and should make the method more generally useful.

15.5.4 High-order probability laws

De Rango *et al.* (1975) have used high-order covariance matrices to show that the maximal determinant rule and the regression equation can be applied successfully to the phase refinement and extension of protein structures. They report that with structure factors calculated from the atomic model of insulin, the use of an order −400 covariance matrix leads to phases with an average error of 15°. The reader should consult references in De Rango *et al.* (1975) for the theory of this technique. The method has also been applied to the data of insulin for phase refinement and phase extension to 1.5 Å. As with the Sayre technique, an improved electron density map is obtained leading to a more straightforward interpretation.

16

MOLECULAR REPLACEMENT

16.1 INTRODUCTION

We have seen in Chapter 2 that many proteins have subunit structures and that the subunits are usually related by symmetry operations such as a two-fold or three-fold axis. How do these proteins crystallize?

Quite often the molecular symmetry axes are used in the symmetry of the crystal, and each crystal asymmetric unit contains one protein subunit. However, even when a protein has several identical subunits related by symmetry axes, more than one protein molecule may pack in the same crystallographic asymmetric unit. For instance, in rhombohedral insulin crystals the dimer (described in Section 2.3, page 54) occupies one asymmetric unit. The two molecules have an approximate two-fold rotation axis relating their positions. However, the two-fold axis is local to the dimer or asymmetric unit, and it is called a local or non-crystallographic two-fold axis. Because the symmetry is non-crystallographic the two molecules have quite different packing relationships with respect to molecules in the next crystal unit cell.

This kind of situation is found in other proteins studied crystallographically including glyceraldehyde phosphate dehydrogenase (Buehner *et al.*, 1974b), asparaginase (Epp *et al.*, 1971) and aldolase (Eagles *et al.*, 1969). Different subunits of these proteins have different crystallographic environments. There are also other occasions when identical or closely related proteins are found in different crystallographic environments. For instance, proteins will sometimes crystallize in different space-groups under slightly different crystallization conditions. Change of pH, change of solvent or precipitating agent or the addition of some special ingredient can easily change the crystal packing. We have seen in Section 3.5 how polymorphic forms are common with a wide range of proteins such as lysozyme, insulin and pepsin. These polymorphic crystals contain essentially the same protein molecules packed in different ways.

Other examples are found with homologous proteins which rarely have the same crystal form. Homologous proteins which are derived by divergent evolution from a single more primitive protein often have a large proportion of their tertiary structure arranged in identical ways. Haemoglobins and myglobins form an homologous series as do chymotrypsin, trypsin and elastase.

These proteins, despite the similarity in tertiary structures, crystallize in quite different space-groups.

In these various situations of identical structures or parts of structures in different crystallographic environments we would expect similarities between their diffraction patterns. This follows from the fact that each diffraction pattern is derived from the Fourier transform of the molecule. If this is so, can we then derive from the diffraction patterns the relative orientations and positions of these molecules? With a knowledge of the structure of one protein in an homologous series, can we find the structures of the others in different crystals? Can we define the positions of the non-crystallographic symmetry axes when there is more than one molecule in the unit cell? Does this imply any restriction on the structure factors which may be used to determine phases and so help determine the crystal structure?

In this chapter we will discuss these questions. All have been the subject of considerable theoretical debate; some have been answered unequivocally in practice.

Here we will begin by considering some theory but we will be primarily concerned with what protein crystallographers have found useful in practice.

Much of this work derived from the paper by Rossmann and Blow (1962) where the location of non-crystallographic symmetry was the main interest. Publications concerning the molecular replacement method have been collected together by Rossmann (1972) in a useful book.

16.2 ORIENTATION OF MOLECULES

Two identical molecules may only differ in orientation and position. It is therefore possible to define a general operation involving only rotation and translation which will bring the equivalent points of electron density in each molecule into coincidence.

If the equivalent points on the two molecules are defined by position vectors X_1 and X_2 referred to a common orthogonal coordinate system, then

$$X_2 = [C] X_1 + d$$

where $[C]$ is a rotation matrix and d a vector defining translation.

We can now find the rotation and translation as a two-stage operation. It is convenient to consider the rotation first.

16.2.1 Direct comparison of Fourier transforms

Intuitively we might expect to define the rotation matrix $[C]$ from inspection of the diffraction patterns. The patterns are derived from the Fourier transform of the molecule and this is illustrated for some simple model systems in Figs. 16.1, 16.2 and 16.3.

16.2.1. DIRECT COMPARISON OF FOURIER TRANSFORMS

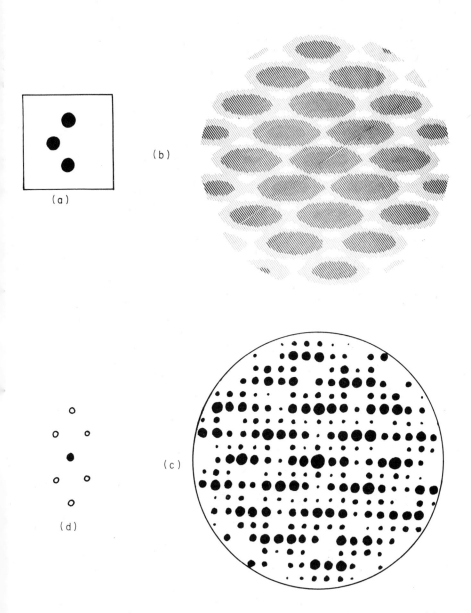

FIG. 16.1 The modulus of the molecular transform of the structure (a) is given in (b). The weighted reciprocal lattice (c) corresponds to the diffraction pattern given by placing the structure in a two-dimensional crystal of unit cell size indicated in (a). The self-Patterson is represented in (d).

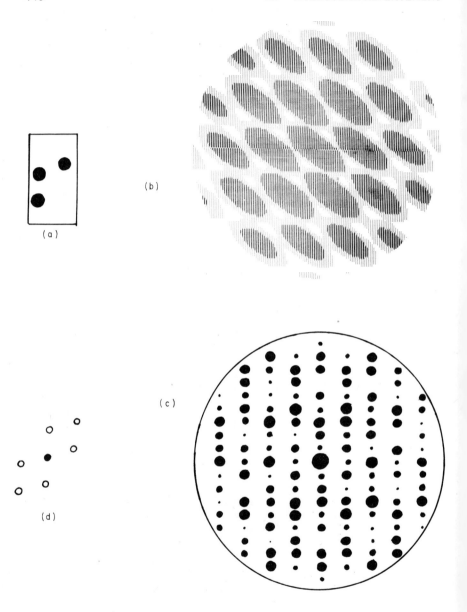

FIG. 16.2 (a) The structure shown in Fig. 16.1 is placed in a different unit cell; (b) represents the modulus of its molecular transform, (c) the weighted reciprocal lattice and (d) the self-Patterson. The relative orientations of the structures in Fig. 16.1 and this figure can be established either by comparing the weighted reciprocal lattice (i.e. the diffraction patterns) or the self-Patterson functions.

16.2.1. DIRECT COMPARISON OF FOURIER TRANSFORMS 447

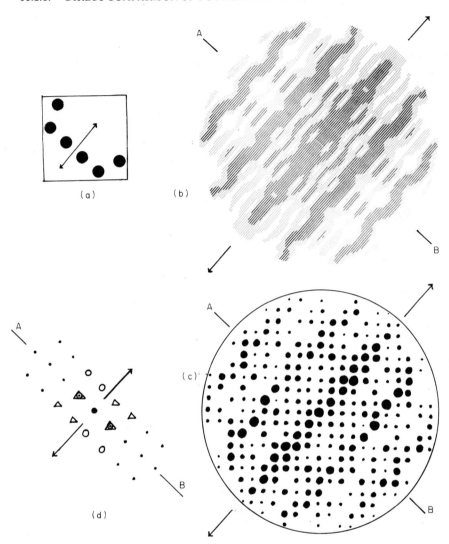

FIG. 16.3 Non-crystallographic symmetry: (a) the structure shown in Fig. 16.1 and 16.2 is placed in a unit cell with a local two-fold axis of symmetry (indicated by an arrow). The modulus of its molecular transform is given in (b); the transorm also has a two-fold axis of symmetry and in addition a mirror plane, AB, perpendicular to the two-fold axis. The weighted reciprocal lattice (c) and the Patterson function (d) show the same symmetry. The Patterson function has many vector peaks in the Harker plane, AB, which give rise to a series of peaks in the direction of the two-fold axis in the molecular transform and the weighted reciprocal lattice. The direction of the two-fold axis can be identified either by the symmetry or by the peaks – spikes – in the transform.

Figure 16.1a shows a hypothetical molecular structure in a cell. Figure 16.1b gives the modulus of the molecular transform in the same orientation and Fig. 16.1c is the diffraction pattern which is related to the Fourier transform sampled at the reciprocal lattice points defined by the crystallographic cell and X-ray wavelength. Figure 16.2 shows the same molecule in a different orientation in a different cell.

The examples given in Figs. 16.1 and 16.2 are both molecules in space-group P1. In each case there is only one molecule in the crystallographic unit cell. With the help of the Fourier transform of the molecule it is quite easy to fix the rotation of the molecule in one cell relative to the other. The principal difficulty is that when the diffraction patterns are rotated into the same relative orientations the diffraction maxima do not necessarily correspond; we need to calculate the values of the Fourier transform at non integral values of the Miller indices. This can be done by interpolation. The problem becomes more complex when there is more than one molecule in the unit cell. Here we have the transforms of the two or more molecules in different orientations plus interference functions between the molecules. A situation of this sort is illustrated in Fig. 16.3. The actual transform is rather difficult to identify although there are simplifying features in symmetry which we shall discuss later.

We may conclude that determination of orientation by direct comparison of Fourier transforms is fairly easy where we compare two crystals each with one molecule in the unit cell. It is even more straightforward when the structure of the molecule is known so that the Fourier transform can be calculated for a sufficiently large cell to overcome the problem of sampling at non-integral reciprocal lattice points and overlap of neighbouring origin peaks. This method has been used successfully by Joynson et al. (1970) to find the position of the lysozyme molecule in the P1 crystal form. They used atomic coordinates for lysozyme found in tetragonal crystals to calculate the Fourier transform of the molecule.

16.2.2 Comparison of Patterson functions

In the case of a unit cell containing more than one molecule, the best approach is through the Patterson function. Any Patterson function will contain sets of peaks representing intramolecular vectors. These are the self vector sets. There will be one set for each molecular orientation found in the crystallographic cell. The peaks of the self-vector sets for the three examples discussed above are shown in Figs. 16.1d, 16.2d and 16.3d. They must all lie within the largest intramolecular distance, r, from the origin of the Patterson function. On the other hand the intermolecular vectors will tend on the average to be longer than the intramolecular ones. These may be termed the cross-vector set. Many will be outside the sphere of radius 'r' around the origin within which the self vector set

16.2.2. COMPARISON OF PATTERSON FUNCTIONS

lies. The relative orientations of the molecules can be determined by comparing the self-Patterson functions which may be considered to be within a sphere of radius 'r'.

However, in cases where the molecule does not approximate to a sphere great care must be taken in choosing the 'r' value. Clearly if the molecule is rod like and sits in a long narrow unit cell some intramolecular distances might exceed one of the cell dimensions. The radius "r" must be somewhat shorter than the shortest cell dimensions, otherwise spurious overlaps will occur with the origin of the next unit cell in Patterson space.

Several modifications to the Patterson function will also improve the comparison. All vectors which are within, say 2 Å, of the origin of the Patterson are best ignored as this region is most likely to contain errors and artifacts. Also it is better to use a "sharpened" Patterson function. This is one which corresponds to vectors between point atoms rather than atoms with a finite sphere of electron density. This hypothetical situation is simulated by increasing the average intensity of the higher $\sin \theta$ value reflections so that the mean value of F_h^2 in any $\sin \theta$ range is the same and does not decrease with increasing $\sin \theta$. This is the same as using E values used for the direct method discussed in Chapter 5, page 142. A Patterson computed with such modified F_h^2 values should have better resolved vector peaks and allow better definition of the relative orientations of the self-vector sets. These operations are discussed fully by Rossmann and Blow (1962) and Nordman (1966).

The next problem is to choose a criterion for the correspondence of the self-vector sets at different orientations. There are several possible techniques. The classic method of Rossmann and Blow (1962) is to compare the values of the product function R where

$$R = \int_U P_2(X_2) P_1(X_1) dX_1$$

for the Patterson P_1 and the rotated Patterson P_2 within the volume U. This will have a maximum value when the two self-vector sets are equivalently oriented. R is known as the rotation function.

Other criteria would be the sum of the two Pattersons "the sum function" or to add up the minimum at each point on either of the Pattersons "the minimum function". These functions both have a maximum at the correct orientation. They are discussed fully by Buerger (1959) and have been applied in various forms by Huber (1969), Nordman (1966) and Tollin (1966).

The computations may be made in either reciprocal or real space. Huber (1959) and Nordmann (1966) have discussed real space techniques. However, for low resolution protein crystal data reciprocal space operations are probably advantageous, and the rotation function of Rossmann and Blow (1962) is generally used.

16.2.3 The rotation function

We now expand the Patterson functions to find the reciprocal space expression for the rotation function. The Patterson function of the first molecule is given by

$$P(\mathbf{X}_1) = \frac{1}{V} \sum_\mathbf{h} |F_\mathbf{h}|^2 \cos(2\pi \mathbf{h} \cdot \mathbf{X}_1)$$

and that of the second by

$$P(\mathbf{X}_2) = \frac{1}{V} \sum_\mathbf{p} |F_\mathbf{p}|^2 \cos(2\pi \mathbf{p} \cdot \mathbf{X}_2)$$

where $\mathbf{X}_1$ and $\mathbf{X}_2$ are the Patterson space vectors of peaks for molecules 1 and 2 such that $\mathbf{X}_2 = \mathbf{X}_1 [C]$ and the $\mathbf{h}$ and $\mathbf{p}$ are reciprocal space vectors so that $\mathbf{p} = [\tilde{C}]\mathbf{h}'$ where $[\tilde{C}]$ is the transpose of $[C]$ and $\mathbf{h}'$ is a non-integral reciprocal lattice vector. Therefore we may rewrite

$$P(\mathbf{X}_2) = \frac{1}{V} \sum_\mathbf{p} |F_\mathbf{p}|^2 \cos(2\pi \mathbf{h}' \cdot \mathbf{X}_1)$$

Now the rotation function $R = \int P(\mathbf{X}_2) P(\mathbf{X}_1) d\mathbf{X}_1$

$$= \frac{1}{V^2} \sum_\mathbf{p} \sum_\mathbf{h} |F_\mathbf{p}|^2 |F_\mathbf{h}|^2 \int_a \cos[2\pi[\mathbf{h}+\mathbf{h}']\mathbf{X}_1] d\mathbf{X}_1$$

The integral is an interference function whose value may be written as

$$(U/V) |G_{\mathbf{h},\mathbf{h}'}|$$

for integration over a volume symmetrical about the Patterson origin.

Thus for overlap of the self vectors of the Patterson function with rotation but without translation, we have

$$R = (U/V^3) \sum_\mathbf{P} \sum_\mathbf{h} |F_\mathbf{p}|^2 |F_\mathbf{h}|^2 G_{\mathbf{h},\mathbf{h}'}$$

If we assume that the protein molecule is approximately spherical then $|G_{\mathbf{h},\mathbf{h}'}|$ is given by

$$|G_{\mathbf{h},\mathbf{h}'}| = \frac{3(\sin 2\pi \mathbf{H} \cdot \mathbf{r} - 2\pi \mathbf{H} \cdot \mathbf{r} \cos 2\pi \mathbf{H} \cdot \mathbf{r})}{(2\pi \mathbf{H} \cdot \mathbf{r})^3}$$

where $\mathbf{H} = \mathbf{h} + \mathbf{h}'$. This function is illustrated in Fig. 16.4. It has a maximum value of 1 and is never greater than 0.086 outside the range $-0.725 < \mathbf{H} \cdot \mathbf{r} < +0.725$. Hence all the terms in the rotation function for which $|\mathbf{H} \cdot \mathbf{r}| > 0.725$ can be neglected.

What does this mean in terms of calculating the rotation function? If we take a protein of diameter 40Å in a cell, say, of 80Å in each direction, then $r = 0.5$.

16.2.3. THE ROTATION FUNCTION

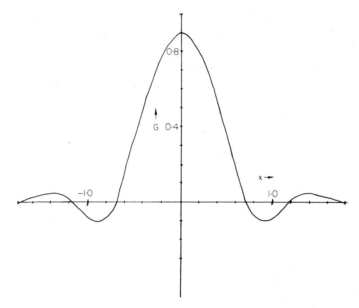

FIG. 16.4 The interference function
$$G = 3(\sin 2\pi x - 2\pi x \cos 2\pi x)/(2\pi x)^3$$
applicable for roughly spherical sub-units. (Rossmann and Blow, 1962)

Therefore we need only consider lattice points where $|\mathbf{h} + \mathbf{h}'|$ is less than 1.5. A cube in reciprocal space containing approximately 27 lattice points satisfies this requirement reasonably well. Tollin and Rossmann (1966) find 27 points give useful results for the systems they have studied. The rotation function can be written as a double summation

$$R = (U/V^3) \sum_\mathbf{p} |F_\mathbf{p}|^2 \{\sum |F_\mathbf{h}|^2 G_{\mathbf{h},\mathbf{h}'}\}$$

The inner summation $\{\sum |F_\mathbf{h}|^2 G_{\mathbf{h},\mathbf{h}'}\}$ need only be performed over those 27 points $\mathbf{h}$ which are close to the non-integral point $-\mathbf{h}'$. It is better that crystal 1 should have close to equal cell dimensions and no systematic absences.

The rotation function is approximately proportional to $|F|^4$ so that its value will be very heavily dominated by the few large products. The smaller intensities will make little contribution to the sum, and are generally omitted in the calculations. This is an extremely important aspect to the calculation which can be very time consuming.

Rotation functions are usually computed by transforming the crystallographic system into a Cartesian coordinate system, and then converting this into a system of Eulerian angles $(\theta_1 \theta_2 \theta_3)$. These three angles define rotation of the Cartesian coordinate system so that any orientation may be achieved. This system is discussed by Rossmann and Blow (1962). It has a very great advantage in describing symmetry operations which take on a simple form for all but cubic groups (Tollin et al., 1966). They are, however, very difficult to visualize and give rise to an unsymmetrical sampling in reciprocal space which must be uneconomical in computing time. Lattman (1972) has suggested a way of overcoming this latter problem but his method destroys the simplicity of the symmetry.

A more easily visualized system is to define the function in terms of a rotation κ, round an axis, the position of which is defined by two spherical coordinates ψ, ϕ. This system is useful for non-crystallographic symmetry as the axis may be made to correspond to the local symmetry axis. The search for, say, a twofold axis is therefore much simplified.

As the rotation function is a very time consuming computation, it is generally initially calculated at $20°$ intervals in the Eulerian system. Higher resolution is required to define the rotation matrix precisely. The resolution which is required is that rotation which moves the most distant reciprocal lattice point (h_{max}) through one reciprocal lattice translation, given approximately by $2\sin^{-1}(1/2h)$. For a cell with 80Å axis and 6Å data this means intervals of $4°36'$ · ($\sin\theta/2 = 0.04$).

Crowther first suggested that the computation of the rotation function may be considerably accelerated by expanding the Patterson density in terms of spherical harmonics rather than Cartesian Fourier components $|F_h|^2$ of the crystal (Crowther, 1971b).

This may be as much as 100 times faster than previous methods while at the same time avoiding the approximations of the conventional method such as truncating the interpolation function $G_{h,h'}$ and by using only large values of $|F_h|^2$.

16.2.4 Non-crystallographic symmetry

As an example let us consider the rotation function on rhombohedral insulin which has non-crystallographic symmetry.

We have described in Chapter 2 how the insulin hexamer has approximate 32 symmetry. The molecules are related together in pairs by a two-fold axis to give dimers. Three dimers are related by a three-fold axis perpendicular to the two fold axes to give a hexamer.

In the rhombohedral crystals the three-fold axis is crystallographic so that one dimer is contained in each asymmetric unit. The position of the non-crystallographic two-fold axis was identified and positioned correctly by Rossmann using a rotation function long before the structure was solved using isomorphous replacement. It lies perpendicular to the three-fold axis at $44°$ to the axis.

16.2.4. NON-CRYSTALLOGRAPHIC SYMMETRY

Figure 16.5a shows the 2Zn insulin hexamer in the crystallographic cell, viewed along the three-fold axis. Figure 5b shows the rotation function at 6Å resolution (Dodson et al., 1966) plotted on a stereogram viewed down the three-fold axis in the same way as 5a. In this case the rotation function is plotted in terms of the two spherical polar coordinates ψ, ϕ with the assumption that there is a two-fold axis, i.e. with $\kappa = 180°$. The peaks in the function correspond to the two-fold axes perpendicular to the three-fold axis and at 60° intervals around it.

It is very instructive to look at the diffraction pattern and the Patterson function derived from it to see how these also reflect the same molecular symmetry. Figure 16.3 demonstrates how the Fourier transform of one molecule is modified by the presence of an exactly similar molecule in a position related by a local two-fold axis which passes through the origin. The modulus of the Fourier transform of any molecule is centrosymmetrical (Friedel's Law; $|F_\mathbf{h}| = |F_\mathbf{\bar{h}}|$) about the origin in reciprocal space and the addition of a two-fold axis passing through the origin will generate a mirror plane of symmetry perpendicular to the two-fold axis. The diffraction pattern of such a pair of molecules will have $2/m$ symmetry. This is shown in Fig. 16.3c. In the case of rhombohedral insulin the effect of adding the centre of symmetry (of the Fourier transform) to a perfect molecular symmetry of 32 is to generate $\bar{3}m$ symmetry in the diffraction pattern. As the molecular symmetry is local and the axis not quite exact, the diffraction pattern will have only approximate $\bar{3}m$ symmetry. The $hk0$ projection (of insulin) with the positions of the expected three-fold and two-fold axes and mirror planes is shown in Fig. 16.5c. The symmetry is there but is not terribly conspicuous!

However, there is a further indication of the two-fold axis in the diffraction pattern. For reflections in reciprocal space along the direction of the two-fold axis the atoms related by the two-fold axis will diffract in phase. These reflections will have an average intensity proportional to half the number of atoms scattering with twice the number of electrons.

This is $N/2 \, (2f)^2 = 2Nf^2$ for a situation of N equal atoms of scattering factor f. For other general reflections the average intensity is Nf^2. Thus these reflections will have double the average intensity of the general reflections. This often manifests itself as "spikes" in the diffraction pattern in the direction of the two-fold axes. This can be seen rather well on the insulin photograph of Fig. 16.5c, and is often seen in the diffraction patterns of the icosahedral viruses.

Indication of the local symmetry can also be seen in the Patterson functions. Figure 16.3d shows the Patterson function of the model system which like the diffraction pattern has $2/m$ symmetry. The Patterson centre of symmetry is added to the two-fold axis between the molecules. There is a Harker plane containing vector peaks between equivalent atoms in the two molecules. This lies perpendicular to the two-fold axis through the origin and is characterized by an unusually high density of peaks. The Patterson function for insulin should have

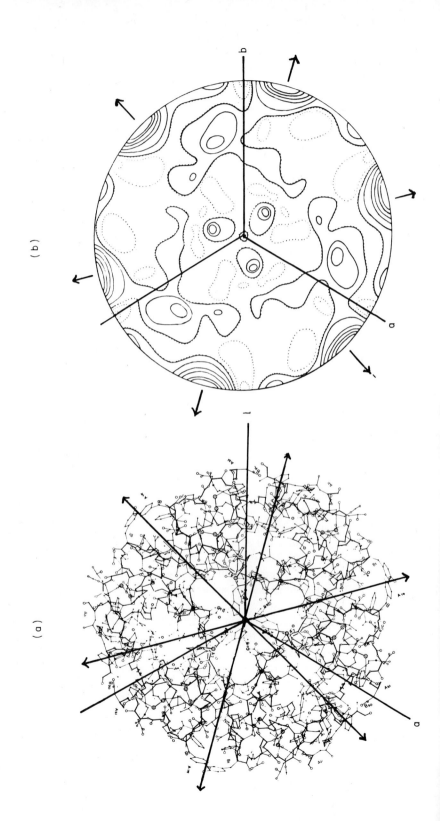

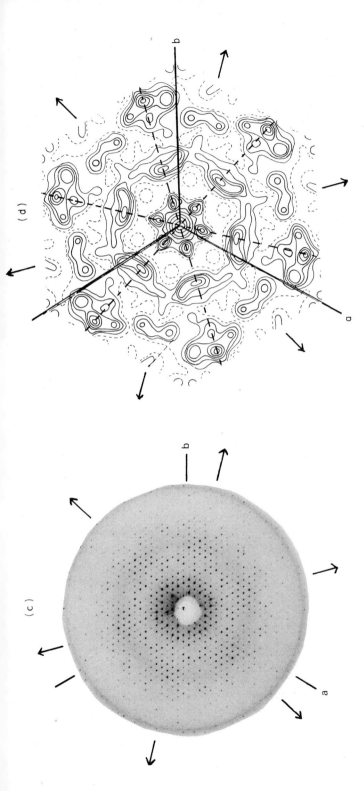

FIG. 16.5 The rotation function: (a) shows the insulin hexamer, the direction of the cell axes **a** and **b** and the local two fold axes. (b) shows the rotation function in the same orientation calculated before the structure was known by Dodson et al., (1966). Peaks correctly indicate the direction of the local two fold symmetry axes. The diffraction pattern (c) and the Patterson function (d) have approximate $\bar{3}m$ symmetry. The directions of the two-fold axes are indicated by spikes in the diffraction pattern. There are peaks on the Harker planes (also mirror planes, indicated by – – – –) in the Patterson function.

$\bar{3}m$ symmetry. In Fig. 16.5d the projection Patterson is shown. Many of the large peaks can be seen to be along the direction of the mirror planes as expected and the Patterson has good mirror plane symmetry. The continuity of the Patterson function makes this a little more obvious than in the $hk0$ projection X-ray photograph where the transform is sampled at reciprocal lattice points. However, they both contain the same information, and are in many ways alternative approaches to the computerized reciprocal space rotation function.

16.3 POSITIONING THE ORIENTED MOLECULE: THE TRANSLATION PROBLEM

16.3.1 Precise and imprecise parameters

With a knowledge of the molecular orientation determined from the self-Patterson, the translation vector **d** may be investigated by considering the cross-Patterson. The cross vectors are intermolecular vectors and as such may be used to find the translation between molecules if we already know their orientations.

It is convenient to consider the translation vector for each atom in terms of two components. One is the vector **t** in the direction of the rotation axis and the second is the vector **s** perpendicular to the axis direction (see Fig. 16.6). Figure 16.6a gives an example for the special case when the rotation is 180°. Figure 16.6b shows the Patterson given by this distribution. The self vector set is not affected by the translation and is the same as in the example of Fig. 16.3 where there is no translation and the two molecules are related by an exact two-fold rotor. The self vector set has $2/m$ symmetry (as was discussed on p. 447).

The cross vector set is disposed about a plane perpendicular to the rotation axis direction and translated **t** along the axis. This plane contains all the vector peaks relating equivalent atoms in the two molecules, and as such it is a Harker plane. It contains an image of the molecule projected along the axis direction and expanded and rotated by an amount which depends on the rotation. For the two-fold axis it will be twice the size and rotated by 180°. Around each Harker peak, the vector peaks form real and inverted images of the molecule. In the special case of the two-fold axis these images will be related by a mirror plane.

This description emphasizes one important point. The vector **t** is the same for all atoms. In the two-fold axis case we get a well-defined Harker plane which is also a mirror plane. On the other hand the vector **s** varies for each atom. It depends on the distance of each atom from the axis of rotation. Superposition of the self vector set on the cross vectors would give a number of positions where some agreement between vector peaks is obtained. These positions would be spread out on the plane perpendicular to rotation axis at a distance **t** along the axis. They correspond to the Harker peak positions. A search conducted in

16.3.1. PRECISE AND IMPRECISE PARAMETERS

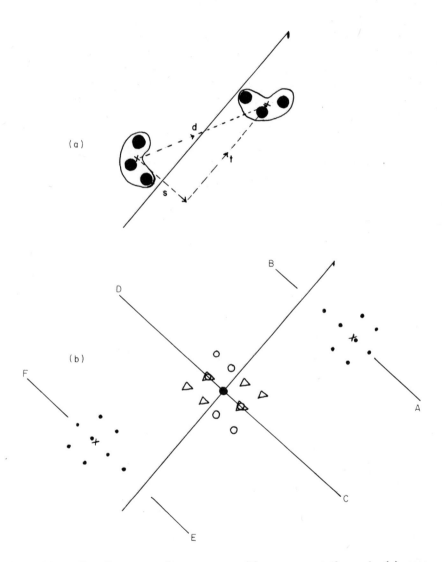

FIG. 16.6 The Patterson of a structure with a screw rotation axis: (a) represents atoms in a two-dimensional model structure. The crosses are the points chosen as the approximate centres of the two molecules which are related by a vector **d**. **d** has components **t** and **s** parallel and perpendicular to the screw rotation axis. (b) shows the vectors arising from the structure in (a). The self vectors of the two molecules are indicated by △ and ○; the cross vectors by ●. Crosses mark the positions of **d** and −**d**. CD is a mirror plane through the origin; AB and EF are Harker planes.

this way would then give precise information on **t** but imprecise information on **s**.

As the individual vectors **s** depend on the atomic arrangement in the molecule, a meaningful interpretation of these vector peaks requires a knowledge of the molecular structure. With this knowledge it is possible to generate a cross vector set and search the Patterson for this arrangement. Conversely we could theoretically analyse the cross vector set to give the molecular structure, but with a protein with many atoms which are unresolved this is not possible.

These arguments show that the translation problem is less tractable than the rotation problem. Without knowledge of the protein structure we can define **t** precisely. From the range of **s** values giving some superposition we can only derive some idea of the size of the protein subunit.

16.3.2 Non-crystallographic two-fold symmetry — the translation function

Even if we know nothing of the molecular structure and therefore cannot calculate a meaningful vector perpendicular to the rotation axis, knowledge of the vector **t** is often important. In the case of non-crystallographic symmetry it allows us to define the translation component of a screw axis. More specifically we can differentiate between a screw and a proper rotation axis.

With aggregated proteins two-fold axes relating the subunits are most common (see Chapter 2) and we will consider only this situation here. This is particularly fortunate as a rotation of 180° is the most straightforward to identify. In this case the two cross vector sets related by the Patterson centre of symmetry can be brought to coincidence by rotating the Patterson function around the axis through the origin by 180° and translating it by 2**t**.

The search is most conveniently carried out by defining a translation function in an analogous way to the rotation function (Rossmann *et al.*, 1964).

Thus we seek to maximize

$$T(\Delta) = \int_{|u|<2r} P(\mathbf{u} + \Delta) P([C]\mathbf{u} - \Delta) d\mathbf{u}$$

where Δ is a translation vector which is independent of the origin, i.e. position, of the rotation axis and relates the "centres" of the two molecules. Rossmann *et al.* (1964) have shown that this function can be expressed as

$$T(\Delta) = \left(\frac{u}{V^3}\right) \sum_{\mathbf{h}} \sum_{\mathbf{p}} |F_{\mathbf{h}}|^2 |F_{\mathbf{p}}|^2 G_{\mathbf{h},\mathbf{p}} \cos\{2\pi(\mathbf{h} - \mathbf{p}) \cdot \Delta\}$$

which is a centrosymmetric Fourier summation with coefficients

$$\{\sum_{\mathbf{p}} |F_{\mathbf{H}+\mathbf{p}}|^2 |F_{\mathbf{p}}|^2 G_{\mathbf{H}+\mathbf{p},\mathbf{p}}\}$$

for the term with indices $\mathbf{H} = \mathbf{h} - \mathbf{p}$.

16.3.2. NON-CRYSTALLOGRAPHIC TWO-FOLD SYMMETRY

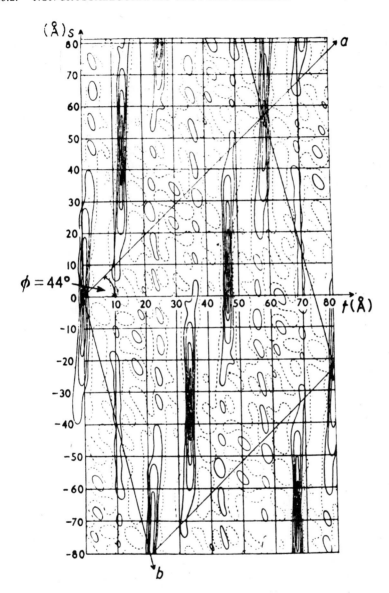

FIG. 16.7 The translation function for insulin hexamers in the hexagonal plane $z = 0$. The self-Patterson has been removed within a sphere of 20 Å. Radius of integration 20 Å, resolution 4.5 Å. Data sharpened by omitting all reflections with spacing greater than 10 Å. The parameters s and t are defined in Fig. 16.6. The function shows that t is defined precisely as zero, i.e. it is a proper two-fold axis of rotation, but s is very imprecisely defined. Peaks at 0,0, $^1/_3$,$^2/_3$ and $^2/_3$,$^1/_3$ correspond to the equivalent hexamers in the hexagonal unit cell. (From Dodson et al., 1966).

This summation is carried out on the Patterson modified to remove the self vector set, i.e. $|u| < 2r$. Like the rotation function, the translation function depends critically on the choice of radius of the vector sets and is improved by sharpening.

The function has been used with insulin to demonstrate that the two-fold rotation identified by the rotation function has no translation component. Figure 16.7 shows the function for zero plane perpendicular to the three fold axis which contains the two-fold axis. As expected the function defines the translation **t** precisely but gives a broad peak perpendicular to the rotation axis. The peak positions correspond to a zero component of translation **t** indicating that the molecules have a proper two-fold axis of rotation, and not a two-fold screw axis. The other peaks 2/3a, 1/3b and 1/3a, 2/3b correspond to two other insulin hexamers in the hexagonal cell.

16.3.3 Positioning a known molecule – the Q function

If the structure of the molecule or part of the molecule is known, the difficulties in defining the translation component perpendicular to the rotation axis can be overcome. The relative positions of the cross vectors can be derived from the molecular structure and search methods can be used to locate the cross vector set in the Patterson.

There are many possible ways of approaching this problem: we could work in real or reciprocal space, or we could use different criteria of fit for our vector sets. Here we will describe one of these methods which is due to Tollin and uses his so-called "Q" function.

The origin of the unit cell will be defined with respect to the symmetry operations, i.e. the screw, and proper rotation axes in the case of protein crystals. The origin is arbitrary when the cell is triclinic, P1 (where there are no symmetry operations). It is fixed perpendicular to the rotation axis in P2 but arbitrary in the direction of the axis. In P222 where there are three orthogonal rotation axes the origin is fixed in all directions. The problem of finding the position of a molecule in a unit cell is then to find its position perpendicular to any rotation axes. For each axis it is a two-dimensional problem. We need to consider only two orthogonal axes and these can be treated independently.

Consider first a molecule of n atoms each at a position defined relative to an arbitrary origin by a vector **r**. Let us place a crystallographic symmetry operator T at $-\mathbf{R}_0$ relative to the same origin. Then there will be a further set of atoms at $T(\mathbf{r}_j + \mathbf{R}_0) - \mathbf{R}_0$. The interatomic vector is given by $\mathbf{r}_j + \mathbf{R}_0 - T,(\mathbf{r}_j' + \mathbf{R}_0)$, each vector will give rise to a peak in the Patterson function. When $j = j'$ the peak will lie on the Harker section.

We can therefore generate precisely a series of relative cross vector peaks for the symmetry operation but their absolute position with respect to the symmetry axis depends on the unknown R_0. Now the sum of the Patterson

16.3.3. POSITIONING A KNOWN MOLECULE – THE Q FUNCTION

density at the ends of the crossvector set is given by

$$Q(R_0) = \sum_h |F_h|^2 \sum_{j=1}^{n} \sum_{j'=1}^{n} \cos 2\pi h[r_j + R_0 - T(r_{j'} + R_0)]$$

and this will be a maximum when the trial cross vector set superpose on the true cross vector peaks. If we then compute $Q(R_0)$ as a function of R_0 it should give a maximum when the cross set is properly positioned relative to the symmetry axis. This function is the Tollin "Q" function (Tollin, 1966).

For a two fold rotation axis parallel to **b**

$$T(xyz) = -x, y, -z$$

This can be substituted in the above equation to give:

$$Q(X_0 Z_0) = \sum_{hkl} |F_{hkl}|^2 \sum_{j=1}^{n} \sum_{j'=1}^{n} \cos[4\pi(hX_0 + lZ_0) + 2\pi(k(x_j + x_{j'}) + k(y_j - y_{j'}) + l(z_j + z_{j'}))]$$

This expression can be rearranged into a form more suitable for calculation. It is clearly a two-dimensional function in X_0 and Z_0, and because of the term in $4\pi(hX_0 + lZ_0)$ it is rather a sharp function. Thus it will not give rise to false peaks due to overlaps. Spurious peaks will however arise whenever two atoms r_1 and r_2 have the same y-coordinate and give rise to a non-Harker peak in the Harker section. Then there will be some value of R_0 for which $T(r_1 + R_0) - (r_2 + R_0) = 0$ and here the Q function will have a value of $\sum_h |F_h|^2$.

The position of this spurious peak can be predicted and it should not be a problem in interpretation of the Q function. Nevertheless in a protein there may be so many cases giving rise to spurious peaks that this is impracticable, and the alternative method of modifying the $|F(h)|^2$ values to remove the Patterson origin is usually adopted (Tollin, 1969). The Q function has been used successfully for a number of proteins but its first application was to position the seal myoglobin molecule using the atomic coordinates of whale myoglobin (Tollin, 1969). Tollin's Q function is closely related to the translation function of Crowther and Blow (1967) where the function

$$T(t) = \int P(u) P(u, t) du$$

is maximized. This finds the best fit of the computed and correctly oriented cross-Patterson $P(u, t)$ with the Patterson $P(u)$. Crowther and Blow (1967) express $T(t)$ in terms of the Fourier transform of the molecule and the Patterson. In Tollin's function we have chosen to define the translation in terms of R_0, a vector relating the coordinate system to the symmetry axis rather than a vector between the molecules **t**. We have also represented the $P(u, t)$ in terms of a series of points in vector space where the function has unit value and putting $P(u, t)$ to zero elsewhere. The Patterson function $P(u, t)$ calculated correctly for the cross vector set would have peaks defined by the product of the scattering factors. The peaks have different weights and would not be points but rather have a finite size dependent on scattering and temperature factors. The success of the Q function shows that here the assumption of point atoms of equal scattering is a reasonable one for proteins.

Nevertheless the Crowther and Blow expression is a more general form of the Q function. As it defines the molecule in terms of the Fourier transform of the molecule rather than in terms of the atomic structure, it might be a more convenient form to use in a comparison where the "known structure" comprises only a low resolution electron density map (Tollin, 1969).

16.4 MOLECULAR REPLACEMENT AND THE CALCULATION OF PHASES

We have outlined techniques for determining the orientation and position of protein molecules either when the structure or part of the structure is known or when it is completely unknown. We have seen that in the latter case the problem is rather intractable. But let us here assume that these operations have been successful and see whether we can use information for the calculation of phases and hence an electron density map.

16.4.1 Known structure

We will consider first a protein of known or partly known structure. If the rotational and translational parameters are also known it is a comparatively straightforward operation to calculate some phases. The structure factors can be calculated for the known part of the structure correctly oriented and positioned in the unit cell. This provides a set of phases which can be used in conjunction with the observed structure amplitudes to compute an electron density map. In the case of seal myoglobin, the coordinates from whale myoglobin have been used to produce a map which compares favourably with the electron density map calculated using the conventional multiple isomorphous replacement method (Tollin, 1969).

16.4.2 Unknown structure

If the structure is unknown, determination of phases is a much more controversial exercise. However, where there is non-crystallographic symmetry or two identical molecules in different cells there are twice the normal number of intensities per subunit volume and there must be some constraints which may be expressed in terms of relations between structure factors. Are these sufficient to determine the phases? A set of equations may be derived by equating the density at the two points where the structures are known to be the same

16.4.2. UNKNOWN STRUCTURE

(Rossmann and Blow, 1963; 1964). This gives rise to equations of the kind

$$|F_\mathbf{p}| \exp[i\alpha_\mathbf{p}] = \frac{u}{V} \sum_{\mathbf{h}-\infty}^{+\infty} |F_\mathbf{h}| \exp\{i\alpha_\mathbf{h}\} \sum_{h=1}^{N} G_{\mathbf{p},\mathbf{h}}$$

where $|F_\mathbf{p}|, \alpha_\mathbf{p}, |F_\mathbf{h}|, \alpha_\mathbf{h}$ are the structure factor amplitudes and their phases at the reciprocal lattice points $\mathbf{p}$ and $\mathbf{h}$ in either the same or different crystals (Main and Rossmann, 1966). $G_{\mathbf{p},\mathbf{h}}$ depends on the rotation and translation parameters relating the two molecules. The problem is to solve these non-linear equations. As with other direct methods (see Section 5.14) this can be approached intuitively by finding from selected combinations of reflections the probabilities of the various estimations. The estimated values can be reinserted in the equations and refined using the other estimated values (Main and Rossmann, 1966, Main, 1967). A similar but perhaps more hopeful approach using linear algebra has been suggested by Crowther (1971b). This has been used by Jack (1973) for direct determination of the phases for the $hk0$ projection obtained from tobacco mosaic virus coat protein at 7Å resolution making use of the 17-fold local symmetry axis of the protein aggregate which is almost parallel to the crystal c axis. Comparison of the phases obtained by this method with those obtained by Gilbert and Klug (1974) using a single isomorphous derivative at 9.5Å resolution showed over 90% agreement for reflections with $(h + k)$ even and 60% for $(h + k)$ odd. In the latter case the heavy atoms of the isomorphous derivative form an almost face-centred lattice in projection and contribute little to the $(h + k)$ odd reflections.

For refinement of the phases the method is really the reciprocal space equivalent of averaging the calculated electron density for the volumes related by the local symmetry followed by a back Fourier transformation to calculate the new phase. Bricogne (1974) has presented a rigorous proof of the formal equivalence between the reciprocal and real space approaches to phase restraint by molecular replacement. The real space method has several advantages; for example the molecular boundaries can be given any desired shape, and the method may be less expensive in computing time. Bricogne has shown that the method is convergent if more than two crystallographically independent copies of the same molecule can be observed.

The molecular replacement methods are probably used most effectively in conjunction with other direct methods such as the tangent formula (Section 5.15) or with isomorphous replacement. For example Buehner et al. (1974a,b) calculated an electron density map using single isomorphous replacement for crystals of lobster glyceraldehyde-3-phosphate dehydrogenase, GAPDH, which contain one tetramer per asymmetric unit. They averaged the electron density over the four equivalent subunits and back transformed to give new phases. The improvement of the phases allowed other derivative heavy atom positions to be found, and was a crucial stage in the structure analysis (see also Argos et al., 1975).

The major difficulty with this method is that the two or more molecules are not in the same crystallographic environments and we have assumed throughout that this does not alter their three-dimensional structures. Although subunits usually have the same general structure and the degree of similarity is sufficient to give good peaks in the rotation and translation functions at low resolution, the structures may be sufficiently different in detail of side chains to give rise to errors at higher resolutions. There is no general rule concerning this. The two-fold symmetry in the chymotrypsin dimer is very much better than in the insulin dimer. The method could work with chymotrypsin but probably would not at 3Å resolution in the case of insulin.

In summary it may be concluded that given good non-crystallographic symmetry the method of molecular replacement may be useful when there are three or more subunits. The results with GAPDH and tobacco mosaic virus coat protein are very encouraging, and indicate that the method can be quite powerful especially if the phase information is combined with information from other methods such as isomorphous replacement.

17

THE USE OF NEUTRONS AND GAMMA-RAYS IN PROTEIN CRYSTALLOGRAPHY

17.1 INTRODUCTION

Although our knowledge of the structure of protein crystals has been obtained largely through the use of X-ray diffraction, it is evident that this technique has a number of shortcomings. X-rays are ionizing radiation and cause extensive crystal damage at normal temperatures. In the method of isomorphous replacement it is difficult to prepare heavy atom derivatives which are strictly isomorphous with the native protein crystals. X-rays are scattered by the electron density and it is impossible to identify light atoms such as hydrogen even at high resolution. Can these difficulties be overcome by using neutrons or gamma rays in diffraction studies? We will first consider neutron diffraction.

17.2 GENERATION OF NEUTRON BEAMS

Neutrons are uncharged particles, but like X-rays they have wave properties. The wavelength, λ, is given by the equation.

$$\lambda = h/mv \qquad (17.1)$$

where h is Plank's constant and m, v are the mass and velocity of the neutrons. If the neutrons make many collisions with atoms at a temperature T in a reactor then their energy will be defined by

$$\tfrac{1}{2}mv^2 = \tfrac{3}{2}kT \qquad (17.2)$$

where k is Boltzmann's constant. Equations 1 and 2 may be combined to give

$$\lambda^2 = h^2/3mkT$$

This means that at $100°C (T = 373°)$, $\lambda = 1.33$ Å, which is of the same order as 1.54 Å the wavelength of copper radiation. This is very fortunate for the

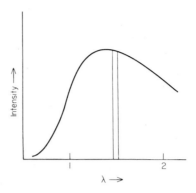

FIG. 17.1 The intensity of a thermal neutron beam as a function of wavelength (Bacon, 1962)

investigation of molecules such as proteins where bondlengths between light atoms lie between 1.2 and 2 Å. Neutrons generated at this temperature are known as "thermal neutrons". Those formed at higher temperatures with shorter wavelength are known as "hot neutrons", while those with longer wavelength are "cold neutrons".

Neutrons are generated in a nuclear reactor and are slowed down by collisions in a moderator of graphite or heavy water. They will have a distribution of velocities following a Maxwellian curve appropriate to the temperature of the moderator. The distribution is illustrated in Fig. 17.1. This shows the intensity versus wavelength distribution for thermal neutrons. The rather broad peak contrasts strongly with the characteristic lines of an X-ray tube shown in Fig. 9.1.

For diffraction experiments it is advantageous to simulate the characteristic radiation beam. This is done in a rather wasteful manner. The neutron beam is reflected from a crystal. The emergent beam is defined by the Bragg equation.

$$\lambda = 2d \sin \theta$$

A wavelength band is selected by a suitable choice of angle θ. Usually a band on the shorter side of the peak is chosen so as to decrease the number of neutrons with half the wavelength some of which would also be reflected at the same angle.

The intensity of the beam will be defined by the angles of divergence before and after the monochromating crystal, and this means that beams of useful intensity tend to be rather broad. Nevertheless 99% of the beam is often wasted. In fact the maximum obtainable monochromatic flux at 1.5 Å is about 10^3 times smaller than the flux obtainable for X-rays of a similar wavelength. Table 17.1 shows some figures for the radiation intensities for the shorter wavelength (and less

17.3. THE SCATTERING OF NEUTRONS BY ATOMS

TABLE 17.I

Comparison of radiation intensities prior to and after diffraction from a single crystal of sperm whale myoglobin. The data represent actual measurements which for comparison reasons were normalized with respect to a single crystal of 1 mm^3 volume. The incident flux in each case represents the flux available within the wavelength range which becomes actually reflected, at the position of the single crystal. The reflected power corresponds to the total number of neutrons or photons which appear per second at the centre of the (603) reflection.

	Radiation	Wavelength [Å]	Peak incident intensity [counts/cm^2/sec]	Reflected power [counts/sec]	Mosaic width [°]
a	X-rays	0.71	2.10^9	4.10^3	0.4
b	Thermal neutrons	1.6	10^7	10	0.3
c	Gamm-rays	0.86	8.10^4	0.1	0.4

a *Mo* X-ray tube with β-filter, 50 kV, 10 mA, distance X-ray tube to crystal 37 cm.
b Brookhaven Natl. Lab. high-flux reactor, germanium monochromator.
c 100 mCi ^{57}Co source with a single emission line of 14.4 keV gamma-radiation; distance source to crystal 20 cm.

intense) *MoKα* radiation and thermal neutrons. This rather low flux is a major disadvantage with neutron diffraction analysis.

17.3 THE SCATTERING OF NEUTRONS BY ATOMS

We have seen that X-rays are scattered by electrons by virtue of their charge which interacts with the incident radiation. The electron density of the atom has roughly the same dimensions as the X-ray wavelength. As a consequence scattered waves from the different parts of the electron density will tend to be increasingly out of phase as the angle between the incident and scattered waves increases. The scattered beam falls off rapidly with increasing angles. We have discussed this fully in section 5.5.1. The X-ray scattering factors are characterized by their proportionality to the number of electrons at zero scattering but have a strong angular dependence for higher angles.

Neutrons are scattered by a quite different process from X-rays. In non-magnetic materials, neutrons are scattered by the nucleus which is very small compared to the neutron wavelength. The scattering for any atom is difficult to calculate as it is a function not only of the nuclear radius but also the

TABLE 17.II

A comparison of the neutron and X-ray scattering factors of certain atoms. (After Bacon, 1962)

Element	Atomic number	Atomic weight of natural element	Specific nucleus	Nuclear spin	Neutrons b_c (10^{-12} cm.)	ζ (barns)	σ (barns)	X-rays f_x, 10^{-12} cm. $\sin\theta = 0$	$(\sin\theta)/\lambda = 0.5$ Å^{-1}
H	1	...	H^1	½	−0.378	1.79	81.5	0.28	0.02
			H^2	1	0.65	5.4	7.6	0.28	0.02
C	6	...	C^{12}	0	0.661	5.50	5.51	1.69	0.48
			C^{13}	½	0.60	4.5	5.5	1.69	0.48
N	7	...	N^{14}	1	0.940	11.0	11.4	1.97	0.53
O	8	...	O^{15}	0	0.577	4.2	4.24	2.25	0.62
Na	11	...	Na^{23}	5/2	0.351	1.55	3.4	3.09	1.14
Mg	12	24.3			0.54	3.60	3.70	3.38	1.35
P	15	...	P^{31}	½	0.53	3.5	3.6	4.23	1.83
S	16	...	S^{32}	0	0.31	1.2	1.2	4.5	1.9
Cl	17	35.5			0.99	12.2	15	4.8	2.0
K	19	39.1			0.35	1.5	2.2	5.3	2.2
Ca	20	40.1			0.49	3.0	3.2	5.6	2.4
			Ca^{40}	0	0.49	3.0	3.1	5.6	2.4
			Ca^{44}	0	0.18	0.4	...	5.6	2.4
Fe	26	55.8			0.96	11.4	11.8	7.3	3.3
			Fe^{54}	0	0.42	2.2	2.5	7.3	3.3
			Fe^{56}	0	1.01	12.8	12.8	7.3	3.3
			Fe^{57}	...	0.23	0.64	2	7.3	3.3

Element	Z	At. wt.	Isotope	Spin	b_c	ζ	σ	f_x	f_x
Co	27	..	Co^{59}	7/2	0.25		6	7.6	3.4
Ni	28	58.7			1.03	13.4	18.0	7.9	3.6
			Ni^{58}	0	1.44	25.9	..	7.9	3.6
			Ni^{60}	0	0.30	1.1	..	7.9	3.6
			Ni^{62}	0	−0.87	9.5	..	7.9	3.6
Cu	29	63.6			0.79	7.8	8.5	8.2	3.8
			Cu^{63}	3/2	0.67	5.7	..	8.2	3.8
			Cu^{65}	3/2	1.11	15.3	..	8.2	3.8
Zn	30	65.4			0.59	4.3	4.2	8.5	3.9
Rb	37	85.5			0.55	3.8	5.5	10.4	5.0
Sr	38	87.6			0.57	4.1	10	10.7	5.2
Pd	46	106.7			0.59	4.4	4.8	12.9	6.5
Ag	47	107.9			0.61	4.6	6.5	13.3	6.7
			Ag^{107}	1/2	0.83	8.7	10	13.3	6.7
			Ag^{109}	1/2	0.43	2.3	6	13.3	6.7
Cd	48	112.4			$0.38 + i0.12$			13.6	6.9
I	53	..	I^{127}	5/2	0.52	3.4	3.8	15.0	7.7
Cs	55	..	Cs^{133}	7/2	0.49	3.0	7	15.5	8.1
Ba	56	137.4			0.52	3.4	6	15.8	8.3
La	57	..	La^{139}	7/2	0.83	8.7	9.3	16.1	8.4
Ce	58	140.25			0.46	2.7	2.7	16.3	8.6
			Ce^{140}		0.47	2.8	2.6	16.3	8.6
			Ce^{142}		0.45	2.6	2.6	16.3	8.6
Pt	78	195.2			0.95	11.2	12	22.0	12.1
Au	79	..	Au^{197}	3/2	0.76	7.3	9	22.2	12.3
Hg	80	200.6			1.3	22	26.5	22.5	12.5
Tl	81	204.4			0.89	10.0	10.1	22.8	12.7
Pb	82	207.2			0.96	11.5	11.4	23.1	12.9
U	92	..	U^{238}		0.85	9.0	..	25.9	14.8

b_c is the neutron coherent scattering amplitude in units of 10^{-12} cm; ζ is the coherent scattering cross section in barns (i.e. 10^{-24} cm^2); σ is the total scattering cross section (including incoherent scattering) in barns; f_x is the X-ray scattering amplitude in units of 10^{-12} cm.

nuclear spin and will be sensitive to the existence of more than one isotope. In fact all atoms scatter more or less equally well, as can be seen from Table 17.II. This contrasts strongly with the dependence of X-ray scattering with the number of extranuclear electrons. As a consequence of the small size of the nucleus, there is very little angular dependence in the scattering of neutrons. Thus the higher angle data should be more easily observable for neutrons than for X-rays, and so theoretically more accurate results are obtainable with neutron diffraction. A plane wave of neutrons can be described by a wave function:

$$\psi = e^{ikz}$$

The scattered wave is usually written as:

$$\psi_s = -\left(\frac{b}{r}\right) e^{ikr}$$

where r is the distance of the point of measurement from the origin at which the nucleus is assumed to be rigidly fixed. b is the scattering length. Like the scattering factor, f, for X-rays (see section 7.2) this is a complex quantity and can be written:

$$b = \alpha + i\beta$$

The imaginary part is large only if the atom has a large absorption component. This occurs when the incident wavelength is close to the resonance wavelength of the atom and this happens infrequently. Cadmium and samarium isotopes are examples; the anomalous scattering caused by the imaginary component can be used in the structure analysis (see Section 7.2). For the moment we will treat the scattering length as if it were completely real. We assume that the incident and resonance energies are well separated and this is a good approximation for most atoms (see Section 7.2). The scattering cross-section is defined as

$$\sigma = \frac{\text{outgoing current of scattered neutrons}}{\text{incident flux}} = 4\pi b^2$$

For a nucleus of zero spin this will be defined mainly by the potential scattering. The potential scattering depends on the radius R and is given by $4\pi R^2$ where the nuclear radius R is $\sim 1.5 \times 10^{-3} A^{1/3}$ cm and A is the mass number. Therefore, the potential scattering increases in proportion to the two-thirds power of mass number. When the nuclear spin (I) is not zero, there are two types of scattering. This is because the compound nucleus (the nucleus plus the neutron) can exist in two states $I + \frac{1}{2}$ or $I - \frac{1}{2}$ which have scattering lengths b^+ and b^-. The total scattering can be considered as the sum of two terms:

$$\sigma = \zeta + \delta$$

ζ is the coherent scattering, i.e. scattering by different nuclei can result in

17.4. CRYSTAL DAMAGE

interference. δ is the incoherent scattering. This can be a nuisance in diffraction experiments as it produces a rather large background so decreasing the precision with which the coherent diffracted beams can be measured.

For nuclei with a spin not zero the coherent and incoherent scattering lengths depend on the relative sizes of b^+ and b^-. If they are of equal size and opposite sign, the coherent scattering will be very small and the incoherent scattering large. This is discussed fully by Bacon (1962). This is unfortunately the case for hydrogen where the coherent scattering is only 2% of the total scattering (see Table 17.II). Some data for the scattering of neutrons by atoms are shown in Table 17.II.

What conclusions can we draw from this concerning neutron diffraction analysis of proteins?

There are several advantages: (1) Neutron diffraction can position light atoms such as hydrogen and deuterium because they have similar scattering lengths to heavier atoms. In fact the scattering length is negative for hydrogen so that the hydrogen atoms would give rise to negative density in a Fourier summation. (2) The scattering lengths of atoms of similar atomic number are often quite different. Thus nitrogen with its greater scattering length can be distinguished from oxygen and carbon. (3) The absorption of neutrons by matter is small and therefore much larger crystals can be used in neutron than in X-ray diffraction. This is very helpful in view of the low flux available from present neutron sources. However, the growth of large crystals is difficult and is usually a rate controlling factor in neutron studies. (4) The anomalous dispersion effect (7.2) is larger for neutrons. We discuss this in Section 17.9.

The disadvantages are: (1) Hydrogen atoms comprise nearly half the atoms in protein crystals. This means that the incoherent scattering will be large and accurate intensity measurements will be difficult to make. For proteins this can be partly overcome by soaking the crystals in heavy water (D_2O) solutions. In this way the hydrogen atoms of the solvent and many of the hydrogens of the protein can be exchanged for deuterium without alteration of the structure. Recently it has become possible to prepare certain completely deuterated proteins, for example by feeding algae on a deuterated broth. (2) The method of isomorphous replacement using platinum, gold, mercury or lead derivatives is probably not useful in neutron analysis as these atoms have scattering lengths of the same order as lighter atoms. However, certain atoms such as Dy^{164} have large scattering lengths; thus in a few cases isomorphous replacement may be helpful (see Section 17.9).

17.4 CRYSTAL DAMAGE

Apart from these factors related to the scattering of neutrons by atoms, there is one outstanding advantage of neutron diffraction. Because thermal neutrons have low energy (0.025 e.v.) compared to X-rays (10,000 e.v.) they are non-ionizing radiation. This means that protein crystals are not damaged by neutrons as they are by electrons or X-rays and complete neutron diffraction data sets can be collected from a single crystal. For instance, Schoenborn has measured 10,000 reflections each for between five and fifteen minutes from a single myoglobin crystal which showed no signs of radiation damage.

17.5 DATA COLLECTION

Neutron beams cannot be measured directly by photographic methods. However, D. Hohlwein (unpublished results) of the Institute Max von Laue-Paul Langevin has developed a photographic method in which the conversion of thermal neutrons to light is achieved by a double layer of ZnS/Ag and Li^6F-powder, each solidified in a plastic matrix, with the photographic film between the two layers.

The difficulty of using photographic methods stimulated research into neutron diffractometers from the beginning of neutron diffraction experiments. Counters of BF_3 containing highly absorbing B^{10} were developed as early as 1950. Some reflections for myoglobin are shown in Fig. 17.2.

However, the low flux of neutron beams make data collection a time consuming process. For instance, early neutron experiments on insulin crystals of 2 mm diameter required counting times of between five and fifteen hours per reflection. The X-ray data of the same resolution need one minute per reflection. Although counting times are smaller on most modern reactors, new techniques for data collection are still an urgent requirement in neutron studies of proteins.

A number of different approaches to this problem are being pioneered. Increase of crystal size may not be possible with proteins of large cell dimensions because of overlap of adjacent reflections. This may be partly avoided by using cold neutrons which as a result of the larger wavelengths give better resolution, and also higher reflectivity, since $I \propto \lambda^3$. There is, however, an upper unit on the wavelength for a given resolution since $d_{min} = \lambda/2 \sin \theta_{max} = \lambda/2$ so for 1.5 Å resolution we need $\lambda < 3$ Å.

Part of the problem of accurate measurement of neutron diffraction intensities is the high background caused by inelastic scattering. Schoenborn *et al.* (1970) have shown that this can be eliminated by placing a monochromator after the beam rather than before it. The pile neutrons can be passed through a beryllium filter which passes neutrons with $\lambda > 3.9$ Å virtually unattenuated, but eliminates radiation. Experiments with a myoglobin crystal are illustrated in Fig. 17.3. In the first case a graphite monochromator was placed before the myoglobin crystal and in the second case, after the myoglobin crystal. There is a large decrease in the background mainly due to the elimination of inelastic scattering.

17.5. DATA COLLECTION

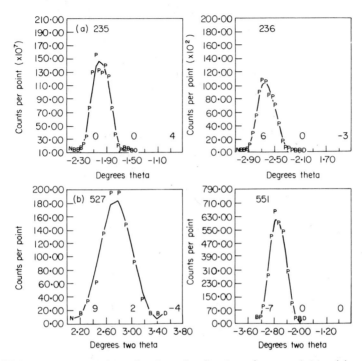

FIG. 17.2 Data scans through selected reflections for myoglobin: (a) ω Scan; (b) $2\theta,\theta$ scan. Data points are indicated by letters: P = peak, B, D and N = background. The solid line represents a "smoothed" integration curve (Schoenborn et al., 1970)

We have seen above that conventional monochromatization techniques are very wasteful, and that a greater efficiency is obtained if a broader spectrum of wavelengths is used in the diffraction experiments. (see Fig. 17.4). A diffraction experiment with an unmonochromatized beam gives a Laue Pattern. A diffracted beam at any angle will comprise many different component parts, each corresponding to a different wavelength and order of diffraction. In the case of X-rays these are very difficult to separate owing to the high speed of the X-ray photons. However, neutrons have a slower speed and the different components can be separated by time of flight techniques. Devices for this type of analysis have been described (Schoenborn et al., 1970; Nunes et al., 1970) and are being developed in several laboratories at the present time. This method uses a much wider range of the reactor spectrum and is therefore more efficient. However, it is probably less useful for anomalous scattering measurements which are strongly wavelength dependent.

Alternatively a modified Laue method can be used (Thomas, 1972). This incorporates a curved or lamellated monochromator system which focuses a

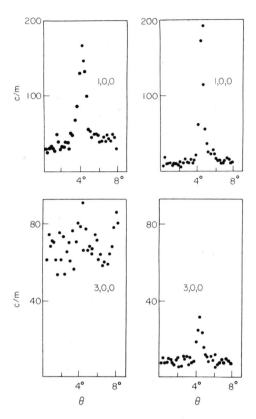

FIG. 17.3 Myoglobin rocking curves ($\lambda = 4.05$ Å) left: with monochromator before the sample, right: with monochromator after the sample. (Schoenborn et al., 1970)

small wavelength band onto the crystal (Fig. 17.5). It is arranged in such a way that different wavelengths come from different directions. It has the advantage of giving a gain in intensity but at the same time produces a good separation of different orders of reflections. This is very important for proteins with large cell dimensions. The Ewald construction is illustrated in Fig. 17.4 (b). Each of the different wavelengths can be represented by an Ewald sphere and these will define a volume in reciprocal space in which the conditions for diffraction are satisfied. All points within the volume will diffract simultaneously. Klar (1972) has developed a neutron diffractometer for measuring these reflections. This is illustrated in Fig. 17.6. The single crystal is surrounded with a metallic sphere on the inside of which are mounted 100 BF_3 – detectors. Each counter can be moved independently within a circular area on the sphere and around an axis which is a radius of the sphere (ϕ). The reflections are measured by scanning or

17.5. DATA COLLECTION

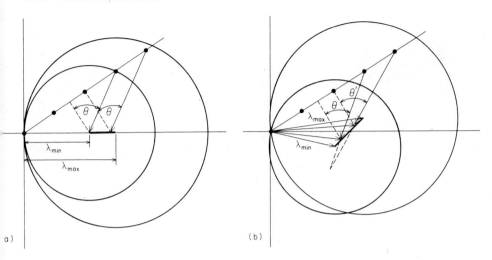

FIG. 17.4 Ewald constructions for classical (a) and modified (b) Laue methods. In the classical methods all wavelengths come from the same direction and there is in principle no limitation of wavelength band; different orders of reflection overlap. In the modified method the different wavelengths of a restricted wavelength band coming from different directions gives a separation of different orders. (Klar, 1972)

by using a wide aperture. Klar estimates that, with one single crystal orientation, between fifty and eighty reflections can be measured simultaneously under favourable conditions.

Position sensitive defectors offer a further alternative for neutron diffraction. U. Arndt and D. Gilmore (unpublished results) have devised a television neutron diffractometer. The diffraction pattern is formed on a flat phosphor screen of 40 cm diameter which can be placed at varying distances from the crystal

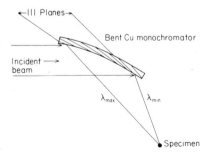

FIG. 17.5 A bent copper monochromator for use with the modified Laue method.

FIG. 17.6 The interior of the sphere with the Eulerian cradle of Klar's single crystal neutron diffractometer with 100 movable detectors.

17.6. NEUTRON FOURIERS

according to resolution required. The resultant light image is demagnified and focused on the photo cathode of the television camera. The image is recorded by a computer. The crystal is rotated at a speed of 1/500° per second. The rate of data collection for a medium size protein is probably comparable to the rate attained by conventional single counter X-ray diffractometers. This type of position sensitive detector could be used in conjunction with the modified Laue method or, after some modification, with the time of flight method.

The different problems in collection of protein neutron data require different solutions. What is suitable for a low resolution data set of a small protein may have disadvantages in the high resolution data collection of a large protein or the measurement of anomalous scattering, or vice versa. As in the case of X-ray diffraction development of various techniques is required.

17.6 NEUTRON FOURIERS

If a protein has alrady been studied at high resolution by X-ray diffraction, the phases needed for the neutron Fourier can be estimated. The calculation of a Fourier map of the protein, using the neutron amplitudes and the phases estimated from the X-ray data may give hydrogen atom positions and allow the distinction between nitrogen and oxygen. This information cannot be obtained from the electron density map.

It is first necessary to make a good estimate of the phases. The phases used in the X-ray Fourier summation will not be very useful as the neutron scattering factors are quite different from the X-ray ones. However, if the positions of the atoms are known, the structure factors for neutron diffraction can be estimated. They will not be good estimates as the positions are badly defined and will probably have an average error of about 0.30 Å, even if they are derived from a good high resolution X-ray density map. Furthermore the hydrogen positions are not known. As these comprise 50% of the scattering matter the calculated structure factors with these atoms omitted may well be quite wrong. Many of the hydrogen atom positions are determined by the positions of the other atoms in the protein and these can be calculated for use in the structure factor calculation. However, this procedure may be a little dangerous if there is any ambiguity in the estimated positions.

Schoenborn *et al.* (1970) have computed a 2 Å Fourier of myoglobin using phases calculated from the non hydrogen atom positions. For this study they used crystals of dimensions 4 mm x 3 mm x 2 mm. These were soaked in mother liquor in which H_2O was replaced by D_2O to reduce the incoherent scattering (see Section 17.3). These crystals were mounted in quartz glass tubes. The diffraction experiments were carried out using a neutron flux at the crystal position of 10^7 neutron/cm²/sec, and wavelengths of 1.2 Å and 1.6 Å. 10,000 reflections were measured with a four circle diffractometer using a counting time of between 5 and 15 minutes per reflection (see Fig. 17.2) 70% of the reflections had $I > 3\sigma$.

The density map indicates that the molecule is well defined. The areas between the molecules show many positive features suggesting that some of the water of crystallization is quite ordered and also that it has been exchanged for D_2O (hydrogen atoms would give negative peaks – see Table 17.II). Hydrogen atoms within the molecule are indicated by negative scattering density. Over half of the hydrogen atom positions can be located from the density map, and some glutamines and histidines can be uniquely positioned as a result of the higher scattering density of nitrogen compared to oxygen and carbon (see Table 17.II) (Schoenborn, 1969, 1971, 1972; Norvell et al., 1975).

17.7 COMPARISION OF X-RAY AND NEUTRON AMPLITUDES

Even if the phases are unknown some information about the molecular organization in complex structures such as lipoproteins or viruses may be available from comparison of the low order reflections obtained from neutron and X-ray diffraction (Schoenborn et al, 1970). Molecular scattering factors of typical organic groups are rather different for neutrons and X-rays. Table 17.III gives the averaged neutron scattering length per atom. Whereas for protein the value is +0.18, for lipid it is -0.03×10^{-12} cm; for X-rays, the values are similar to each other. Thus a comparison of the low order reflections from X-ray and neutron studies of a protein lipid complex may give some information about the positions of the protein and lipid. This has already proved useful in the study of nerve membranes (Caspar and Kirschner, 1971) but it may have applications in protein complexes with lipid and other organic materials.

17.8 ISOMORPHOUS REPLACEMENT WITH NEUTRON DIFFRACTION

As most atoms have similar neutron scattering length, the use of isomorphous replacement in neutron studies in the conventional manner is not so straight forward. "Heavy atoms" such as platinum and mercury have scattering lengths comparable with that of nitrogen. The addition of these atoms would therefore make little measurable difference to the neutron intensities.

An exception to this is dysprosium (Dy^{164}) which has a very large coherent scattering amplitude (4.94×10^{-12} cm). S. Mason (unpublished results) has suggested that it may be easily substituted in proteins in a similar manner to other rare earths, as discussed in Section 8.6.1. It may be possible to measure the isomorphous differences and use them for phase determination.

A less conventional technique is to exploit the isomorphous differences caused by deuteration of subunits in crystals or protein complexes. Most

TABLE 17.III
Average neutron scattering factors for groups of atoms found in biological structures. (From Schoenborn et al., 1970).

Molecule	Averaged neutron Scattering length per/atom (10^{-12} cm^1)
H_2O	−0.06
D_2O	0.60
PO_3	0.57
Protein	0.18
Lipid	−0.03

protein complexes can be dissociated and then reassembled in a biologically active form. For a bacterial complex this can be achieved by growing the bacteria on a D_2O medium. Complexes can be formed with specific subunits deuterated by providing the relevant components for reassembly. Comparison of the neutron diffraction of series of such specifically deuterated complexes should then allow a computation of the intersubunit distances and thus the molecular organization (Engelman and Moore, 1972).

17.9 ANOMALOUS SCATTERING OF NEUTRONS

We have seen in Section 7.2 that the presence of a natural or resonance frequency in the scattering density gives rise to anomalous scattering for X-rays. This anomalous scattering has two components: (i) a real (dispersion) component which has values at all wavelengths and changes sign at the resonance frequency and (ii) an imaginary (absorption) component which has values only at the resonance frequency. Anomalous scattering also occurs with neutrons in an analogous way, and this is illustrated in Fig. 17.7. We can thus use anomalous scattering of neutrons to simulate isomorphism (Peterson and Smith 1962; Sikka, 1969) to solve the phase problem (Section 7.3). Alternatively we can use the inequality of $F_{PH}(+)$ and $F_{PH}(-)$ resulting from the absorption component to calculate the phases (Peterson and Smith, 1962; Ramaseshan, 1964; Dale and Willis, 1966; MacDonald and Sikka, 1969).

Stable isotopes with a resonance wavelength of between 0.5 and 2.0 Å include Gd^{157} and Cd^{113}. Others such as Lu^{176} and U^{233} are unstable and therefore not suitable for diffraction studies (see Table 17.IV).

The dependence of the real and imaginary parts of the scattering length on the wavelength is illustrated in Fig. 17.7. It can be seen that the imaginary part is only large in the vicinity of the resonance. It is therefore necessary to work close to $\lambda = 0.678$ Å to measure anomalous scattering with Cd^{113}. This wavelength is

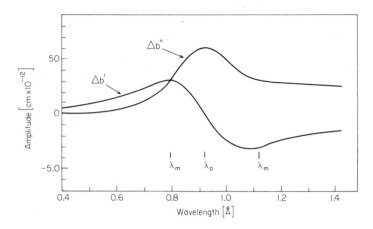

FIG. 17.7 Wavelength dependence of the resonant contributions to the real (b') and the imaginary (b'') components of the scattering amplitude for the case of the thermal neutron resonance in Sm^{149}. The resonance is centred at $\lambda_0 = 0.92$ Å; the peaks of the dispersion curve appear at the positions indicated by λm.

rather inconvenient for two reasons. First, it corresponds to a less intense part of the neutron beam as the thermal neutron flux peaks at about 1 Å. Secondly, the short wavelength necessitates small crystals for structures with large cell dimensions in order to resolve the reflections. This would further decrease the intensities of the diffracted beams.

Sm^{149} has a resonance wavelength at 0.915 and is therefore more suitable for large crystal cells. Gd^{157} has an even larger resonance absorption at 1.61 Å. As samarium and gadolinium can be relatively easily introduced into protein crystals (see Section 8.6.1), these isotopes seem to be most useful in neutron anomalous scattering measurements of proteins.

Neutron diffraction measurements could be taken at two different wavelengths, one close to the resonance wavelength and one much further away. The measurements taken at a distance from the resonance scattering should, within experimental error, obey Friedel's Law and these are used as the native set. The Bijvoet differences can be used to calculate a $(\Delta F_{ano})^2$ Patterson function from which the coordinates of anomalous scattering atoms can be determined (see Section 11.3.2). The structure factors of the anomalous scatterers can then be estimated and used in conjunction with the native and derivative amplitudes to estimate phases in the same way considered for X-ray diffraction (see Chapter 12). The best strategy for data collection is not clear. Measurement of two sets of Friedel pairs or of three sets at different wavelengths with no Friedel pairs (i.e. to simulate two isomorphous derivatives) may be better alternatives. In all cases the advantage of neutron measurements is that native and derivative

TABLE 17.IV

Some properties of resonant scattering nuclei (at resonance) with resonances for neutrons of $\lambda > 0.6$ Å.[a]

Isotope	Half life	$\lambda_{resonance}$ (Å)	b_{real} (10^{-12} cm)	$b_{imaginary}$ (10^{-12} cm)	σ_{scat} (barns)	σ_{abs} (barns)
Gd^{155}	Stable	1.747	0.806	1.68	43.9	5.86×10^4
Gd^{157}	Stable	1.614	0.809	7.16	652.1	2.31×10^5
Xe^{135}	15.3 min	0.987	0.769	214.5	5.8×10^5	3.3×10^6
Sm^{149}	Stable	0.915	0.795	6.051	467.0	1.1×10^5
Lu^{176}	3.75 hr	0.756	0.840	0.962	20.5	1.5×10^4
U^{233}	1.6×10^5 yr	0.688	0.923	0.003	10.7	45.0
Cd^{113}	Stable	0.678	0.725	4.507	26.1	6.08×10^4

[a]From Stehn, J. R., Goldberg, M. D., Mayurno, B. A., Wiener-Chasman, R., *Neutron Cross Sections, BNL 325*, Brookhaven National Laboratory (1964).

data are measured from the same crystal and so there is no problem caused by lack of isomorphism in the phasing step.

Preliminary studies of a cadmium myoglobin derivative by Schoenborn have demonstrated that meaningful Bijvoet differences can be measured despite the necessity of small crystals and Mason has studied crystals of Gd^{157}–insulin which give good anomalous scattering. These techniques appear to have a very great potential.

17.10 GAMMA–RAY RESONANCE

Gamma-rays, which are emitted in the radioactive decay of certain isotopes, can also be diffracted by crystals, and are therefore potentially useful in determining crystal structures.

Gamma-radiation is diffracted by two different mechanisms. First, the electronic shells of all atoms give rise to a scattered wave in an analogous way to X-rays. Secondly, certain nuclei may have a resonance frequency which corresponds to the gamma-radiation frequency and these will give rise to anomalous scattering. As with X-rays and neutrons, the anomalous scattering will have a real (n') and imaginary (n'') component.

The imaginary or absorption component will show a sharp peak corresponding to the absorption wavelength, λ_0, of the nuclear transition, while the real or dispersion component will change sign at λ_0 but have measurable values at wavelengths some distance from this. As there is a sharp single resonance the dispersion curve will be antisymmetrical about λ_0 and in this way differs from the situation of X-ray anomalous scattering where there is a continuum of absorption frequencies (Chapter 7.). Theoretically it is possible to solve the phase problem using anomalous scattering techniques with gamma rays which are equivalent to those that we have already described for X-ray and neutron diffraction; but in practice there are many difficulties.

Studies have so far employed the 14.4 keV gamma-radiation in ^{57}Fe, which is emitted on the radiation decay of ^{57}Co. This radiation has a wavelength of 0.86 Å and therefore is suited for structure studies using large cells. The resonance scattering amplitudes are several hundred times that of X-rays or neutrons and therefore anomalous scattering or resonance techiques can be used with fewer anomalous scatterers, but the brightness of the available radiation sources is much smaller for gamma rays than even for neutrons and are about 10^{-5} of fluxes obtainable with ordinary X-ray tubes. However, before we consider in more detail the problems which are evident in the use of gamma-radiation, let us survey its advantages which justify the further study of this technique.

Mossbauer (1973) has underlined several useful aspects of gamma rays: (1) The phase determination for any Bragg reflection can be calculated using the

17.10. GAMMA-RAY RESONANCE

same crystal by measuring the scattered intensity at three different frequencies. This is particularly easy to achieve with gamma rays as the frequency can be varied by using the Doppler effect, i.e. by moving the source towards or away from the crystal. Measurements of the diffracted amplitudes can then be made at different wavelengths in the region of λ_0 in order to simulate isomorphism (see section 7.3) and to solve the phase problem. (2) In contrast to the use X-rays and neutrons, the frequency changes required with the gamma-ray method are extremely small (of the order of 10^{-8} eV). Thus there are no differences of absorption of the radiation diffracted by the normal scatterers, the counter will not change in sensitivity with the different radiation frequencies and there will be no change with frequency of the geometry of the diffraction experiment. (3) Although gamma-rays show a well-known and easily accountable radioactive decay, they show no longtime fluctuations in their intensity. Gamma-ray sources have better stability than X-ray generators or neutron reactors. (4) Iron occurs naturally in many proteins such as haemoglobin, myoglobin and cytochromes. The introduction of the required resonant nucleus, ^{57}Fe, then is an isotope exchange problem which should be more straight-forward than the usual preparation of heavy atom derivatives required by X-ray analysis (Chapter 11).

However, the crystal specimens, the gamma reaction detectors and in particular the gamma ray sources present very real problems, and we must now discuss these and outline possible remedies.

17.10.1 Crystal specimen

The gamma ray lines reside above a general background, and the fraction of the gamma-line intensity which occurs at the resonance frequency is known as the Debye–Waller factor or the f-factor. In the case of the 14.4 keV line, the f-factor has a value of about 90% for metallic iron or ionic iron compounds. However, for iron-containing proteins such as myoglobin the f-factor may be as low as 1%, at T = 300°K because the iron is bound to the rest of the molecule by comparatively weak bonds. However, this can be overcome by cooling the crystals to T = 185°K (Parak and Formanek, 1971) which is achieved by the application of external hydrostatic pressures of up to 2.5 k bar during the cooling period. This induces a different ice phase, and avoids cracking of the crystals which would normally occur on freezing the solvent of crystallization. Thus gamma-ray resonance techniques are feasible for proteins but have to be carried out at very unphysiological conditions.

17.10.2 Gamma-radiation detectors

The most efficient counting method uses area detectors which can discriminate between different Bragg reflections for the crystal at one orientation given by a divergent incident beam. These detectors must combine a high energy resolution with a high detection efficiency. The usual multiwire and position sensitive counter system with a planar geometry cannot be used. On the other hand, solid state Si(Li) – counters with many elements (surface areas of 2.5 mm^2) have a moderate position and high energy resolution and have high counting efficiency (Mossbauer, 1973).

17.10.3 Gamma radiation sources

The weakness of the source remains the biggest problem. Clearly there is no advantage in increasing the area of the gamma-ray source beyond the dimensions of the crystal, and efforts have concentrated on increasing the activity per unit area (Mossbauer, 1973). The thickness of the source is limited by self absorption problems inside the source itself. This is partly due to the generation of ^{57}Fe on radioactive decay of ^{57}Co which then absorbs the radiation so that it has less and less chance of leaving the source. Methods are being studied which allow the ^{57}Fe to be removed easily at intervals from the active ^{57}Co, but this is proving to be a difficult technical problem.

So far phase determination using gamma-ray anomalous scattering has been achieved only for a small molecule, $K_3Fe(CN)_6$ (Parak, 1971). However, studies are underway on crystals of myoglobin. It is estimated that a 3 Å resolution study would require a total measuring time of about one year using a crystal of 1 mm^3, a carrier-free source of 150 mC strength spread over the area of 1 mm^2 and an area detector. This makes the method potentially very hopeful, and it may very well be more useful than neutron diffraction in solving the phase problem in protein crystallography.

18

ELECTRON MICROSCOPY

18.1 INTRODUCTION

The electron microscope has undoubtedly contributed more than any other technique to our understanding of cellular morphology and of the organization of intracellular organelles. The ability to form a direct image of suitably prepared specimens at very high magnifications, has been exploited in almost all areas in the biological and materials sciences. Nevertheless the image formed in the electron microscope suffers from severe limitations. Firstly, although the nominal resolving power of an electron microscope is of the order of 3 Å, the resolution obtainable for biological specimens is usually no better than 15 Å. Secondly, although biological structures are three-dimensional, the image formed in an electron microscope is a two-dimensional projection of the structure. Thirdly, for biological molecules there are considerable problems associated with contrast and specimen preservation. These limitations have meant that the contribution of electron microscopy to the study of protein structure has been relatively meagre compared with the wealth of detail supplied by high resolution X-ray crystallographic studies. Observations on protein molecules in the electron microscope have been limited to observations of molecular size and shape. For example, at a time when there was considerable controversy over the biochemical evidence for the subunit structure of aldolase, electron microscopy showed convincingly that the molecule was composed of four subunits (Penhoet et al., 1967). To this day the reliable imaging of an individual protein molecule as small as molecular weight 160,000 by conventional electron microscopy represents a great achievement. A more recent example of determination of molecular sub-structure, is given by the work on the $C1q$ complex of the complement system of the immune response. Electron microscopy of negatively stained specimens has revealed a "flower-pot" arrangement of six strands with heads joined to a thicker compact structure at the base (Shelton et al., 1972; Knobel et al., 1975). Examination of protein crystals coupled with image analysis and reconstruction techniques has provided slightly more detailed information. For example the somewhat unusual shapes of phosphoglycerate kinase (Eagles, 1970) and γG1 immunoglobulin (Labaw and Davies, 1971), which are now known from X-ray studies, were observed in electron micrographs of crystals of these proteins. Electron microscopy can produce results from very small amounts of microcrystals. Because of this it is often possible to deduce some information on the overall shape of a protein in advance of material suitable for X-ray studies, as for example in the recent work on the lac-repressor protein (Steitz et al., 1974). Further there is the exciting possibility that the image of the molecule seen in the electron microscope can be used to calculate an approximate set of low resolution phases. These phases could then aid an interpretation of the X-ray diffraction pattern. A promising start along these lines has been made by De Rosier and Oliver (1971) for the core protein of the α-ketoglutarate dehydrogenase complex and for phasing of electron diffraction patterns by Unwin and Henderson (1975).

In spite of the considerable achievements described in these examples, they represent the next best type of study in the absence of suitable large crystals and detailed X-ray work. Information on subunit structure, molecular size and shape can be obtained much more precisely (although admittedly less directly and with considerably more labour) by X-ray diffraction studies. Nevertheless there are numerous proteins and protein complexes which, because of their size or availability, will never be amenable to X-ray diffraction experiments and it is for these problems that we seek reliable high resolution images from the electron microscope. The answer to this quest must lie in new developments both in terms of instruments and in terms of specimen preparation. So far biological specimens have been examined in the dry state and dehydration obviously has a deleterious effect on conformation. The development of an environmental cell for use with the high voltage electron microscope seems likely to overcome this problem. Radiation damage by the electron beam is especially important for biological molecules and the severity of this effect is only just beginning to be understood through systematic studies. Contrast forms another problem. In order to provide sufficient contrast, biological molecules are stained with heavy metal salts. Staining is often achieved by trial and error methods with very little specificity. There is scope for improvements in the staining technique. The new high resolution scanning transmission electron microscope (STEM) permits the location and indentification of individual heavy atom clusters and its development is therefore of tremendous significance for the study of biological macromolecules.

In this chapter we first discuss the scope and limitations of present day conventional electron microscopy so that the protein crystallographer may assess the usefulness of this technique and the type of contribution which it can make both towards the solution of his particular protein problem and in the context of the wider biological implications of his work. There are many textbooks available on this subject. Those especially recommended are G. A. Meek (Practical Electron Microscopy for Biologists: Wiley–Interscience, 1970), C. E. Hall (Introduction to Electron Microscopy: McGraw Hill, 1966) and G. M. Haggis (The Electron Microscope in Molecular Biology: Longman, 1970). We then discuss some of the more recent developments in electron microscopy which are especially relevant for protein crystallography and which may enable the ultimate potential of the instrument to be fully realized. An account of some of these developments is to be found in the publication of a symposium, organized by Huxley and Klug, in *Phil-Trans. Roy. Soc. Lond.* (1971) **B261** pl-230 and published as a book "New Developments in Electron Microscopy" (eds. H. Huxley and A. Klug, Royal Society, 1971).

18.2 THE PRINCIPLE AND DESIGN OF THE ELECTRON MICROSCOPE

The basic difference between the formation of an image in an X-ray diffraction experiment and the formation of an image in the electron microscope is that in the latter case it is possible to view the image directly without the need for phase computation and Fourier analysis. The small negatively charged electrons can be focused using either electrostatic fields or (as in the case of most conventional transmission microscopes) magnetic fields.

18.2.1 The wavelength of the electron

From the deBroglie theory of the wave-particle duality of fast moving particles, the wavelength associated with an electron which has been accelerated through a potential difference of V volts is given by:

$$\lambda = \frac{h}{\sqrt{2me(V + 10^{-6}V^2)}} = \frac{12.3}{\sqrt{V + 10^{-6}V^2}} \text{ Å}$$

where the symbols have their usual meaning in quantum mechanics. The wavelengths for voltages commonly used in electron microscopy are shown in Table 18.I. At 50 kV the electron is travelling with a speed approximately one third the speed of light and at voltages greater than this the relativistic term, which is represented by the $10^{-6} V^2$ term in the above equation, becomes important. The wavelength of the electron is very, very much shorter than the wavelength of X-rays, and in principle it might be expected that the electron microscope should be capable of giving images of much higher resolution than encountered in X-ray diffraction. That this is not the case is due to two factors. Firstly there is the impossibility of designing lenses which are capable of focusing widely scattered electrons and secondly there are almost insurmountable difficulties associated with specimen preparation for the electron microscope, especially for specimens of biological materials.

TABLE 18.I
Wavelength of electrons for accelerating voltages commonly used in the electron microscope.

V k volts	λ Å
50	0.054
80	0.042
100	0.037
1000	0.0087

18.2.2 The formation of the image

The arrangement of lenses in the electron microscope is shown in Fig. 18.1. The electrons are emitted by a hot tungsten filament and are accelerated towards the anode through the appropriate voltage. The electron gun incorporates a bias shield around the filament which is held at a slight negative potential with respect to the filament. This has the dual purpose of focusing the electrons towards the anode and of controlling the emission current. The condenser lens system is used to control the intensity of illumination and the area of the specimen illuminated. The specimen scatters the electrons and the scattered electrons are collected by the objective lens to form an intermediate magnified image. The image is further magnified by the intermediate and projector lenses. The final image is either viewed on the fluorescent screen or the screen is swung out of the way and the image recorded on a photographic plate. Since electrons are scattered by air, the entire system must be kept under vacuum at a pressure of less than 10^{-5} τ and the microscope column must contain the necessary airlocks so that specimens, photographic plates, or new filaments can be inserted while the rest of the column is maintained at high vacuum. The objective aperture is also shown in Fig. 18.1; its purpose is discussed later.

A ray diagram illustrating the formation of a high magnification image with the three image forming lenses is shown in Figs. 18.2 and 18.3a. The magnification range covered is generally between 5,000 and 200,000. The final magnification is changed by varying the strength of the intermediate lens. This alters the position of the object plane of the intermediate lens and consequently the current in the objective lens must also be changed (i.e. the lens focused) to compensate whenever the magnification is changed. In most electron microscopes the projector lens magnification is kept fixed.

18.2.3 Electron diffraction

In the examination of crystalline specimens, both the real image and the electron diffraction pattern can be recorded. The arrangement of the lenses for electron

18.2.3. ELECTRON DIFFRACTION

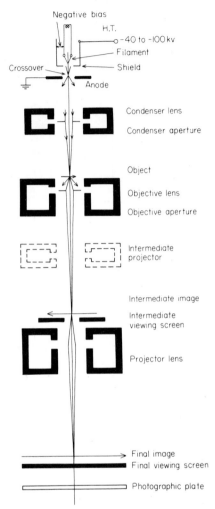

FIG. 18.1 The arrangement of the lenses in the electron microscope. (from Hall, 1966)

diffraction is shown in Fig. 18.3b. The electron diffraction pattern is formed in the back focal plane of the objective lens. The intermediate lens is therefore adjusted so that this plane becomes the object plane of the intermediate lens and the image of the diffraction pattern formed by the intermediate lens falls in the object plane of the projector lens. This is equivalent to turning the intermediate lens to zero magnification so that the rays which pass through the specimen parallel to the axis of the microscope are brought to a focus on the fluorescent screen. In order to observe the high order terms of the diffraction pattern the objective aperture must be removed. In this mode of operation the whole area of

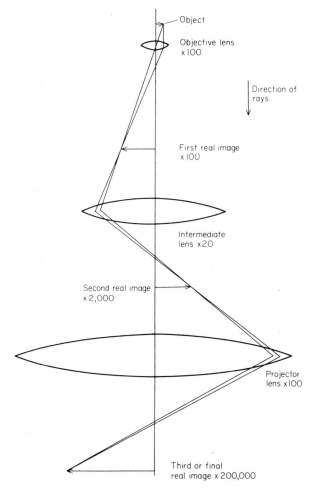

FIG. 18.2 A ray diagram showing the formation of an image by a three lens system. (from Meek, 1970)

the specimen illuminated contributes to the diffraction pattern. In many cases with microcrystals we may wish to examine only a very small area (a few micron in diameter, say) of the specimen. Selection of such a small area is difficult to perform at the specimen but is readily achieved by placing a field limiting aperture in the image plane of the objective (Fig. 18.3b). The specimen is magnified about ×100 at this stage and hence apertures of reasonable physical dimensions can be used. It is possible to turn from the selected area electron diffraction pattern of the specimen to the magnified image merely by adjusting the intermediate lens (Fig. 18.3c).

18.2.4. RESOLVING POWER

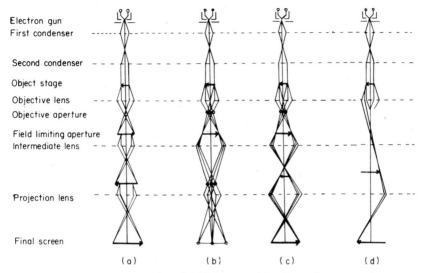

FIG. 18.3 Ray diagrams for (a) high magnification electron microscopy, (b) electron diffraction (selected area), (c) selected area electron microscopy and (d) dark field electron microscopy.

In this arrangement for electron diffraction the maximum observable spacing is usually less than 15 Å. Diffraction from repeat spacings greater than this is obscured by the main beam. In some microscopes longer camera lengths, which allow the resolution of longer spacings, are achieved with an additional projector lens, sometimes referred to as a diffraction lens. Alternatively it is possible to modify the conventional microscope for low angle diffraction (Glaeser and Thomas, 1969). The double condenser lens system is used to focus a minimum beam spot in a plane just below the specimen. The condenser aperture is replaced by a very small aperture (25 μ) whose size determines the area of the specimen illuminated. The objective and intermediate lenses are switched off and the projector lens alone is used to focus a magnified image of the electron diffraction pattern on the screen. With this system camera lengths suitable for most protein crystals can be achieved. Unfortunately because of alignment problems, it is not easy to turn between the image and the diffraction pattern. The use of the electron diffraction pattern as a sensitive indication of radiation damage is discussed in Section 18.4.

18.2.4 Resolving power

The ultimate resolving power of the electron microscope is governed by the performance of the objective lens. Most objective lenses are of the polepiece design. They consist of a solenoid which is encased in a soft iron sheath except for a small gap on the inside of the lens which is filled with a non-magnetic spacer. When a current is passed through the solenoid a strong magnetic field is developed at the gap and this is concentrated into a very short distance (a few mm) and brought as close as possible to the axis of the microscope by means of the cylindrical, tapering iron polepieces. The strength of the magnetic field and hence the focal length of the lens is dependent upon, although not quite linearly related to, the current through the solenoid. This means that the magnification, focus, and illumination are easily controlled by adjusting the appropriate lens current controls. In spite of considerable ingenuity and improvements in recent years in the design polepieces, magnetic lenses always exhibit the defect of spherical aberration. This means that the strength of the lens is greater away from the axis of the microscope so that electrons which are widely scattered are brought to a focus at a different position from the electrons which travel close to the axis. The result is that rays from a point object are spread over a radius Δr in the image plane. Magnetic lenses are always convergent and hence there is no way of compensating for spherical aberration analogous to the combination of convergent and divergent lenses in the optical microscope.

It is possible to show that a magnetic lens is able to focus to the extent that Sin α may be approximated by α, where α is the semi-angle of the cone of scattered electrons. Since

$$\sin \alpha = \alpha - \frac{\alpha^3}{3!} + \frac{\alpha^5}{5!} + \dots$$

The radius Δr is approximated by the first order correction term:

$$\Delta r = C_s \alpha^3$$

where C_s is the spherical aberration coefficient. By careful design of polepieces and excitation current, a lens with $C_s = 0.5$ mm with a focal length of 1 mm has been achieved but for most microscopes $C_s \simeq 2.8$ mm and $f = 3$ mm.

In order to minimize effects from spherical aberration α needs to be made as small as possible, but in order to improve resolutions, the Abbē criteria (d min $= 0.5\lambda/\alpha$) which is based on diffraction theory, requires α to be as large as possible. These two conflicting requirements result in an optimum value for α which is given by:

$$C_s \alpha^3 = \frac{0.5\lambda}{\alpha}$$

i.e.

$$\alpha = \left\{ \frac{0.5\lambda}{C_s} \right\}^{1/4}$$

$$\simeq 5.6 \times 10^{-3} \text{ radians for } C_s = 2.8 \text{ mm}$$
$$V = 50 \text{ kV}$$

$$\therefore d_{min} = 0.6\, C_s^{1/4} \lambda^{3/4}$$

$$\simeq 5\ \text{Å}$$

In order to limit the rays to those which subtend an angle α close to the optimum value, a small objective aperture is inserted between the specimen and the objective lens (see Fig. 18.1). An objective aperture of 50 μ placed 3 mm from the specimen subtends an angle of 8×10^{-3} radians at the specimen.

Thus spherical aberration forms the main limitation to the theoretical resolving power of the electron microscope. Since C_s is involved in the equation as the ¼th power, any improvement needs to be of an order of magnitude in order to make a significant improvement in the resolution. Some advantage may be obtained by working at higher voltages (i.e. shorter wavelengths), but the gains are small since spherical aberration also increases with voltage. The ultimate theoretical resolution achieved with the very best microscopes is of the order of 2 Å.

18.2.5 The interactions of electrons with matter

However, this nominal resolution can only be achieved with ideal, ultra-thin specimens. For a specimen of any appreciable thickness chromatic aberration effects become important. The interaction of electrons with matter can be divided into three distinct processes: (i) No interaction. Over 90% of the incident electrons pass straight through the specimen and do not interact at all. (ii) Elastic scattering. The incident electron is scattered by the screened Coulombic potential which exists between the positively charged nuclei of the atoms of the specimen and the negatively charged incoming electron. Since the mass of the proton is 2,000 times that of the electron, the electron is scattered with essentially no loss of energy. It is these elastically scattered electrons which are most important in image formation in the conventional transmission electron microscope. The Coulombic scattering of electrons by matter is approximately a million times stronger than the corresponding interaction of X-rays with matter. This means that appreciable electron scattering can take place with very thin films and intensities of the scattered rays recorded within seconds rather than hours. (iii) Inelastic scattering. The fast moving incident electron can also interact with the slow moving atomic electrons of the specimen. Some energy will be transferred between the two and the scattered electron will continue on its way having suffered a loss of energy, and thus a change of wavelength. Electrons of different wavelengths are brought to a focus at different positions and this gives rise to a blurring of the image known as chromatic aberration. The inelastically scattered electrons contain significant information about the specimen, but in the conventional transmission microscope this information cannot be used. Only the high resolution scanning electron microscope (Section 18.7) has the ability to provide both spatial and energy resolution. In order to minimize chromatic aberration, very thin specimens are required. For graphite the energy loss suffered by the electrons is of the order of 20 V. With a typical lens of chromatic aberration coefficient of C_c = 2 mm, the radius of the chromatic disc is approximately 8 Å at 100 kV for a 100 Å layer of carbon. A rough rule which governs the relationship between the thickness of the specimen and the resolution is:

resolution $\simeq$ $1/10$ × thickness of specimen at 100 kV.

The main criteria is the "mass-thickness", the product of physical thickness and density. At higher voltages, such as the million volt electron microscope, energy loss due to inelastic scattering is much less severe, and thicker specimens can be used (Section 18.5).

18.2.6 Contrast

Finally, we need to consider the role of contrast in image formation. There is little point in striving after the best resolving power if the details in the specimen cannot be distinguished through lack of contrast. In the light microscope contrast is mostly achieved by differential absorption of light by different parts of the object. In the electron microscope with ultra-thin specimens, contrast is almost solely achieved by scattering. Suppose an area of the specimen that contains an accumulation of heavy atoms scatters electrons. Most of these scattered electrons will be stopped by the objective aperture and the corresponding area in the image plane will appear dark on the fluorescent screen and relatively transparent on the photographic plate. A part of the specimen containing a low concentration of matter will provide little scattering and this area will appear bright on the screen and dark on the photographic plate. Thus the objective aperture serves two purposes. It restricts scattered rays to those which pass nearly parallel to the axis and which can be focused by the objective lens and it provides a mechanism for contrast by blocking out the widely scattered electrons. In fact in some cases it may be advisable to sacrifice resolution by using a smaller objective aperture than is strictly demanded in order to enhance contrast.

The other source of contrast familiar to electron microscopists is known as "defocus contrast" or phase contrast. This effect is due to the formation of Fresnel fringes about any parts of the specimen where there is a rapid change in mass thickness. Fresnel fringes arise from interference effects in the out of focus position. If the objective aperture is removed and the objective focus control turned rapidly from one side of focus to the other, a marked increase in contrast is observed on either side of focus which is at its greatest in the slightly underfocused position. Nearly every biological electron microscopist takes advantage of this effect in order to improve the contrast in his electron micrographs. Care and judgement are needed however in order to avoid loss of

18.2.8. PHASE CONTRAST MICROSCOPY

resolution and introduction of spurious images. The effects of spherical aberration, aperture size and defocus contrast on resolution are discussed further in Section 18.6 on image analysis.

In biological specimens the problems of contrast are likely to be especially acute. Biological molecules are composed almost entirely of carbon, nitrogen, oxygen and sulphur, and are usually examined either supported on a thin carbon film or embedded in a plastic resin. Thus there is essentially little difference in mass thickness between the specimen and the background. For this reason it is necessary to stain biological materials with heavy metal stains. The incorporation of a stain inevitably leads to some modification of the structure, especially since relatively high mass thicknesses are required in order to provide sufficient contrast over noise due to fluctuations in background. It is not surprising therefore that in recent years there has been an investigation into instrumental methods of providing contrast. Both dark field and phase contrast electron microscopy have been developed, the basic principles of which have their analogies in light microscopy.

18.2.7 Dark field electron microscopy

In dark field electron microscopy, the image is formed solely with the scattered electrons and the main beam of unscattered electrons is prevented from reaching the image plane. This is achieved either with conical illumination, by the insertion of a precisely aligned circular stop in the condenser aperture, or by tilting the main beam so that the undeflected beam is stopped by the objective aperture, or by shifting the objective aperture itself to one side so that only scattered electrons are allowed through (Fig. 18.3d). This last method is the easiest to use, since it requires only a small alteration from the normal bright field conditions, but it necessarily requires that the objective lens accepts more widely scattered electrons which may result in some loss of resolution due to spherical aberration. Exposure times may be up to thirty times as long with dark field microscopy, because fewer electrons are reaching the image plane. Hence an extremely stable electron microscope is required. With this technique Ottensmeyer and his colleagues (Henkelman and Ottensmeyer, 1972; Whiting and Ottensmeyer, 1972) have reported the visualization of single atoms of several heavy metal complexes and of single heavy atoms complexed with specific bases in DNA strands.

18.2.8 Phase contrast microscopy

Phase contrast electron microscopy has been pioneered by Unwin (1971, 1972) who has shown that a suitable electrostatic phase plate can be produced by placing a small aperture, with a thin, poorly conducting thread spanning its diameter, at the back focal plane of the objective lens. The uniformly intense beam of unscattered electrons causes the central portion of the thread to become positively charged and the resulting electrostatic field produces the necessary phase shift between scattered and unscattered electrons. The best material for the thread was found to be a gold plated spider's thread. Ironically even with the phase plate in place, it was found necessary to enclose the specimen (tobacco mosaic virus) in a film of stain, in order to prevent it from disintegrating in the electron beam. Images of very high quality have been obtained. These are felt to be a more realistic representation of the specimen than have hitherto been obtained in normal bright field conditions, especially since they lead to Fourier transforms which are in better agreement with the X-ray diffraction results than the normal bright field images. The resolution in the phase contrast electron micrographs extends to 8.5 Å, a limit which is probably set by specimen damage. The technique requires critical adjustment of the phase plate and stringent focusing conditions.

18.3 PREPARATION OF BIOLOGICAL SPECIMENS

The preparation of biological specimens for the electron microscope must satisfy very rigorous requirements. Firstly, the specimen must be preserved in some way so that it will withstand desiccation in the high vacuum of the microscope column and the bombardment by a high flux of electrons; secondly, the specimen must be thin so that sufficient electrons can reach the image plane without significant energy loss; and thirdly, the specimen must be stained in some way to provide contrast. There are four basic techniques of specimen preparation which are applicable to the examination of protein crystals: thin sections; negative staining; shadowing; and freeze fracture.

18.3.1 Thin sections

In this method, which is borrowed from the histologists and widely used for examining soft tissues, the material is embedded in a plastic resin which is then cut into thin sections with an ultramicrotome. The specimen is first fixed in order to preserve the structure. Typically the material is treated with a 1% glutaraldehyde solution which reacts rapidly, cross-linking the protein probably by reaction with the lysine residues, and renders the specimen insoluble. It is customary to follow at this stage with post-fixation with osmium tetroxide solution which provides additional stabilizing cross-links and some degree of heavy metal staining. Because most resins are not miscible with water, it is then necessary to dehydrate the specimen by transferring it through a series of successive water–alcohol mixtures until it is in absolute alcohol. The material is then embedded in an appropriate epoxy-resin, such as epon or araldite, and the blocks left to harden in an oven at 60°C for 48 hours. The blocks are afterwards trimmed and thin sections cut with an ultra-microtome. The thickness of the

18.3.2. NEGATIVE STAINING

sections is judged according to the interface colours observed when the section is floating on water. These are given approximately by:

Dark grey	less than	400 Å
Grey		400–500 Å
Silver		500–700 Å
Gold		700–900 Å
Purple	greater than	900 Å

For work with biological specimens, a thickness of around 500 Å is usually sought. The sections are transferred to an electron microscope grid (typical dimensions: diameter 3.05 mm; 200 holes per square inch, 50 μ thick) and stained by soaking for periods between 1 to 30 minutes in solutions of uranyl acetate, or lead citrate, or phosphotungstic acid. It is generally assumed that the type of staining which results from this procedure is positive staining (that is the specimen itself takes up the stain) but Labaw and Davies (1971) report that with crystals of γGl immunoglobulin the stain appeared mostly to penetrate the spaces between molecules. Often double staining techniques are used by soaking the specimen first in one stain and then in a second stain. If the material appears reluctant to take up the stain at this stage, it is sometimes worthwhile to include a brief staining procedure after the osmium tetroxide fixation step.

Because of the problems in cutting and handling very thin sections, it is difficult to achieve resolutions better than 25 Å with this technique. Protein crystals are often hard to cut and some experimentation with different resins and hardeners may be needed in order to match the hardness of the block to the specimen. The dehydration steps in alcohol are time consuming and may lead to structural artefacts even with fixed specimens.

The great advantage of this method of specimen preparation for protein crystallography is that it is possible to examine the same crystals in the electron microscope that have been used in the X-ray diffraction experiments. Further the crystals can be oriented in the block so that sections in preferred orientations may be cut. Precise orientations can then be achieved with a tilting stage in the electron microscope.

The preparation of specimens by this method is lengthy and involves several steps where artefacts might be introduced. New advances in cryobiology might well prove relevant. The specimen is rapidly frozen to liquid nitrogen temperatures, and thin slices cut with a cryomicrotome and examined in the microscope still at low temperature. Many problems still remain to be overcome with this approach, however.

18.3.2 Negative staining

The method, described rather loosely as negative staining, is most suitable for high resolution work with particulate material, such as protein crystals, viruses, ribosomes and protein molecules, where the specimen itself is already very thin.

For the examination of protein crystals by this method, the smallest possible microcrystals are required; the sort of crystals which are barely visible under the light microscope and which might excite the biochemist when first observed as a "sheen" during the protein preparation. The requirement for very small crystals means that some progress towards structural information can be made with the electron microscope in advance of X-ray diffraction studies. An indication of the suitability of the crystals for X-ray work might be obtained for example, although there is always likely to be some uncertainty concerning the precise similarity between the micro-crystals and larger crystals.

A support film is needed in order to prevent the particulate material slipping through the holes in the grid. Each laboratory has its own preferred method of preparing such films. A holey carbon film has been found most successful for protein crystals. This is prepared by covering the grid with a thin (approximately 200 Å) film of plastic, such as formvar, in which a few drops of water are finely dispersed. A thin layer of carbon (approximately 20 Å) is evaporated onto the plastic film and the plastic is subsequently removed by exposing the grid to the vapour of a suitable solvent such as chloroform. The carbon layer then provides the necessary support film for the particulate material. Carbon has the advantages that it is stable in the electron beam, it conducts electricity and hence reduces charging effects at the specimen, and it conducts heat from the specimen to the grid bars. The idea behind the droplets of water in the formvar solution is that these should give rise to very small holes in the plastic film. In fortunate cases the specimen can be stretched over a hole and therefore be observed without the carbon background.

In the preparation of specimens, a drop of a suspension of the specimen is placed on the grid and most of the surplus liquid drawn off with filter paper. A drop of a dense salt solution is then added to the grid, surplus fluid again drawn off, and the grid dried rapidly under a lamp. Typical stains are uranyl acetate or uranyl formate and phosphotungstic acid. The principle behind negative staining rests on the assumption that these salts rapidly penetrate the spaces between molecules in the specimen and then set in a glass-like structure so that the protein molecules are preserved in a matrix of electron dense material. On examination of the grid in the electron microscope, amplitude contrast is provided by the scattering of the electrons by the stain and the protein molecules are observed as relatively transparent holes in the stain. The success of salts like uranyl acetate or formate in this method is mainly attributed to the fact that they lose water quicker than the protein molecules in the drying process and hence help to preserve the structure of the hydrated protein. The uranyl salts dry down to amorphous deposits with little granular or crystalline structure and this deposit is likely to be more resistant to beam damage than the biological molecule. An improvement in resistance to damage has been noted in the use of mixed salts such as uranyl formate and aluminium formate. Phosphotungstic acid is thought to be an effective stain because the many polar

18.3.2. NEGATIVE STAINING

atoms on its periphery impose an ordered structure on adjacent water molecules and leave the structure beyond it undistorted so that water can evaporate with little damage to the surface of the molecule. An example of a negatively stained protein crystal is shown in Figs. 18.4 and 18.5.

The term "negative staining" is somewhat arbitrary and the process is invariably accompanied by some degree of positive staining, depending upon

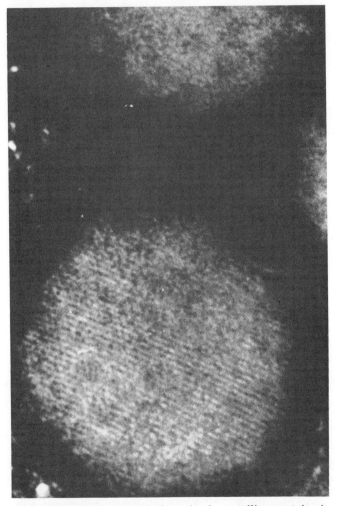

FIG. 18.4 An example of a negatively stained crystalline protein. An electron micrograph of β-granules of the rat showing the storage of insulin in a form which resembles the packing of the insulin molecules in the rhombohedral 2-zinc crystals (from Greider et al., 1969). Negative stain: 2% phosphotungstic acid; periodicity 50 Å.

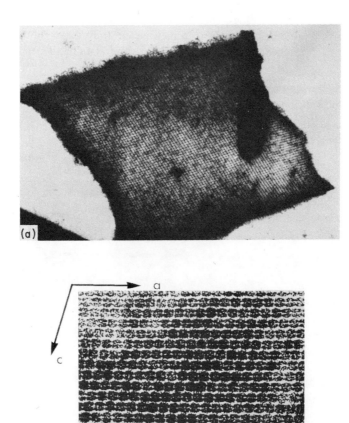

FIG. 18.5 Negatively stained phosphorylase **b** crystals. Drops of the crystalline suspension were placed on a carbon coated grid and after about 1 minute a drop of unbuffered 1% uranyl acetate was applied. Excess solution was quickly removed with filter paper. (a) A single crystal, dimensions approximately $1\ \mu \times 1\ \mu$. (b) An enlarged area of the crystalline array. The protein is black. The

18.3.2. NEGATIVE STAINING

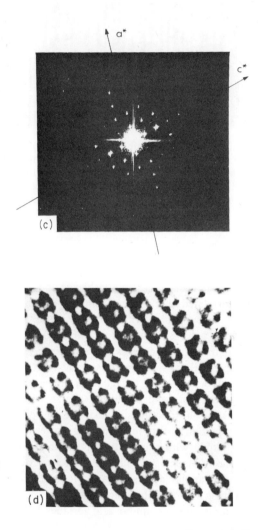

repeat distance of the cell edges is $a = c = 125$ Å. $\beta = 107°$. (c) The optical diffraction pattern of the area shown in (b). The resolution extends to about 23 Å. (d) Optically filtered image. The protein molecules (black) are seen to be composed of four subunits. (From Eagles and Johnson, 1972)

FIG. 18.6 An example of positive staining. (a) An electron micrograph of s.l.s. collagen positively stained with uranyl acetate and phosphotungstic acid. (b) A diagram of the distribution of charged amino acids along an chain. A dark block, three residues wide, is centred on the positions of arginine, lysine, glutamic acid and aspartic acid. A constant axial residue translation is assumed and appears to be justified by the close correspondence between the patterns of figures (a) and (b). (From Doyle et al., 1974)

conditions of pH etc. Collagen forms an outstanding example of a material in which positive staining has been successfully achieved and moreover the staining pattern has been correlated with the chemical structure. Under certain conditions the collagen micro-fibrils (which are composed of three polypeptide chains each 1000 amino acids long wound around each other to form a triple helix) can be induced to align themselves in parallel arrays known as segment long spacing (s.l.s.). When these specimens are stained with uranyl acetate, the negatively charged amino acids (i.e. the carboxylate groups of glutamic and aspartic acid) take up the stain and produce a characteristic banding pattern. Likewise on staining with phosphotungstic acid, the phosphotungstate is taken up by the basic groups and a different banding pattern is observed. These banding patterns are visible with this particular specimen because the ordered array of collagen microfilaments provides a sufficient number of identical staining sites. When the amino acid sequence of the collagen polypeptide is aligned with the electron micrographs there is an extremely good correlation between banding patterns and the amino acid residues (Fig. 18.6) (Traub and Piez, 1971; Doyle et al., 1974).

18.3.3 Shadowing

In this technique contrast and preservation are achieved by casting a thin metal film over the particulate material. The method is especially useful for revealing shape and surface structure and some estimate of the third dimension of the specimen (i.e. its thickness) can be made from the known angle of shadowing.

18.3.4. FREEZE FRACTURE

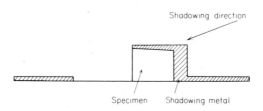

FIG. 18.7   Metal shadowing.

The principle of the method is shown in Fig. 18.7. A thin film of a metal such as platinum or a platinum–iridium alloy is evaporated at an angle onto the specimen in a vacuum. The metal accumulates on the near side of the particle leaving a clear area on the side which is shielded by the particle. When printed in reverse contrast, electron micrographs of shadowed specimens give the impression that particles are illuminated by a setting sun; their three-dimensional appearance is most striking. The method is not suitable for high resolution work because of the size of the metal granules. Nevertheless it has been skillfully applied to the study of macro molecular complexes such as myosin, polysomes and the tangled thread of DNA released from T2 bacteriophage. One of the earliest electron micrographs of a protein crystal was recorded from a specimen prepared in this way (Hall, 1950).

18.3.4 Freeze fracture

The techniques of freeze fracture and freeze etching (Moor and Muhlethaler, 1963; Moor, 1971; Bullivant, 1973) also reveal surface details but the specimen is prepared in such a way as to show a completely different surface from that shown by conventional shadowing techniques. The specimen is preserved in its natural state by vitrification; that is very rapid freezing (approximately $10°C/s$) to a temperature of $-100°C$. A reduction in the normal critical freezing rate for vitrification is achieved by the prior addition to the specimen of an antifreeze agent such as glycerol. The specimen, still held at the very low temperature, is then transferred to a vacuum chamber and fractured by a blow with a knife edge. Specimens tend to fracture along the line of weakest bonding at low temperatures and so surfaces are revealed which are quite different to those that are usually seen. There is good evidence, for example, that membranes fracture along the central hydrophobic plane of the lipid bilayer (Branton, 1969). After fracturing, the surface must be quickly replicated, by shadowing with platinum – carbon at an angle fo 45° and coating with carbon in a direction normal to the specimen surface. The avoidance of any contamination or formation of ice crystals is crucial at this stage. The specimen is subsequently removed from the vacuum, and the replica floated off into a cleaning liquid in order to remove all traces of the adhering specimen. The cleaned replicas are then examined in the electron microscope and reveal details of the fracture surface.

The details of the outer surface of a particle may also be studied by introducing an etching step between the fracture and replication steps. This involves leaving the specimen for a short time (approximately 30 s to a minute) at $-100°C$ to allow sublimation of the water molecules around the outer surface of the particle. Because of the danger of contamination during etching, special precautions are necessary to eliminate any condensable gases from the vacuum chamber.

The outstanding application of freeze fracture techniques has been in the study of the structure of membranes. The method has not yet found widespread application to the study of protein crystals but it is anticipated that interesting results may emerge in the future.

18.4 RADIATION DAMAGE

Observations on specimens in the electron microscope take place in the context of an extremely hostile environment; the specimen is stained with a heavy metal, dried by a high vacuum, and bombarded with a high flux of fast moving electrons. In order to realize the ultimate instrumental resolving power of the electron microscope, which in principle can provide information at the 5 Å level, progress is required on all three of these unfavourable aspects, but especially on the problem of radiation damage. In the brief discussion of the phase contrast electron microscope, we have seen that the resolution was limited by the degree of specimen preservation. The severity of radiation damage in the electron microscope may be appreciated when we consider that at 100 kV a single electron passing through an organic specimen sustains approximately one inelastic collision per $0.1\ \mu$ thickness. Each collision yields an energy of 32 eV (the energy of a C–C single bond is of the order of 5 eV). At a magnification of 10,000 times, a current density of at least 10^{-2} amp/cm^2 is required, and from these values it has been calculated (Stenn and Bahr, 1970) that the energy absorbed by one gram of specimen in a 1s exposure is sufficient to bring 20 g of ice to steam. Systematic studies on the consequences of this large transfer of energy to the specimen have only begun in the last decade.

Absorption of energy by a molecule leads to the formation of excited or ionized species. Transitions of electrons between quantum levels in an excited atom may be accompanied by X-radiation which is characteristic of the atom and this type of "damage" is turned to good account in X-ray microanalyser electron microscopes. A more severe consequence of the impact of the electron beam with the specimen leads to the loss of an orbital electron and the formation of ions or radicals, which are highly reactive and whose subsequent reactions alter the chemical nature of the specimen. From studies with model

18.4. RADIATION DAMAGE

compounds it is known that C–H bonds are more susceptible than C–C bonds (i.e. there is no simple relationship between damage and bond energy although there may be some correlation between damage and the ease with which broken fragments can diffuse apart and so resist repair). Aromatic structures are more resistant to damage and their resistance is directly related to the extent of delocalization of the π electron orbitals (Pullman and Pullman, 1963). Amino acids are susceptible to bond scission at the α–C bonds. The scission of bonds in polymeric structures introduces unsaturation into the remaining portion of the chain which may attempt repair through cross-linking and subsequent alteration in conformation. Although the precise details of changes brought about by radiation damage will depend on the three-dimensional conformation of the individual molecule, we may confidently predict that biological molecules will be significantly altered or destroyed by a high radiation dose. Low temperature stages designed for use at liquid helium temperatures would not in principle be expected to reduce events leading to primary bond scission but nevertheless may prove advantageous in preventing the disruptive consequences of these events due to the trapping of radicals and a reduction in the thermal motion of the molecule. It is interesting that heating effects seem to have little role in radiation damage (at least for specimens in the dry state). Glaeser (1971) has demonstrated a reciprocal relationship between specimen life time and current density at the specimen, which indicates the absence of any dose-rate effect, such as heating, as a cause of damage.

The response to beam damage of three typical crystalline biological specimens, valine, adenosine and uranyl acetate stained catalase, has been investigated by Glaeser (1971) in one of the most detailed studies on radiation damage to date. Low angle electron diffraction patterns were observed and times for total damage estimated from the times for complete fading of the pattern. These times were then correlated with the radiation dose in terms of the current density passing through the specimen. Electron diffraction patterns can be recorded with a considerably lower flux of electrons than that required for the recording of the final image since all unit cells of the crystal contribute to the diffraction pattern but in the formation of the image each unit cell is viewed individually. The use of the criterion of fading of the diffraction pattern is likely to give an overestimate of the radiolytic yield of the specimen since it reflects a breakdown in the order of the crystal and not necessarily damage to each and every molecule in the crystal itself. Valine proved extremely susceptible to radiation damage. The diffraction pattern faded after 15 sec with a current of 10^{-4} amps/cm^2 at 80 kV. This corresponds to a dose of approximately 8×10^{15} electrons/cm^2 and from this value and a calculation of the linear energy transfer for 80 kV electrons a radiolytic yield of eight molecules damaged per 100 eV of energy absorbed was estimated. Adenosine, probably because of the aromatic nature of part of the molecule, proved considerably more resistant to radiation damage. At a similar current density of

10^{-4} amps/cm² at 80 kV, the pattern faded after 100 secs, corresponding to a flux of approximately 6×10^{16} electrons/cm² and a radiolytic yield of 1 molecule destroyed per 90 eV of absorbed energy. The damage reported for the stained catalase crystals was of a fundamentally different nature. No change in the pattern was observed for fluxes up to 1×10^{17} electrons/cm² but thereafter the high angle terms, corresponding to spacing less than 25 Å, disappeared and changes in intensities of the remaining reflections were observed. It appears as if a significant portion of matter within the unit cell changes to a more stable state as a result of irradiation and this is likely to cause some uncertainty as to the relationship between the feature observed in the final image and the original object. This work suggests that unstained biological specimens are extremely susceptible to radiation damage and that, although staining provides some protection, there is likely to be an alteration of the stain and the specimen structures in the electron beam. Indeed direct evidence for changes in the distribution of stain has been obtained by Unwin (1974).

What sort of electron fluxes do we need in order to "see" molecular detail? If we assume that the total number of electrons arriving at the image plane is a random process, then the statistical fluctuations for a given image point receiving n electrons is $\sqrt{n}$ where n is given by

$$n = jtd^2$$

where j is the current density through the object for an exposure time t, d^2 is the area of the picture element, and we assume that all the electrons passing through the object reach the image plane. For a low contrast image to be resolved, the inherent contrast must exceed the statistical fluctuations by the minimum acceptable signal-to-noise ration S/N. This leads to the inequality (Glaeser, 1971)

$$C \geqslant S/N \times \frac{\sqrt{n}}{n} = \frac{S/N}{\sqrt{jtd^2}}$$

where C denotes the inherent image contrast which is defined as the difference in image intensity between two points divided by the local average in image intensity. Hence in order to record some object feature with a contrast C at a resolution d, the integrated flux density jt must be greater than or equal to $(S/N)^2/C^2 d^2$.

If we assume a value of $C/N = 5$, the value usually quoted for visual perception of structure, and an inherent image contrast of $C = 0.1$, the minimum resolution attainable with a flux density of 10^{17} electrons/cm² (the maximum permissible dose for stained catalase crystals) is of the order of 16 Å. In order to resolve a spacing of 5 Å, a current density of 10^{18} electrons/cm² would be required, but this would totally damage the specimen.

Thus radiation damage leads to a paradoxical situation. In order to "see" detail, a relatively high flux of electrons is required; in order to preserve the structure from beam damage, a low flux is needed. It is possible that image intensifiers could permit the formation of images with extremely low fluxes but their use is still controversial and they would not overcome the essentially statistical nature of the problem. In the case of crystals, however, it may be possible to use sub-optimally exposed electron micrographs and to extract the relevant data from the noise by image reconstruction procedures (Section 18.6). The use of protective agents such as those discussed in Chapter 9 (Section 9.3.5) for X-radiation damage may prove useful. Higher voltages also offer a further method for minimizing damage. (Fast electrons interact less than slow electrons.) Calculations and experimental data suggest that improvements in the lifetimes of the specimen of between 3 to 10 times can be expected by raising the voltage from 50 kV to 1000 kV. Radiation damage is likely to prove the main barrier to the achievement of ultra-high resolution in the electron microscope.

Damage to crystalline specimens in the electron microscope occurs both because of the interaction of the electrons with the specimen and because of the disordering effect of the high vacuum. Negative staining alleviates these deleterious effects to some extent, as has been discussed previously. More recently however a significant decrease in radiation damage has been achieved by embedding the specimen in ice (Taylor and Glaeser, 1974). Alternatively, Unwin and Henderson (1975) have shown that high resolution images (9 Å) of protein crystals can be obtained by replacing the aqueous medium of the crystals by another liquid such as glucose which has similar chemical and physical properties but which is non-volatile. By using this technique to prevent dehydration damage and by recording images with very low doses to reduce radiation damage, Unwin and Henderson were able to produce electron micrographs for which high resolution information could be extracted by quantitative computer processing techniques (Section 18.6).

18.5 HIGH VOLTAGE ELECTRON MICROSCOPY

The major advantages of using higher voltages (>100 kV) in electron microscopy are the increase of penetration of the electrons, which allows the use of thicker specimens, and a reduction in radiation damage (which has been discussed in the previous section). The decrease in wavelength of the high voltage electrons should produce an improvement in the theoretical resolving power of the instrument but this is not realized in practice due to an increase in spherical aberration and problems associated with electrical and mechanical stability. In fact the nominal resolving power of most high voltage microscopes is of the same magnitude or slightly worse than the resolving power of conventional microscopes. The disadvantage associated with the increasing penetration of the fast moving electrons is that contrast also falls off markedly with increasing voltage, although this has not proved a noticeable draw-back with stained biological specimens so far.

The use of thicker specimens has two implications for protein crystallography. Firstly let us consider the depth of field of the electron microscope. Because the objective aperture restricts the scattered rays to those within a cone of very small semi-angle α, the depth of field is very large, and usually much larger than the thickness of the specimen. This means that both the upper and lower surfaces of the specimen will be in focus at the same time, resulting in an image which is essentially a two-dimensional projection of the specimen. Depth of field, D_{fi}, is given by

$$D_{fi} = \frac{\lambda}{NA^2}$$

where λ is the wave-length of the electrons and NA is the numerical aperture of the objective lens. For 50 kV electrons and a numerical aperture of 10^{-3}, $D_{fi} = 5.4\ \mu$; for 1000 kV electrons ($\lambda = 0.0087$ Å) and numerical aperture of 10^{-3}, $D_{fi} = 0.87\ \mu$. Thus the use of thicker specimens, which may be of greater thickness than the depth of focus, allows the possibility of optical sectioning through the specimen whereby different planes of the object are brought into focus in turn. This technique has been applied to the three-dimensional structure of chromosomes, 1.5 μ thick.

Secondly, the use of high voltages permits the use of an environmental stage in which the specimen can be kept in a humid environment. It is well known from X-ray diffraction studies that protein crystals lose their crystallinity and become disordered on drying and this is one of the major limitations in the production of very high resolution images with the conventional electron microscope. The most successful environmental cell for high resolution work is of the aperture design. The specimen consists of four co-linear apertures, centred on the axis of the microscope column with the specimen positioned between the inner two 75 μ diameter apertures. The separation of these two apertures is of the order of 1.5 mm. A pressure gradient is maintained across the apertures by having a source of water vapour feeding the specimen and differentially pumping, by means of an additional diffusion pump, on the low pressure side of the apertures. In this way the specimen is kept moist, but the water vapour is not allowed to contaminate the rest of the microscope column. With this environmental cell, Matricardi *et al.* (1972) have recorded the electron diffraction of a catalase crystal which extends to a resolution of 2 Å. The voltage was 200 kV and the authors were careful to keep the maximum electron dose below 0.6×10^{16} electrons/cm^2 in order to avoid radiation damage. With beam exposures greater than 1.3×10^{16} electrons/cm^2 the pattern disappeared. Although the calculated reconstruction of an image from its electron diffraction pattern requires a detailed knowledge of the electron scattering processes, which are only approximately understood for protein crystals, the design and construction of the environmental cell represents one of the most significant advances for the study of protein crystals in the electron microscope. The reader is referred to the reviews of Joy (1973) and Parsons et al. (1974) for more detailed accounts of electron microscope observations on aqueous biological specimens.

High voltage electron microscopes have been in use for a number of years – the 1500 kV electron microscope in Toulouse was commissioned in 1960. There are now a number of commercial instruments available but because of their cost and their size (the lenses alone weigh a ton), high voltage electron microscopes are unlikely to become general laboratory instruments. Instead it is likely that there will be a number of centres where a communal microscope will serve the surrounding area.

18.6 IMAGE ANALYSIS

An electron micrograph of a regular object, such as a protein crystal, contains many views of the repeating unit. If the specimen is truly regular then these units should all be the same, although in the electron micrograph their appearances may vary due to the effects of noise arising from perturbations of structure, variability of staining, granularity of the support film etc. It was natural that electron microscopists should seek a method which would enable such periodic specimens to be accurately analysed and which would allow the effects of noise to be filtered out.

The optical diffractometer (Klug and Berger, 1964) provides the most direct approach to this problem. A schematic diagram of the instrument has already been shown in Chapter 5 (Fig. 5.3). Monochromatic light from a laser emerges from a small aperture which is located at the front focal plane of the condenser lens C. The electron micrograph is placed in position M and is illuminated with parallel light. The variations in optical density in the electron micrograph give rise to scattered rays and these rays are collected by lens S to form an image of the diffraction pattern in plane D. The diffraction pattern may either be recorded on a photographic film placed in position D or, alternatively, the diffracted rays are allowed to continue and are brought to a focus by the lens system O to form an image of the electron micrograph in the image plane I.

The first stage of this process, the formation of the diffraction pattern, is relatively straightforward and may be accomplished with lenses of moderate quality. The diffraction pattern is most informative, especially to those trained in X-ray crystallography. Often periodicities are revealed which are difficult to detect by eye in the original micrograph. Because all the regular repeating units that are contained in the area illuminated contribute to the diffraction pattern, more precise estimates of the repeat distances, the crystal unit cell and its symmetry are obtained from the diffraction pattern than from the micrograph, once the optical diffractometer has been carefully calibrated. An objective estimate of the resolution of the micrograph is given from a measurement of the maximum observed spacing in the diffraction pattern. Finally specimen drift and objective lens astigmatism give rise to characteristic features which are most easily recognized in the diffraction pattern of the granular carbon support film.

The next stage, the optical filtering of the noise and the reconstruction of the filtered image, require very careful alignment of the optical diffractometer and very good quality lenses. In order to see how the filtering process works we note that the regular portions of the electron micrograph contribute to the diffraction lattice in the diffraction pattern while the irregular fluctuations due to noise contribute to the overall background scattering. Therefore by placing a mask, with holes drilled to correspond to the diffraction lattice, in the plane of the diffraction pattern, those diffracted rays which have arisen from the noise are prevented from reaching the image plane. The mask has to be carefully made to correspond precisely with the observed lattice. Those diffracted rays which are allowed through the mask are collected by the lens system O to form a filtered image. Fig. 18.5 illustrates the steps involved in this process. The optically filtered image still contains many of the features present in the original micrograph, but the periodic nature of the lattice and the shape of the individual protein subunits are much clearer.

Optical filtering and image reconstruction are simple to perform but suffer the disadvantage that artefacts can be introduced due to the imperfections of the lenses. This problem is especially serious for the lens system O which has to collect the widely diverging rays of the magnified diffraction pattern. The problem may be overcome by the use of the computer to calculate the Fourier transforms (De Rosier and Klug, 1968). A selected area of the electron micrograph is first digitized by means of an automatic densitometer, such as that used for measuring the optical densities in X-ray diffraction photographs. The Fourier transform is calculated, preferably by means of a fast Fourier transform routine (Cooley and Tukey, 1965), and this gives numerical values for the structure factor amplitudes and their phases of the diffraction pattern. The noise is selectively removed and the image reconstructed by the inverse Fourier transformation. Note that in these calculations there is no "phase problem" because we have the structure already in the electron micrograph.

The technique of image reconstruction has been extended to cover the three-dimensional reconstruction of particle images (De Rosier and Klug, 1968; De Rosier and Moore, 1970; Hoppe, *et al.*, 1974). The method is based on the projection theorem, familiar to X-ray crystallographers, which states that the two-dimensional Fourier transform of a plane projection of a three-dimensional density distribution is identical to the corresponding central section of the three-dimensional transform normal to the direction of view. The three-dimensional transform can therefore be built up section by section using transforms of different views of the object and the three-dimensional recon-

18.6. IMAGE ANALYSIS

struction of the image produced by the inverse Fourier transformation, as in the computation of a three-dimensional Fourier synthesis in an X-ray diffraction experiment. This procedure has been successfully applied to the structure of a regular spherical virus (Crowther, 1971) and to helical decorated actin filaments (Moore et al., 1970). In both cases because of the symmetrical nature of the object, a single micrograph was sufficient to provide the necessary views. For other objects different views have to be obtained by tilting the specimen, or by recording several micrographs with the object in different orientations. This method has been used by Hoppe et al., (1974) to provide a three-dimensional reconstruction of individual negatively stained yeast fatty acid synthetase molecules.

In an earlier section we noted that contrast could be achieved both by amplitude contrast and by phase contrast effects which arise from contributions from defocusing and spherical aberration of the objective lens. Phase contrast provides the predominant factor at medium and high resolutions. The effects of phase contrast are displayed much more simply in the Fourier transform of the image than in the image itself and hence they may be readily analysed with the aid of the optical diffractometer or computer transforms. In the wave theory of image formation, the effects of spherical aberration and defocusing are attributed to a phase shift ($\chi(\alpha)$) which the scattered electron wave undergoes at the diffraction plane of the microscope. This is a function of scattering angle α and is given by:

$$\chi(\alpha) = \frac{2\pi}{\lambda}(-\tfrac{1}{4}C_s\alpha^4 + \tfrac{1}{2}\Delta f \alpha^3)$$

where C_s is the spherical aberration coefficient and Δf is the defocus (positive for underfocus). The transform of the phase contrast image, T_{ph}, is related to the true transform of the object, $T(\alpha,\phi)$, by

$$T_{PH} = -T(\alpha,\phi)A(\alpha)f(\alpha)\sin\chi(\alpha)$$

where α is the angle of scatter as before, ϕ is the azimuthal co-ordinate, $f(\alpha)$ is the atomic scattering factor for electrons and $A(\alpha)$ is the objective aperture function ($A(\alpha) = 1$ if $\alpha \leqslant$ defining aperture angle; $A(\alpha) = 0$ elsewhere). The expression shows that the true transform is modulated by three factors, the most important of which is $\sin\chi(\alpha)$, which is known as the transfer function (Hanzen, 1971). The transfer function will have the following effects on terms corresponding to a spatial resolution of λ/α: (i) if $\sin\chi(\alpha)$ is close to 1, then these terms will contribute their full weight to the Fourier transform and details corresponding to this resolution will be faithfully imaged; (ii) if $\sin\chi(\alpha)$ is close to zero, then terms of this spatial resolution will be removed from the transform and will not contribute to the final image; (iii) if $\sin\chi(\alpha)$ changes sign then terms corresponding to this resolution will contribute to the final image with reverse contrast leading to the danger of spurious images. We therefore require

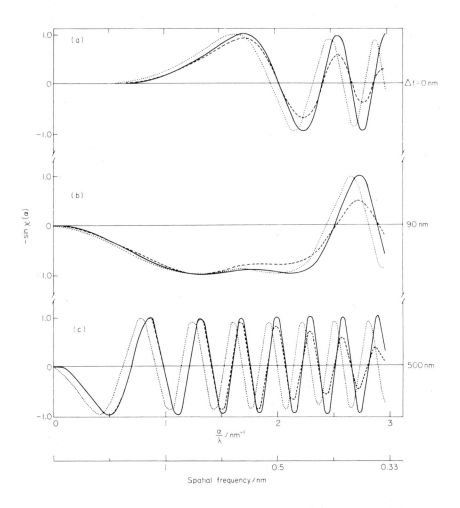

FIG. 18.8 The phase-contrast transfer function, $-\sin\chi(\alpha)$, plotted as a function of α/λ, in nm^{-1}, for $\lambda = 0.0042$ nm, $C_s = 1.3$ mm, and for the indicated values of Δf. A negative value of this function implies that the corresponding region of the object transform is contributing to the image with normal contrast, i.e. a subtraction from the background electron intensity over regions of high mass density. The solid curves are for pure phase contrast. The dashed curves are corrected for the effects of chromatic aberration with normal electrical instabilities, by averaging over a range of Δf of ± 20 nm. The dotted curves are corrected for the effects of the partial coherence of the electron source, assuming a 100 μm diameter condenser aperture. (from Erickson and Klug, 1971)

18.7. TRANSMISSION SCANNING ELECTRON MICROSCOPE (STEM)

that Sin $\chi(\alpha)$ be relatively invariant over the range of resolutions in which we are interested.

Erikson and Klug (1971) have calculated curves for sin $\chi(\alpha)$ for different degrees of underfocus and their results are reproduced in Fig. 18.8. It is seen that 900 Å underfocus would be a favourable choice for imaging details between 20 to 5 Å resolution, while a greater degree of underfocus (5000 Å) would be more appropriate for details in the 50 to 20 Å range. The fact that Sin $\chi(\alpha)$ approaches zero as α approaches zero is not important since at low resolution features will be faithfully produced by amplitude contrast. The effects of the transfer function on real images were also examined by Erickson and Klug (1971). A series of electron micrographs at different degrees of underfocus for uranyl acetate stained catalase crystals were taken and three of these together with their optical diffraction patterns are shown in Fig. 18.9. At $\Delta f = 0$, phase contrast is zero for the effective resolution range within which the specimen is preserved and the contrast in the image arises solely from amplitude contrast. It is seen that the high resolution terms in the optical diffraction pattern are weak or missing. At $\Delta f = 5400$ Å, many of the high resolution components are enhanced and the micrograph contains more detail and better contrast. At $\Delta f = 14,500$ Å (i.e. 1.45 μ), the effects of the rapidly oscillating transfer function in the gross underfocus condition become apparent. Since these oscillations occur within the spatial frequency range of the catalase transform, significant deterioration of the image results. This detailed study thus provides an insight into one of the most important features of image formation. It is comforting to find that Erickson and Klug (1971) conclude that "the moderate underfocused micrographs used in most biological microscopy are valid images, frequently the best possible in terms of resolution and contrast, with no artefacts in the low and medium range of interest".

In the same study Erickson and Klug (1971) also investigated the degree of multiple scattering (i.e. the Renninger effect) in the thin protein crystal. Although there was some evidence that the intensity of one of the reflections was affected, it appears that such effects are likely to be small for very thin specimens.

18.7 THE HIGH RESOLUTION TRANSMISSION SCANNING ELECTRON MICROSCOPE (STEM)

The development of the high resolution transmission scanning electron microscope represents a significant departure from conventional electron microscopy and one of the most important steps forward in recent years for the study of biological molecules. The interaction of a fast moving electron beam with matter results in both elastic and inelastic scattering of the incident beam. The ratio of the scattering cross sections for the elastically (σe) and inelastically (σi) scattered

514 18. ELECTRON MICROSCOPY

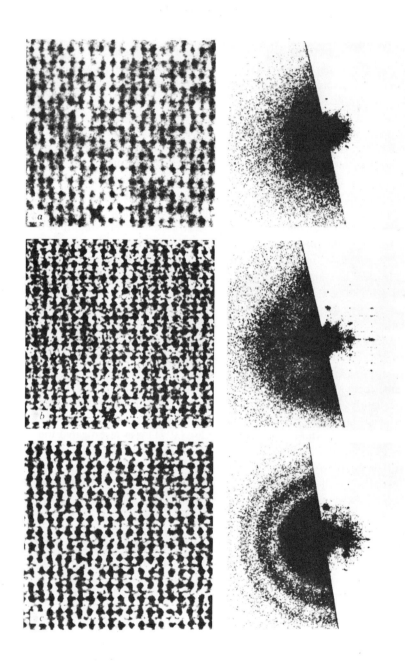

18.7. TRANSMISSION SCANNING ELECTRON MICROSCOPE (STEM)

electrons is given, very approximately, by:—

$$\frac{\sigma i}{\sigma e} = \frac{25}{Z}$$

where Z is the atomic number of the scattering atom in the specimen. This relationship suggests that if the scattered electrons could be separated by a velocity filter on the basis of their energy, then a composite image formed from the differences in intensities of the elastically and inelastically scattered electrons should contain both spatial and analytical information of the atoms in the specimen. In the conventional transmission electron microscope the entire area of the specimen is illuminated by the electron beam and the scattered electrons focused by the objective lens so that all points of the image are generated simultaneously. Although in principle it might be possible to attach a spectrometer below the specimen in order to separate the elastically scattered from the inelastically scattered electrons, the technical problems associated with simultaneous energy and spatial resolution are almost insuperable. In the transmission scanning electron microscope a fine beam of electrons is scanned point by point over the area of the specimen and the intensity of the resultant transmitted or secondary electrons used to modulate the intensity of a synchronously scanned cathode ray tube to form an image of the specimen. The spatial resolution is provided before the beam hits the specimen and is limited by the diameter of the beam spot. Since there are no lenses after the specimen, it is relatively simple to select the image forming electrons on the basis of their energy loss by placing a spectrometer below the specimen.

A high resolution scanning electron microscope with just such an energy analyser has been designed and built by Crewe and his colleagues in Chicago (Crewe et al., 1969; Crewe, 1970). A schematic diagram of their instrument is shown in Fig. 18.10. The major difficulty in achieving high resolution with the scanning electron microscopes arises from the requirement for a very small beam spot of great intensity. The electron beam from a hot tungsten filament, which is suitable for the conventional transmission electron microscope, is nowhere near

FIG. 18.9 The image of a negatively stained catalase crystal and the corresponding optical diffraction pattern at different values of defocusing; (a) in focus, (b) 540 nm under focus, (c) 1450 nm under focus. The apparent lattice spacings in the micrographs are 6.55 nm in the horizontal and 8.9 nm in the vertical direction. The reciprocal lattice spacings in the diffraction pattern are $1/6.55$ nm^{-1} and $1/17.8$ nm^{-1} (corresponding to the true vertical repeat of 2×8.9 nm). The left half of the optical diffraction pattern has been overexposed to show the low intensity spots and the effect of the transfer function on the noise pattern. The highest resolution diffraction spot, indices (1,7), corresponds to a reciprocal spacing of $1/2.4$ nm^{-1}. (from Erickson and Klug, 1971)

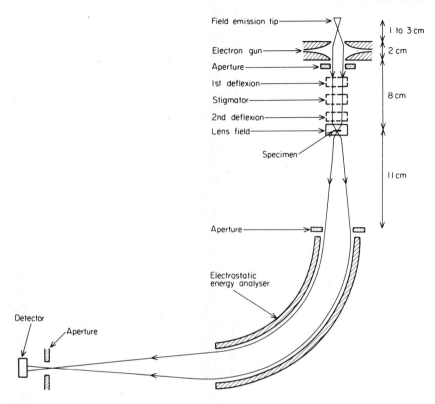

FIG. 18.10 A schematic diagram of the scanning transmission electron microscope. (from Crewe 1970)

bright enough for high resolution scanning work. For this reason, Crewe and his colleagues have developed their own field emission electron gun. In this device electrons are pulled from a finely etched tungsten point by a high voltage (2kV). An ultra-high vacuum ($10^{-10} \tau$) is required in order to prevent contamination and flash over. This type of gun can produce a 100 Å spot with a current of 10^{-10} A at the specimen, which is many orders of magnitude brighter than that produced by a hot tungsten filament. The beam is focused by condenser lenses to produce a spot of 5 Å diameter at the specimen.

With this instrument visualization of single heavy atom clusters and of single strands of uranyl acetate stained DNA with clusters of stain 20 Å apart have been reported (Crewe, 1971). In addition detailed analysis of the energy loss spectra provides information on the electronic structure of the molecule. Energy losses in the range 0 – 70 eV correspond to interactions with outer shell orbital electrons, while more substantial losses (> 285 eV) give information concerning the K shell transitions. The energy loss spectra of 25 keV electrons traversing

18.7. TRANSMISSION SCANNING ELECTRON MICROSCOPE (STEM)

thin films (approximately 350 Å thick) of the nucleic acid bases adenine, thymine and uracil have been examined (Isaacson, 1972). The fine structure of these spectra are characteristic of molecular structure (for example there are characteristic differences in the spectra of uracil and thymine, which differ merely by the presence of an additional methyl group in thymine). In the region where the energy loss spectra overlaps with conventional ultra-violet spectroscopy (< 10 eV) there is good agreement between the results. Variation in energy loss spectra also provides a sensitive monitor of radiation damage. Maximum dosages determined by this method are comparable with those obtained by other methods and again indicate the severe effects of radiation damage on biological molecules.

The possibility of obtaining both spatial resolution of the order of 5 Å and contrast according to atom or molecule-type in the high resolution scanning electron microscope has tremendous implications for the study of biological macromolecules.

NOTE ADDED IN PROOF
THE STRUCTURE OF A MEMBRANE BOUND PROTEIN

Since this chapter was written, Henderson and Unwin [*Nature* (1975) **257**, 28, and Unwin and Henderson (1975)] have achieved a notable breakthrough in the analysis of periodic arrays of biological molecules. By combining data from low dose images and electron diffraction patterns recorded with a conventional transmission electron microscope they have obtained a 3D image at 7 Å resolution of the purple membrane. This specialized membrane functions as a light driven hydrogen ion pump. 75% of its mass is composed of identical protein molecules (MW 26,000) arranged in a crystal lattice (P3) that is one unit cell thick (45 Å). Using the glucose technique to alleviate dehydration damage and low doses (<0.5 e/Å^2) to reduce beam damage (Section 18.4), electron micrographs of unstained membranes were recorded in which the only source of contrast was weak phase contrast produced by defocussing. In these almost featureless micrographs, the weak signal was extracted from noise by averaging over thousands of unit cells by means of Fourier transforms. The computed transforms were used to provide the phases (suitably corrected for contrast transfer function (Section 18.6) by reference to optical transforms of high dose images) while the amplitudes of the structure factors were obtained by measurement of intensities in the electron diffraction patterns. A 3D Fourier synthesis was computed using these data obtained for various tilts of the membrane. The map shows that the protein is globular, that it extends to both sides of the membrane and that it contains 7 α-helices packed 10–12 Å apart and 35–40 Å in length, running normal to the plane of the membrane. The molecules are grouped around the 3-fold axis with a space 20 Å diameter at the centre which is filled with lipid. Thus, for the first time we know the structure of a membrane-bound protein *in situ*: a most significant step forward.

19

ACHIEVEMENTS OF PROTEIN CRYSTALLOGRAPHY

It is evident from our discussions in previous chapters that, although some techniques in protein crystallography are well established, many are still under development and have yet to prove their usefulness. Protein crystallography is still a very lively science, and this is also reflected in its achievements which year by year take advantage of new developments and push back the frontiers of our knowledge in new areas.

In the early days there was great excitement in following the folding of each polypeptide chain and noting its helices and sheets. Although there are too few known protein structures for us to attempt a classification system in terms of a finite number of basic archetypes, certain general principles have emerged, as we have seen in Chapter 2, and it is possible to rationalize the folding of the polypeptide chain in terms of weak non-bonded interactions (hydrogen bonds, van der Waals interactions and hydrophobic bonds). Many proteins are organized into separate domains and among these domains certain recurring patterns have been recognized, such as for example the co-enzyme binding domain that occurs in all four of the dehydrogenases studied at high resolution and in some other proteins or the antiparallel β-pleated sheet sandwich domain that forms the common structural feature of the variable and constant halves of the light and heavy chains of human and mouse myeloma immunoglobulins. These structural homologies among polypeptide chains of widely different sequences form one of the most exciting features of present-day protein crystallography. Yet we still do not know enough about the energetics of weak interactions nor about the role of the solvent molecules for us to be able to predict a three-dimensional structure from its chemical sequence. One of the most important problems in protein structure is the nature of the code that relates primary amino acid sequence to tertiary structure. More basic knowledge is needed both from theoretical chemistry and from a greater number of and more precise structure determinations.

The results concerning the relationship between protein structure and function have proved more revolutionary although at the same time the lack of self-sufficiency of the X-ray method in this field is obvious. In order to fully understand the behaviour of biological molecules we require collaboration of chemists and crystallographers in the design and execution of experiments based on the knowledge of the protein structure and the constraints that this structure imposes on possible mechanism. Happily there is now a fruitful interplay of ideas among different scientists and the multidisciplinary approach is proving most productive. In the paragraphs below we discuss some of the highlights of the achievements of protein crystallography in the interpretation of function in terms of structure.

Almost all chemical reactions in living systems are brought about by enzymes. The rate of a chemical reaction depends on the free energy of the transition state through which the reactants must pass to be transformed into products. For years it had been realized that the powerful properties of enzymes must be due to the very special way in which the polypeptide chain is folded to provide a site which is or can be made complementary to that of a transition state structure and within which there is a critical arrangement of groups of atoms which are endowed by virtue of the protein environment to bring about catalysis. Thus the enzyme stabilizes a transition state and acts as a reversible energy store. Within the last decade we have seen how this is accomplished for a few enzymes.

The specificity of an enzyme may be elucidated fairly readily from binding studies on various inhibitors and substrate and transition state analogues and these studies also form the basis of proposals for mechanistic pathways. As a result of crystallographic and chemical experiments the mechanism of action of one enzyme, lysozyme (Blake *et al.*, 1967a,b; Phillips, 1967; Imoto *et al.*, 1972) is known in detail (Figs. 19.1–3) and the essential features of the mechanisms of ribonuclease (Richards and Wyckoff, 1971), chymotrypsin (Blow *et. al.*, 1969; Blow, 1971) carboxypeptidase (Lipscomb *et al.*, 1970; Hartsuck and Lipscomb, 1971) and alcohol dehydrogenase (Eklund *et al.*, 1974) are also understood.

The nature of the active site varies from enzyme to enzyme. In the case of lysozyme the active site is a well-defined cleft (see Fig. 19.1), but in the case of chymotrypsin it is a more shallow depression on the enzyme surface. A well-defined cleft has the advantage of giving higher binding and greater specificity, but if the substrate is also a macromolecule some alteration of the substrate conformation may be required before binding can take place. A shallow depression will make the active site groups more available for action on a native molecule. We need to understand how the binding of the substrate to the active site can result in a favouring of a transition state conformation or configuration. The free energy of the transition state depends on entropic factors such as the proximity of the reactants and their relative orientations as well as factors such as strain in the system which contribute to the enthalpy. Thus in ribonuclease, the proximity of two histidines to the scissile bond and their precise orientation allows them to act one as an acid the other as a base in the hydrolysis of phosphate ester linkages in ribonucleic acids. In lysozyme there is general acid-nucleophilic attack by two acid groups which are correctly

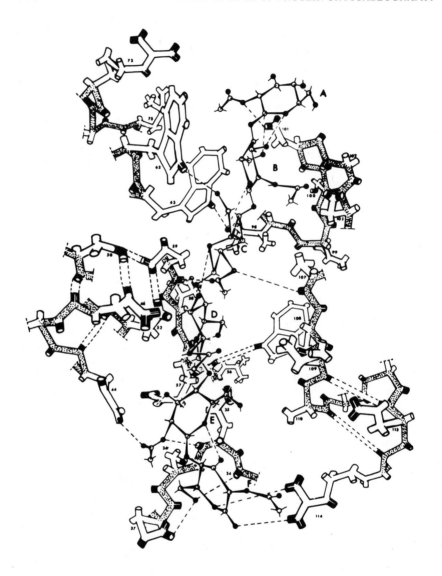

FIG. 19.1 The active site of lysozyme showing the binding of hexa-N-acetyl-chitohexose. The three sugar residues in the upper half of the diagram are bound in the way observed in crystallographic experiments for tri-N-acetylchitotriose. The mode of binding of the three in the lower half has been derived from model building. The positions of atoms in the enzyme are those observed in the native structure: no allowance has been made for the small shifts that are seen on binding the trisaccharide or of the possibility that additional changes occur when substrate is bound. (From Blake et al., 1967a)

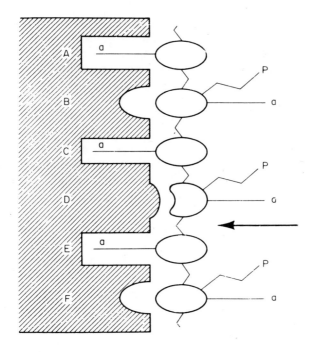

FIG. 19.2 A schematic diagram showing the specificity of lysozyme for hexasaccharide substrates. The substrate is composed of alternate residues of N-acetylglucosamine and N-acetylmuramic acid. Six sub-sites A-F on the enzyme bind the sugar residues. Alternate sites interact with the acetamido side chains (a) and these sites are unable to accommodate the N-acetylmuramic acid residues with their lactyl side chains (P). Site D cannot bind a sugar without distortion and the β(1–4) glycosidic linkage between residues D and E that is cleaved is shown by the arrow (from Imoto et al., 1972).

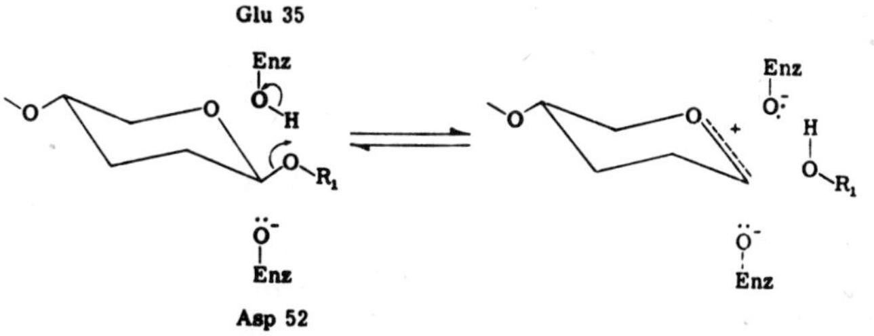

FIG. 19.3 The mechanism for bond rearrangement in lysozyme catalysis showing general acid attack by glutamic acid 35 and the enzyme bound carbonium ion which is favoured and stabilized by the distortion of the sugar and the nucleophile aspartic acid 52.

positioned adjacent to a β-1-4 glycosidic linkage, in which one of the sugar rings is stericly distorted towards the transition state (see Figs. 19.2 and 19.3). In chymotrypsin and related enzymes the hydrolysis is brought about by nucleophilic attack combined with general acid catalysis, in which a "charge relay system" leads to electronic strain and distortion of the substrate towards the transition state of the reaction (Fig. 2.3c). In alcohol dehydrogenase the reaction is brought about by electrophilic catalysis mediated by the active site zinc in which the zinc-bound water plays a crucial role. In most cases it appears that the enzyme is itself distorted or undergoes an "induced fit" when the substrate is bound and in some cases such as carboxypeptidase this may involve quite considerable movements (14Å) of some of the catalytic groups.

Thus both the specificity of the active site which is determined by the overall conformation of the enzyme and the possibility of induced conformational changes are of critical importance in the positioning and orientation of the reactants and catalytic groups, and also in the achievement of "steric" or "electronic" strain. Knowledge of the structures of enzymes can indicate how these factors change the entropy and enthalpy of the transition state, leading to a transition state of lower free energy and a consequent rate enhancement by the enzyme.

For years haemoglobin has held a special fascination for chemists, biochemists and physiologists. The protein is readily available since it is the main component of the red blood cells and is of prime physiological importance in the transport of oxygen from the lungs via the arteries to the tissues and in the transport of carbon dioxide through the veins back to the lungs. Mammalian haemoglobin is composed of four subunits, which are identical in pairs, and each subunit contains one iron atom situated at the centre of a haem group. It was established by Conant as long ago as 1923 that the iron remains in the ferrous state during the reversible uptake of oxygen whereas in solution without the protein the haem iron is rapidly oxidized to the ferric state. The binding curve of

---

FIG. 19.4 Diagrammatic sketch showing one possible sequence of steps in the reaction of haemoglobin with oxygen. Deoxy-haemoglobin with all salt-bridges intact and with one molecule of diphosphoglycerate clamped between the two β-chains. At steps 1–2 and 2–3 the α-chains are oxygenated. The tyrosine pockets are narrowed, the tyrosines expelled and the salt-bridges with the partner α-chains are broken. In step 3–4 the quaternary structure clicks to the oxy form, accompanied by expulsion of diphosphoglycerate, and breakage of the salt links between subunits $\alpha_1-\beta_2$ and $\alpha_2-\beta_1$. Note that the internal salt-bridges of the β chains are intact and their haems too narrow to admit ligands. At steps 4–5 and 5–6 the β subunits react in turn, accompanied by widening of the haem pockets, narrowing of the tyrosine pockets and rupture of the internal salt links from the C-terminal histidines to aspartates-94. *Note:* The change in quaternary structure could in fact take place at any stage of the reaction, and the co-operative mechanism is independent of the sequence in which the individual subunits react (from Perutz, 1970).

19. ACHIEVEMENTS OF PROTEIN CRYSTALLOGRAPHY

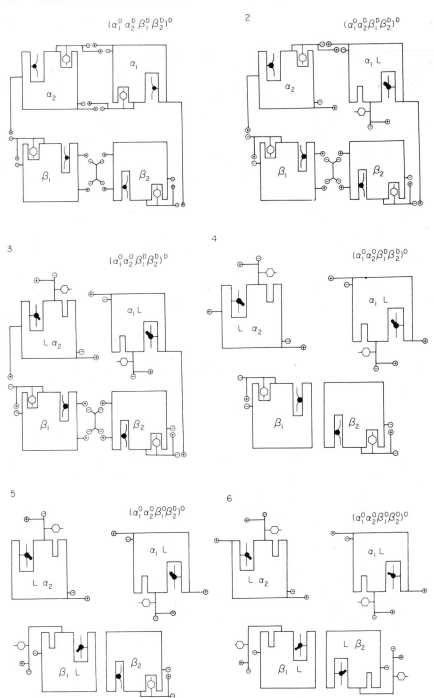

oxygen to haemoglobin exhibits a characteristic sigmoid shape which shows that the oxygen affinity depends on the number of oxygen molecules already combined. The oxygen affinity is also sensitive to concentrations of other metabolites, such as hydrogen ions and organic phosphates, which are not related to the substrate. Haemoglobin therefore poses special problems; firstly as to how the environment of the protein controls the properties of the iron atom and secondly as to how the binding of oxygen at one haem group influences the oxygen affinity of the adjacent haem group. The structural studies have provided keys to the answers to these questions. The analysis of haemoglobin, and related monomeric proteins has shown that the haem is located in a hydrophobic pocket on the protein surface which provides a medium of low dielectric constant in which oxygenation is favoured and oxidization discouraged (Perutz et al., 1965; see Fig. 2.20). The reversible combination of oxygen with the haem group is known, from chemical studies, to result in a change in the electronic state of iron which is characterized by a transition from a high spin paramagnetic state to a low spin diamagnetic state. In haemoglobin, the cooperative effects have been shown by Perutz's X-ray studies (Perutz, 1970) to be triggered by small movements of the iron atom, relative to the porphyrin ring, which take place on the uptake of oxygen and accompany the spin state transition of the iron. These shifts are transmitted to other parts of the molecule so that constraints which hold the protein in the deoxy state are relaxed and it is therefore easier for the next subunit to take up oxygen (Fig. 19.4). So far haemoglobin is the only allosteric control protein whose structure is known in detail. Such proteins, for which a small change in concentrations of metabolites results in large changes in biological activity, occur at key points in all metabolic pathways. It will be interesting to understand the molecular mechanics of their reactions and interactions.

In aerobic cells the free energy contained in foodstuffs is converted via enzymatic oxidation into free energy of the molecule of adenosine triphosphate (ATP). The final common pathway by which all electrons derived from different fuels of the cell are transferred to oxygen involves a series of electron-transferring molecules called the cytochromes. The haem iron in each of the cytochromes can accept an electron in its oxidized form and become reduced. The reduced iron, in turn, can donate its electron to the next carrier and so on until the last enzyme gives up its electron to molecular oxygen. The proteins do not occur singly in the soluble portion of the cell cytoplasm but are fixed in well organized assemblies in the mitochondria, the sub-cellular organelles which are the power-houses of the cell. Other electron transferring proteins are found elsewhere in cells and also in bacteria and plants and these proteins may be classed together under the heading redox proteins. The structures of redox proteins known to date (Table 19.1) include the haem proteins, cytochrome c, cytochrome c_2, and cytochrome b_5 and the non-haem iron-sulphur proteins rubredoxin, high potential iron protein and ferredoxin and a flavoprotein, flavodoxin. The iron containing proteins exhibit a wide range of redox potentials

TABLE 19.I
Proteins whose structures have been solved at high resolution

Protein and Reference[x]	Species	Symmetry of oligomeric aggregate	Resolution Å
1. Enzymes			
Acid proteinase Jenkins et al. (1976)	E. parasitica		3.0
Adenylate kinase Schulz et al. (1973)	Porcine muscle		3.0
Alcohol dehydrogenase Branden et al. (1973) Eklund et al. (1974)	Horse liver	2	2.4
Carbonic anhydrase C Liljas et al. (1972)	Human		2.0
Carbonic anhydrase B Kannan et al. (1975)	Human		2.3
Carboxypeptidase A Lipscomb et al. (1970)	Bovine		2.0
Carboxypeptidase B Schmid & Herriott (1975)	Bovine		2.8
α-chymotrypsin Blow (1971) Tulinsky et al. (1973)	Bovine	2	2.0 2.8
γ-chymotrypsin Segal et al. (1971)	Bovine		2.7
Chymotrypsinogen Freer et al. (1970)	Bovine		2.5
Elastase Shotton et al. (1971)	Porcine		3.5
Glyceraldehyde −3 phosphate dehydrogenase Buehner et al. (1974a,b) Wonacott et al. (1975)	Lobster B. Stereothermophilus	222 222	3.0 2.7
Hexokinase Fletterick et al. (1975)	Yeast		2.7
Lactate dehydrogenase Adams et al. (1973)	Dogfish M4 apo	222	2.0
Adams et al. (1973)	Dogfish M4 ternary	222	3.0
Lysozyme Imoto et al. (1972)	Hen Egg White		2.0
Banyard et al. (1973) Matthews and Remington (1974)	Human Bacteriophage T4		2.5

TABLE 19.I (*continued*)
Proteins whose structures have been solved at high resolution

Protein and Reference[x]	Species	Symmetry of oligomeric aggregate	Resolution Å
Malate dehydrogenase Hill et al. (1972)	Porcine	2	2.5
Nuclease Cotton et al. (1971)	Staphylococcus Aureus		2.0
Papain Drenth et al. (1971a,b,c)	Papaya latex		2.8
Pepsin Andreeva et al. (1976)	Porcine		2.7
Phosphoglycerate Kinase Blake and Evans (1974)	Horse		3.0
Bryant et al. (1974)	Yeast		3.5
Phosphoglycerate mutase Campbell et al. (1974)	Yeast	222	3.5
Phospholipase A_2 Drenth et al. (1976)	Porcine		3.0
Protease S/GPB Delbaere et al. (1975)	S. griseus		2.8
Rhodanese Smit et al. (1973)	Bovine liver	2	3.9
Ribonuclease A Kartha et al. (1967)	Bovine		2.0
Carlisle et al. (1974)			2.5
Ribonuclease S Richards and Wyckoff (1971)	Bovine		2.0
Subtilisin BPN' Wright et al. (1969)	B. amyloliquefaciens		2.0
Subtilisin Novo Drenth et al. (1971a,b,c)	B. Subtilis		2.8
Superoxide dismutase Richardson et al. (1975)	Bovine		3.0
Thermolysin Colman et al. (1972a)	B. Termoproteolyticus		2.3

TABLE 19.I (continued)
Proteins whose structures have been solved at high resolution

Protein and Reference[x]	Species	Symmetry of oligomeric aggregate	Resolution Å
Triosephosphate isomerase Banner et al. (1975)	Chicken	2	2.5
Trypsin Stroud et al. (1974)	Bovine		2.7
Trypsin Inhibitor Deisenhofer et al. (1973)	Bovine Pancreas		1.5
Trypsin-trypsin inhibitor complex Rühlmann et al. (1973)	Bovine		2.6
Huber et al. (1974)			1.9
Trypsin-Soybean trypsin inhibitor complex Sweet et al. (1974)	Porcine		2.6
Tyrosine t-RNA Synthetase Blow et al. (1975)			
2. Globins			
Haemoglobin Love et al. (1971)			2.5
Huber et al. (1970)	Blood worm		
Perutz et al. (1968)	Chironomus		2.5
Bolton and Perutz (1970)	Horse oxy	2	2.8
Muirhead and Greer (1970)	Horse deoxy	2	2.8
Hendrickson and Love (1971)	Human deoxy	2	3.5
	Lamprey		2.0
Myoglobin Watson (1969)	Sperm whale		1.4
3. Redox Systems			
Cytochrome b_5 Mathews et al. (1971)	Calf liver		
Cytochrome c Dickerson et al. (1971)	Horse ferri		2.8
Dickerson et al. (1971)	Bonito ferri		2.8
Tanaka et al. (1973)	Bonito ferro		2.3
Takano et al. (1971)	Tuna ferri		2.8
Tanaka et al. (1973)	Tuna ferro		2.4
Cytochrome c_2 Salemme et al. (1973)	R. rubrum		2.0
Ferredoxin Adman et al. (1973)	P. aerogenese		2.0

TABLE 19.I *(continued)*
Proteins whose structures have been solved at high resolution

Protein and Reference[x]	Species	Symmetry of oligomeric aggregate	Resolution Å
Flavodoxin Andersen et al. (1972)	Clostridium MP		1.9
Watenpaugh et al. (1972)	D. Vulgaris		2.0
High Potential Iron protein Carter et al. (1972, 1974b)	Chromatium vinosium		2.0
Rubredoxin Watenpaugh et al. (1973)	C. pasteurianum		1.5
Thioredoxin S$_2$ Holmgren et al. (1975)	E. coli	2	2.8
4. Hormones			
Glucagon Sasaki et al. (1975) Blundell (1975)	Porcine	3	3.0
Insulin Blundell et al. (1972)	Porcine	32	1.9
Peking Insulin Research Group (1973)			1.8
Peking and Shanghai Research Group (1974)			
5. Immunoglobulins			
Bence-Jones Fragment Au. Fehlhammer et al. (1975)	Human		2.5
Bence-Jones Fragment Rhe Wang et al. (1975)	Human	2	3.0
Bence–Jones protein Mcg Schiffer et al. (1973)	Human	2	3.5
Bence–Jones protein REJ Epp et al. (1974)	Human	2	2.8
Fab fragment New Poljak et al. (1973, 1974) Chen and Poljak (1974)	Human		2.8
Fab fragment McPC 603 Segal et al. (1974)	Mouse		3.1
F$_c$ fragment Deisenhofer et al. (1976)	Human		3.5
6. Others			
Bacteriochlorophyll protein Fenna & Matthews (1976)	C. limicola	3	2.8

19. ACHIEVEMENTS OF PROTEIN CRYSTALLOGRAPHY

TABLE 19.I (*continued*)
Proteins whose structures have been solved at high resolution

Protein and Reference[x]	Species	Symmetry of oligomeric aggregate	Resolution Å
Concanavalin A Jack et al. (1971)	Jack bean	222	2.8
Hardman and Ainsworth (1972)			2.4
Edelman et al. (1972)			2.0
Myogen Kretsinger and Nockolds (1973)	Carp		1.9
Prealbumin Blake et al. (1974)		222	2.5

[x]Data taken from Liljas and Rossmann (1974) with additions. References refer to the most comprehensive description on the structure and are often not the first report.

(E = + 0.32 V for cytochrome c_2, + 0.21 V for cytochrome c, + 0.02 V for cytochrome b_5; − 0.57 V for rubredoxin, + 0.35 V for high potential iron protein and − 0.45 V for ferredoxin). One of the most interesting problems concerns the mechanism by which one electronic state of the iron atom is stabilized with respect to the other.

The particular properties of a metal in a metal–protein system are likely to depend most critically on such small deviations from standard geometry caused by the protein environment. It is essential to have definite structural evidence for such deviations. However where the redox protein is involved as part of an electron transport system, a full explanation of the properties of the protein requires an understanding of the interactions with neighbouring proteins in the chain and poses problems concerned with biological organization of multicomponent systems which are far more complex than the individual proteins so far studied by X-ray crystallography.

The structural studies on proteins have contributed to our knowledge of evolutionary processes. In several cases proteins from different species have been studied such as hen and human lysozymes, horse, bonito and tuna fish cytochrome c, human, horse, insect, lamprey and blood worm haemoglobins (Table 19.I). Chemical analysis of proteins from different species indicate that the proteins form a homologous series; they have evolved from a common ancestral enzyme. Structural studies show that the amino acid changes for the most part have had little effect on structure especially in the region of the active site. A series of haemoglobin and myoglobin structures is shown in Fig. 19.5. The overall structure and location of the haem have been highly conserved during evolution.

Nature, having once hit upon a good way of achieving a desired end, has taken care not to change it too much from species to species.

An interesting sequel to these studies is provided by the structure determination of a bacterial cytochrome, cytochrome c_2, from the simple photosynthetic bacterium *Rhodospirillum rubrum* (Salemme et al., 1973). This protein, which is used in photosynthetic electron transfer, is clearly but distantly related to mammalian cytochromes showing 30% identical amino acids to human cytochrome *c*. The construction of a phylogenetic tree based on a comparison of the sequences of cytochrome *c* from 36 different species (Dayhoff *et al.*, 1972) suggest that the bacteria diverged from higher organisms about 2,600 million years ago. The structure of bacterial cytochrome c_2 is similar to the structure of horse cytochrome *c* especially in the region around the haem group. These observations lend support to the proposals (derived from other evidence) that the mitochondria, the sub-cellular organelles within which the cytochrome c_2 are located, share a common origin with the bacteria.

Among higher organisms, a series of related enzymes are often observed where each enzyme has a slightly different specificity and physiological purpose but shares a common mechanism with the other enzymes; the serine proteases, chymotrypsin, trypsin and elastase, form a good example. These enzymes, so called because of their uniquely reactive serine residue which is involved in catalysis, are formed in the pancreas and secreted into the small intestine where they break down ingested proteins. Each cleaves the polypeptide chain specifically on the carboxyl side of certain amino acids. Chemical studies have shown that these enzymes have fairly similar amino acid sequences indicating a common evolutionary origin. It has been estimated that the duplications which gave rise to the separate genes for these enzymes took place only some 1,300 million years ago. It was not surprising therefore to find that their structures are also closely similar, especially with respect to the critical arrangement of ionizable groups at the active site, (see Fig. 2.3b) termed the "charge relay system" which is responsible for the activation of the serine residue. The structural studies show, however, how the changes at certain points in the sequence lead to changes at the substrate binding site which account for the differing specificities of the enzymes (Fig. 19.6). Certain bacteria also contain a serine proteinase, subtilisin, but chemical studies suggest that this enzyme is not

---

FIG. 19.5 A comparison of different myoglobin and haemoglobin structures showing the "myoglobin fold" which is relatively constant between mammalian, insect and annelid worm globins. Only α-carbon atoms are shown. (a) α chain of horse met haemoglobin; (b) β chain of horse met haemoglobin; (c) sperm whale myoglobin; (d) lamprey haemoglobin; (e) *chironomus* haemoglobin; (f) *glycera* haemoglobin (from Love *et al.*, 1971).

19. ACHIEVEMENTS OF PROTEIN CRYSTALLOGRAPHY 531

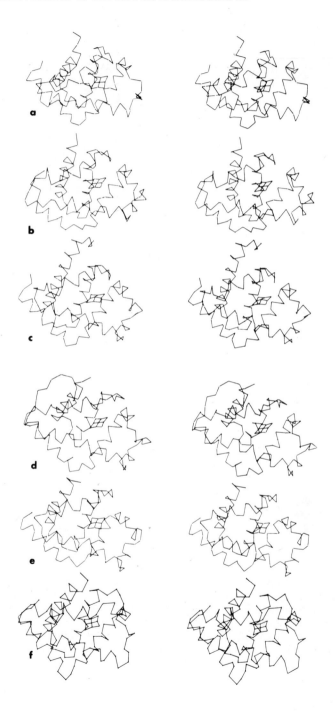

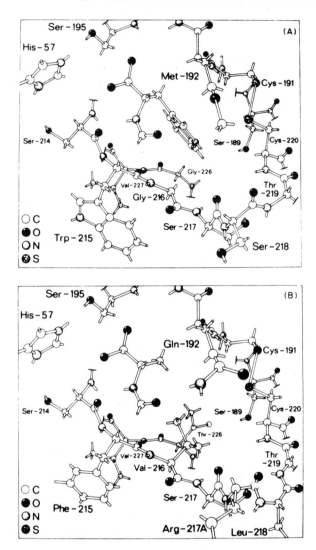

FIG. 19.6 Perspective drawings of the active sites of (a) chymotrypsin with bound formyl-L-tryptophan (b) elastase with bound formyl-L-alanine arranged in a similar conformation. Access to the binding pocket for aromatic residues in chymotrypsin is blocked by valine 216 and threonine 226 in elastase and this explains the specificity of elastase for small non-polar amino acids. In trypsin residues 216 and 226 are glycine but residue 189 is aspartic acid, and this explains the specificity of trypsin for basic amino acids (from Hartley and Shotton, 1970).

related to the mammalian counterparts. X-ray studies have demonstrated that the overall folding of the polypeptide chain is indeed very different from chymotrypsin but that the enzyme contains the same type of charge relay system for the activation of the serine. Thus we appear to have an example of convergent evolution whereby the same mechanism has been developed within different molecular frameworks and via different evolutionary pathways for mammals and bacteria.

A further unexpected relationship has recently emerged from the structural studies on dehydrogenases, enzymes that require the co-factor nicotinamide-adenine-dinucleotide (NAD) for their activity. These enzymes contain a common structural feature which is composed of about 150 amino acids arranged in six parallel strands of β-pleated sheet enveloped by four α-helices. This specially stable structure forms part of the adenine binding site for the co-factor in the dehydrogenases and we have described it in Chapter 2 (see Fig. 2.26). The structure has also been observed in other proteins which are not related to the dehydrogenases, but which do have an affinity for the adenine nucleotide such as phosphoglycerate kinase, adenylate kinase and flavodoxin.

These observations have led to the suggestion that the nucleotide binding domains of the dehydrogenases and kinases have evolved from a common ancestral gene. However comparisons of the amino acid sequences have shown that there is little homology among the primary structures as judged by usual standards and that if indeed the domain structure has evolved from a common origin then the ancestral structure must have been present during precellular evolution (i.e. about 3,200 million years ago) (Rossmann *et al.*, 1974). Alternatively it has been suggested (Fletterick *et al.*, 1975) that there may be a limited number of ways of forming super-secondary structure and that the alternating sheet/helix arrangement represents a specially stable structure. It is worth noting that a similar domain structure to that observed for the dehydrogenases occurs in phosphoglyceratemutase (a protein which shows no affinity for adenine nucleotides) while in hexokinase (a protein which does bind adenine nucleotides) the domain structure is only superficially similar to that of the dehydrogenases. Clearly these results and those from other proteins such as triosephosphate-isomerase and subtilisin, which also show an ordered pattern of sheet and helix, must be taken into account in the debate on the evolutionary or thermodynamic significance of these structures which is now in progress.

The assumption that functional parts of a molecule remain invariant during evolution has been turned to advantage in the study of insulin. The structure of this important hormone (see Fig. 2.24) has been solved independently first in Oxford and now in Peking and forms perhaps the major definite evidence for an understanding of the molecular details of the action of insulin. Very little is known of the active site nor of the means by which the hormone interacts with its receptor in the target tissue. A tentative identification of the active region has been approached by considering the natural variations of different insulins and chemically modified insulins in relation to their biological activity (Blundell et al., 1971). Only a small number of surface residues are found to be invariant and the X-ray model shows these residues to be clustered together. The biological activity is very susceptible to chemical modification of these residues and also depends critically on the conservation of the general three-dimensional arrangement of the hormone structure and this region in particular. For these reasons, this part of the hormone is thought to be involved in the interaction with the receptor. However, not all polypeptide hormones have globular structures. It appears that, in the crystal, glucagon has a mainly helical structure but has no hydrophobic core (Blundell, 1975). In solution at high dilutions it probably exists as an equilibrium population of conformers, and only achieves the helical conformation on association with other glucagon molecules or with the receptor. Structural studies of these hormones are providing blue-prints for pharmaceutical chemists who are struggling to make analogues available for medical treatment.

An understanding of the structural basis of the immune response and the manner in which diversity of specificity is accomplished within similar molecular frameworks is emerging from X-ray studies on immunoglobulins and their fragments. These studies are likely to form the greatest contribution by protein crystallography to medicine. Immunoglobulins are proteins with specific antibody activity or with structural features which closely resemble those of antibodies. They are multichain proteins made up of two light (L) and two heavy (H) chains linked by non-covalent interactions and by disulphide bonds (Fig. 19.7). Polypeptide chains of antibodies in an individual vary greatly in primary structure in their amino terminal regions but are relatively invariant elsewhere. Each light and heavy chain may thus be divided into a variable region at the amino terminal portion of the chain (the V_L and V_H regions) and a constant region (C_L and C_H). These Y shaped molecules can be cleaved by certain enzymes into two different types of fragments; the Fab fragment which contains the antigen binding site and consists of whole light chain and the amino terminal half (Fd portion) of the heavy chain; and the Fc fragment which can be crystallized and mediates complement-fixing and cell-binding activities of the molecule.

Immunoglobulins, even those directed against small molecules, are heterogeneous and in order to obtain sufficient supplies of pure homogeneous immunoglobulins for crystallographic studies it has been necessary to work with the immunoglobulins which are produced in large amounts by certain plasma cell tumours (myelomas) of man and mouse. Myeloma proteins have not been shown to be antibodies in the sense that they have been induced by stimulation with known antigens but they have the same basic structure as normal immunoglobulins and each myeloma protein appears to be a unique example of one of the many different immunoglobulin molecules (see review by

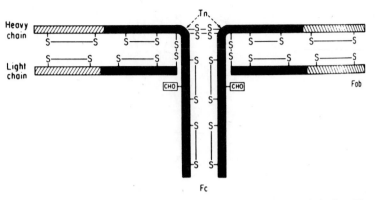

FIG. 19.7 A model of the structure of human γG1 immunoglobulin. The site of cleavage by trypsin (Tn) to produce *Fab* and *Fc* fragments is shown by the arrows. Interchain and intrachain di-sulphide bonds are indicated, and CHO marks the approximate position in the heavy chain of the carbohydrate moiety. The variable regions of light and heavy chains are indicated by slanted lines (from Edelman and Gall, 1969).

Edelman and Gall, 1969). Some myeloma proteins have been shown to precipitate with antigens and this precipitation may be prevented by binding of haptens. The structures of several immunoglobulin fragments are known: (1) the Fab fragment of a human myeloma immunoglobulin IgG New; (2) the Fab fragment of a mouse immunoglobulin IgA (K) McPC603; (3) the Mcg Bence–Jones protein (a human λ light chain dimer); and (4) the Bence–Jones protein REI (a human KV_L dimer) (Table 19.I). These structures have caused great excitement because again a common domain structure has emerged. The Fab molecules consist of four globular subunits which correspond to the variable (V_L, V_H) and constant (C_L and C_H) regions of the light (L) and heavy (H) chains, arranged in a tetrahedral conformation (Fig. 19.8). These globular subunits share a basic structure composed of a sandwich of two anti-parallel β-pleated sheets surrounding a tightly packed hydrophobic core (Fig. 19.9). The domains are covalently linked in pairs (V_L to C_L and V_H to C_H) by linear stretches of polypeptide chain ("switch" regions). The domain structure is well conserved except at those regions which show hypervariability in sequence. Comparison of the V_L domains of mouse and human immunoglobulins reveals a closer homology in structure than between the V_H and V_L regions of mouse immunoglobulin which suggests (Segal et al., 1974) that the V_L and V_H regions diverged earlier than rodents and primates. The active site (identified by hapten binding studies) occurs at a shallow cavity formed by the hypervariable regions of the V_L and V_H chains. For IgG New (Poljak et al., 1974) this region contains a high density of aromatic side chains which explains specificity for haptens that include benzene and naphthalene aromatic rings (Fig. 19.10). Structural

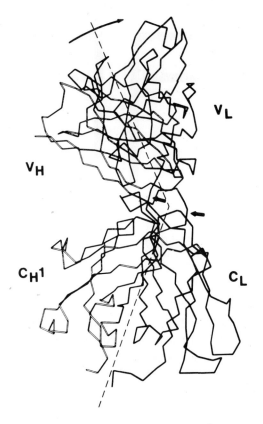

FIG. 19.8 A view of the α-carbon backbone of *Fab'* New. The diagram shows the V and C domains, the L chain (open line), the *Fd'* chain and the local, approximate 2-fold axes (broken lines) which relate the V_L to the V_H subunit and C_L to the C_H subunit. The two short arrows indicate the switch region of both chains. The longer arrow indicates a possible relative motion of the V and C domains (from Poljak *et al.*, 1974).

diversity of the V regions is localized to the hypervariable loops at the V_H and V_L chains and could give rise to the remarkable breadth of specificity in the immune system. (See review by Davies *et al.*, 1975).

Out of the bewildering complexity of biological systems some order is beginning to emerge. For the future there are many exciting prospects. There will certainly be a great many protein structures solved and it is anticipated that these determinations will show an increase in accuracy as new ways are found to collect high resolution data and to refine structures. The future significance for biology of these new structures is impossible to predict but there are several areas where breakthroughs are likely to be achieved. The physiological control of

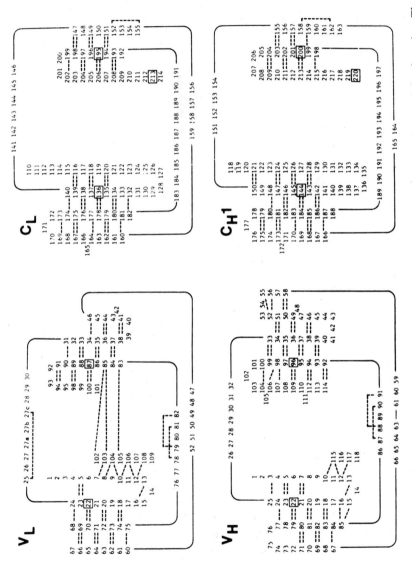

FIG. 19.9. Diagram of the hydrogen bonds between main chain atoms for the V_L, C_L, V_H, and C_H domains. The two sets of β-pleated sheet in each domain come together to form a sandwich so that the four domain structures are similar to each other.

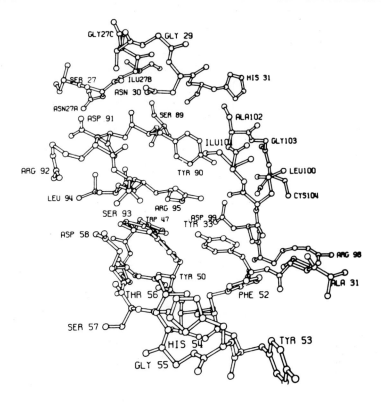

FIG. 19.10 View of some of the amino acid residues at the active site of Fab' New. Residues number 27 to 31 and 89 to 95 correspond to the V_L chain; other residue numbers correspond to the V_H chain. (Poljak et al., 1974)

enzyme activity is often mediated by small changes in the concentrations of metabolites which bear no structural relation to the substrate of the enzyme. This is the case for aspartate transcarbamylase (ATCase) which is the first enzyme in a series leading to the synthesis of a base and which is inhibited by the base. The structures of this and other allosteric proteins are likely to emerge in the near future and will give us a better understanding of the molecular basis of control. In living organisms chemical reactions seldom occur in isolation but are usually part of a metabolic pathway which allows the controlled conversion of one product into another. A concerted effort is underway on the enzymes of at least one metabolic pathway, glycolysis; the pathway is responsible for the conversion of glycogen and glucose to pyruvate with the controlled release and storage of energy. These studies should enrich our understanding of the inter-relationships of the enzymes, especially those enzymes which are adjacent to each other in the pathway where the product of one reaction becomes the substrate for the next. Already the structures of four adjacent glycolytic enzymes are known, (triosephosphate isomerase, glyceraldehyde-3-phosphate dehydrogenase, phosphoglycerate kinase and phosphoglyceromutase). In some metabolic systems the catalysis of a co-ordinated sequence of reactions is brought about by multi-enzyme complexes. These multi-component systems allow the transfer of the product of the reaction of one enzyme to the next enzyme without the need for the intermediate metabolite (which may be unstable) to go into a common pool. X-ray and electron microscope studies on crystals of some of these large complexes are now underway (De Rosier and Oliver, 1971). These studies and also the more advanced studies on some of the small viruses [see, for example, the discussion of the structure of viruses in Cold Spring Harbour Symposium (1971) 36, 433–510] illustrate the usefulness of the combined X-ray and electron microscope approach for large complexes. The methods of protein crystallography have recently been successfully applied to the determination of the tertiary structure of yeast phenylalanine transfer RNA (Suddath et al., 1974; Kim et al., 1974; Robertus et al., 1974; Klug et al., 1974). The structure confirms the existence of four lengths of double helical structure which were predicted from the nucleic acid sequence, but shows that these are actually organized in pairs to give two helices in a T shape. The structure explains the existence of certain invariant bases in the sequence which take part in tertiary interactions, and suggest the three-dimensional arrangement of codon, ribosome, synthetase and amino acid to which the t-RNA must bind during protein synthesis.

This brief discussion indicates some of the areas where answers to biological questions are likely to be provided by X-ray analysis. However, there are many other areas where crystalline material of multi-molecular complexes is unlikely to be available and yet detailed molecular knowledge is required. The problems which are of special importance today concern the study of membranes and the study of protein–nucleic acid interactions. For membranes, we wish to understand the organization of proteins within the membrane that is responsible for mediating the signals from molecules on the outside of the cell or organelle to those within it. In genetics we wish to know the structures of the protein–nucleic acid complexes in the cell nucleus that lead to the control of the expression of the gene and likewise the organizations of the proteins and nucleic acids on the ribosomes in the cell cytoplasm that lead to the control of the translation of the genetic message. These problems are being approached by a wide variety of biochemical, chemical, genetic and immunological techniques and the structures of some of the complexes can be visualized directly in the electron microscope (albeit after severe preparative treatment and in a hostile environment). The contribution of protein crystallography to these and other, wider, biological problems must be to provide a precise knowledge of the structures of the individual components and the principles upon which these structures are based so that, in the same way as the small molecule crystal structure determinations gave us the structures of the amino acids and peptide bond upon which our interpretation of protein structures is based, so the structures of the proteins will form a foundation for an understanding of very much larger complexes. Protein crystallography has come of age.

REFERENCES

REFERENCES

Abrahamsson, S. (1966). *J. Sci. Instr.*, **43**, 931.
Adams, M. A. (1968). D. Phil. Thesis., University of Oxford.
Adams, M. J., Collier, E., Dodson, G., Hodgkin, D. C. and Ramaseshan, S. (1966). Int. Union Cryst. 7th Meeting Abstr. A165.
Adams, M. J., Blundell, T. L., Dodson, E., Dodson, G. G., Vijayan, M., Baker, E. N., Harding, M. M., Hodgkin, D. C., Rimmer, B. and Sheet, S. (1969). *Nature Lond*, **224**, 491.
Adams, M. J., McPherson, A., Rossmann, M. G., Schevitz, R. W. and Wonacott, A. J. (1970a). *J. Molec. Biol.*, **51**, 31.
Adams, M. J., Ford, G. C., Koekoek, R., Lentz, P. J., McPherson, A., Rossmann, M. G., Smiley, I. E., Schevitz, R. W. and Wonacott, A. J. (1970b). *Nature*, **227**, 1098.
Adams, M. J., Buehner, M., Chandrasekhar, K., Ford, G. C., Hackert, M. L., Liljas, A. and Rossmann, M. G. (1973). *Proc. Nat. Acad. Sci. USA*, **70**, 1968.
Adman, E. T., Sieker, L. C. and Jensen, L. H. (1973). *J. Biol. Chem.*, **248**, 3987.
Akervall, K. and Strandberg, B. (1971). *J. Molec. Biol.*, **62**, 625.
Alden, R. A., Wright, C. S. and Kraut, J. (1970). *Phil. Trans. Roy. Soc. Lond.*, **B257**, 119.
Alexander, L. E. and Smith, G. S. (1962). *Acta Cryst.*, **15**, 983.
Alexander, L. E. and Smith, G. S. (1964). *Acta Cryst.*, **17**, 447.
Amzel, L. M., Poljak, R. J., Saul, R., Varga, J. M. and Richards, F. F. (1974). *Proc. Nat. Acad. Sci. USA*, **71**, 1427.
Anderson, R. D., Apgar, P. A., Burnett, R. M., Darling, G. D., Lequesne, M. E., Mayhew, S. G. and Ludwig, M. L. (1972). *Proc. Nat. Acad. Sci. USA*, **69**, 3189.
Andreeva, N. S., Federov, A. A., Gushina, A. W., Shutskever, N. E., Riskulov, R. R. and Volinova, T. V. (1976). Unpublished results.
Argos, P. and Mathews, F. S. (1973). *Acta Cryst.*, **B29**, 1604.
Argos, P., Ford, G. C. and Rossmann, M. G. (1975). Acta Crysta., **A31**, 499.
Arndt, U. W. (1968). *Acta Cryst.*, **B24**, 1355.
Arndt, U. W. and Phillips, D. C. (1961). *Acta Cryst.*, **14**, 807.
Arndt, U. W. and Willis, B. T. M. (1966). "Single Crystal Diffractometry", Cambridge University Press.
Arndt, U. W., North, A. C. T. and Phillips, D. C. (1964). *J. Sci. Instr.*, **41**, 421.
Arndt, U. W., Crowther, R. A. and Mallett, J. F. W. (1968). *J. Phys. Sci. Instr.*, **1**, 510.

Arndt, U. W., Gilmore, D. J. and Boutle, S. H. (1972). *Adv. Electronics Electron Phys.*, **33B**, 1069.
Arndt, U. W., Champness, J. N., Phizackerley, R. P. and Wonacott, A. J. (1973). *J. Appl. Cryst.*, **6**, 457.
Arnon, R., Shapira, E. (1969). *J. Biol. Chem.*, **244**, 1033.
Arnone, A., Bier, C. J., Cotton, F. A., Day, V. W., Hazen, E. E., Richardson, D. C., Richardson, J. S., Yonath, A. (1971). *J. Biol. Chem.*, **246**, 2302.
Avey, H. P., Shall, S. (1969). *J. Molec. Biol.*, **43**, 341.
Bacon, G. E. (1962). Neutron Diffraction, Oxford Univ. Press.
Bailey, J. L. (1967). Techniques of Protein Chemistry 2d. ed., American Elsevier Publishing Company, New York.
Banaszak, L. J., Watson, H. C. and Kendrew, J. C. (1965). *J. Molec. Biol.*, **12**, 130.
Banner, D. W. (1972). D. Phil. Thesis. University of Oxford.
Banner, D. W., Bloomer, A. C., Petsko, G. A., Phillips, D. C., Pogson, C. I., Wilson, I. A., Corron, P. H., Furth, A. J., Milman, J. D., Offord, R. E., Priddle, J. D. and Waley, S. G. (1975). *Nature, Lond.*, **255**, 609.
Banyard, S. H., Blake, C. C. F. and Swan, I. D. A. (1973). *In* "Lysozyme" (eds. Osserman, E. F., Canfield, R. E. and Beychok, S.), Academic Press, New York, London and San Francisco.
Barrett, A. N. and Zwick, M. (1971). *Acta Cryst.*, **A27**, 6.
Barry, C. D. and North, A. C. T. (1971). *Cold Spr. Harb. Symp. Quant. Biol.*, **36**, 577.
Barry, C. D., Ellis, R. A., Graesser, S. M. and Marshall, G. R. (1969). *In* "Pertinent Concepts in Computer Graphics" (eds. Faiman M. and Nievergelt J.), University of Illinois Press, Urbana, Illinois.
Beddell, C. R. (1970). D. Phil. Thesis, University of Oxford.
Beddell, C. R. and Blake, C. C. F. (1970). *In* "Chemical Reactivity and Biological Role of Functional Groups in Enzymes". Biochemistry Society Symposium, 31, (ed. Smellie, R. M. S.) Academic Press. p. 157.
Benesch, R. and Benesch, R. E. (1956). *J. Am. Chem. Soc.*, **78**, 1597.
Benisek, W., and Richards, F. M. (1968). *J. Biol. Chem.*, **243**, 4267.
Bernal, J. D. (1969). "Science in History", Vol. III, Pelican Books p. 871.
Bernal, J. D. and Crowfoot, D. C. (1934). *Nature, Lond.*, **133**, 794.
Berreman, D. W. (1955). *Rev. Sci. Instr.*, **26**, 1048.
Bijvoet, J. M. (1949). *Proc. Koninkl. Ned. Akad. Wetenschap (B)*, **52**, 313.
Bijvoet, J. M. (1954). *Nature, Lond.*, **173**, 888.
Birktoft, J. J. and Blow, D. M. (1973). *J. Molec. Biol.*, **68**, 187.
Birktoft, J. J., Blow, D. M., Henderson, R. and Steitz, T. A. (1970). *Phil. Trans. Roy. Soc. London*, **B257**, 67.
Blake, C. C. F. (1968). *Adv. Protein Chem.*, **23**, 59.
Blake, C. C. F. and Evans, P. R. (1974). *J. Molec. Biol.*, **84**, 585.
Blake, C. C. F. and Phillips, D. C. (1962). in "Biological Effects of Ionising Radiation at the Molecular Level". IAEA Symposium, Vienna, 1962 p. 183.
Blake, C. C. F. and Swan, I. D. A. (1971). *Nature, New Biol.*, **232**, 12.
Blake, C. C. F., Fenn, R. H., North, A. C. T., Phillips, D. C. and Poljak, R. J. (1962). *Nature, Lond.*, **196**, 1173.
Blake, C. C. F., Koenig, D. F., Mair, G. A., North, A. C. T., Phillips, D. C. and Sarma, V. R. (1965). *Nature, Lond.*, **206**, 757.
Blake, C. C. F., Johnson, L. N., Mair, G. A., North, A. C. T., Phillips, D. C. and Sarma, V. R. (1967a). *Proc. Roy. Soc.*, **B167**, 378.

Blake, C. C. F., Mair, G. A., North, A. C. T., Phillips, D. C. and Sarma, V. R. (1967b). *Proc. Roy. Soc.*, **B167**, 365.
Blake, C. C. F., Evans, P. R. and Skopes, R. K. (1973). *Nature New Biol.*, **235**, 195.
Blake, C. C. F., Geisow, M. J. Swan, I. D. A., Rerat, C. and Rerat, B. (1974). *J. Molec. Biol.*, **88**, 1.
Blow, D. M. (1958). *Proc. Roy. Soc.*, **A247**, 302.
Blow, D. M. (1971). *In* "The Enzymes", Vol. III, 3rd edition, (Boyer, P. D. Boyer, ed.), p. 185, Academic Press, New York, London and San Francisco.
Blow, D. M. and Crick, F. H. C. (1959). *Acta Cryst.*, **12**, 794.
Blow, D. M. and Matthews, B. W. (1973). *Acta Cryst.*, **A29**, 56.
Blow, D. M. and Rossmann, M. G. (1961). *Acta Cryst.*, **14**, 1195.
Blow, D. M., Birktoft, J. J. and Hartley, B. S. (1969). *Nature, Lond.*, **221**, 337.
Bluhm, M. M., Bodo, G., Dintzis, H. M., Kendrew, J. C. (1958). *Proc. Roy. Soc., Lond.*, **A246**, 369.
Blundell, T. L. (1970). Unpublished Results.
Blundell, T. L. (1975). *New Scientist*, **67**, 662.
Blundell, T. L. and Johnson, L. N. (1972). *In* "Chemical Crystallography", (ed. Robertson J. M.) Butterworths, London p. 199.
Blundell, T. L., Cutfield, J. F., Dodson, G. G., Dodson, E., Hodgkin, D. C., Mercola, D., Vijayan, M. (1971a). *Nature, Lond.*, **231**, 506.
Blundell, T. L., Dodson, E. J., Dodson, G. G., Hodgkin, D. C. and Vijayan, M. (1971b). *Comtemporary Phys.*, **12**, 209.
Blundell, T. L., Hodgkin, D. C., Dodson, E., Dodson, G. G. and Vijayan, M. (1971c). *Recent. Prog. Horm. Res.*, **27**, 1.
Blundell, T. L., Hodgkin, D. C., Dodson, G. G. and Mercola, D. A. (1972). *Adv. Protein Chem.*, **26**, 279.
Bode, W. and Schwager, P. (1975). *FEBS Lett.*, **56**, 139.
Bodo, G., Dintzis, H. M., Kendrew, J. C. and Wyckoff, H. W. (1959). *Proc. Roy. Soc.*, **A253**, 70.
Bokhoven, C., Schoone, J. C. and Bijvoet, J. M. (1951). *Acta Cryst.*, **4**, 275.
Bolton, W. and Perutz, M. F. (1970). *Nature, Lond.*, **228**, 551.
Borras-Cuesta, F. (1972). D. Phil. Thesis, Oxford.
Boyes-Watson, I., Davidson, E. and Perutz, M. F. (1947). *Proc. Roy. Soc. Ser.*, **A191**, 83.
Braams, R. and Swallow, A. J. (1968). *In "Advances in Chemistry. Radiation Chemistry"* (ed. R. F. Gould) **81**, 464.
Bradshaw, R. A., Ericsson, L. H., Walsh, K. A. and Neurath, H. (1969). *Proc. Nat. Acad. Sci. USA*, **63**, 1389.
Bragg, W. L. (1913). *Proc. Camb. Phil. Soc.*, **17**, 43.
Bragg, W. L. (1962). "The Crystalline State", Vol. I. General survey. Third Edition. Bell: London. *See also* "The Development of X-ray Analysis" Bell: London (1975).
Bragg, W. L. (1969). *Acta Cryst.*, **A25**, 1.
Bragg, W. L. and Perutz, M. F. (1952). *Proc. Roy. Soc.*, **A213**, 425.
Bragg, W. L., James, R. W. and Bosanquet, C. M. (1921). *Phil. Mag.*, **42**, 1.
Branden, C. I. and Zeppezauer, E. (1974). Unpublished results.
Branden, C. I., Eklund, H., Nordstrom, B., Boiwe, T., Soderlund, G., Zeppezauer, E., Ohlsson, I. and Akeson, A. (1973). *Proc. Nat. Acad. Sci. USA*, **70**, 2439.
Brandenburg, D. (1969). Hoppe-Seyler's *Z. Physiol. Chem.*, **350**, 741.

Branton, D. (1969). *Ann. Rev. Plant. Physiol.*, **20**, 209.
Bricogne, G. (1974). *Acta Cryst.*, **A30**, 395.
Brooks, B. C. and Dick, W. F. L. (1951). "Introduction to Statistical Methods". Heinemann.
Bryant, T. N., Watson, H. C. and Wendell, P. L. (1974). *Nature, Lond.*, **247**, 14.
Buehner, M., Ford, G. C., Moras, D., Olsen, K. W. and Rossmann, M. G. (1974). *J. Molec. Biol.*, **82**, 563.
Buehner, M., Ford, G. C., Moras, D., Olsen, K. W. and Rossmann, M. G. (1973). *Proc. Nat. Acad. Sci.*, **70**, 3052.
Buerger, M. J. (1942). "X-ray Crystallography". John Wiley, New York.
Buerger, M. J. (1959) "Vector Space", John Wiley, New York.
Buerger, M. J. (1963) "Elementary Crystallography", John Wiley, New York.
Buerger, M. J. (1964) "The Precession Method", John Wiley, New York.
Bullivant, S. (1973). *In* "Some Biological Techniques in Electron Microscopy", p. 123 (ed. Parsons, D. F.) Academic Press, New York, London and San Francisco.
Bunn, C. W. (1961). "Chemical Crystallography", Second Edition. Oxford University Press.
Bunn, C. W., Moews, P. C. and Baumber, M. E. (1971). *Proc. Roy. Soc. Lond.*, **B178**, 245.
Burbank, R. D. (1964). *Acta Cryst.*, **17**, 434.
Busing, W. R. and Levy, H. A. (1957). *Acta Cryst.*, **10**, 180.
Campbell, J. W., Watson, H. C. and Hodgson, G. I. (1974). *Nature, Lond.*, **250**, 301.
Carlise, C. H., Gorinsky, B. A., Mazumdar, S. K., Palmer, R. A., Yeates, D. G. R. (1974). *J. Molec. Biol.*, **85**, 1.
Carter, C. W., Freer, S. T., Xuong, N. H., Alden, R. A. and Kraut, J. (1971). *"Cold Spring Harbor Symposium"*, **36**, 381.
Carter, C. W., Kraut, J., Freer, S. T. and Alden, R. A. (1974a). *J. Biol. Chem.*, **249**, 6339.
Carter, C. W., Kraut, J., Freer, S. T., Xuong, N. H., Alden, R. A. and Bartsch, R. G. (1974b). *J. Biol. Chem.*, **249**, 4212.
Caspar, D. and Kirschner, D. (1971). *Nature New Biol.*, **231**, 46.
Chambers, J. L., Christoph, G. G., Krieger, M., Hay, L. and Stroud, R. M. (1974). *Biochem. Biophys. Res. Commun.*, **59**, 70.
Chen, B. L. and Poljak. (1974). *J. Biol. Chem.*, **13**, 1295.
Chothia, C. (1973). *J. Molec. Biol.*, **75**, 295.
Chothia, C. and Janin, J. (1975). *Nature, Lond.*, **256**, 705.
Collins, D. M., Cotton, F. A., Hazen, E. E., Meyer, E. F. and Morimoto, C. W. (1975). *Science, NY*, **190**, 1047.
Colman, P. M. and Matthews, B. W. (1971). *J. Molec. Biol.*, **60**, 163.
Colman, P. M., Jansonius, J. N. and Matthews, B. W. (1972a). *J. Molec. Biol.*, **70**, 701.
Colman, P. M., Weaver, L. H. and Matthews, B. W. (1972b). *Biochem. Biophys. Res. Commun.*, **46**, 1999.
Cooley, J. W. and Tukey, J. W. (1965). Mathematics of Computation, **19**, No. 90, 297.
Cork, C., Fehr, D., Hamlin, R., Vernon, W., Xuong, N. H. and Perez-Mendez, V. (1974). *J. Appl. Cryst.*, **7**, 319; *Acta Cryst.*, **A31**, 701.
Coster, D., Knol, K. S. and Prins, J. A. (1930). *Z. Phys.*, **63**, 345.
Cotton, F. A., Bier, C. J., Day, V. W., Hazen, E. E. Jr., Larsen, S. (1971). *Cold Spring Harb. Symp. Quant. Biol.*, **36**, 243.

Coulter, C. L. (1971). *Acta Cryst.*, **B27**, 1730.
Cox, E. G. and Shaw, W. F. B. (1930). *Proc. Roy. Soc.*, **127**, 71.
Crewe, A. V. (1970). *Q. Rev. Biophys.*, **3**, 137.
Crewe, A. V. (1971). *Phil. Trans. Roy. Soc.*, **B261**, 61.
Crewe, A. V., Isaacson, M. and Johnson, D. (1969). *Rev. Sci. Instr.*, **40**, 241.
Crick, F. H. C. and Magdoff, B. (1956). *Acta Cryst.*, **9**, 901.
Cromer, D. T. and Liberman, D. (1970). *J. Chem. Phys.*, **53**, 1891.
Crowfoot, D. M. (1935). *Nature, Lond.*, **135**, 591.
Crowfoot, D., Bunn, C. W., Rogers-Low, B. W. and Turner Jones, A. (1949). *In* "The Chemistry of Pencillin", p. 310 (eds. Clarke, H. T., Johnson, J. R. and Robinson, R.) Oxford University Press.
Crowther, R. A. (1971a). *Phil. Trans. Roy. Soc.*, **B261**, 221.
Crowther, R. A. (1971b). *In* "Molecular Replacement Method", (ed. Rossmann, M. G.) Gordon and Breach, New York.
Crowther, R. A. and Blow, D. M. (1967). *Acta Cryst.*, **23**, 544.
Cruickshank, D. W. J. (1949). *Acta Cryst.*, **2**, 65.
Cullis, A. F., Muirhead, H., Perutz, M. F., Rossmann, M. G. and North, A. C. T. (1961). *Proc. Roy. Soc.*, **A265**, 15.
Cullis, A. F., Muirhead, H., Perutz, M. F., Rossmann, M. G. and North, A. C. T. (1962). *Proc. Roy. Soc.*, **A265**, 161.
Czok, R. and Bucher, T. H. (1960). *Adv. Protein Chem.*, **15**, 337.
Dale, D. H. and Willis, B. T. M. (1966). *AERE – Report*, **R5195**.
Dale, D., Hodgkin, D. C. and Venkatesan, K. (1963). *In* "Crystallography and Crystal Perfection", p. 237, (ed. Ramachandran, G. N.) Academic Press, London, New York and San Francisco.
Darnall, D. W. and Birnbaum, E. R. (1970). *J. Biol. Chem.*, **245**, 6484.
Darwin, C. G. (1914). *Phil. Mag.*, **27**, 315.
Dauben, C. H. and Templeton, D. H. (1955). *Acta Cryst.*, **8**, 841.
Davies, D. R. and Doctor, B. P. (1971). "Procedures in Nucleic Acid Research", Harper and Row, New York.
Davies, D. R. and Segal, D. M. (1971). "Methods in Enzymology", **22**, 266.
Davies, D. R., Padlan, E. A. and Segal, D. M. (1975). *Ann. Rev. Biochem.*, **44**, 639.
Dayhoff, M. O., Park, C. M. and McLaughlin, P. J. (1972). *In* "Atlas of Protein Sequence and Structure 1972" (ed. Dayhoff M. O.) Vol. 5 pp. 7–16. Natl. Biomedical Res. Fnd., Washington.
Debye, P. (1914). *Annl. Phys.* **43**, 49.
Deisenhofer, H. and Steigemann, W. (1974). *In* "2nd. Int. Research Conference on Proteinase Inhibitors", (eds. Fritz, H., Tschesche, H., Green, L. J. and Truscheir, E.), Springer-Verlag, New York. (in press).
Delbaere, L. T. J., Hutcheon, W. L. B., James, M. N. G. and Thiessen, W. E. (1975). *Nature, Lond.*, **257**, 758.
De Rango, C., Mauguen, Y. and Tsoucaris, G. (1975). *Acta Cryst.*, **A31**, 227.
DeRosier, D. J. and Klug, A. (1968). *Nature, Lond.*, **217**, 130.
DeRosier, D. J. and Moore, P. B. (1970). *J. Molec. Biol.*, **52**, 355.
DeRosier, D. and Oliver, R. M. (1971). *Cold Spring Harb. Symp. Quant. Biol.*, **36**, 199.
Diamond, R. (1966). *Acta Cryst.*, **21**, 253.
Diamond, R. (1969). *Acta Cryst.*, **A25**, 43.
Diamond, R. (1971). *Acta Cryst.*, **A27**, 436.
Diamond, R. (1974a). *J. Molec. Biol.*, **82**, 371.
Diamond, R. (1974b). *Private Communication*.

Dickerson, R. E. and Geiss, I. (1969). "Structure and Action of Proteins", Harper Row, New York.
Dickerson, R. E., Kendrew, J. C. and Strandberg, B. E. (1961). *Acta Cryst.*, **14**, 1188.
Dickerson, R. E., Kopka, M. L., Varnum, J. C. and Weinzierl, J. E. (1967). *Acta Cryst.*, **23**, 511.
Dickerson, R. E., Weinzierl, J. E. and Palmer, R. A. (1968). *Acta Cryst.*, **B24**, 997.
Dickerson, R. E., Eisenberg, D., Varnum, J. and Kopka, M. L. (1969). *J. Molec. Biol.*, **45**, 77.
Dickerson, R. E., Takano, T., Eisenberg, D., Kallai, O. B., Samson, L., Cooper, A. and Margoliash, E. (1971). *J. Biol. Chem.*, **246**, 1511.
Dixon, M. and Webb, E. C. (1961). *Adv. Protein Chem.*, **16**, 197.
Dobler, M., Dover, S. D., Laves, K., Binder, A. and Zuber, H. (1972). *J. Molec. Biol.*, **71**, 785.
Dodson, E. J. and Vijayan, M. (1971). *Acta Cryst.*, **B27**, 2402.
Dodson, E., Harding, M. M., Hodgkin, D. C. and Rossmann, M. G. (1966). *J. Molec. Biol.*, **16**, 227.
Dodson, E. J., Evans, P. R. and French, S. (1975). *In:* "Anomalous Scattering" p. 423 (Eds. Ramaseshan, S. and Abramsons, S. C.), Munksgaard, Copenhagen.
Dodson, E. J., Isaacs, N. W. and Rollett, J. S. (1976). *Acta Cryst.*, **A32**, 311.
Doyle, B. B., Hulmes, P. J. S., Miller, A., Parry, D. A. D., Piez, K. A. and Woodhead-Galloway, J. (1974). *Proc. Roy. Soc. Lond.*, **B187**, 37.
Drenth, J., Jansonius, J. N. and Wolthers, B. G. (1967). *J. Molec. Biol.*, **24**, 449.
Drenth, J., Jansonius, J. N., Koekoek, R., Marrink, J., Munnik, J. and Wolthers, B. G. (1962). *J. Molec. Biol.*, **5**, 398.
Drenth, J., Jansonius, J. N., Koekoek, R., Swen, H. H. and Wolthus, B. G. (1968). *Nature, Lond.*, **218**, 929.
Drenth, J., Jansonius, J. N., Koekoek, R. and Wolthers, B. G. (1971a). *Adv. Protein Chem.*, **25**, 79.
Drenth, J., Hol, W. G. J., Jansonius, J. N. and Koekoek, R. (1971b). *Cold Spring Harb. Symp. Quant. Biol.*, **36**, 107.
Drenth, J., Jansonius, J. N., Koekoek, R. and Wolthus, B. G. (1971c). *In* "The Enzymes", Vol. III, 3rd Edition, (ed. Boyer, P. D.), p. 484, Academic Press, New York, London and San Francisco.
Drenth, J., Enzing, C., Kalk, K. H. and Vessies, J. (1975). Unpublished results.
Eagles, P. A. M. (1970). D. Phil. Thesis, University of Oxford.
Eagles, P. A. M. and Johnson, L. N. (1972). *J. Molec. Biol.*, **64**, 693.
Eagles, P. A. M., Johnson, L. N., Joynson, M. A., McMurray, C. H. and Gutfreund, H. (1969). *J. Molec. Biol.* **45**, 533.
Ebert, M. and Swallow, A. J. (1965). *In* "Advances in Chemistry. The Solvated Electron", (ed. Gould, R. F.) **50**, 289.
Edelman, G. A. and Gall, W. E. (1969). *Ann. Rev. Biochem.*, **38**, 415.
Edelman, G. M., Cunningham, B. A., Reeke, G. N., Becker, J. W., Waxdal, M. L. and Wang, J. L. (1972). *Proc. Nat. Acad. Sci. USA*, **69**, 2580.
Edman, P. (1967). *Eur. J. Bio. Chem.*, **1**, 80.
Ehrenberg, W. and Spear, W. E. (1951). *Proc. Phys. Soc.*, **B64**, 67.
Einarsson, R., Wallen, L. and Zeppezauer, M. (1972). *Chemica Scripta*, **2**, 84.
Einarsson, R., Eklund, H., Zeppezauer, E., Boiwe, T. and Branden, C-I. (1974). *Eur. J. Biochem.*, **49**, 41.

REFERENCES

Eisenberg, D. (1970). *In* "The Enzymes", Vol. I, 3rd edition, (ed. Boyer, P. D.), Academic Press, New York, London and San Francisco.
Eisenberg, D., Heidner, E. G., Goodkin, P., Dastoor, M., Weber, B. H, Wedler, F. and Bell, J. D. (1971). *Cold Spring Harb. Symp. Quant Biol.*, **36**, 291.
Eklund, H., Nordstrom, B., Zeppezauer, E., Soderlund, G., Ohlsson, I., Boiwe, T. and Brandon, C-I. (1974). *FEBS Letters*, **44**, 200.
Eklund, H., Nordstram, B., Zeppezauer, E. S., Soderlund, G., Ohlsson, I., Boiwe, T. and Branden, C. (1976). *J. Molec. Biol.*, (in press).
Ely, K. R., Girling, R. L., Schiffer, M., Cunningham, D. E. and Edmundson, A. E. (1973). *Biochem. J.*, **12**, 4233.
Engelman, D. M. and Moore, B. P. (1972). *Proc. Nat. Acad. Sci. USA*, **69**, 1997.
Epp, O., Steigemann, W., Formanek, H. and Huber, R. (1971). *Eur. J. Biochem.*, **20**, 432.
Epp, O., Colman, P., Fehlhammer, H., Bode, W., Schiffer, M., Huber, R. and Palm, W. (1974). *Eur. J. Biochem.*, **45**, 513.
Erickson, H. P. and Klug, A. (1971). *Phil. Trans. Roy. Soc.*, **B261**, 105.
Evans, P. R. (1973). D. Phil. Thesis. University of Oxford.
Ewald, P. P. (1921). *Z. Kristollagr. Miner*, **56**, 129.
Fehlhammer, H. *et al.* (1975). *Biophys. Struct. Mechanism*, **1**, 139.
Fenna, R. E. and Matthews, B. W. (1975). *Nature, Lond.*, **258**, 573.
Fermi, G. (1970). *J. Molec. Biol.*, **46**, 211.
Fletterick, R. J., Bates, D. J. and Steitz, T. A. (1975). *Proc. Nat. Acad. Sci. USA*, **72**, 38.
Ford, L. O., Johnson, L. N., Machin, P. A., Phillips, D. C. and Tjian, R. (1974). *J. Molec. Biol.*, **88**, 349.
Freer, S. T., Kraut, J., Robertus, J. D., Wright, H. T. and Xuong, N. H. (1970). *Biochem.*, **9**, 1997.
Freer, S. T., Alden, R. A., Carter, C. W. and Kraut, J. (1974). *J. Biol. Chem.*, **250**, 46.
Foltmann, B. (1959). *Acta Chem. Scand.*, **13**, 1927.
Friedrich, W., Knipping, P. and Von Laue, M. (1912). *Proc. Bavarian Acad. Sci.*, p. 303. Reprinted in *Naturwissenschaften*, (1952) **39**, 367.
Furnas, T. C. and Harker, D. (1955). *Rev. Sci. Instr.*, **26**, 449.
Gassmann, J. and Zechmeister, K. (1972). *Acta Cryst.* **A28**, 270.
Geisow, M. S. (1975). Private Communication.
Gilbert, P. F. C. and Klug, A. (1974). *J. Molec. Biol.* **86**, 193.
Glaeser, R. M. (1971). *J. Ultrastruct. Res.*, **36**, 466.
Glaeser, R. M. and Thomas, G. (1969). *Biophys. J.*, **9**, 1073.
Glasstone, S. (1974). "Text Book of Physical Chemistry", Macmillan and Co. Ltd., London.
Goodwin, R. P. (1969). *Springer Tracts in Modern Physics*, (ed. Hohler, G.), **51**, 1.
Gould, R. W., Bates, S. R. and Sparks, C. J. (1968). *Appl. Spect.*, **22**, 549.
Green, A. A. (1931). *J. Biol. Chem.*, **93**, 495.
Green, A. A. (1932). *J. Biol. Chem.*, **95**, 47.
Green, A. A. and Hughes, W. L. (1955). *In* "Methods in Enzymology", Vol. 1, Academic Press, New York, London and San Fransisco.
Green, D. W., Ingram, V. M. and Perutz, M. F. (1954). *Proc. Roy. Soc.*, **A225**, 287.
Greider, M. H., Howell, S. L. and Lacy, P. E. (1969). *J. Cell. Biol.*, **41**, 162.
Gunnarsson, P. O. and Petterson, G. (1974). *Eur. J. Bio. Chem.*, **43**, 479.

Gunnarsson, P. O., Pettersson, G. and Zeppezauer, M. (1974). *Eur. J. Bio. Chem.,* **43**, 479.
Haas, D. J. and Rossmann, M. G. (1970). *Acta Cryst.,* **B26**, 998.
Hagman, L. O., Larson, L. O. and Kierkegaard, P. (1969). *Int. J. Prot. Res.,* **1**, 283.
Hall, C. E. (1950). *J. Biol. Chem.,* **185**, 749.
Hall, C. E. (1966). "Introduction to Electron Microscopy", McGraw Hill.
Hamilton, W. C., Rollett, J. S. and Sparks, R. A. (1965). *Acta Cryst.,* **18**, 129.
Hanzen, K. J. (1971). *Adv. Opt. Elect. Microsc.,* **4**, 1.
Harding, M. M. (1962). D. Phil. Thesis, Oxford University.
Hardman, K. (1975). Private Communication.
Hardman, K. D. and Ainsworth, C. F. (1972). *Biochem.,* **11**, 4910.
Harker, D. (1956). *Acta Cryst.,* **9**, 1.
Hart, R. G. (1961). *Acta Cryst.,* **14**, 1194.
Hart, G. (1975). *Acta Cryst.,* (in press).
Hartley, B. and Shotton, D. (1970). *In* "The Enzymes", Vol. III, 3rd edition (ed. Boyer, P.), Academic Press, New York, London and San Francisco.
Hartshorne, N. H. and Stuart, A. (1960). "Crystals and the Polarising Microscope", Third Edition. Arnold, London.
Hartsuck, J. A. and Lipscomb, W. N. (1971). *In* "The Enzymes", Vol. III, 3rd edition (ed. Boyer, P. D.), p. 1, Academic Press, New York, London and San Francisco.
Hauptman, H. and Karle, J. (1953). "The Solution of the Phase Problem. I The Centrosymmetric Crystal". ACA Monograph No. 3. Ann Arbor, Michigan. Edwards Brothers.
Haurowitz, F. (1950). "Chemistry and Biology of Proteins", P. 91 Academic Press, New York, London and San Fransisco.
Henderson, R. C. (1970). *J. Molec. Biol.,* **54**, 341.
Henderson, R. C. and Moffat, J. K. (1971). *Acta Cryst.,* **B27**, 1414.
Hendrickson, W. A. (1971). *Acta Cryst.,* **B27**, 1472.
Hendrickson, W. A. and Karle, J. (1973). *J. Biol. Chem.,* **248**, 3327.
Hendrickson, W. A. and Lattman, E. E. (1970). *Acta Cryst.,* **B26**, 136.
Hendrickson, W. A. and Love, W. E. (1971). *Nature New Biol.,* **232**, 197.
Henkelman, R. M. and Ottensmeyer, F. P. (1972). *Proc. Nat. Acad. Sci.,* **68**, 3000.
Henry, N. F. M., Lipson, H. and Wooster, W. A. (1951). "Interpretation of X-ray Diffraction Photographs", Macmillan, London.
Herriott, J. R., Sieker, L. C., Jensen, L. H. and Lovenberg, W. (1970). *J. Molec. Biol.,* **50**, 391.
Herriott, J. R., Watenpaugh, K. D., Sieker, L. C. and Jensen, L. H. (1973). *J. Molec. Biol.,* **80**, 423.
Hess, G. P. and Rupley, J. A. (1971). *Ann. Rev. Biochem.,* **40**, 1013.
Hill, E., Tsernoglou, D., Webb, L. and Banaszak, L. J. (1972). *J. Molec. Biol.,* **72**, 577.
Hirs, C. H. W. (1967). *In* "Methods in Enzymology", Vol. XI, Academic Press, New York, London and San Francisco.
Hjerten and Mosbach. (1962). *Anal. Biochem.,* **3**, 109.
Hodgkin, D. C. and Riley, D. P. (1968). *In* "Structural Chemistry and Molecular Biology", Freeman, San Francisco. p. 15.
Hodgman, C. D. (1949). "Handbook of Chemistry and Physics", Chemical Rubber Publishing Co. Cleveland, Ohio.

Hol, W. G. J. (1971). Ph. D. Thesis. Rijksuniversiteit te Groningen.
Holmes, K. C. and Blow, D. M. (1966). "The Use of X-ray Diffraction in the study of Protein and Nucleic Acid Structure", Interscience. New York.
Holmgren, A., Sodeberg, B.-O., Eklund, H. and Branden, C-I. (1975). *Proc. Nat. Acad. Sci. USA*, **72**, 2305.
Hönl, H. (1933). *Z. Phys.*, **84**, 1.
Hoppe, W. (1975). Unpublished results.
Hoppe, W. and Gassmann, J. (1968). *Acta Cryst.*, **B24**, 97.
Hoppe, W., Gassmann, J. and Zechmeister, K. (1970). *In* "Crystallographic Computing", p. 26, (ed. Ahmed, F. R.), Copenhagen, Munkgaard.
Hoppe, W., Gassmann, J., Hensmann, N., Schramm, H. J. and Stur, M. (1974). *Hoppe–Seylers Z. Physiol. Chem.*, **355**, 1483.
Huber, R. (1969). *In* "Crystallographic Computing Proc.", p. 96, (ed. Ahmed, F. R.) Munksgaard, Copenhagen.
Huber, R. and Kopfman, G. (1969). *Acta Cryst.*, **A25**, 143.
Huber, R., Epp, O. and Formanek, H. (1969a). *J. Molec. Biol.*, **42**, 591.
Huber, R., Epp, O. and Formanek, H. (1969b). *Naturwissenschaften*, **56**, 362.
Huber, R., Epp, O. and Formanek, H. (1970). *J. Molec. Biol.*, **52**, 349.
Huber, R., Epp, O., Steigemann, W. and Formanek, H. (1971). *Eur. J. Biochem.*, **19**, 42.
Huber, R., Kukla, D., Bode, W., Schwager, P., Bartels, K., Deisenhofer, J. and Steigemann, W. (1974). *J. Molec. Biol.*, **89**, 73.
Huxley, H. E. (1953). *Acta Cryst.*, **6**, 457.
Imoto, T., Johnson, L. N., North, A. C. T., Phillips, D. C. and Rupley, J. A. (1972). *In* "The Enzymes", Vol. VII, 3rd edition, p. 665, (ed. Boyer P. B.), Academic Press, New York, London and San Francisco.
Insulin Research Group, Peking. (1974). *Academia Sinica, Scientia Sinica*, **XVII**, 79.
Isaacson, M. (1972). *J. Chem. Phys.*, **56**, 1803.
Jack, A. (1973). *Acta Cryst.*, **A29**, 545.
Jack, A., Weinzierl, J. and Kalb, A. J. (1971). *J. Molec. Biol.*, **58**, 389.
Jacobi, T. H., Ellis, R. A. and Fritsch, J. M. (1972). *J. Molec. Biol.*, **72**, 589.
Jackoby, W. B. (1971). *In* "Methods in Enzymology", Vol. XII p. 248, Academic Press, New York, London and San Francisco.
James, R. W. (1957). "The Optical Principles of The Diffraction of X-rays The Crystalline State", Vol. II. Bell, London. See also *Acta Cryst.* (1948), **1**, 132.
Jansen, E. F., Nutting, M. D. F., Jang, R. and Balls, A. K. (1949). *J. Biol. Chem.*, **179**, 189.
Jeffrey, J. W. and Rose, K. M. (1964). *Acta Cryst.*, **17**, 343.
Jenkins, J. A., Tickle, I. J., Blundell, T. L. and Ungaretti, L. (1976). Unpublished results.
Jensen, L. (1971). Private Communication.
Johnson, L. N., Madsen, N. B., Mosley, J. and Wison, K. S. (1974). *J. Molec. Biol.*, **90**, 703.
Joy, R. T. (1973). *Adv. Opt. Elect. Microsc.*, **5**, 297.
Joynson, M. A., North, A. C. T., Sarma, V. R., Dickerson, R. E. and Steinrauf, L. K. (1970). *J. Molec. Biol:* **50**, 137.
Kannan, K. K., Liljas, A., Waara, I., Bergsten, P. C., Lövgren, S., Strandberg, B., Bengtsson, U., Carlbom, U., Fridborg, K., Jarup, L. and Petef, M. (1971). *Cold Spring Harb. Symp. Quant. Biol.*, **36**, 221.

Kannan, K. K., Notstrand, B., Fridborg, K., Lövgren, S., Ohlsson, A. and Petef, M. (1975). *Proc. Nat. Acad. Sci. USA*, **72**, 51.
Karle, J. and Hauptman, H. (1956). *Acta Cryst.*, **9**, 635.
Karle, J. and Karle, I. L. (1966). *Acta Cryst.*, **21**, 849.
Kartha, G. (1965). *Acta Cryst.*, **19**, 883.
Kartha, G. (1969). Private Communication.
Kartha, G. and Parthasarathy, R. (1965a). *Acta Cryst.*, **18**, 745.
Kartha, G. and Parthasarathy, R. (1965b). *Acta Cryst.*, **18**, 749.
Kartha, G., Bello, J. and Harker, D. (1967). *Nature, Lond.*, **213**, 862.
Katayama, C., Sakabe, N. and Sakabe, K. (1972). *Acta Cryst.*, **A28**, 293.
Kendrew, J. C. (1962). *Brookhaven Symp. Biol.*, **15**, 216.
Kendrew, J. C., Bodo, G., Dintzis, H. M., Parrish, R. G., Wykoff, H. and Phillips, D. C. (1958). *Nature, Lond.*, **181**, 662.
Kendrew, J. C., Dickerson, R. E., Strandberg, B. E., Hart, R. G., Davies, D. R., Phillips, D. C. and Shore, V. C. (1960). *Nature, Lond.*, **185**, 442.
Kendrew, J. C., Watson, H. C., Strandberg, B. E., Dickerson, R. E., Phillips, D. C. and Shore, V. C. (1961). *Nature, Lond.*, **190**, 666.
Kestleman, H. (1968). *In* "Exploring University Mathematics 2", (ed. Hardiman, N. J.) Pergamon Press, Oxford.
Kim, S. H., Suddath, F. L., Quigley, G. J., McPherson, A., Sussman, J. L., Wang, A. H. J., Seeman, N. C. and Rich, A. (1974a). *Science, New York*, **185**, 435.
Kim, S. H., Sussman, J. L., Suddath, F. L., Quigley, G. J., McPherson, A., Wang, A. H. J., Seeman, N. C. and Rich, A. (1974b). *Proc. Nat. Acad. USA*, **71**, 4970.
King, M. V. (1954). *Acta Cryst.*, **7**, 601.
King, M. V. (1955). *J. Molec. Biol.*, **11**, 549.
Klar, B. (1972). *Acta Cryst.*, S250.
Klug, A. and Berger, J. E. (1964). *J. Molec. Biol.*, **10**, 565.
Klug, A., Ladner, J. and Robertus, J. D. (1974). *J. Molec. Biol.*, **89** 511.
Knobel, H. R., Williger, W. and Isliker. (1975). *Eur. J. Immunol.*, **5**, 78.
Kopfmann, G. and Huber, R. (1968). *Acta Cryst.*, **A24**, 348.
Korn, A. H., Feairheller, S. H. and Filachione, E. M. (1972). *J. Molec. Biol.*, **65**, 525.
Kornicker, and Vallee, B. L. (1969). *Ann. NY Acad, Sci.*, **153**, 689.
Kraut, J. (1968). *J. Molec. Biol.*, **37**, 225.
Kraut, J., Sieker, L. C., High, D. F. and Freer, S. T. (1962). *Proc. Nat. Acad. Sci. USA*, **48**, 1417.
Kraut, J. G., Strahs, G. and Freer, S. T. (1968). *In* "Structural Chemistry and Molecular Biology", (eds. Rich, A. and Davidson, N.) Freeman, San Francisco, California.
Kretsinger, R. H. (1968). *J. Molec. Biol.*, **31**, 315.
Kretsinger, R. H. and Nockolds, C. (1973). *J. Biol. Chem.*, **248**, 3313.
Kretsinger, R. H., Watson, H. C. and Kendrew, J. C. (1968). *J. Molec. Biol.*, **31**, 305.
Kretsinger, R. H., Nockolds, C. E., Coffee, C. J. and Bradshaw, R. A. (1971). *Cold Spring Harb. Symp. Quant. Biol.*, **36**, 217.
Labaw, L. W. and Davies, D. R. (1971). *J. Biol. Chem.*, **246**, 3760.
Lang, A. R. (1954). *Rev. Sci. Instr.*, **25**, 1039.
Lattman, E. E. (1972). *Acta Cryst.*, **B28**, 1065.
Leach, S. J., Nermethy, G. and Scheraga, H. A. (1966). *Biopolymers*, **4**, 887.
Legendre, A. M. (1806). Nouvelles methods pour le determination des orbites des cometes. Courcier, Paris 1806 P72.

Lehninger, A. L. (1972). "Biochemistry", Worth Publishers Inc., New York.
Levinthal, C. (1966). *Sci. Am.*, **214**, 42.
Levinthal, C. C., Barry, C. D., Ward, S. A. and Zwick, M. (1968). *In* "Emerging Concepts in Computer Graphics", (eds. Nievergelt, J. and Secrest, D.), W. A. Benjamin, New York.
Levitt, M. (1974). *J. Molec. Biol.*, **82**, 393.
Levitt, M. and Lifson, S. (1969). *J. Molec. Biol.*, **46**, 269.
Levitt, M. and Warshel, A. (1975). *Nature, Lond.*, **253**, 694.
Liljas, A. (1971). Ph.D Thesis, Acta Universitatis Upsaliensis Weilands Tryckeri, Uppsala.
Liljas, A. and Rossmann, M. G. (1974). *Ann. Rev. Biochem.*, **43**, 475.
Liljas, A., Kannan, K. K., Bergsten, P. C., Waara, I., Fridborg, K., Strandberg, B., Carlborn, U., Jarup, L., Lovgren, S. and Petef, M. (1972). *Nature New Biol.*, **235**, 131.
Lipscomb, W. N., Coppola, J. C., Hartsuck, J. A., Ludwig, M. L., Muirhead, H., Searl, J. and Steitz, T. A. (1966). *J. Molec. Biol.*, **19**, 423.
Lipscomb, W. N., Hartsuck, J. A., Reeke, G. N., Quiocho, F. A., Bethge, P. H., Ludwig, M. L., Steitz, T. A., Muirhead, H. and Coppola, J. C. (1968). *Brookhaven Symp. Biol.*, **21**, 24.
Lipscomb, W. N., Hartsuck, J. A., Quiocho, F. A. and Reeke, G. N. (1969). *Proc. Nat. Acad. Sci. USA*, **64**, 28.
Lipscomb, W. N., Reeke, G. N., Hartsuck, J. A., Quiocho, F. A. and Bethge, P. H. (1970). *Phil. Trans. Roy. Soc. London.*, **B257**, 177.
Lipson, H. and Cochran, W. (1966). "The Determination of Crystal Structures", 3rd edition, Bell and Sons, London.
Lipson, H. and Taylor, C. A. (1958). "Fourier Transforms and X-ray Diffraction", Bell, London.
Love, W. E. (1971). Private Communication.
Love, W. E., Klock, P. A., Lattman, E. E., Padlan, E. A., Ward, K. B. and Hendrickson, W. A. (1971). *Cold Spring Harb. Symp. Quant. Biol.*, **36**, 349.
Low, B. W. and Berger, J. E. (1961). *Acta Cryst.*, **14**, 82.
Low, B. W. and Richards, F. M. (1952). *J. Am. Chem. Soc.*, **74**, 1660.
Ludwig, M. L., Anderson, R. D. Apgar, P. A., Burnett, R. M., LeQuesne, M. E. and Mayhew, S. G. (1971). *Cold Spring Harb. Symp. Quant. Biol.*, **36**, 369.
Luzzati, V. (1952). *Acta Cryst.*, **5**, 802.
Luzzati, V. (1953). *Acta Cryst.*, **6**, 142.
MacDonald, A. C. and Sikka, S. K. (1969). *Acta Cryst.*, **B25**, 1804.
Mahler, H. R. and Cordes, E. H. (1968). "Biological Chemistry", Harper and Row, New York.
Main, P. (1967). *Acta Cryst.*, **23**, 50.
Main, P. and Rossmann, M. G. (1966). *Acta Cryst.*, **21**, 67.
Marsh, D. J. and Petsko, G. A. (1973). *J. Appl. Crystallog.*, **6**, 76.
Marsh, R. E., Corey, R. B. and Pauling, L. (1955). *Acta Cryst.*, **8**, 710.
Matricardi, V. R., Moretz, R. C. and Parsons, D. F. (1972). *Science, NY.*, **177**, 268.
Matthews, B. W. (1966a). *Acta Cryst.*, **20**, 230.
Matthews, B. W. (1966b). *Acta Cryst.*, **20**, 82.
Matthews, B. W. (1968). *J. Molec. Biol.*, **33**, 491.
Matthews, B. W. (1974a). *In* "The Proteins", (eds. Neurath, M. and Hill, R. L.) 3rd edition, Academic Press, New York, London and San Francisco.
Matthews, B. W. (1974b). *J. Molec. Biol.*, **82**, 513.
Matthews, B. W. (1974c). Private Communication.

Matthews, B. W. and Remington, S. H. (1974). *Proc. Nat. Acad. Sci. USA.*, **71**, 4178.
Matthews, B. W. and Weaver, L. H. (1974). *Biochemistry*, **13**, 1719.
Matthews, B. W., Colman, P. M., Jansonius, J. N., Titani, K., Walsh, K. A. and Neurath, H. (1972a). *Nature New Biol.*, **238**, 41.
Matthews, B. W., Klopfenstein, C. E. and Colman, P. M. (1972b). *J. Phys. Sci. Instr.*, **5**, 353.
Matthews, B. W., Jansonius, J. N., Colman, P. M., Schoenborn, B. P. and Dupourque, D. (1972c). *Nature New Biol.*, **238**, 37.
Matthews, B. W., Sigler, P. B., Henderson, R. and Blow, D. M. (1967). *Nature, Lond.*, **214**, 652.
Mathews, F. S., Argos, P. and Levine, M. (1971). *Cold Spring Harb. Symp. Quant. Biol.*, **36**, 387.
Meek, G. A. (1970). "Practical Electron Microscopy for Biologists", Wiley Interscience, London.
Merlin, J. F. and Lipsky, S. (1962). *In* "Biological Effects of Ionising Radiation at the Molecular Level", IAEA Symposium, Vienna, 1962 p. 73.
Minor, T. C., Milch, J. R. and Reynolds, G. T. (1974). *J. Appl. Cryst.*, **7**, 323.
Mitchell, C. M. (1957). *Acta Cryst.*, **10**, 475.
Moncrief, J. W. and Lipscomb, W. N. (1966). *Acta Cryst.*, **21**, 322.
Moor, H. (1971). *Phil. Trans. Soc. Lond.*, **B261**, 121.
Moor, H. and Muhlethaler, K. (1963). *J. Cell. Biol.*, **17**, 609.
Moore, P. B., DeRosier, D. J. and Huxley, H. E. (1970). *J. Molec. Biol.*, **50**, 279.
Morimoto, H. and Uyeda, R. (1963). *Acta Cryst.*, **16**, 1107.
Mosley, J. (1973). D. Phil. Thesis, University of Oxford.
Mossbauer, R. L. (1973). *Naturwissenschaften.*, **60**, 493.
Muirhead, H. and Greer, J., (1970). *Nature, Lond.*, **228**, 516.
Muirhead, H., Cox, J. M., Mazzarella, L. and Perutz, M. F. (1967). *J. Molec. Biol.*, **28**, 117.
Muller, A. (1927). *Proc. Roy. Soc.*, **A117**, 30.
Muller, A. (1929). *Proc. Roy. Soc.*, **125**, 507.
Muller, A. (1962). *In* "Biological Effects of Ionising Radiation at the Molecular Level", IAEA Symposium, Vienna. p. 61.
Muller, A. and Clay, R. E. (1939). *J. Inst. Elect. Engrs.*, **84**, 261.
Nagano, K. (1974). *J. Molec. Biol.*, **8**, 337.
Neidle, S. (1973). *Acta Cryst.*, **B29**, 2645.
Neurath, H., Bradshaw, R. A., Ericsson, L. H., Babin, D. R., Petra, P. H. and Walsh, K. A. (1968). *Brookhaven Symp. Biol.*, **21**, 1.
Nockolds, C. E. and Kretsinger, R. H. (1970). *J. Phys. Sci. Instr.*, **3**, 842.
Nordman, C. E. (1966). *Trans Am. Cryst. Assoc.*, **2**, 29.
Norne, J.-E., Bull, T. E., Einarsson, R., Lindman, B. and Zeppezauer, M. (1973). *Chemica Scripta.*, **3**, 142.
North, A. C. T. (1959). *Acta Cryst.*, **12**, 512.
North, A. C. T. (1964). *J. Sci. Instr.*, **41**, 42.
North, A. C. T. (1965). *Acta Cryst.*, **18**, 212.
North, A. C. T. and Phillips, D. C. (1968). *Prog. Biophys.*, **1**,
North, A. C. T., Phillips, D. C. and Mathews, F. S. (1968). *Acta Cryst.*, **A24**, 351.
Norvell, J. C., Nunes, A. C. and Schoenborn, B. P. (1975). *Science, NY*, **190**, 568.
Nunes, A. C., Nathans, R. and Schoenborn, B. P. (1971). *Acta Cryst.*, A27, 284.
Ohlsson, I., Nordstrom, B. and Branden, C-I. (1974). *J. Molec. Biol.*, **89**, 339.
Oesterhelt, D., Bauer, H. and Lynen, F. (1969). *Proc. Nat. Acad. Sci. USA.*, **63**, 1377.

Osserman, E. F., Cole, S. J., Swan, I. D. A. and Blake, C. C. F. (1969). *J. Molec. Biol.*, **46**, 211.
Padlan, E. A. and Love, W. E. (1968). *Nature, Lond.*, **220**, 376.
Parak, F. (1971). *Z. Physik.*, **244**, 456.
Parak, F. and Formanek, H. (1971). *Acta Cryst.*, **A27**, 573.
Parratt, L. G. (1959). *Rev. Sci. Instr.*, **30**, 297.
Parsons, D. F., Matricardi, V. R., Moretz, R. C. and Turner, J. N. (1974). *Adv. Biol. Med. Phys.*, **15**, 161.
Patterson, A. L. (1934). *Phys. Rev.*, **46**, 372.
Pauling, L. and Corey, R. B. (1951). *Proc. Nat. Acad. Sci. USA.*, **37**, 729.
Pauling, L., Corey, R. B. and Branson, H. R. (1951). *Proc. Nat. Acad. Sci. USA.*, **37**, 205.
Peerdeman, A. F., Bommel, A. J. V. and Bijvoet, J. M. (1951). *Proc. Acad. Sci. Amst.*, **54**, 16., *Nature, Lond.*, **165**, 271.
Peking Insulin Research Group. (1973). *Scientia Sinica.*, **16**, 136.
Peking and Shanghai Insulin Research Group. (1974). *Scientia Sinica.*, **17**, 752.
Penhoet, E., Kochman, M., Valentine, R. and Rutter, W. J. (1967). *Biochemistry.*, **6**, 2940.
Pepinsky, R. and Okaya, Y. (1956). *Proc. Nat. Acad. Sci. USA.*, **42**, 286.
Perutz, M. F. (1946). *Trans. Faraday Soc.*, **42B**, 187.
Perutz, M. F. (1956). *Acta Cryst.*, **9**, 867.
Perutz, M. F. (1970). *Nature, Lond.*, **228**, 726, 734.
Perutz, M. F., Kendrew, J. C. and Watson, H. C. (1965). *J. Molec. Biol.*, **13**, 669.
Perutz, M. F., Muirhead, H., Cox, J. M. and Goaman, L. G. C. (1968). *Nature, Lond.*, **219**, 131.
Peterson, S. W. and Smith, H. G. (1961). *Phys. Rev.*, **6**, 7.
Petsko, G. A. (1973). D. Phil. Thesis, Oxford University.
Petsko, G. A. (1975). *Acta Cryst.*, (in press).
Petsko, G. A. and Williams, R. J. P. (1975) (unpublished results).
Phillips, C. S. G. and Williams, R. J. P. (1966). "Inorganic Chemistry", Vol. II, Oxford University Press.
Phillips, D. C. (1964). *J. Sci. Instr.*, **41**, 123.
Phillips, D. C. (1966). *Adv. Res. Diff. Meth.*, **2**, 75.
Phillips, D. C. (1967). *Proc. Nat. Acad. Sci. USA*, **57**, 493.
Phillips, F. C. (1963). "An Introduction to Crystallography", 3rd edition, Longmans, London.
Phillips, J. C., Wlodawar, A., Yevitz, M. M. and Hodgson, K. O. (1976). *Proc. Nat. Acad. Sci. USA.* **73**, 128.
Piearson, R. J. (1963). *J.A.C.S.*, **85**, 3533.
Poljak, R. J., Amzel, L. M., Avey, H. P., Becka, L. W., Goldstein, D. J. and Humphrey, R. L. (1971). *Cold Spring Harb. Symp.*, **36**, 421.
Poljak, R. J., Amzel, L. M., Avey, H. P., Chen, B. L. Phizackerley, R. P. and Saul, F. (1973). *Proc. Nat. Acad. Sci.*, **70**, 3305.
Poljak, R. J., Amzel, L. M., Chen, B. L., Phizackerley, R. P. and Saul, F. (1974). *Proc. Nat. Acad. Sci. USA.*, **71**, 3440.
Pullman, B. and Pullman, A. (1963). Quantum Biochemistry, Interscience: New York.
Quiocho, R. A. and Richards, F. M. (1964). *Proc. Nat. Acad. Sci. US.*, **52**, 833.
Quiocho, F. A., McMurray, C. H. and Lipscomb, W. N. (1972). *Proc. Nat. Acad. Sci. USA*, **69**, 2850.
Ramachandran, G. N. (1963). *In* "Advanced Methods of Crystallography", (ed.

Ramachandran, G. N.) Academic Press, New York, London and San Francisco.
Ramachandran, G. N. and Kolaskar, A. S. (1973). *Bio. Chem. Bio. Phys. Acta.*, **303**, 385.
Ramachandran, C. N. and Raman, S. (1956). *Current Sci. India.*, **25**, 348.
Ramachandran, G. N. and Raman, S. (1959). *Acta Cryst.*, **12**, 957.
Ramachandran, G. N. and Sasisekharan, V. (1968). *Adv. Protein Chem.*, **23**, 283.
Ramachandran, G. N. and Srinivasan, R. (1961). *Nature, Lond.*, **190**, 159.
Ramachandran, G. N. and Srinivasan, R. (1970). "Fourier Methods in Crystallography", Interscience, Wiley, New York.
Ramachandran, G. N., Ramakrishnan, C. and Sasisekharan, V. (1963). *J. Molec. Biol.*, **7**, 955.
Ramakrishnan, C. and Ramachandran, G. N. (1965). *Bio. Phys. J.* **909**.
Raman, S. (1959). *Acta Cryst.*, **12**, 964.
Ramaseshan, S. (1963). *In* "Advanced Methods of Crystallography", (ed. Ramachandran, G. N.) p. 67 Academic Press, New York, London and San Francisco.
Ramaseshan, S. (1966). *Current Sci. India.*, **35**, 87.
Ramaseshan, S. and Venkatesan, K. (1957). *Current Sci. India.*, **26**, 352.
Ramussen, S. E. and Henriksen, K. (1970). *J. Appl. Cryst.*, **3**, 100.
Rasse, D., Warme, P. K. and Scheraga, H. A. (1974). *Proc. Nat. Acad. Sci. USA.*, **71**, 3736.
Reeke, G. (1974). Private Communication.
Reeke, G. N. and Lipscomb, W. N. (1969). *Acta Cryst.*, **B25**, 2614.
Reeke, G. N., Hartsuck, J. A., Ludwig, M. L., Quiocho, F. A., Steitz, T. A. and Lipscomb, W. N. (1967). *Proc. Nat. Acad. Sci. USA.*, **58**, 220.
Richards, F. M. (1968). *J. Molec. Biol.*, **37**, 225.
Richards, F. M. and Knowles, J. R. (1968). *J. Molec. Biol.*, **37**, 231.
Richards, F. M. and Wyckoff, H. W. (1971). *In* "The Enzymes", Vol. IV (ed. Boyer, P. D.) Academic Press, New York, London and San Francisco.
Richardson, J. S., Thomas, K. A., Rubin B. H. and Richardson, D. C. (1975). (to be published).
Robertson, J. M. and Woodward, I. (1937). *J. Chem. Soc.*, **219**.
Robertus, J. D., Ladner, J. E., Finch, J. T., Rhodes, D., Brown, R. S., Clark, B. F. C. and Klug, A. (1974). *Nature, Lond.*, **250**, 546.
Rollett, J. S. (1965). *In* "Computing Methods in Crystallography", (ed. Rollett, J. S.), Pergamon Press.
Rosenbaum, G., Holmes, K. C. and Witz, J. (1971). *Nature, Lond.*, **230**, 434.
Rossmann, M. G. (1960). *Acta Cryst.*, **13**, 221.
Rossmann, M. G. (1961). *Acta Cryst.*, **14**, 383.
Rossmann, M. G. (1972). "Molecular Replacement Method", Gordon and Breach, New York.
Rossmann, M, G., Jeffrey, B. A., Nain, P. and Warren, S. (1967). *Proc. Nat. Acad. Sci. USA*, **57**, 515.
Rossmann, M. G. and Blow, D. M. (1962). *Acta Cryst.*, **15**, 24.
Rossmann, M. G. and Blow, D. M. (1963). *Acta Cryst.*, **16**, 39.
Rossmann, M. G. and Blow, D. M. (1964). *Acta Cryst.*, **17**, 1474.
Rossmann, M. G., Blow, D. M., Harding, M. M. and Collier, E. (1964). *Acta Cryst.*, **17**, 338.
Rossmann, M. G., Adams, M. J., Buehner, M., Ford, G. C., Hachert, M. L., Lentz, P. J. Jr., McPherson, A. Jr., Schevitz, R. W. and Smiley, I. E. (1971). *Cold Spring Harb. Symp. Quant. Biol.*, **36**, 179.

REFERENCES

Rossmann, M. G., Jeffrey, B. A., Main, P. and Warren, S. (1967). *Proc. Natn. Acad. Sci. USA*, **57**, 515.
Rossmann, M. G., Adams, M. J., Buehner, M., Ford, G. C., Hackert, M. L., Liljas, A., Rao, S. T., Banaszak, L. J., Hill, E., Tsernoglou, D., Webb, L. (1973). *J. Molec. Biol.*, **76**, 533.
Rossmann, M. G., Moras, D. and Olsen, K. W. (1974). *Nature, Lond.*, **250**, 194.
Rühlmann, A., Kukla, D., Schwager, P., Bartels, K. and Huber, R. (1973). *J. Molec. Biol.*, **77**, 417.
Salemme, F. R. and Fehr, D. G. (1972). *J. Molec. Biol.*, **70**, 697.
Salemme, F. R., Freer, S. T., Xuong, N. G. H., Alden, R. A. and Kraut, J. (1973). *J. Biol. Chem.*, **248**, 3910.
Sanger, F. and Thompson, E. O. P. (1953). *Bio. Chem. J.*, **53**, 353.
Sanger, F. and Tuppy, H. (1951). *Biochem. J.*, **49**, 463.
Sasaki, K., Dockcrill, S., Adamnak, D., Tickle, I. J. and Blundell, T. L. (1975). *Nature, Lond.*, **257**, 751.
Sayre, D. (1952). *Acta Cryst.*, **5**, 60.
Sayre, D. (1972). *Acta Cryst.*, **A28**, 210.
Sayre, D. (1974). *Acta Cryst.*, **A30**, 180.
Schevitz, R., Navia, M., Bantz, D., Cornich, G., Rosa, J., Rosa, M. and Sigler, P. (1972). *Science NY.*, **177**, 429.
Schiffer, M., Girling, R. L., Ely, K. R. and Edmundson, A. E. (1973). *Biochemistry*, **12**, 4233.
Schlicktkrull, J. (1958). "Insulin Crystals", Munksgaard, Copenhagen.
Schmid, M. F. and Herriott, J. R. (1975). *Acta Cryst.*, **A31**, 530.
Schoenborn, B. P. (1969). *Nature, Lond.*, **224**, 143.
Schoenborn, B. P. (1971). *Cold Spring Harb. Symp. Quant. Biol.*, **36**, 569.
Schoenborn, B. P. (1972). "Harwell School on Neutron Beams in Research". Unpublished.
Schoenborn, B. P., Watson, H. C. and Kendrew, J. C. (1965). *Nature, Lond.*, **207**, 28.
Schoenborn, B. P., Nunes, A. C. and Nathans, R. (1970). *Ber. Bunsenges physik chem.*, **74**, 1202.
Schulz, G. E., Elzinga, M., Marx, F. and Schirmer, R. H. (1974). *Nature, Lond.*, **250**, 120.
Schulz, G. E., Biederman, K., Kabsch, W. and Schirmer, R. H. (1973). *J. Molec. Biol.*, **80**, 857.
Schwager, P., Bartels, K. and Huber, R. (1973). *Acta Cryst.*, **A29**, 291.
Scott, R. A. and Scheraga, H. A. (1966). *J. Chem. Phys.*, **45**, 2091.
Segal, D. M., Powers, J. C., Cohen, G. H., Davies, D. R., Wilcox, P. E. (1971). *Biochemistry*, **10**, 3728.
Segal, D. M., Padlan, E. A., Cohen, G. H., Rudikoff, S., Potter, M., Davies, D. R. (1974). *Proc. Nat. Acad. Sci. USA*, **71**, 4298.
Shall, S. and Barnard, E. A. (1969). *J. Molec. Biol.*, **41**, 237.
Shelton, E., Yonemasu, K. and Straid, R. M. (1972). *Proc. Nat. Acad. Sci. USA*, **69**, 65.
Shotton, D. and Watson, H. C. (1970a). *Nature, Lond.*, **225**, 811.
Shotton, D. and Watson, H. C. (1970b). *Phil. Trans. Roy. Soc. Lond.*, **B257**, 111.
Shotton, D. M., White, N. J., Watson, H. C. (1971). *Cold Spring Harb. Symp. Quant. Biol.*, **36**, 91.
Sigler, P. B. and Blow, D. M. (1965). *J. Molec. Biol.*, **12**, 17.
Sigler, P. B., Jeffery, B. A., Matthews, B. W. and Blow, D. M. (1966). *J. Molec. Biol.*, **15**, 175.

Sigler, P. B., Blow, D. M., Matthews, B. W. and Henderson, R. (1968). *J. Molec. Biol.,* **35**, 143.
Sikka, S. K. (1969). *Acta Cryst.,* **A25**, 539.
Sim, G. A. (1959). *Acta Cryst.,* **12**, 813.
Sim, G. A. (1960). *Acta Cryst.,* **13**, 511.
Singh, A. K. and Ramaseshan, S. (1966). *Acta Cryst.,* **21**, 279.
Smit, J. D. G., Ploegman, J. H., Kalk, K. H., Jansonius, J. N., Drenth, J. (1973). *IsraelJ. Chem.,* **12**, 287.
Smocka, G. E., Birnbaum, E. R. and Darnall, D. (1971). *Biochemistry,* **10**, 4556.
Snape, K. (1974). D. Phil. Thesis, University of Oxford.
Sokolovsky, M., Riordan, J. F. and Vallee, B. L. (1967). *Biophys. Biochem. Res. Commun.,* **27**, 20.
Sperling, R. and Steinberg, I. Z. (1974). *Biochemistry,* **13**, 2007.
Sperling, R., Burstein, Y. and Steinberg, I. Z. (1969). *Biochemistry,* **8**, 3810.
Srinivasan, R. (1961). *Acta Cryst.,* **14**, 607.
Srinivasan, R., Sarma, V. R. and Ramachandran, G. N. (1963). *In* "Crystallography and Crystal Perfection", (ed. Ramachandran, G. N.) p. 85, Academic Press, London, New York and San Francisco.
Steinberg, I. Z. and Sperling, R. (1967). *In* "Conformation of Biopolymers" (ed. Ramachandran, G. N.) p. 215, Academic Press, New York, London and San Francisco.
Steiner, L. A. and Blumbery, P. M. (1971). *Biochemistry,* **10**, 4725.
Steinrauf, L. K. (1963). *Acta Cryst.,* **16**, 317.
Steitz, T. A. (1968). *Acta Cryst.,* **B24**, 504.
Steitz, T. A., Richmond, T. J., Wise, D. and Engelman, D. (1974). *Proc. Nat. Acad. Sci. USA,* **71**, 593.
Stenn, K. and Bahr, G. F. (1970). *J. Ultrastruct. Res.,* **31**, 526.
Stout, G. H. and Jensen, L. H. (1968). "X-ray Structure Determination", Macmillan, New York.
Strahs, G. and Kraut, J. (1968). *J. Molec. Biol.,* **35**, 503.
Stroud, R. M., Kay, L. M. and Dickerson, R. E. (1971). *Cold Spring Harb. Symp. Quant. Biol.,* **36**, 125.
Stroud, R. M., Kay, L. M. and Dickerson, R. E. (1974). *J. Molec. Biol.,* **83**, 185.
Stryer, L., Kendrew, J. C. and Watson, H. C. (1964). *J. Molec. Biol.,* **8**, 96.
Stubbs, G. (1972). D. Phil. University of Oxford.
Suddath, F. L., Quigley, G. J., McPherson, A., Sneden, D., Kim, J. J., Kim, S. H. and Rich, A. (1974). *Nature, Lond.,* **248**, 20.
Sweet, R. M., Wright, H. T., Janin, J., Chothia, C. H. and Blow, D. M. (1974). *Biochemistry,* **13**, 4212.
Takano, T., Swanson, R., Kallai, O. B. and Dickerson, R. E. (1971). *Cold Spring Harb. Symp. Quant. Biol.,* **34**, 397.
Tanaka, N., Yamane, T., Tsukihara, T., Ashida, T., Kakudo, M. (1973). 28th Annual Meeting of Chemical Society of Japan.
Tanaka, N., Yamane, T., Tsukihara, T., Ashida, T. and Kakudo, M. (1975). *J. Biochem.,* **77**, 147.
Taylor, A. (1949). *J. Sci. Instr.,* **26**, 225.
Taylor, C. A. and Lipson, H. (1964). "Optical Transforms", G. Bell and Sons, London.
Taylor, K. A. and Glaeser, R. M. (1974). *Science, NY.,* **186**, 1036.
Ten Eyck, L., Weaver, L. H. and Matthews, B. W. (1976). *Acta Cryst.,* **A32**, 349.
Theorell, H. (1932). *Biochem. J. 2.,* **252**, 1.
Thomas, V. P. (1972). *J. Appl. Cryst.,* **5**, 83.

REFERENCES

Thomson, A. J., Williams, R. J. P. and Reslova, S. (1972). *Struct. Bonding*, **11**, 1.
Tickle, I. J. (1975). *Acta Cryst.*, **B31**, 329.
Tilander, B., Strandberg, B. and Friborg, K. (1965). *J. Molec. Biol.*, **12**, 740.
Timkovich, R. and Dickerson, R. E. (1972). *J. Molec. Biol.*, **72**, 199.
Timkovich, R. and Dickerson, R. E. (1973). *J. Molec. Biol.*, **79**, 39.
Tollin, P. (1966). *Acta Cryst.*, **21**, 613.
Tollin, P. (1969). *Acta Cryst.*, **A25**, 376.
Tollin, P. and Rossmann, M. G. (1966). *Acta Cryst.*, **21**, 872.
Tollin, P., Main, P., Rossmann, M. G. (1966). *Acta Cryst.*, **20**, 404.
Traub, W. and Piez, K. A. (1971). *Adv. Protein Chem.*, **25**, 243.
Tsernoglou, D., Hill, E. and Banaszak, L. J. (1972). *J. Molec. Biol.*, **69**, 75.
Tulinsky, A. (1974). Private Communication.
Tulinsky, A., Vandlen, R. L., Morimoto, C. N. Mani, N. V., Wright, L. H. (1973). *Biochemistry*, **12**, 4185.
Unwin, P. N. T. (1971). *Phil. Trans. Roy. Soc. Lond.*, **B261**, 95.
Unwin, P. N. T. (1972). *Proc. Roy. Soc.*, **A329**, 327.
Unwin, P. N. T. (1974). *J. Molec. Biol.*, **87**, 657.
Unwin, P. N. T. and Henderson, R. (1975). *J. Molec. Biol.*, **94**, 425.
Wang, B. C., Yoo, C. S. and Sax, M. (1975). *Acta Cryst.*, **A31**, S30.
Waser, J. (1951). *Rev. Sci. Instr.*, **22**, 563.
Watenpaugh, K., Sieker, L. C., Herriott, J. R. and Jensen, L. H. (1971). *Cold Sping Harb, Symp. Quant. Biol.*, **36**, 359.
Watenpaugh, K. D., Sieker, L. C., Jensen, L. H., Legall, J. and Dubourdieu, M. (1972). *Proc. Nat. Acad. Sci. USA*, **69**, 3185.
Watenpaugh, K. D., Sieker, L. C., Herriott, J. R. and Jensen, L. H. (1973). *Acta Cryst.*, **B29**, 943.
Watson, G. N. (1972) "The Theory of Bessel Functions", Cambridge University Press.
Watson, H. C. (1969). *Progr. Stereochem.*, **4**, 299.
Watson, H. C., Kendrew, J. C. and Stryer, L. (1964). *J. Molec. Biol.*, **8**, 166.
Watson, H. C., Shotton, D. M., Cox, J. M. and Muirhead, H. (1970). *Nature, Lond.*, **225**, 806.
Weber, B. H. and Goodkin, P. E. (1970). *Arch. Biochem. Biophys.*, **141**, 489.
Weinzierl, J. E., Eisenberg, D. and Dickerson, R. E. (1969). *Acta Cryst.*, **B25**, 380.
Whiting, R. F. and Ottensmeyer, F. P. (1972). *J. Molec. Biol.*, **67**, 173.
Whittaker, E. J. W. (1953). *Acta Cryst.*, **6**, 218.
Whiley, D. C., Evans, D. R., Warren, S. G., McMurray, C. H., Edwards, B. F. P., Franks, W. A. and Lipscomb, W. N. (1971). *Cold Spring Harb. Symp. Quant. Biol.*, **36**, 285.
Wilson, A. J. C. (1949). *Acta Cryst.*, **2**, 318.
Wilson, A. J. C. (1950). *Acta Cryst.*, **3**, 397.
Winkler, F. K. and Dunitz, J. D. (1971). *J. Molec. Biol.*, **59**, 169.
Witz, J. (1969). *Acta Cryst.*, **A25**, 30.
Wonacott, A. J. (1974). Private Communication.
Woolfson, M. M. (1961). "Direct Methods in Crystallography", Oxford University Press, London.
Woolfson, M. M. (1970) "Introduction to X-ray Crystallography", Cambridge University Press, Cambridge.
Wooster, W. A. (1964). *Acta Cryst.*, **17**, 878.
Wright, C. S., Alden, R. A. and Kraut, J. (1969). *Nature, Lond.*, **221**, 233.

Wyckoff, H. W., Doscher, M., Tsernoglou, D., Inagami, T., Johnson, L. N., Hardman, K. D., Allewell, N. M., Kelly, D. M. and Richards, F. M. (1967a). *J. Molec. Biol.*, **27**, 563.

Wyckoff, H. W., Hardman, K. D., Allewell, N. M., Inagami, T., Tsernoglou, D., Johnson, L. N. and Richards, F. M. (1967b). *J. Biol. Chem.*, **242**, 3984.

Wyckoff, H. W., Tsernoglou, D., Hanson, A. W., Knox, J. R., Lee, B. and Richards, F. M. (1970). *J. Biol. Chem.*, **245**, 305.

Xuong, N-H. (1969). *J. Phys. Sci. Instr.*, **2**, 485.

Xuong, N-H. and Freer, S. T. (1971). *Acta Cryst.*, **B27**, 2380.

Xuong, N-H., Kraut, J., Sely, O., Freer, S. T. and Wright, C. S. (1968). *Acta Cryst.*, **B24**, 289.

Zaloga, G. and Sarma, V. R. (1974). *Nature, Lond.*, **251**, 551.

Zeppezauer, M., (1971). *Adv. Enzymol.*, **22**, 253.

Zeppezauer, E., Soderburg, B. O., Branden, C. I., Akeson, A., Theorell, H. (1967). *Acta Chem. Scand.*, **21**, 1099.

Zeppezauer, M., Eklund, H. and Zeppezauer, E. S. (1968). *Arch Biochem Biophys.*, **126**, 564.

Zwick, M. (1968). Ph.D. Thesis, M.I.T.

SUBJECT INDEX

A

Absolute configuration
 of amino acids, 19
 determination using anomalous scattering, 180, 373
Absolute scale, 333
Absorption by crystal, 249–250
 by air, 250, 317
 by X-ray film, 315
Absorption correction of intensities
 diffractometer, 324
 photographic techniques, 320
Acid proteinase from *E. parasitica*, 525
Adenylate kinase, 7, 525
Alcohol dehydrogenase, 6, 519, 523
Aldolase, 9, 485
Alkaline phosphatase, 8
Amino acids, 18
Amylase, 8
Anomalous dispersion
 to simulate isomorphous changes, 168
Anomalous scattering, 165
 and absolute configuration, 180, 373
 absorption component, 167
 in calculation of phases, 177, 371
 in correlation of origins, 361
 difference, 171, 341
 dispersion component, 167
 in finding heavy atom positions, 338
 Fourier syntheses, 175
 of gamma rays, 481
 of neutrons, 479
 in Pattersons, 345
 simple theory, 165
Antibodies, 5, 534
Apoenzyme, 58

Area detectors, 241, 283
Aspartate transcarbamoylase, 7, 539
Asymmetric unit, 84
Atomic scattering factors, X-rays, 120
 imaginary part, 167
 real part, 166
Atomic scattering factors, neutrons, 467, 468, 478

B

Background corrections, for intensity measurements, 314, 321
Bacteriochlorophyll protein, 528
Beta sheet, 31
Bijvoet difference, 171
Blow and Crick treatment of errors in isomorphous replacement, 364
Bragg's Law, 109
Bravais Lattices, 90–93

C

Calculated phases in difference maps, 417
Carbonic anhydrase, 9, 41, 525
Carboxypeptidase A, 8, 11, 519, 522, 525
Carboxypeptidase B, 525
Carrier and storage proteins, 5
Catalase, 6
 beam damage in electron microscope, 505
Chymotrypsin, 8, 11
 difference Fouriers, 414
 heavy atom derivatives, 188, 345
 mechanism of action, 519, 522, 525, 533
 secondary structure, 41
Chymotrypsinogen, 4, 525
Coenzyme, 58

Cofactor, 57
Collagen, 502
Collimation, 256–257
Complement, electron microscopy of Clq, 485
Compton scattering, 119
Computer graphics, 399
Concanavalin A, 5, 392, 528
Conformation, 28
Conformational changes
  detection in difference Fouriers, 411–416
Contrast in electron microscope, 494–496
Contrast transfer function in electron microscopy, 511–513
Cox and Shaw oblique incidence correction, 316
Crosslinking protein crystals, 199
Crystal
  mounting, 80
  optical properties, 97
  solvent of, 79
  spacegroups, 94
  symmetry, 83
Crystallization, 1, 3, 59
  batch methods, 69
  equilibrium dialysis techniques, 71
  hot box technique, 70
  microdialysis cells, 72
  solvent of, 79
  vapour diffusion techniques, 75
Cylindrical polar co-ordinates, 257–259
Cytochrome b_5, 524, 527
Cytochrome c, 524, 527, 530
Cytochrome c_2, 524, 527, 530

D

Darwin's formula, 246–247
Dark field electron microscopy, 495
Data collection, X-rays, 240, 310
  neutrons, 472
Data processing, 311
Densitometers, 313
  precision of, 318
Deoxyribonucleic acid, 495
Denaturation of proteins, 49
Depth of field, in electron microscope, 508
Depth of focus, in electron microscope, 508

Development of film, 283–289
Difference Fouriers, 404
  errors, 409
  detection of movements of atoms, 411
  peak heights, 407
Diffractometers, X-ray, 284–309
  neutron, 472
Direct Methods, 140–144
  in finding heavy atom positions, 351
  in refinement of protein crystal phases, 436

E

Elastase, 9, 525
Electron, wave lengths of, 487
  diffraction, 488–491
  interactions with matter, 493–494
Electron density equation, 134, 381
Electron density maps, 381
  display of, 390
  high resolution, 389
  interpretation, 16, 399
  low resolution, 387
  medium resolution, 389
  resolution, 381
Electron microscope, principles and design, 487–496
  high voltage, 507–509
  preparation of specimens for, 496–504
  resolving power, 491–493
Electron microscopy, 485
Energy loss spectra, 516–517
Enolase, 9
Environmental stage in electron microscope, 508–509
Enzymes, 3
  classification, 6
  mechanism, 519–522
Equivalent reflections, comparison of, 331
Errors, treatment of in method of isomorphous replacement, 364–371
  mean square error in electron density map, 367–369
Evolution, relevance of structural studies to, 529–534
Ewald construction, 132

SUBJECT INDEX

F

Fatty acid synthetase, 511
Ferredoxin, 527
Fibrous proteins, 3
Figure of merit
    definition, 368
    variation with $\sin^2\theta/\lambda^2$, 371
Film characteristics, 310–313
Film development, 283–284
Flavodoxin, 374, 527, 528
Flow cell, 238
Four circle diffractometer, 285–289
Free radicals, radiation damage and, 251–252
Freeze fracture, 503
Friedel's law, 133–134
    breakdown due to anomalous scattering, 170

G

Gamma-ray resonance, 481
Glutaraldehyde
    use in crosslinking crystals, 199
    use in electron microscopy, 496
Glyceraldehyde-3-phosphate dehydrogenase, 4, 6, 525
Glucagon, 5, 57, 528
Glide plane, 93
Glycolysis, 528

H

Haem, 57
Haemoglobin, 5
    biological activity, 522, 527, 531
    tetramer, 54
Hard ions and ligands, 192
Harker construction
    anomalous scattering, 178, 180
    double isomorphous replacement, 161
    single isomorphous replacement, 157
Heating effects in radiation damage, 253–254
Heavy atom binding to proteins, 192
    effect of pH, 195
    effect of salts and buffers, 196
    effect of concentration, 198
    effect of temperature, 198
    lability, 194
    thermodynamic stability, 192
Heavy atom derivatives, preparation of, 14, 183
    gold, 221
    iodination, 235
    iridium, 230
    labelled specific inhibitors, 188
    lanthanides, 202
    lead, 202
    masked metal ions, 234
    mercury, 205, 206, 211, 212
    in metalloproteins, 184
    osmium, 230
    platinum, 221
    replacement of amino acids, 191
    silver, 219
    thallium, 202
    uranyl compounds, 200
    use of flow cell, 238
Heavy atom positions, correlation of origins, 360
Heavy atom positions, determination of, 15, 337
    from centric zones, 338
    use of difference Fouriers, 349
    using isomorphous differences, 338
    using anomalous differences, 338
    using combination coefficients, 339
Heavy atom positions, refinement of, 352
Helix, alpha, 32
    π, 33
    3_{10}, 32
    pitch, 31
Hendrickson–Lattman method for phase probability distributions, 375–380
Hexokinase, 7, 525, 534
High potential iron potential, 527
    refinement, 432
High resolution transmission scanning electron microscope (STEM), 513–517
High voltage electron microscope, 507–509
Holoenzyme, 58
Homologous proteins, 4, 443
Hormones, 5, 534
Hydrogen bond, 22
Hydrophilic groups, 22, 23
    bonding, 49

I

Image analysis in electron microscopy, 509–513
Immunoglobulin, 485, 497, 528, 534–538
Inclination geometry, 291–295
Indicatrix, 102
Inhibitors of enzyme, 58
Insulin, 5, 334, 527, 528, 534
   dimers, 55
   electron density maps, 386, 393
   hexamers, 56
   non-crystallographic symmetry, 453
   primary structure of, 24
   secondary structure of, 34
   tertiary structure of, 45
Ion pair, 22, 23
Isoelectric point, 50
Isomorphous differences, 158
   in acentric zones, 159, 338
   in centric zones, 159, 338
   in determination of heavy atom positions, 338
   in correlation of origins, 360
   Pattersons, 343
   treatment of errors, 340
Isomorphism, lack of, 163
   tests for, 162
Isomorphous replacement, 11, 151
   with anomalous scattering, 180
   in centric zones, 159
   double phased synthesis, 157
   β-isomorphous synthesis, 157
   multiple, 160
   neutron diffraction, 478
   single, 153

K

α-ketoglutarate dehydrogenase complex, 486

L

Lac repressor, 486
"Lack of closure" error, 365–366, 371–373
Lactate dehydrogenase, 4, 6
   in evolution, 525, 533
   quaternary structure, 53
   secondary structure, 41
   spacegroup, 90
Laue equations, 126
Least squares method of refinement, 144–148
Linear diffractometer, 291, 302
   setting errors, 298
Lorentz correction, 247
   camera geometries, 319
   diffractometer geometries, 323
Low temperature device for protein crystals, 253
Luzzati factor in difference Fouriers, 407
Lysozyme, Bacteriophage T4, 525
Lysozyme, hen egg white, 8, 11
   biological activity, 519, 525
   crystallization, 69
   data collection, 303,
   difference Fouriers, 404, 412, 414
   electron density, 135, 382, 388, 390
   refinement, 426
   structure, 40
   X-ray data, 249

M

Malate dehydrogenase, 6, 525
   structure, 42, 44, 52, 53
Microdialysis cells, 72
Mirror plane, 86
Model building
   in computer, 423
   in optical comparator, 392
Molecular replacement, 443
   and calculation of phases, 462
Molecular transforms, 112, 123
   in finding orientation of molecules, 444
Molecular weights, determination of, 104–106
Monochromators, X-ray, 245–246
   neutron, 466
Myogen, 528
Multiple counter diffractometry, 299–309
Myoglobin, 404, 527, 531
   secondary structure, 40, 41

N

Negative staining, 497–502
Neutron diffraction, 465
Neutron Fouriers, 477
Nicotinamide adenine dinucleotide, NAD, 58

SUBJECT INDEX

Non-crystallographic symmetry, 443, 452
  calculation of phases, 462
Non-linear response of X-ray film, 311–313, 314–315
Nuclease, 9, 525
Nucleation of crystals, 66

O

Oblique incidence absorption correction, 315
Oligomeric proteins, 4
Optical comparator, 392
Optical diffractometer, 110, 509–510
Optical properties of crystals, 97–104
Ordinate analysis, 289–290, 307
Oscillation camera, 274–283

P

Papain, 9, 526
Patterson function, 137–140
  anomalous differences, 345
  combined differences, 347
  defining molecular orientation, 448
  even and odd, 176
  isomorphous differences, 343
  sum differences, 347
Pepsin, 8, 525
Peptide, link, 25
  cis, 27
  trans, 27
Phase problem, 136–137
Phases, calculation of, 16, 363
Phase contrast electron microscopy, 495–496
Phenyl-tRNA synthetase, 9
Phosphoglucomutase, 7
Phosphoglycerate kinase, 4, 7, 249, 307–308, 485, 526
Phosphoglycerate mutase, 9, 526, 534
Phospholipase A_2, 525
Phosphorylase, 7, 249, 375
Phosphotungstic acid, use of in electron microscopy, 497–502
Photographic techniques of data collection
  neutrons, 472
  X-rays, 260, 310
pK, 19
Polarization of scattered X-rays, 119, 247
  correction for, 319

Polarizing microscope, 98
  use in protein crystallography, 103
Position sensitive detectors, X-rays, 241, 283
  neutrons, 475
Prealbumin, 528
Precession Camera, 266–274
Primary structure, 25
  chemical determination of, 25
  from electron density maps, 400
Profile analysis method, 290–291
Prosthetic groups, 58
Pyruvate dehydrogenase, 6
Pyruvate kinase, 7

Q

Q-function, 460
Quartz plate, 101
Quaternary structure, 51

R

R value, 147–148
  in determination of heavy atom positions, 355, 357, 358
  of protein structure factors, 428
Radiation damage
  in electron microscope, 504
  by neutrons, 471
  by X-rays, 351
Radiation damage correction
  in diffractometer methods, 330
  in photographic methods, 321
Radiation intensities of X-rays, neutrons and gamma-rays, 467
Ramachandran plot, 30
Rational indices, law of, 84
Reciprocal lattice, 130–132
Reciprocal unit cell dimensions, 130–131
Redox proteins, 4
Refinement
  energy, 431
  free atom reciprocal space, 429
  gradient-curvature method, 427
  of heavy atom positions, 353
  high potential iron protein, 432
  least squares, 17, 144
  linked atom real space, 423
  lysozyme, 426
  of protein atom positions, 423
  rubredoxin, 434, 439
  trypsin inhibitor, 435

trypsin—trypsin inhibitor, 432
 use of difference Fouriers, 428
Refinement of phases
 direct methods, 436
 real space direct methods, 438
 Sayre method, 439
 tangent formula, 437
 Tsoucaris—De Rango method, 442
Reflections, number to be measured, 248–249
Reflection, width of, 254–256
Rennin, 8
Resolution, 117, 381
Rhodanese, 7, 526
Ribonuclease, 9, 11, 519, 526, 40
Rotation axis symmetry operation, 86
Rotation camera, 274–283
Rotation function, 450
Rubredoxin
 biological activity, 524, 527, 528
 electron density maps, 394
 refinement of atomic positions, 434
 refinement of phases, 439
 sequence, 400

S

Salting in, 61
Salting out, 61
Sayre equation, 142, 436
Scaling, different sets of data, 331–332
 of derivative to native data, 333–336
Screenless precession photography, 270–274
Screenless photography, 269–283
Screw axis, 93
Secondary structure, 31
Sequence of proteins, 25
 chemical determination, 25
 from electron density maps, 400
Shadowing, 502–503
Sim weighting in difference maps, 417–419
Soft ions and ligands, 192
Solvated electron, 252
Solubility of proteins, 60
 and ionic strength, 61
 organic solvents, 66
 pH and counter ions, 64
 temperature, 65

Solvent content of protein crystals, 104
Space groups, 94–97
Space lattice definition, 83
Spherical aberration, in electron microscope, 492
Standard deviation, of measurements of intensities by photographic methods, 317–318
 by diffractometer methods, 321–323
Structure factor equation, 127
Subtilisin, 9, 526, 534, 41
Superoxide dismutase, 520
Supersecondary structure, 49
Symmetry operations, 86
Symmetry, non-crystallographic, 443
Synchrotron radiation, 244–245, 309

T

Tangent formula, 144, 436
Temperature factor, 121, 333–334
Ternary complex, 58
Tertiary structure, 45
Thermal neutrons, 466
Thermolysin, 9, 526, 48, 400
Thin sections in electron microscopy, 496–497
Thioredoxin S_2, 527
Thomson scattering of X-rays, formula, 119
Transfer function, in electron microscopy, 511–513
Triose phosphate isomerase, 4, 9, 526,– 534
t-RNA, 539
Translation function
 Crowther and Blow, 461
 Rossman and Blow, 456
Trypsin, 8, 526
Trypsin inhibitor, 526
 refinement, 435
Trypsin—trypsin inhibitor complex
 biological role, 526, 527
 data collection by screenless precession photography, 273
 refinement, 432

U

Unit cell definition, 83–84
 dimensions for different lattices, 130–131

SUBJECT INDEX

W

Water molecules, compensation for displaced water molecules in difference maps, 416–417
Wave nature of X-rays, 107
Wave length of electron, 487
  of neutrons, 465
Width of reflection, 254–256
Wilson plot, 334

Wooster effect, 314
Wyckoff, method of measuring reflection, 291

X

X-rays
  production of, 242
  wavelength, 243

Molecular Biology

An International Series of Monographs and Textbooks

Editors

BERNARD HORECKER

Roche Institute of Molecular Biology
Nutley, New Jersey

NATHAN O. KAPLAN

Department of Chemistry
University of California
At San Diego
La Jolla, California

JULIUS MARMUR

Department of Biochemistry
Albert Einstein College of Medicine
Yeshiva University
Bronx, New York

HAROLD A. SCHERAGA

Department of Chemistry
Cornell University
Ithaca, New York

HAROLD A. SCHERAGA. Protein Structure. 1961

STUART A. RICE AND MITSURU NAGASAWA. Polyelectrolyte Solutions: A Theoretical Introduction, *with a contribution by Herbert Morawetz.* 1961

SIDNEY UDENFRIEND. Fluorescence Assay in Biology and Medicine. Volume I—1962. Volume II—1969

J. HERBERT TAYLOR (Editor). Molecular Genetics. Part I—1963. Part II—1967

ARTHUR VEIS. The Macromolecular Chemistry of Gelatin. 1964

M. JOLY. A Physico-chemical Approach to the Denaturation of Proteins. 1965

SYDNEY J. LEACH (Editor). Physical Principles and Techniques of Protein Chemistry. Part A—1969. Part B—1970. Part C—1973

KENDRIC C. SMITH AND PHILIP C. HANAWALT. Molecular Photobiology: Inactivation and Recovery. 1969

RONALD BENTLEY. Molecular Asymmetry in Biology. Volume I—1969. Volume II—1970

JACINTO STEINHARDT AND JACQUELINE A. REYNOLDS. Multiple Equilibria in Protein. 1969

DOUGLAS POLAND AND HAROLD A. SCHERAGA. Theory of Helix-Coil Transitions in Biopolymers. 1970

JOHN R. CANN. Interacting Macromolecules: The Theory and Practice of Their Electrophoresis, Ultracentrifugation, and Chromatography. 1970

WALTER W. WAINIO. The Mammalian Mitochondrial Respiratory Chain. 1970

LAWRENCE I. ROTHFIELD (Editor). Structure and Function of Biological Membranes. 1971

ALAN G. WALTON AND JOHN BLACKWELL. Biopolymers. 1973

WALTER LOVENBERG (Editor). Iron-Sulfur Proteins. Volume I, Biological Properties—1973. Volume II, Molecular Properties—1973. Volume III, in preparation

A. J. HOPFINGER. Conformational Properties of Macromolecules. 1973

R. D. B. FRASER AND T. P. MACRAE. Conformation in Fibrous Proteins. 1973

OSAMU HAYAISHI (Editor). Molecular Mechanisms of Oxygen Activation. 1974

FUMIO OOSAWA AND SHO ASAKURA. Thermodynamics of the Polymerization of Protein. 1975

LAWRENCE J. BERLINER (Editor). Spin Labeling: Theory and Applications. 1976

T. BLUNDELL AND L. JOHNSON. Protein Crystallography. 1976

in preparation

HERBERT WEISSBACH AND SIDNEY PESTKA (Editors). Molecular Mechanisms of Protein Biosynthesis